Proceedings of the 5th International Conterence "Computational Mechanics and Virtual Engineering" COMEC 2013

Transilvania University of Braşov

24 - 25 October 2013, Braşov, Romania

György SZEIDL, Chairman
Petre P. TEODORESCU, Co-Chairman
Sorin VLASE, President
Michael M. DEDIU, Editor

DERC Publishing House
Tewksbury (Boston), Massachusetts, U. S. A.

American Mathematical Society
2010 Mathematical Subject Classification: 65-xx, 70-xx, 74-xx, 76-xx, 82-xx

Library of Congress Cataloging in Publication Data

Computational Mechanics and Virtual Engineering, the 5th International Conference, 24- 25 October 2013, Braşov, Romania / György SZEIDL, Chairman, Petre P. TEODORESCU , Co-Chairman, Sorin Vlase, President, Michael M. Dediu, Editor
    p. cm. – (Proceedings of the 5th International Conference "Computational Mechanics and Virtual Engineering" COMEC 2013)
    Includes bibliographical references

ISBN-13: 978-1-939757-11-1

PREFACE

After receiving many favorable comments and having interesting discussions regarding the Proceedings of the 4th International Conference "Advanced Composite Materials Engineering" COMAT 2012, we are pleased to present these Proceedings of the 5th International Conference "Computational Mechanics and Virtual Engineering" COMEC 2013. This conference was dedicated to the great Professor Teodorescu, who supported these conferences for many years. I want to express my respect and reverence to this extraordinary Professor, who had astonishing accomplishments, kindness, patience, dedication and optimism. Professor Petre P. Teodorescu will be deeply missed by all his former students, by colleagues, by those who know him from the books, written by him, or written by others about him, and by many others.

Computational mechanics and virtual engineering (CMVG) are increasingly important areas of engineering science, dedicated to the use of computational methods and devices to characterize, simulate and predict physical events and engineering systems governed by the laws of mechanics, and to integrating geometric models, analysis, simulation, optimization and decision making tools within a computer-generated environment that accelerates product development. **CMVG** include new models of physical and biological systems based upon quantum, molecular and biological mechanics, and there is a vast potential for future growth and applicability. Some examples of techniques used by CMVG are numerical methods for the analysis of the nonlinear continuum response of materials, linear and finite deformation elasticity, inelasticity and dynamics, numerical formulation and algorithms (like variational formulation and variational constitutive updates, finite element discretization, error estimation, constrained problems, time integration algorithms and convergence analysis), and parallel computer implementation of algorithms. The applications of CMVG are in manufacturing, aerospace, civil engineering structures, communication, geotechnics, flow problems, automotive engineering, predictive surgery, transportation, geo-environmental modeling, biomechanics, electromagnetism, medicine, prediction of natural physical events, metal forming and numerous other fields.

These Proceedings of the 5th International Conference "Computational Mechanics and Virtual Engineering" COMAT 2013 include 58 papers, which analyze many important practical applications. The topics range from numerical simulations regarding the determination of stresses in all plies of various composite laminates, structural synthesis for redundant industrial robots with more than 6 axes, and the sound attenuation in a sonic composite with point defects studied using a new method that combines the features of the cnoidal method and the genetic algorithm, to a set of dual vectors based methods for rigid body displacement and motion parameterization, vibrations of heterogeneous curved beams subjected to a radial force at the crown point, and computation of the longitudinal tensile break stress of multiphase composite materials with short fibers as reinforcement.

I wish to thank all the conference participants for their interesting presentations, and Mrs. Sophia Dediu for her assistance in preparing this volume.

There is, certainly, much more that can be said about computational mechanics and virtual engineering than we presented here. We hope that the papers included here will provide ideas for our audience, and will stimulate more research, development and applications.

We look forward to receiving comments and suggestions from our readers.

Michael M. Dediu

Boston, USA, December 12, 2013

Previously published in this series:

1. Ioan Goia, *Mechanics of Materials*
2. Ionel Staretu, *Gripping Systems*
3. Proceedings of the 4th International Conference "Advanced Composite Materials Engineering" COMAT 2012

The 5th International Conference
"Computational Mechanics and Virtual Engineering "
COMEC 2013
24- 25 October 2013, Braşov, Romania

STRESSES IN VARIOUS COMPOSITE LAMINATES FOR GENERAL SET OF APPLIED IN-PLANE LOADS

H. Teodorescu-Draghicescu[1], S. Vlase[2]
[1]Transilvania University of Brasov, Braşov, ROMANIA, e-mail draghicescu.teodorescu@unitbv.ro,
[2]svlase@unitbv.ro

Abstract: *This paper presents the numerical simulations regarding the determination of stresses in all plies of various composite laminates such as anti-symmetric, symmetric cross-ply, symmetric angle-ply, and balanced angle-ply laminates for a general set of in-plane loads.*

We used the basic laminate theory considering that in all plies, the fibers and matrix properties, fibers content, and ply thickness are the same.

The following input data have been used: stack of plies, fibers axis of each ply at specified angle to reference direction, axial and transverse Young's moduli, shear modulus, both for fibers and matrix, axial-transverse, Poisson ratio for fibers and matrix.

The output data include: longitudinal, transverse, and shear stresses in all plies at angles varying between 0° and 90°, to reference direction.

Keywords: composite laminates, anti-symmetric, symmetric cross-ply, symmetric angle-ply, balanced angle-ply.

1. INTRODUCTION

The fibers-reinforced polymer matrix composite materials are heterogeneous and anisotropic materials, and, therefore, their mechanics is more complex than that of conventional materials.

The basic element of a composite laminate structure is the individual layer (called lamina) unidirectional reinforced with fibers inserted into a resin system (called matrix). The basic assumptions in the description of the interaction between fibers and matrix in a unidirectional reinforced lamina subject to tensile loads are [1]:

- Both fibers and matrix act as a linear elastic material;
- Initially, the lamina presents no residual stresses;
- The loads are applied parallel or perpendicular to the fibers direction;
- The matrix presents no voids and failures;
- The bond between fibers and matrix is perfect;
- The fibers are uniformly distributed in the matrix.

For in-plane stress condition by superimposing three loadings $\sigma_\parallel$, $\sigma_\perp$ and $\tau_\#$, following law of elasticity can be given:

$$
\begin{bmatrix} \varepsilon_{II} \\ \varepsilon_{\perp} \\ \gamma_{\#} \end{bmatrix} = \begin{bmatrix} \dfrac{1}{E_{II}} & -\dfrac{\upsilon_{II\perp}}{E_{\perp}} & 0 \\[2mm] -\dfrac{\upsilon_{\perp II}}{E_{II}} & \dfrac{1}{E_{\perp}} & 0 \\[2mm] 0 & 0 & \dfrac{1}{G_{\#}} \end{bmatrix} \cdot \begin{bmatrix} \sigma_{II} \\ \sigma_{\perp} \\ \tau_{\#} \end{bmatrix} \tag{1}
$$

where the second matrix is called the compliances matrix.

The transverse contraction coefficients $\upsilon_{II\perp}$ and $\upsilon_{\perp II}$ are not independent of each other. If we assume the existence of small deformations and a linear elastic behavior of the composite material than, between the coefficients of transverse contraction υ and the Young's moduli E, there is the following relationship:

$$
\frac{\upsilon_{\perp II}}{E_{II}} = \frac{\upsilon_{II\perp}}{E_{\perp}} \tag{2}
$$

The relation (2) is called the Maxwell-Betti law. Therefore, the unidirectional reinforced lamina can be described by four basic elasticity terms: E_{II}, $E_{\perp}$, $\upsilon_{\perp II}$ and $G_{\#}$. If desired, the expression (1) in terms of stresses versus strains can be written as following:

$$
\begin{bmatrix} \sigma_{II} \\ \sigma_{\perp} \\ \tau_{\#} \end{bmatrix} = \begin{bmatrix} \dfrac{E_{II}}{1-\upsilon_{\perp II}\cdot\upsilon_{II\perp}} & \dfrac{\upsilon_{II\perp}\cdot E_{\perp}}{1-\upsilon_{\perp II}\cdot\upsilon_{II\perp}} & 0 \\[3mm] \dfrac{\upsilon_{\perp II}\cdot E_{II}}{1-\upsilon_{\perp II}\cdot\upsilon_{II\perp}} & \dfrac{E_{\perp}}{1-\upsilon_{\perp II}\cdot\upsilon_{II\perp}} & 0 \\[3mm] 0 & 0 & G_{\#} \end{bmatrix} \cdot \begin{bmatrix} \varepsilon_{II} \\ \varepsilon_{\perp} \\ \gamma_{\#} \end{bmatrix}, \tag{3}
$$

where the second matrix is called the stiffness matrix.

Expressing the strains versus stresses lead to the advantage of computing the compliances as a function of lamina's basic elastic properties. These basic elastic properties are also called technical constants. These constants can be determined from the lamina's micromechanics using the fibers and matrix elastic properties [1]:

$$
E_{II} = E_F \cdot \varphi + E_M \cdot (1-\varphi), \tag{4}
$$

$$
\upsilon_{\perp II} = \varphi \cdot \upsilon_F + (1-\varphi) \cdot \upsilon_M, \tag{5}
$$

$$
E_{\perp} = \frac{E_M}{1-\upsilon_M^2} \cdot \frac{1+0{,}85\cdot\phi^2}{(1-\phi)^{1{,}25} + \dfrac{\phi\cdot E_M}{(1-\upsilon_M^2)\cdot E_F}} \tag{6}
$$

$$
\upsilon_{II\perp} = \upsilon_{\perp II} \cdot \frac{E_{\perp}}{E_{II}} \tag{7}
$$

$$
G_{\#} = G_M \cdot \frac{1+0{,}6\cdot\phi^{0{,}5}}{(1-\phi)^{1{,}25} + \phi\cdot\dfrac{G_M}{G_F}} \tag{8}
$$

Regarding the lamina being in the stress plane state, its strains can be expressed versus stresses using the transformed components of the compliances matrix:

$$\begin{bmatrix} \varepsilon_{xx} \\ \varepsilon_{yy} \\ \gamma_{xy} \end{bmatrix} = \begin{bmatrix} c_{11} & c_{12} & c_{13} \\ c_{12} & c_{22} & c_{23} \\ c_{13} & c_{23} & c_{33} \end{bmatrix} \cdot \begin{bmatrix} \sigma_{xx} \\ \sigma_{yy} \\ \tau_{xy} \end{bmatrix} \tag{9}$$

These transformed compliances can be computed by following [1]:

$$c_{11} = \frac{\cos^4\alpha}{E_{II}} + \frac{\sin^4\alpha}{E_\perp} + \frac{1}{4} \cdot \left(\frac{1}{G_{\#}} - \frac{2 \cdot \upsilon_{\perp II}}{E_{II}} \right) \cdot \sin^2 2\alpha, \tag{10}$$

$$c_{22} = \frac{\sin^4\alpha}{E_{II}} + \frac{\cos^4\alpha}{E_\perp} + \frac{1}{4} \cdot \left(\frac{1}{G_{\#}} - \frac{2 \cdot \upsilon_{\perp II}}{E_{II}} \right) \cdot \sin^2 2\alpha, \tag{11}$$

$$c_{33} = \frac{\cos^2 2\alpha}{G_{\#}} + \left(\frac{1}{E_{II}} + \frac{1}{E_\perp} + \frac{2 \cdot \upsilon_{\perp II}}{E_{II}} \right) \cdot \sin^2 2\alpha, \tag{12}$$

$$c_{12} = \frac{1}{4} \cdot \left(\frac{1}{E_{II}} + \frac{1}{E_\perp} - \frac{1}{G_{\#}} \right) \cdot \sin^2 2\alpha - \frac{\upsilon_{\perp II}}{E_{II}} \cdot \left(\sin^4\alpha + \cos^4\alpha \right), \tag{13}$$

$$c_{13} = \left(\frac{2}{E_\perp} + \frac{2 \cdot \upsilon_{\perp II}}{E_{II}} - \frac{1}{G_{\#}} \right) \cdot \sin^3\alpha \cdot \cos\alpha - \left(\frac{2}{E_{II}} + \frac{2 \cdot \upsilon_{\perp II}}{E_{II}} - \frac{1}{G_{\#}} \right) \cdot \cos^3\alpha \cdot \sin\alpha, \tag{14}$$

$$c_{23} = \left(\frac{2}{E_\perp} + \frac{2 \cdot \upsilon_{\perp II}}{E_{II}} - \frac{1}{G_{\#}} \right) \cdot \cos^3\alpha \cdot \sin\alpha - \left(\frac{2}{E_{II}} + \frac{2 \cdot \upsilon_{\perp II}}{E_{II}} - \frac{1}{G_{\#}} \right) \cdot \sin^3\alpha \cdot \cos\alpha. \tag{15}$$

In the case in which the stresses are expressed versus strains, the relations are:

$$\begin{bmatrix} \sigma_{xx} \\ \sigma_{yy} \\ \tau_{xy} \end{bmatrix} = \begin{bmatrix} r_{11} & r_{12} & r_{13} \\ r_{12} & r_{22} & r_{23} \\ r_{13} & r_{23} & r_{33} \end{bmatrix} \cdot \begin{bmatrix} \varepsilon_{xx} \\ \varepsilon_{yy} \\ \gamma_{xy} \end{bmatrix}, \tag{16}$$

where r_{ij} represent the transformed components of the stiffness matrix.

A laminate composite structure is considered formed of N unidirectional reinforced laminae subjected to a general set of in-plane loads. The elasticity law of a unidirectional reinforced K lamina is:

$$\begin{bmatrix} \sigma_{xx\,K} \\ \sigma_{yy\,K} \\ \tau_{xy\,K} \end{bmatrix} = \begin{bmatrix} r_{11K} & r_{12K} & r_{13K} \\ r_{12K} & r_{22K} & r_{23K} \\ r_{13K} & r_{23K} & r_{33K} \end{bmatrix} \cdot \begin{bmatrix} \varepsilon_{xx\,K} \\ \varepsilon_{yy\,K} \\ \gamma_{xy\,K} \end{bmatrix}, \tag{17}$$

where r_{ijK} represents the transformed stiffness, σ_{xxK} and σ_{yyK} are medium stresses of a K lamina on x and y axes, τ_{xyK} represent the medium shear stress according to x-y coordinate system. The laminate balance equations are:

$$n_{xx} = \underline{\sigma}_{xx} \cdot t = \sum_{K=1}^{N} \left(\sigma_{xxK} \cdot t_K \right) = \sum_{K=1}^{N} n_{xxK}, \tag{18}$$

$$n_{yy} = \underline{\sigma}_{yy} \cdot t = \sum_{K=1}^{N} \left(\sigma_{yyK} \cdot t_K \right) = \sum_{K=1}^{N} n_{yyK}, \tag{19}$$

$$n_{xy} = \underline{\tau}_{xy} \cdot t = \sum_{K=1}^{N} \left(\tau_{xyK} \cdot t_K \right) = \sum_{K=1}^{N} n_{xyK} \ , \tag{20}$$

where n_{xx} and n_{yy} are the normal forces, n_{xy} is the shear force, $\underline{\sigma}_{xx}$ and $\underline{\sigma}_{yy}$ represent the normal stresses, $\underline{\tau}_{xy}$ is the shear stress of the composite laminate, t_K and t are the K lamina thickness respective the laminate thickness, n_{xxK} and n_{yyK} are normal forces on the unit length of the K lamina and n_{xyK} is the in-plane shear force on the unit length of the K lamina. With relations (17)-(20) the composite laminate elasticity law can be obtained:

$$\begin{bmatrix} \underline{\sigma}_{xx} \\ \underline{\sigma}_{yy} \\ \underline{\tau}_{xy} \end{bmatrix} = \begin{bmatrix} \sum_{K=1}^{N}\left(r_{11K} \cdot \frac{t_K}{t} \right) & \sum_{K=1}^{N}\left(r_{12K} \cdot \frac{t_K}{t} \right) & \sum_{K=1}^{N}\left(r_{13K} \cdot \frac{t_K}{t} \right) \\ \sum_{K=1}^{N}\left(r_{12K} \cdot \frac{t_K}{t} \right) & \sum_{K=1}^{N}\left(r_{22K} \cdot \frac{t_K}{t} \right) & \sum_{K=1}^{N}\left(r_{23K} \cdot \frac{t_K}{t} \right) \\ \sum_{K=1}^{N}\left(r_{13K} \cdot \frac{t_K}{t} \right) & \sum_{K=1}^{N}\left(r_{23K} \cdot \frac{t_K}{t} \right) & \sum_{K=1}^{N}\left(r_{33K} \cdot \frac{t_K}{t} \right) \end{bmatrix} \cdot \begin{bmatrix} \varepsilon_{xx} \\ \varepsilon_{yy} \\ \gamma_{xy} \end{bmatrix} , \tag{21}$$

and from these relations, the composite laminate stiffness can be determined:

$$\underline{r}_{ij} = \sum_{K=1}^{N} \left(r_{ijK} \cdot \frac{t_K}{t} \right). \tag{22}$$

In these conditions, the composite laminate elasticity law becomes:

$$\begin{bmatrix} \underline{\sigma}_{xx} \\ \underline{\sigma}_{yy} \\ \underline{\tau}_{xy} \end{bmatrix} = \begin{bmatrix} \underline{r}_{11} & \underline{r}_{12} & \underline{r}_{13} \\ \underline{r}_{12} & \underline{r}_{22} & \underline{r}_{23} \\ \underline{r}_{13} & \underline{r}_{23} & \underline{r}_{33} \end{bmatrix} \cdot \begin{bmatrix} \varepsilon_{xx} \\ \varepsilon_{yy} \\ \gamma_{xy} \end{bmatrix} , \tag{23}$$

where $\underline{r}_{ij}$ are functions of the basic elastic properties of each lamina $E_{\|K}$, $E_{\perp K}$, $\upsilon_{\perp\|K}$, $G_{\#K}$ and of the fibers disposal angle. Analogue to stresses, a strains analysis can be carried out. From relation (23) the strains ε_{xx}, ε_{yy} and γ_{xy} can be computed. The individual strains of each lamina can be determined through transformation as following:

$$\begin{bmatrix} \varepsilon_{\|K} \\ \varepsilon_{\perp K} \\ \gamma_{\#K} \end{bmatrix} = \begin{bmatrix} cos^2 \alpha_K & sin^2 \alpha_K & sin\,\alpha_K \, cos\,\alpha_K \\ sin^2 \alpha_K & cos^2 \alpha_K & -sin\,\alpha_K \, cos\,\alpha_K \\ -2 sin\,\alpha_K \, cos\,\alpha_K & 2 sin\,\alpha_K \, cos\,\alpha_K & \left(cos^2 \alpha_K - sin^2 \alpha_K \right) \end{bmatrix} \cdot \begin{bmatrix} \varepsilon_{xxK} \\ \varepsilon_{yyK} \\ \gamma_{xy} \end{bmatrix} . \tag{24}$$

Finally, from the strains presented in relation (24), the stresses in each individual lamina can be computed:

$$\sigma_{\|K} = \frac{E_{\|K}}{1-\upsilon_{\perp\|K} \cdot \upsilon_{\|\perp K}} \cdot \varepsilon_{\|K} + \frac{\upsilon_{\perp\|K} \cdot E_{\perp K}}{1-\upsilon_{\perp\|K} \cdot \upsilon_{\|\perp K}} \cdot \varepsilon_{\perp K} , \tag{25}$$

$$\sigma_{\perp K} = \frac{\upsilon_{\perp\|K} \cdot E_{\perp K}}{1-\upsilon_{\perp\|K} \cdot \upsilon_{\|\perp K}} \cdot \varepsilon_{\|K} + \frac{E_{\perp K}}{1-\upsilon_{\perp\|K} \cdot \upsilon_{\|\perp K}} \cdot \varepsilon_{\perp K} , \tag{26}$$

$$\tau_{\#K} = G_{\#} \cdot \gamma_{\#} . \tag{27}$$

To carry out the prediction regarding the failure of the individual laminae, a break criterion is usually used. The system of coordinates $\|$-$\perp$-z represents the local system of coordinates and is applied to each individual lamina. The x-y-z system of coordinates represents the global system of coordinates and is usually applied to the entire laminate. Some experimental results obtained on various composite laminates subjected to a wide range of loadings are presented in references [2], [3], [7-11] as well as numerical simulations to predict the elastic properties of some fibers-reinforced composite laminates are given in papers [4-6].

2. COMPUTATIONAL MICROMECHANICS OF FOUR LAMINATES

In order to compute stresses in each individual lamina of a composite laminate, four examples of laminates have been chosen. These laminates are:

- Anti-symmetric laminate with following plies sequence: [30/0/0/-30]; symmetric cross-ply laminate with plies distribution: [90/0/0/90];
- Symmetric angle-ply laminate with plies sequence: [30/-30/-30/30];
- Balanced angle-ply laminate with following plies distribution: [30/30/-30/-30].

The computational method is based on the approach presented in reference [12]. For all types of laminates, the Tenax IMS65 carbon fibers have been taken into account as well as Huntsman XB3585 epoxy resin with following input data:

- Matrix axial and transverse Young's modulus: 3.2 GPa;
- Fibers axial Young's modulus: 290 GPa;
- Fibers transverse Young's modulus: 4.8 GPa;
- Matrix axial-transverse Poisson ratio: 0.3;
- Fibers axial-transverse Poisson ratio: 0.05;
- Matrix axial-transverse shear modulus: 1.15 GPa;
- Fibers axial-transverse shear modulus: 4.2 GPa;
- Fibers volume fraction: 0.51;
- Applied normal stress in x-direction: 2000 MPa;
- Applied normal stress in y-direction: 200 MPa;
- Applied shear stress in x-y plane: 100 MPa;
- Off-axis loading system: between 0° and 90°.

For these types of laminates, stresses in each lamina have been computed. Example of stresses in some plies in case of some considered laminates subjected to a general set of in-plane loads are presented in tables 1-2 and stresses distributions according to different off-axis loading angles are visualized in figures 1-8.

Table 1: Example of computational stresses in first ply in case of [30/0/0/-30] anti-symmetric laminate

Off-axis loading system	Normal stress $\sigma_\parallel$ [GPa]	Normal stress $\sigma_\perp$ [GPa]	Shear stress $\tau_\#$ [GPa]
0°	2179.08820396	-0.71697426	-35.75065488
10°	939.39407259	36.90536763	-49.96934150
20°	122.51085318	121.05810845	-26.13182244
30°	-173.03328210	241.59118570	32.88674569
40°	88.40865122	383.96653138	119.96785257
50°	875.30289753	531.01157756	224.60823152
60°	2092.73839739	664.99052153	334.18670852
70°	3593.87446208	769.74352531	435.48650202
80°	5197.65192780	832.63583067	516.28936191
90°	6710.63156295	846.08169739	566.84927077

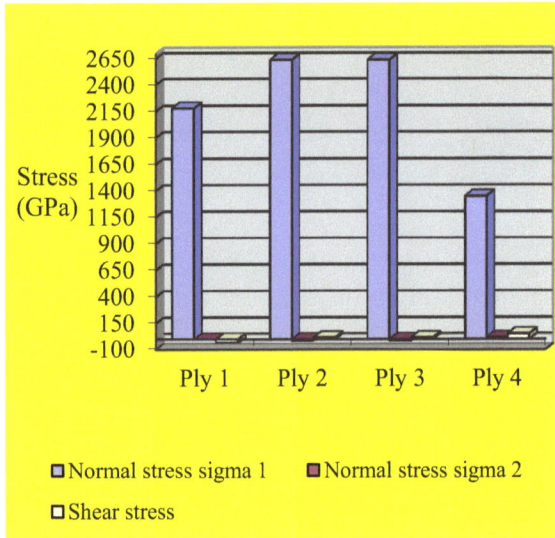

Figure 1: Stresses in plies' laminate [30/0/0/-30]

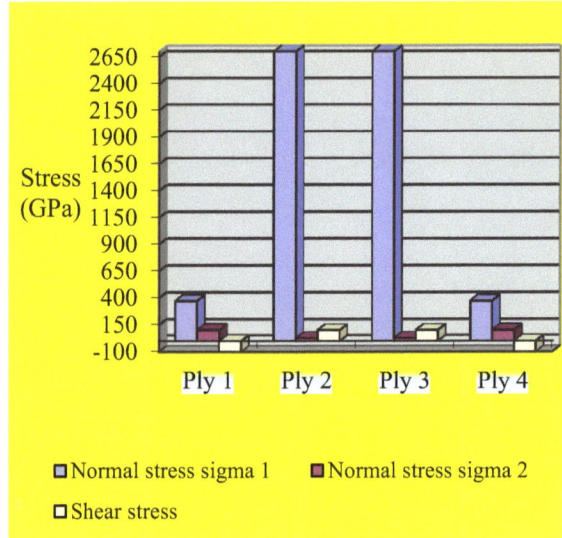

Figure 2: Stresses in plies' laminate [90/0/0/90]

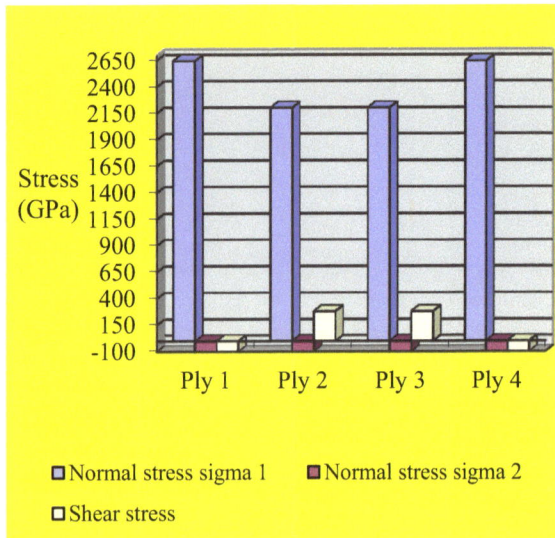

Figure 3: Stresses in plies' laminate [30/-30/-30/30]

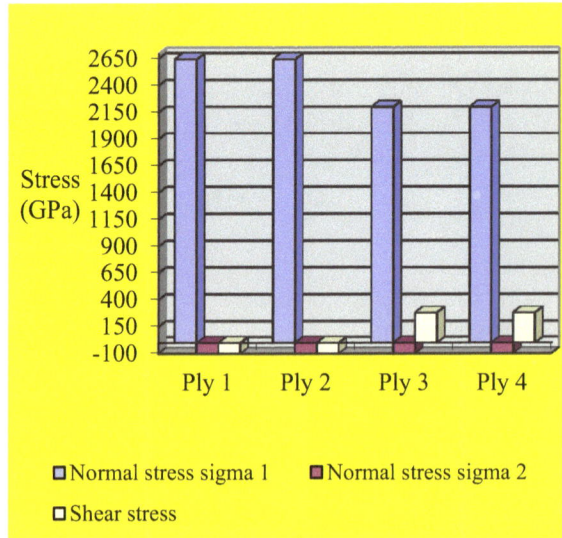

Figure 4: Stresses in plies' laminate [30/30/-30/-30]

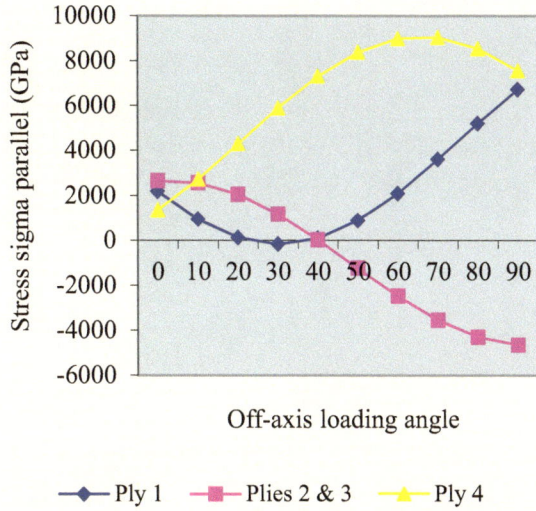

Figure 5: Stress $\sigma_{\parallel}$ in plies' laminate [30/0/0/-30]

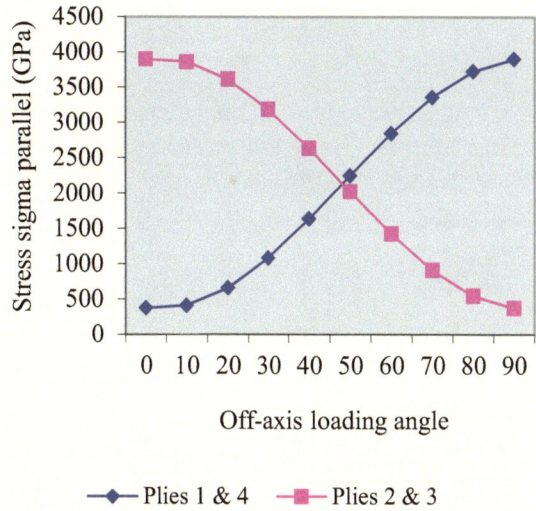

Figure 6: Stress $\sigma_{\parallel}$ in plies' laminate [90/0/0/90]

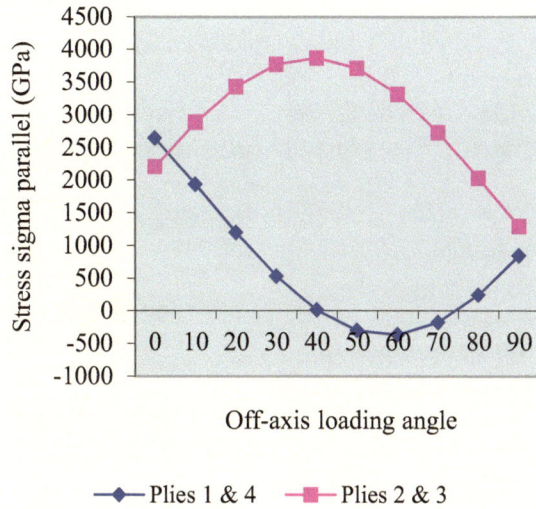

Figure 7: Stress $\sigma_{\parallel}$ in laminate [30/-30/-30/30]

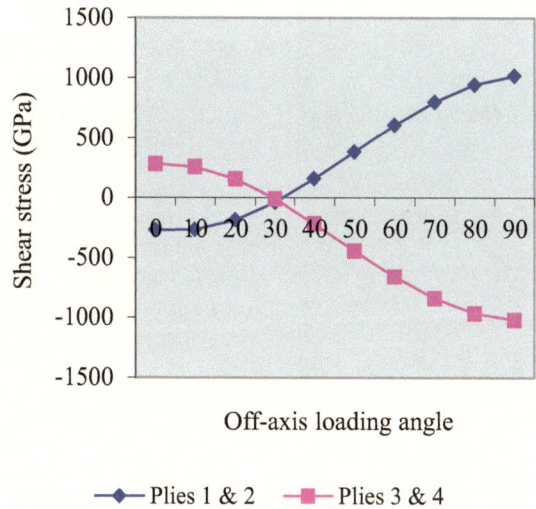

Figure 8: Shear stress in laminate [30/30/-30/-30]

Table 2: Example of computational stresses in first ply in case of [90/0/0/90] symmetric cross-ply laminate

Off-axis loading system	Normal stress $\sigma_{\parallel}$ [GPa]	Normal stress $\sigma_{\perp}$ [GPa]	Shear stress $\tau_{\#}$ [GPa]
0°	373.01363977	103.92906445	-100.00000000
10°	412.30478474	103.07095551	213.84886691
20°	659.32319009	97.67613434	501.90440441
30°	1084.27479052	88.39529598	729.42286341
40°	1635.90415140	76.34784651	868.96215995
50°	2247.67663063	62.98688614	903.69179548
60°	2845.80343841	49.92394388	829.42286342
70°	3358.14165431	38.73460334	655.11329304
80°	3722.89572819	30.76846414	401.78739109
90°	3896.07093555	26.98636024	100.00000002

3. CONCLUSION

A strong anisotropy at all considered laminates can be noticed. Stresses along fibers direction are up to ten times greater than those transverse to fibers direction. The theoretical approach can be used to determine computational stresses in various composite laminates, stresses that can be compared with experimental results obtained by different methods.

REFERENCES

[1] Schürmann H., Konstruiren mit Faser-Kunststoff-Verbunden, Springer, 2005.
[2] Teodorescu-Draghicescu H., Vlase S., Scutaru M.L., Serbina L., Calin M.R., Hysteresis Effect in a Three-Phase Polymer Matrix Composite Subjected to Static Cyclic Loadings, Optoelectron. Adv. Mater. – Rapid Comm. 5(3), 273, 2011.
[3] Vlase S., Teodorescu-Draghicescu H., Motoc D.L., Scutaru M.L., Serbina L., Calin M.R., Behavior of Multiphase Fiber-Reinforced Polymers Under Short Time Cyclic Loading, Optoelectron. Adv. Mater. – Rapid Comm. 5(4), 419, 2011.
[4] Vlase S., Teodorescu-Draghicescu H., Calin M.R., Serbina L., Simulation of the Elastic Properties of Some Fibre-Reinforced Composite Laminates Under Off-Axis Loading System, Optoelectron. Adv. Mater. – Rapid Comm. 5(4), 424, 2011.
[5] Teodorescu-Draghicescu H., Stanciu A., Vlase S., Scutaru M.L., Calin M.R., Serbina L., Finite Element Method Analysis Of Some Fibre-Reinforced Composite Laminates, Optoelectron. Adv. Mater. – Rapid Comm. 5(7), 782, 2011.
[6] Teodorescu-Draghicescu H., Vlase S., Homogenization and Averaging Methods to Predict Elastic Properties of Pre-Impregnated Composite Materials, Computational Materials Science, 50(4), 1310, Elsevier, Feb. 2011.
[7] Stanciu A., Teodorescu-Draghicescu H., Vlase S., Scutaru M.L., Calin M.R., Mechanical Behavior of CSM450 and RT800 Laminates Subjected to Four-Point Bend Tests, Optoelectron. Adv. Mater. – Rapid Comm. 6(3-4), 495, 2012.
[8] Vlase S., Teodorescu-Draghicescu H., Calin M.R., Scutaru M.L., Advanced Polylite composite laminate material behavior to tensile stress on weft direction, J. Optoelectron. Adv. Mater., 14(7-8), 658, 2012.
[9] Teodorescu-Draghicescu H., Scutaru M.L., Rosu D., Calin M.R., Grigore P., New Advanced Sandwich Composite with twill weave carbon and EPS, J. Optoelectron. Adv. Mater., 15(3-4), 199, 2013.
[10] Modrea A., Vlase S., Teodorescu-Draghicescu H., Mihalcica M., Calin M.R., Astalos C., Properties of Advanced New Materials Used in Automotive Engineering, Optoelectron. Adv. Mater. – Rapid Comm. 7(5-6), 452, 2013.
[11] Vlase S., Purcarea R., Teodorescu-Draghicescu H., Calin M.R., Szava I., Mihalcica M., Behavior of a new Heliopol/Stratimat300 composite laminate, Optoelectron. Adv. Mater. – Rapid Comm. 7(7-8), 569, 2013.
[12] Hull D., Clyne T.W., An Introduction to Composite Materials, CUP, 1996.

ON THE SONIC COMPOSITES WITH DEFECTS

Iulian Girip[1], Rodica Ioan[1,2], Mihaela Alexandra Popescu[1], Ligia Munteanu[1], Veturia Chiroiu[1]
[1]Institute of Solid Mechanics, Bucharest,
[2]University Spiru Haret, Bucharest

Dedicated to the memory of Prof. Petre P. Teodorescu (1929-2013).

Abstract: The sound attenuation in a sonic composite with point defects is studied using a new method that combines the features of the cnoidal method and the genetic algorithm. Acoustic scatterers are composed by piezoceramic hollow spheres of functionally graded materials - the Reddy and cosine graded hollow spheres.

This method enables to obtain the dispersion relation for defect modes, and the prediction of the evanescent nature of the modes inside the band-gaps.

Key-Words: Sonic composites, sound attenuation, defects, evanescent waves, full band-gaps.

1. INTRODUCTION

A sonic composite is a finite size periodic array composed of scatterers embedded in a homogeneous material [1-3]. A great number of applications based on the sonic composites are explained by the existence of the band-gaps into the acoustic filters, acoustic barriers or wave guides [4-7]. The generation of large band-gaps is due to the superposition of multiple reflected waves within the array according to the Bragg's theory, and consequently, it is connected with a large acoustic impedance ratio between the scatterers' material and the matrix' material, respectively.

The band-gaps correspond to the Bragg reflections that occur at different frequencies inverse proportional to the central distance between two scaterers. The waves are reflected completely from this periodic array in the frequency range where all partial band-gaps for the different periodical directions overlap.

This makes sharp bends of the wave-guide in the sonic composite. The evanescent waves characterized by complex wave numbers, are distributed across the boundary of the waveguide into the surrounding composite by several times the lattice constant [8].

Recent experimental and theoretical results [9-11] show that the presence of defects in sonic composite is related to the generation of localized modes in the vicinity of the point defect with a significant evanescent behavior of the waves outside the defect point. This means that the evanescent modes are related to the existence of the band-gaps where no real wave number exists. The authors have revealed that the level of the sound is higher inside the vicinity of the defect point than into the composite. Recent works show the calculation of complex band structures for photonic crystals [12-14], using the explicit matrix formulation and the approximation of the supercells. This technique enables to be extended to the 2D complete sonic composites, as well as in sonic composites with point defects.

The goal of this paper is to propose an alternative method for obtaining the band structures of the 3D sonic composites without/ with point defects. The point defects are vacancies or foreign interstitial atoms which are supported by the interfaces between the hollow spheres and the matrix. The proposed method is used to simulate a sonic plate composed of an array of acoustic scatterers which are piezoceramic hollow spheres embedded in an epoxy matrix. The scatterers are made from functionally graded materials with radial polarization, which support the Reddy and cosine laws [15-17]. Readers are also referred to [18-20] for vibrations of solid spheres of functionally graded materials and for the wave propagation in functionally graded materials.

The proposed approach is based on the theory of piezoelectrics coupled with the cnoidal method and a genetic algorithm. For a single sphere made from a functionally graded material, the free vibration problem was analyzed in [21-25].

2. THEORY

The sonic composite is consisting of an array of acoustic scatterers embedded in an epoxy matrix. The acoustic scatterers are hollow spheres made from a nonlinear isotropic piezoelectric ceramic, while the matrix is made from a nonlinear isotropic epoxy resin (Figure 1). The sonic plate consists of 72 local resonators of diameter a. A rectangular coordinate system $Ox_1x_2x_3$ is $Ox_1x_2x_3$ employed. The origin of the coordinate system is located at the left end, in the middle plane of the sample, with the axis Ox_1 in-plane and normal to the layers and the axis Ox_3 out-plane and normal to the plate. The length of the plate is l, its width is d, while the diameter of the hollow sphere is a and its thickness is $e > a$. In order to avoid unphysical reflections from the boundaries of the specimen, we have implemented the absorbing boundary conditions in the x_1-direction, at $x_1 = 0$ and $x_1 = l$. A transducer and a receiver are located at $x_1 = b$ and $x_1 = l - b$, respectively. The role of the transducer is to inject into the plate the plane monochromatic waves propagating in the x_1-direction.

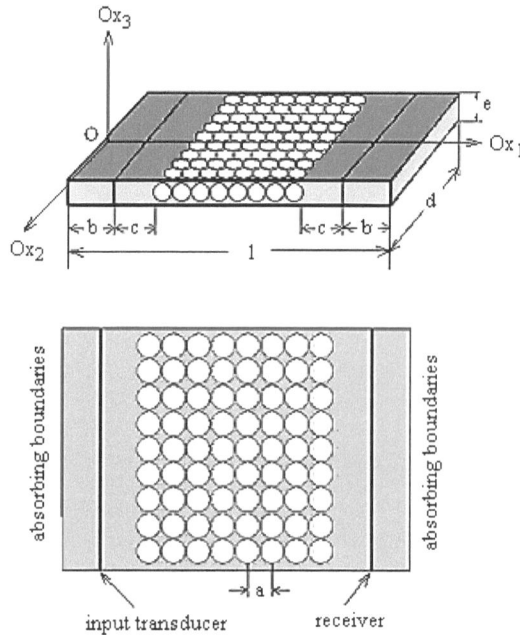

Figure 1: Sketch of the sonic composite.

The governing equations are given by [2, 3]

$$rA_{,r} = MA, \quad rB_{,r} = PB,$$

(1)

where

$$B = [\Sigma_{rr}, \Sigma_2, G, w, \Lambda_r, \phi]^T,$$

$$M = \begin{bmatrix} -2 & -C_{66}(\nabla^2 + 2) + r^2\rho\dfrac{\partial^2}{\partial t^2} \\ C_{44}^{-1} & 1 \end{bmatrix},$$

with $\nabla^2 = \dfrac{\partial^2}{\partial\theta^2} + \cot\theta\dfrac{\partial}{\partial\theta} + \csc^2\theta\dfrac{\partial^2}{\partial\varphi^2}$. It should be noted that equation $(1)^1$ is related to two state variables,

namely $A = [\Sigma_1, F]^T$, while equations $(1)^2$ are related to six state variables $\Sigma_{rr}, \Sigma_2, G, w, \Lambda_r, \phi$

$$\Sigma_{rr} = r\sigma_{rr} = C_{13}S_{\theta\theta} + C_{13}S_{\varphi\varphi} + C_{33}S_{rr} + f_{33}r\phi_{,r}$$

$$\Lambda_r = rD_r = f_{31}S_{\theta\theta} + f_{31}S_{\varphi\varphi} + f_{33}S_{rr} - \zeta_{33}r\phi_{,r},$$

$$u_\theta = -\csc\theta F_{,\varphi} - G_{,\theta}, \quad u_\varphi = F_{,\theta} - \csc\theta G_{,\varphi}, \quad u_r = w,$$

$$\Sigma_{r\theta} = -\csc\theta\Sigma_{1,\varphi} - \Sigma_{2,\theta}, \quad \Sigma_{r\varphi} = \Sigma_{1,\theta} - \csc\theta\Sigma_{2,\varphi}. \tag{2}$$

where σ_{ij} is the stress tensor, ϕ is the electric potential, D_i is the electric displacement vector, C_{ij} are the elastic constants, $C_{66} = (C_{11} - C_{12})/2$, f_{ij} are the piezoelectric constants f_{ij}, ζ_{ij} are the dielectric constants, and $i = r, \theta, \varphi$. The elastic, piezoelectric and dielectric constants are arbitrary functions of the radial coordinate r.

On denoting the components of the strain tensor and displacement vector by ε_{ij} and u_i, $i = r, \theta, \varphi$, respectively.

The nonzero components of the matrix P are given by

$$P_{11} = 2\beta - 1, \quad P_{12} = \nabla^2, \quad P_{13} = k_1\nabla^2, \quad P_{14} = -2k_1 + r^2\rho\dfrac{\partial^2}{\partial t^2},$$

$$P_{15} = 2P_{25} = -P_{64} = 2\gamma, \quad P_{21} = \beta, \quad P_{22} = -2, \quad P_{23} = k_2\nabla^2 - 2C_{66} + r^2\rho\dfrac{\partial^2}{\partial t^2},$$

$$P_{24} = -k_1, \quad P_{32} = C_{44}^{-1}, \quad P_{33} = P_{34} = -P_{55} = 1, \quad P_{36} = C_{44}^{-1}f_{15}, \quad P_{41} = \alpha^{-1}\zeta_{33}, \quad P_{43} = \beta\nabla^2,$$

$$P_{44} = -2\beta, \quad P_{45} = \alpha^{-1}f_{33}, \quad P_{52} = C_{44}^{-2}f_{15}\nabla^2, \quad P_{56} = k_3\nabla^2, \quad P_{61} = \alpha^{-1}f_{33}, \quad P_{63} = \gamma\nabla^2, \quad P_{65} = -\alpha^{-1}C_{33}, \tag{3}$$

where

$$\alpha = C_{33}\zeta_{33} + f_{33}^2, \quad \beta = \alpha^{-1}(C_{13}\zeta_{33} + f_{31}f_{33}), \quad \gamma = \alpha^{-1}(C_{13}f_{33} - C_{33}f_{31}),$$

$$k_1 = 2(C_{13}\beta + f_{31}\gamma) - (C_{11} + C_{12}), \quad k_2 = 0.5k_1 - C_{66}, \quad k_3 = \zeta_{11} + f_{15}^2 C_{44}^{-1}.$$

Consider now two piezoceramic hollow spheres with the ratio of the inner and outer radii ξ_0. Two laws represent the functionally graded property of the material. The first one is the Reddy law given by [15-17]

$$M = M_p\mu^\lambda + M_z(1 - \mu^\lambda), \tag{4}$$

where μ is the gradient index [22], M_p and M_z are material constants of two materials, namely PZT-4 and ZnO. The case $\mu = 0$ corresponds to a homogeneous PZT-4 hollow sphere and $\mu \to \infty$, to a homogeneous ZnO hollow sphere. The second law is expressed as

$$M = M_p\cos\mu + M_z(1 - \cos\mu). \tag{5}$$

At the interfaces between the hollow spheres and the matrix, sharp periodic boundary conditions for the displacement and traction vectors are added [29] for sonic composites without/with defects. In addition, in the case of the point defects situated in the interfaces between the hollow spheres and the matrix, a new equation must be added to reflect the dynamic of the concentration of the defects.

The distribution of the defects of the concentration $c(r)$ is characterized by the diffusive contribution $-D\nabla c$, where D is the diffusion coefficient, and the forced contribution ηcF where η is the mobility and D the driving force which acts upon the defects. The mobility and the diffusivity are related through the Nernst-Einstein relation $D = \eta kT$ where k is the Boltzmann's constant and T the absolute temperature.

The rate of the change of the concentration in the presence of the source $S(r,t)$ is [26]

$$\frac{\partial c}{\partial t} = -\nabla \left(-D\nabla c \right) + S(r,t), \tag{6}$$

or

$$\frac{\partial c}{\partial t} = D \left(\nabla^2 c - \frac{1}{kT} \left(F\nabla c + c\nabla F \right) \right) + S(r,t). \tag{7}$$

The source $S(r,t)$ represents the rate at which the defects are created at any point.

3. THE BEHAVIOR OF THE SONIC COMPOSITE WITH DEFECTS

Consider a plate with the length $l = 18\text{cm}$ and width $d = 11\text{cm}$, while the diameter of the hollow sphere and its thickness are $a = 10.5\text{mm}$ and $e = 12\text{mm}$, respectively, and $\xi_0 = 0.3$. The numerical results are carried out for the following constants [2]:

for PZT-4

$$C_{11} = 13.9 \times 10^{10} \text{N/m}^2, \quad C_{12} = 7.8 \times 10^{10} \text{N/m}^2, \quad C_{13} = 7.4 \times 10^{10} \text{N/m}^2,$$
$$C_{33} = 11.5 \times 10^{10} \text{N/m}^2, \quad C_{44} = 2.56 \times 10^{10} \text{N/m}^2, \quad f_{15} = 12.7 \text{C/m}^2, \quad f_{31} = -5.2 \text{C/m}^2,$$
$$f_{33} = 15.1 \text{C/m}^2, \quad \zeta_{11} = 650 \times 10^{-11} \text{F/m}, \quad \zeta_{33} = 560 \times 10^{-11} \text{F/m}, \quad \rho = 7500 \text{kg/m}^3,$$

for ZnO

$$C_{11} = 20.97 \times 10^{10} \text{N/m}^2, \quad C_{12} = 12.11 \times 10^{10} \text{N/m}^2, \quad C_{13} = 10.51 \times 10^{10} \text{N/m}^2,$$
$$C_{33} = 21.09 \times 10^{10} \text{N/m}^2, \quad C_{44} = 4.25 \times 10^{10} \text{N/m}^2, \quad f_{15} = -0.59 \text{C/m}^2, \quad f_{31} = -0.61 \text{C/m}^2,$$
$$f_{33} = 1.14 \text{C/m}^2, \quad \zeta_{11} = 7.38 \times 10^{-11} \text{F/m}, \quad \zeta_{33} = 7.83 \times 10^{-11} \text{F/m}, \quad \rho = 5676 \text{kg/m}^3,$$

and for epoxy-resin

$$\lambda^e = 42.31 \times 10^9 \text{N/m}^2, \quad \mu^e = 3.76 \times 10^9 \text{N/m}^2, \quad A^e = 2.8 \times 10^9 \text{N/m}^2, \quad B^e = 9.7 \times 10^9 \text{N/m}^2,$$
$$C^e = -5.7 \times 10^9 \text{N/m}^2, \text{ and } \rho^e = 1170 \text{ kg/m}^3.$$

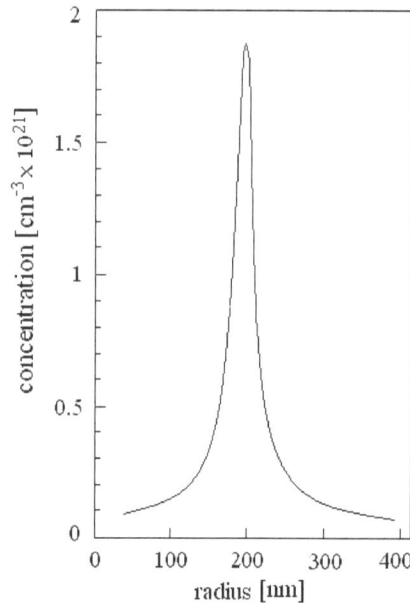

Figure 2: Variation of the concentration with respect to the radius of Σ which encloses the point defect.

The independent sets of equations (1) yield two independent classes of free vibrations. The first class does not involve the piezoelectric or dielectric parameters, being identical to the one for the corresponding spherically isotropic elastic sphere. The second class depends on the piezoelectric or dielectric parameters. With the increase of the gradient index μ, the natural frequencies increase for all modes and functionally graded laws, the variation being more significant when $\mu \leq 10$. For $\mu \to \infty$ the variation of natural frequencies is not significant with respect to those of $\mu = 10$. It is seen that for a piezoceramic hollow sphere, the piezoelectric effect consists of increasing the values for the natural frequencies in both classes of vibrations. If $\xi = 2r/a$ increases, the natural frequencies increase for the first class of vibrations and decrease for the second class.

The propagation of sound is characterized by the superposition of multiply reflected waves. Featuring of the length scale a, the structure of the full band-gap can be better understood by representing the linear band structure (dispersion curve).

The simulation is carried out for a variation of the defects concentration represented in Figure 2 with respect to the radius of Σ which encloses the point defect (vacancies or foreign interstitial atoms), for 300K. In this figure only the effect of diffusion is shown, without any stress gradient, after 1000sec.

Figure 3 plots the dispersion curve including the first partial band-gaps for the composite without defects and for the composite with defects ($c = 1.5 \times 10^{21} \text{cm}^{-3}$). The reduced units for the frequency are $\omega a / 2\pi c_0$, with c_0 the speed of sound in air. We see that the point defects confine acoustic waves in localized modes and in consequence the band-gaps are larger than in the case of the complete composite.

The guided waves are accompanied by evanescent waves which extend to the periodic array of the scatterers surrounding the wave-guide. Using the Joannopoulus representation [8] for the bad-gap structure, Figure 4 presents the band structure with the evanescent modes with exponential decay for the sonic composite without defects. The modes present purely imaginary wave vectors. The central grey region is the full band-gap ranged between 8.02 kHz and 8.72 kHz, given by the real part of the wave vector constrained in the first Brillouin zone for each frequency. The left region represents the imaginary part of the wave vector for longitudinal direction frequency (tension/compression), while the right region is the imaginary part of the wave vector for transverse direction frequency (shear). The red lines represent the imaginary part of the wave vector of the evanescent modes inside the bad-gap.

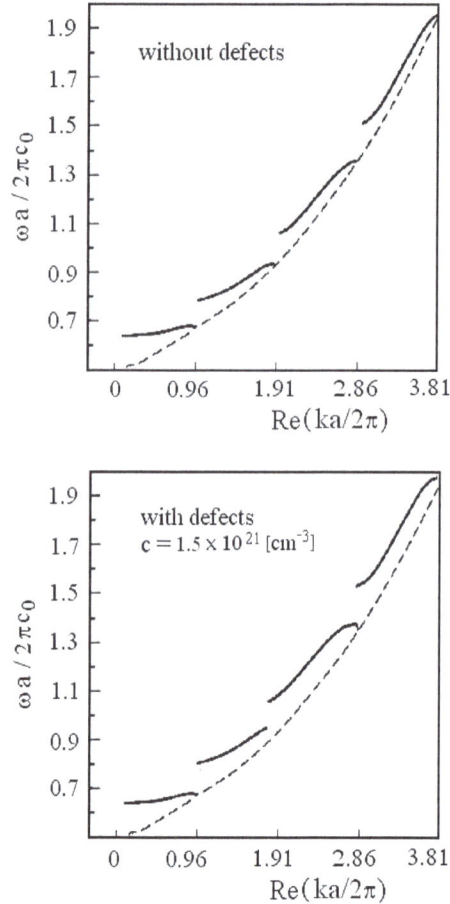

1.   **Figure 3:** Linear dispersion for sonic composite without/with defects.

If we want to have a full band-gap, we must have structures with band-gaps for both longitudinal and transverse waves in the same frequency region.

The difference in the sound velocities between transverse and longitudinal modes causes partial gaps at different frequencies. If the mechanical contrast is small, these partial gaps are narrow and do not overlap. As mechanical contrast increases, the partial gaps widen and begin to overlap in the same frequency region leading to the appearance of a full band-gap independent of the polarization.

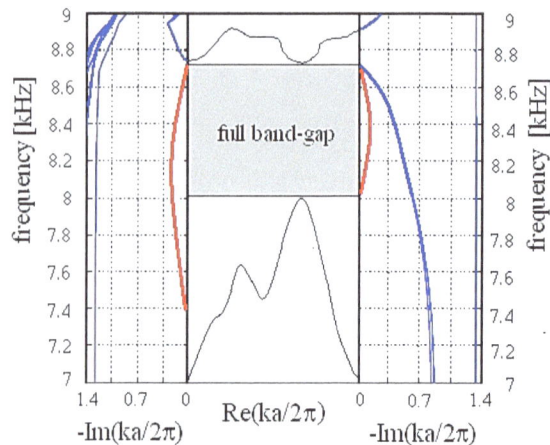

Figure 4: Band structure for the sonic composite without defects in the case of Reddy law.

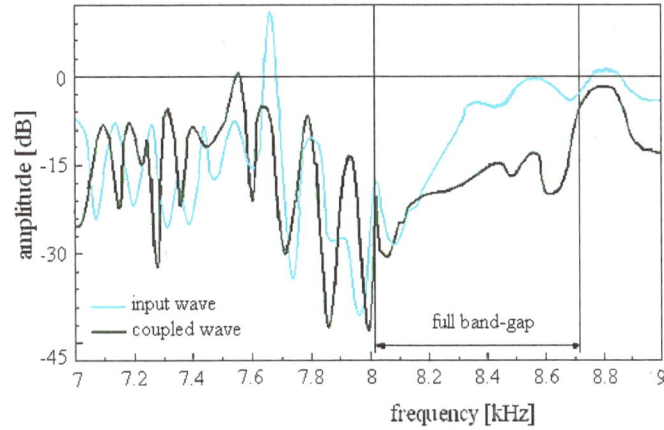

2. **Figure 5:** The input and coupled waves for sonic composite without defects in the case of Reddy law.

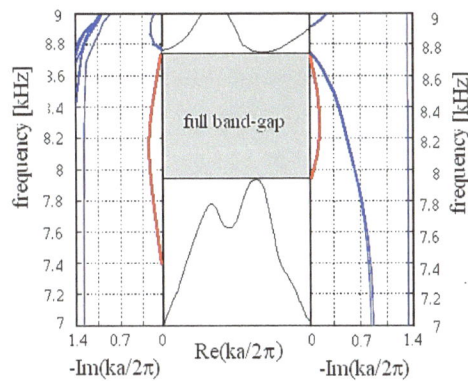

Figure 6: Band structure for the sonic composite with defects ($c = 1.5 \times 10^{21} \, \text{cm}^{-3}$) in the case of Reddy law.

It is strongly expected that mode coupling waves arise between adjacent wave-guides. The output of the coupled modes is compared with the input waves, as shown in Figure 5 in the case of Reddy law and $\sigma_e = 2.2 \text{kPasm}^{-2}$, Figure 6 presents the band structure for the sonic composite with defects ($c = 1.5 \times 10^{21} \, \text{cm}^{-3}$) in the case of the Reddy law. We observe that this time the portion widens from 7.95 kHz to 8.76 kHz. The coupled and the input waves are shown in Figure 7 for $\sigma_e = 2.2 \text{kPasm}^{-2}$. The difference from the composite without defects consists only in the size and structure of the full band-gaps.

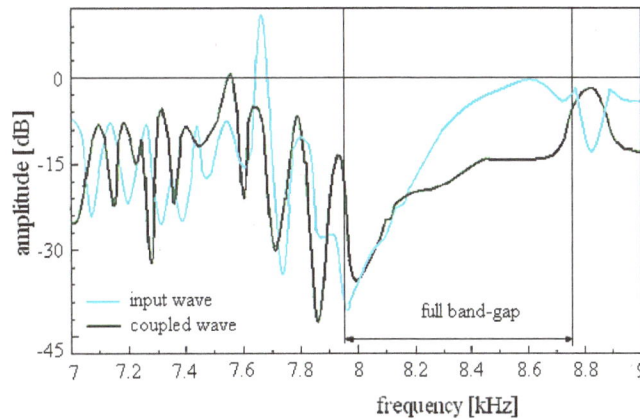

3. **Figure 7:** The input and coupled waves for sonic composite with defects ($c = 1.5 \times 10^{21} \, \text{cm}^{-3}$) in the case of Reddy law.

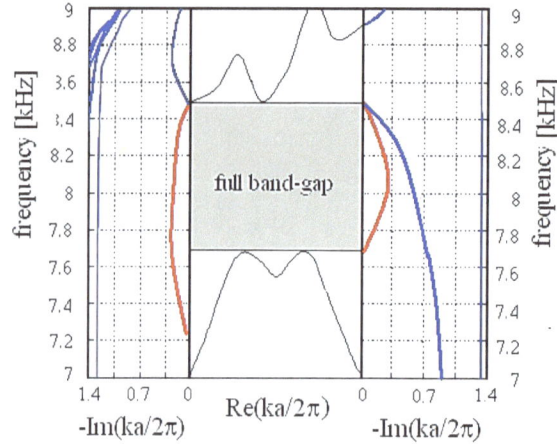

Figure 8: Band structure for the sonic composite with defects ($c = 1.5 \times 10^{21} \text{cm}^{-3}$) in the case of the cosine law.

Figure 8 presents the band structure for the sonic composite with defects ($c = 1.5 \times 10^{21} \text{cm}^{-3}$) in the case of the cosine law. We observe that the full band-gap has the same length but has undergone a translation, i.e. to 7.67 kHz and 8.48 kHz.

The length of the full band-gap as function of the concentration of the point defects is represented in Figure 9, for both Reddy law and cosine law, respectively. For Reddy law, the length increases with concentration and shows a flat portion starting to the concentration of 1.5 with a slight decrease of concentration at around 0.88 kHz. For the cosine law, the length shows two flat zones and a visible drop of concentration around 0.64 kHz.

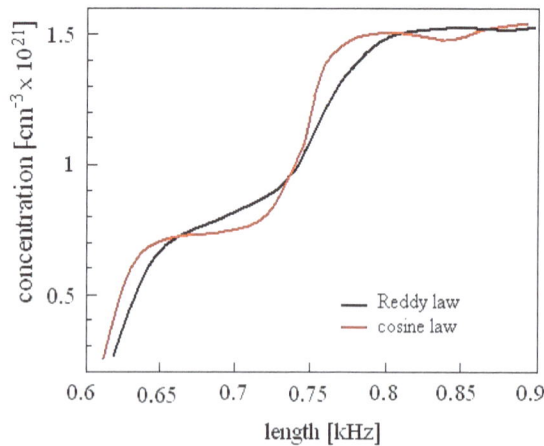

Figure 9: The length of the full band-gap as function of the concentration of the point defects.

6. CONCLUSIONS

Analytical and numerical solutions and verification to real results are presented in this paper for propagation of waves in sonic composites without/with defects. The point defects are vacancies or foreign interstitial atoms which are situated in the interfaces between the hollow spheres and the matrix. It is shown that the point defects confine acoustic waves in localized modes, and the defect modes are created within the band-gaps of the composite. The localization of the sound in the defect regions leads to increase in the intensity of the acoustic wave's interactions. Such localizations increase the length of the full band-gap frequency for sonic composites with defects by 15-20% compared with the values for similar composites without defects.

The scattering problem inside the sonic composite is solved by a method which combines the cnoidal method with a genetic algorithm. The reason for choosing the cnoidal method lies in the facts that the governing equations are reduced to Weierstrass equations with polynomials of higher-order, for a change of variable $t \rightarrow x - ct$, with c a constant. The solutions are expressed as a sum of the linear and the nonlinear superposition of cnoidal vibrations, respectively

ACKNOWLEDGEMENT

The authors gratefully acknowledge the financial support of the National Authority for Scientific Research ANCS/UEFISCDI through the project PN-II-ID-PCE-2012-4-0023, Contract nr.3/2013.

REFERENCES

[1]   M. Hirsekorn, P.P. Delsanto, N.K. Batra, P. Matic, *Modelling and simulation of acoustic wave propagation in locally resonant sonic materials*, Ultrasonics, 42, 231–235, 2004.

[2]   L. Munteanu, V. Chiroiu, *On the dynamics of locally resonant sonic composites*, European Journal of Mechanics-A/Solids, 29(5), 871–878, 2010.

[3]   L. Munteanu, *Nanocomposites* , editura Academiei, Bucharest, 2012.

[4]   J.V. Sánchez-Pérez, D. Caballero, R. Mártinez-Sala, C. Rubio, J. Sánchez-Dehesa, F. Meseguer, J. Llinares, F. Gálvez, *Sound attenuation by a two-dimensional array of rigid cylinders*, Phys. Rev. Lett. 80(24), 5325-5328, 1998.

[5]   J.V. Sánchez-Pérez, C. Rubio, R. Martínez-Sala, R. Sánchez-Grandia, V. Gómez, *Acoustic barrier based on periodic arrays of scatterers*, Appl. Phys. Lett. 81, 5240-5242, 2002.

[6]   A. Khelif, A. Choujaa, B. Djafari-Rouhani, M. Wilm, S. Ballandras, V. Laude, *Trapping and guiding of acoustic waves by defect modes in a full-band-gap ultrasonic crystal*, Phys. Rev. B 68, 214301, 2003.

[7]   A. Khelif, M. Wilm, V. Laude, S. Ballandras, and B. Djafari-Rouhani, *Guided elastic waves along a rod defect of a two-dimensional photonic crystal*, Phys. Rev. E 69(6), 067601, 2004.

[8]   J.D. Joannopoulus, S.G. Johnson, J.N. Winn, R.D. Meade, *Photonic Crystals. Molding the Flow of Light*, Princeton University Press, 2008.

[9]   V. Romero-García, J.V. Sánchez-Pérez, L.M. Garcia-Raffi, *Evanescent modes in sonic crystals: Complex relation dispersion and supercell approximation*, Journal of Applied Physics, 108(4), 108-113, 2010.

[10]   F. Wu, Z. Hou, Z. Liu, T. Liu, *Point defect states in two-dimensional phononic crystals*, Phys. Lett. A 292, 198, 2001.

[11]   Y. Zhao, L.B. Yuan, *Characteristics of multi-point defect modes in 2D photonic crystals*, J. Phys. D: Appl. Phys. 42(1), 015403, 2009.

[12]   T. Miyashita, *Full band gaps of sonic crystals made of acrylic cylinders in air-numerical and experimental investigations*, Japanese Journal of Applied Physics, 41, 3170-1-3175, 2002.

[13]   V. Laude, Y. Achaoui, S. Benchabane, A. Khelif, *Evanescent Bloch waves and the complex band structure of phononic crystals*, Phys. Rev. B 80, 092301 (2009).

[14]   R. Sainidou, N. Stefanou, *Guided and quasiquided elastic waves in photonic crystal slabs*, Phys. Rev. B 73, 184301, 2006.

[15]   J.N. Reddy, *A Generalization of Two-Dimensional Theories of Laminated Composite Laminate*, Comm. Appl. Numer. Meth., 3, 173–180, 1987.

[16]   J.N. Reddy, C.F. Liu, *A higher-order theory for geometrically nonlinear analysis of composite laminates*, NASA Contractor Report 4056, 1987.

[17] J.N. Reddy, C.M. Wang, S. Kitipornchai, *Axisymmetric bending of functionally graded circular and annular plates*, Eur. J. Mech., A/Solids 18, 185–199, 1999.

[18] B.M. Singh, J. Rokne, R.S. Dhaliwal, *Vibrations of a solid sphere or shell of functionally graded materials*, European Journal of Mechanics-A/Solids, 27, 3, 460–468, 2008.

[19] B. Collet, M. Destrade, G.A. Maugin, *Bleustein-Gulyaev waves in some functionally graded materials*, European Journal of Mechanics-A/Solids, 25, 5, 695–706, 2006.

[20] A. Berezovski, J. Engelbrecht, G.A. Maugin, *Numerical simulation of two-dimensional wave propagation in functionally graded materials*, European Journal of Mechanics-A/Solids, 22, 2, 257–265, 2003

[21] V. Chiroiu, L. Munteanu, *On the free vibrations of a piezoceramic hollow sphere*, Mech. Res. Comm., Elsevier, 34, 2, 123–129, 2007.

[22] W.Q. Chen, L.Z. Wang, and Y. Lu, *Free vibrations of functionally graded piezoceramic hollow spheres with radial polarization*, J. Sound Vibr., 251, 1, 103–114, 2002.

[23] W. Q. Chen, *Vibration theory of non-homogeneous, spherically isotropic piezoelastic bodies*, J. Sound Vibr., 229, 833–860, 2000.

[24] M. Mihailescu, V. Chiroiu, *Advanced mechanics on shells and intelligent structures,* Editura Academiei, Bucharest, 2004.

[25] R.A. Toupin, *Piezoelectric relations and the radial deformation of a polarized spherical shell*, J. Acoust. Soc. Am., 31, 315–318, 1959.

[26] Britton, D.T., Harting, M., *The influence of strain on point defect dynamics*, Advanced Engineering Materials, 4(8), 628-635, 2008.

(c) Dragos Catalin

VIBRATIONS OF HETEROGENEOUS CURVED BEAMS SUBJECTED TO A RADIAL FORCE AT THE CROWN POINT

György Szeidl[1], László Kiss[1]

[1] Department of Mechanics, University of Miskolc, Miskolc, HUNGARY

Gyorgy.szeidl@uni-miskolc.hu; mechkiss@uni-miskolc.hu

VIBRATIONS OF HETEROGENEOUS CURVED BEAMS SUBJECTED TO A RADIAL FORCE AT HE CROWN POINT

György Szeidl[1], László Kiss[1]

[1]Department of Mechanics, University of Miskolc, Miskolc, HUNGARY
gyorgy.szeidl@uni-miskolc.hu, mechkiss@uni-miskolc.hu

Abstract: This paper is concerned with the vibrations of heterogenous curved beams under the assumption that the load on the beam is perpendicular to the centerline. It is assumed that (a) the radius of curvature is constant and (b) the Young modulus and Poisson number depend on the cross sectional coordinates only. We have the following objectives: (1) to determine the Green function matrices for pinned beams, fixed beams and beams fixed at one end and pinned at the other end provided that the beam is subjected to a radial load; (2) to clarify how the load affects the natural frequencies if the beam is subjected to a radial force (a vertical force) at the crown point; (3) to develop such a numerical model which makes possible to determine how the natural frequencies are related to the load. We shall present the computational results in a graphical format.

Keywords: curved beams, heterogeneous material, natural frequency as a function of the load, Green function matrices

1. INTRODUCTION

Curved beams are used in various practical applications. We can mention, for example, arch bridges, roof structures, and stiffeners in aerospace applications. We remark that research into the mechanical behavior of curved beams began in the 19th century – see book [1, 1944] by Love for further details. The free vibrations of curved beams have been extensively investigated. Survey papers on the vibrations of these beams were published by Markus and Nanasi [2, 1981], Laura and Maurizi [3, 1987] as well as Chidampram and Lessia [4, 1993]. It may be worth citing the PhD thesis by Szeidl [5, 1975] which clarifies within the framework of the linear theory how the extensibility of the centerline affects the free vibrations and stability of circular beams subjected to a constant radial load (dead load). Solutions for the natural frequencies in [5, 1975] were computed by utilizing different numerical models. One of them is based on the use of the Green function matrix of the corresponding boundary value problem. Unfortunately the results of this work have not been published in English.

Paper by [6, 2005] Lawther attacks the problem how a prestressed state of the body affects the natural frequencies a more form. He studies finite dimensional multiparameter eigenvalue problems and comes to the conclusion that for multiparameter problems, the eigenvalue part of the solution is described by interaction curves in an eigenvalue space, and every such eigenvalue solution has an associated eigenvector. If all points on a curve have the same eigenvector then the curve is necessarily a straight line, but the converse is far more complex.

In the light of Lawther's results there arises the question how the natural frequencies change if a curved beam is subjected to a radial (vertical) load at the crown point. During our investigations we shall assume that the curved beam is made of heterogenous, isotropic and linearly elastic material. As regards heterogeneity it is assumed that the elastic parameters can be varied arbitrarily over the beam cross section but they are independent of the coordinate perpendicular to the cross section. Under these assumptions our main objectives are as follows: (1) derivation of those boundary value problems which make it possible to clarify how the radial load affects the natural frequencies of the beam; (2) determination of the Green function matrices that can be used to reduce the eigenvalue problems set up for the natural frequencies (which depend on the load) to eigenvalue problems governed by systems of Fredholm integral equations; (3) to reduce the eigenvalue problem to an algebraic one which can be solved numerically. The corresponding computational results are then presented in a graphical format.

The paper is organized into eight sections. Section 2 is a summary of the governing equations. After having defined the Green function matrices we reduced the eigenvalue problems to be solved to eigenvalue problems governed by Fredholm integral equations in Section 3. As regards the solution algorithm Section 4 provides an outline. Calculation of the Green function matrices is detailed in Section 5. Relationships between the axial strain on the centerline and the load are presented in Section 6 which also contains formulae for the critical value of the axial strain. Computational result are shown in Section 7. The last section is a conclusion.

2. GENERALIZATIONS OF SOME CLASSICAL FORMULAE

On the basis of article [7, 2012] the most important formulae are all gathered in this section. Figure 1. shows a part of the beam with the applied curvilinear coordinate system ($\xi = s, \eta, \zeta$) and a pinned-pinned beam. Observe that the coordinate line $\xi = s$ coincides with the so-called (E-weighted) centerline. Point C determines the location of the centerline uniquely via equation

$$S_{e\eta} = \int_A E(\eta, \zeta)\zeta\, \mathrm{d}A = 0 .\tag{1}$$

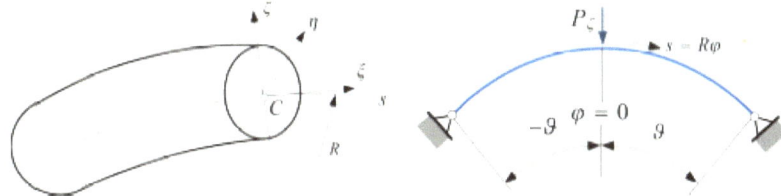

Figure 1. (a) The coordinate system, (b) Pinned-pinned beam

Here $S_{e\eta}$ is the E-weighted first moment of the cross section with respect to the axis η. By assumption the cross section of the beam is symmetric with respect to the axis ζ and the Young modulus depends on the cross sectional coordinates only: $E(\eta, \zeta) = E(-\eta, \zeta)$.

For our latter considerations let us introduce the integrals

$$A_e = \int_A E(\eta, \zeta)\mathrm{d}A , \qquad I_{e\eta} = \int_A E(\eta, \zeta)\zeta^2\mathrm{d}A \tag{2}$$

which are referred to as the E-weighted area and the E-weighted moment of inertia.

In what follows we shall separate the load-induced, and otherwise time-independent, mechanical quantities from those which belong to the vibrations of the loaded beam. The latter ones are actually the time-dependent increments and are uniformly denoted by a subscript $_b$. Let u_o, w_o and R be the tangential and radial displacements and the radius of the centerline, respectively. The connection between the coordinate line s and the angle coordinate φ is defined by $s = R\varphi$.

The axial strain $\varepsilon_{o\xi}$ and the rotation $\psi_{o\eta}$ on the centerline can be given in terms of u_o and w_o as

$$\varepsilon_{o\xi} = \frac{\mathrm{d}u_o}{\mathrm{d}s} + \frac{w_o}{R}, \qquad \psi_{o\eta} = \frac{u_o}{R} - \frac{\mathrm{d}w_o}{\mathrm{d}s} .\tag{3}$$

On the basis of the principle of virtual work, after some here omitted manipulations, we obtain that the axial force N and the bending moment M should satisfy equilibrium equations

$$\frac{\mathrm{d}N}{\mathrm{d}s} + \frac{1}{R}\left[\frac{\mathrm{d}M}{\mathrm{d}s} - \left(N + \frac{M}{R}\right)\psi_{o\eta}\right] + f_t = 0 ,\tag{4a}$$

$$\frac{\mathrm{d}}{\mathrm{d}s}\left[\frac{\mathrm{d}M}{\mathrm{d}s} - \left(N + \frac{M}{R}\right)\psi_{o\eta}\right] - \frac{N}{R} + f_n = 0 \tag{4b}$$

where the intensity of the distributed loads on the centerline in the tangential and normal directions are denoted by f_t and f_n, respectively.

The axial force N and the bending moment M are related to the deformations via the Hook law:

$$N = \frac{I_{e\eta}}{R^2}\varepsilon_{o\xi} - \frac{M}{R}, \qquad M = -I_{e\eta}\left(\frac{\mathrm{d}^2 w_o}{\mathrm{d}s^2} + \frac{w_o}{R^2}\right) , \qquad N + \frac{M}{R} = \frac{I_{e\eta}}{R^2}\varepsilon_{o\xi}, \quad \text{where} \quad m = \frac{A_e R^2}{I_{e\eta}} - 1 .\tag{5}$$

For our later considerations we introduce dimensionless displacements and a notational convention for the derivatives:

$$U_o = \frac{u_o}{R}, \qquad W_o = \frac{w_o}{R}; \qquad (\ldots)^{(n)} = \frac{\mathrm{d}^n(\ldots)}{\mathrm{d}\varphi^n}, \quad n = 0, 1, 2, \ldots .\tag{6}$$

Upon substitution of the Hooke law (5) and then the kinematical quantities (3) into equilibrium equations (4) we arrive at a system of differential equations:

$$\begin{bmatrix} 0 & 0 \\ 0 & 1 \end{bmatrix}\begin{bmatrix} U_o \\ W_o \end{bmatrix}^{(4)} + \begin{bmatrix} -m & 0 \\ 0 & 2 - m\varepsilon_{o\xi} \end{bmatrix}\begin{bmatrix} U_o \\ W_o \end{bmatrix}^{(2)} +$$

$$+ \begin{bmatrix} 0 & -m \\ m & 0 \end{bmatrix}\begin{bmatrix} U_o \\ W_o \end{bmatrix}^{(1)} + \begin{bmatrix} 0 & 0 \\ 0 & 1 + m(1 - \varepsilon_{o\xi}) \end{bmatrix}\begin{bmatrix} U_o \\ W_o \end{bmatrix} = \frac{R^3}{I_{e\eta}}\begin{bmatrix} f_t \\ f_n \end{bmatrix} .\tag{7}$$

As regards the increments in the axial strain and in the rotation the corresponding formulae have a structure similar to that of equations (3):

$$\varepsilon_{mb} = \varepsilon_{o\xi\, b} + \psi_{o\eta}\psi_{o\eta\, b}, \qquad \psi_{o\eta\, b} = \frac{u_{ob}}{R} - \frac{dw_{ob}}{ds}, \qquad \varepsilon_{o\xi\, b} = \frac{du_{ob}}{ds} + \frac{w_{ob}}{R}. \tag{8}$$

Further it can be verified that the differential equations the increments in the axial force and in the bending moment should satisfy assume the forms

$$\frac{d}{ds}\left(N_b + \frac{M_b}{R}\right) - \frac{1}{R}\left(N + \frac{M}{R}\right)\psi_{o\eta\, b} + f_{tb} = 0, \tag{9a}$$

$$\frac{d^2 M_b}{ds^2} - \frac{N_b}{R} - \frac{d}{ds}\left[\left(N + \frac{M}{R}\right)\psi_{o\eta\, b} + \left(N_b + \frac{M_b}{R}\right)\psi_{o\eta}\right] + f_{nb} = 0. \tag{9b}$$

Since the process is a dynamical one – there are no changes in the original loads – it follows that the increments f_{tb} and f_{nb} are in fact forces of inertia, that is,

$$f_{tb} = -\rho_a A \frac{\partial^2 u_{ob}}{\partial t^2}, \qquad f_{nb} = -\rho_a A \frac{\partial^2 w_{ob}}{\partial t^2}. \tag{10}$$

Here A is the area and ρ_a is the averaged density over the cross section. In addition the Hooke law for the increments in the inner forces yields

$$N_b = \frac{I_{e\eta}}{R^2} m \varepsilon_{o\xi\, b} - \frac{M_b}{R}, \qquad M_b = -I_{e\eta}\left(\frac{d^2 w_{ob}}{ds^2} + \frac{w_{ob}}{R^2}\right), \qquad N_b + \frac{M_b}{R} = \frac{I_{e\eta}}{R^2} m \varepsilon_{o\xi\, b}. \tag{11a}$$

A comparison of equations (8), (9) and (11) results in the equations of motion

$$\begin{bmatrix} 0 & 0 \\ 0 & 1 \end{bmatrix}\begin{bmatrix} U_{ob} \\ W_{ob} \end{bmatrix}^{(4)} + \begin{bmatrix} -m & 0 \\ 0 & 2 - m\varepsilon_{o\xi} \end{bmatrix}\begin{bmatrix} U_{ob} \\ W_{ob} \end{bmatrix}^{(2)} +$$
$$+ \begin{bmatrix} 0 & -m \\ m & 0 \end{bmatrix}\begin{bmatrix} U_{ob} \\ W_{ob} \end{bmatrix}^{(1)} + \begin{bmatrix} 0 & 0 \\ 0 & 1 + m(1 - \varepsilon_{o\xi}) \end{bmatrix}\begin{bmatrix} U_{ob} \\ W_{ob} \end{bmatrix} = \frac{R^3}{I_{e\eta}}\begin{bmatrix} f_{tb} \\ f_{nb} \end{bmatrix}. \tag{12}$$

Observe that during the formal derivations we linearized the problem: (a) we neglected the quadratic term $\varepsilon_{o\xi}\varepsilon_{o\xi\, b}$ in (9a); (b) we took the inequalities $\varepsilon_{o\xi\, b} \gg (\varepsilon_{o\xi\, b}\psi_{o\eta})^{(1)}$ and $1 \gg \varepsilon_{o\xi}$ into account in (9b) when we utilized the Hooke law.

If we assume harmonic vibrations and denote the dimensionless displacement amplitudes by $\hat{U}_{ob}$ and $\hat{W}_{ob}$ then we have

$$\begin{bmatrix} 0 & 0 \\ 0 & 1 \end{bmatrix}\begin{bmatrix} \hat{U}_{ob} \\ \hat{W}_{ob} \end{bmatrix}^{(4)} + \begin{bmatrix} -m & 0 \\ 0 & 2 - m\varepsilon_{o\xi} \end{bmatrix}\begin{bmatrix} \hat{U}_{ob} \\ \hat{W}_{ob} \end{bmatrix}^{(2)} +$$
$$+ \begin{bmatrix} 0 & -m \\ m & 0 \end{bmatrix}\begin{bmatrix} \hat{U}_{ob} \\ \hat{W}_{ob} \end{bmatrix}^{(1)} + \begin{bmatrix} 0 & 0 \\ 0 & 1 + m(1 - \varepsilon_{o\xi}) \end{bmatrix}\begin{bmatrix} \hat{U}_{ob} \\ \hat{W}_{ob} \end{bmatrix} = \lambda \begin{bmatrix} \hat{U}_{ob} \\ \hat{W}_{ob} \end{bmatrix}; \quad \lambda = \rho_a A \frac{R^3}{I_{e\eta}}\alpha^2 \tag{13}$$

where λ is the eigenvalue sought and α is the natural frequency of the arch.

For an unloaded beam – i.e. when $\varepsilon_{o\xi} = 0$ – we get back the equations which govern the free vibrations – compare equation

$$\begin{bmatrix} 0 & 0 \\ 0 & 1 \end{bmatrix}\begin{bmatrix} \hat{U}_{ob} \\ \hat{W}_{ob} \end{bmatrix}^{(4)} + \begin{bmatrix} -m & 0 \\ 0 & 2 \end{bmatrix}\begin{bmatrix} \hat{U}_{ob} \\ \hat{W}_{ob} \end{bmatrix}^{(2)} + \begin{bmatrix} 0 & -m \\ m & 0 \end{bmatrix}\begin{bmatrix} \hat{U}_{ob} \\ \hat{W}_{ob} \end{bmatrix}^{(1)} +$$
$$+ \begin{bmatrix} 0 & 0 \\ 0 & m + 1 \end{bmatrix}\begin{bmatrix} \hat{U}_{ob} \\ \hat{W}_{ob} \end{bmatrix} = \lambda \begin{bmatrix} \hat{U}_{ob} \\ \hat{W}_{ob} \end{bmatrix} \tag{14}$$

to equation (11) in [8, 2013]. Depending on the supports applied the above system should be associated with appropriate boundary conditions. The left side of equation (13) can be rewritten in the form

$$\mathbf{K}\left[y\left(\varphi\right), \varepsilon_{o\xi}\right] = \overset{4}{\mathbf{P}} y^{(4)} + \overset{2}{\mathbf{P}} y^{(2)} + \overset{1}{\mathbf{P}} y^{(1)} + \overset{0}{\mathbf{P}} y^{(0)}, \qquad y = \begin{bmatrix} \hat{U}_{ob} \\ \hat{W}_{ob} \end{bmatrix}. \tag{15}$$

It is easy to see that the operator $\mathbf{K}$ is self-adjoint.

Differential equations (14) (or which is the same equations (15)) and the boundary conditions valid for pinned-pinned beams, fixed-fixed beams and beams pinned at one end and fixed at the other end constitute three eigenvalue problems.

Observe that the i-th eigenfrequency α_i in these eigenvalue problems depend on the magnitude of the concentrated force P_ζ, or what is the same, on the dimensionless load $\mathcal{P} = P_\zeta R^2 \vartheta / (2 I_{e\eta})$ through the axial strain: $\varepsilon_{o\xi} = \varepsilon_{o\xi}(\mathcal{P})$. We also remark that the heterogeneity appears in the formulation via the parameters m and ρ_a.

3. THE GREEN FUNCTION MATRIX

Observe that differential equations (15) are degenerated since the matrix $\overset{4}{\mathbf{P}}$ has no inverse. Let $\mathbf{r}(\varphi)$ be a prescribed inhomogeneity. Consider the boundary value problems defined by

$$\mathbf{K}(\mathbf{y}) = \sum_{\nu=0}^{4} \overset{\nu}{\mathbf{P}}(\varphi)\mathbf{y}^{(\nu)}(\varphi) = \mathbf{r}(\varphi), \qquad \overset{3}{\mathbf{P}}(\varphi) = 0 \tag{16}$$

and the boundary conditions valid for pinned beams:

$$\hat{U}_{ob}(-\vartheta) = 0 \quad \hat{W}_{ob}(-\vartheta) = 0 \quad \hat{W}_{ob}^{(2)}(-\vartheta) = 0 \quad | \quad \hat{U}_{ob}(\vartheta) = 0 \quad \hat{W}_{ob}(\vartheta) = 0 \quad \hat{W}_{ob}^{(2)}(\vartheta) = 0 , \tag{17a}$$

for fixed beams:

$$\hat{U}_{ob}(-\vartheta) = 0 \quad \hat{W}_{ob}(-\vartheta) = 0 \quad \hat{W}_{ob}^{(1)}(-\vartheta) = 0 \quad | \quad \hat{U}_{ob}(\vartheta) = 0 \quad \hat{W}_{ob}(\vartheta) = 0 \quad \hat{W}_{ob}^{(1)}(\vartheta) = 0 , \tag{17b}$$

and for beams fixed at the left end and pinned at the right end:

$$\hat{U}_{ob}(-\vartheta) = 0 \quad \hat{W}_{ob}(-\vartheta) = 0 \quad \hat{W}_{ob}^{(1)}(-\vartheta) = 0 \quad | \quad \hat{U}_{ob}(\vartheta) = 0 \quad \hat{W}_{ob}(\vartheta) = 0 \quad \hat{W}_{ob}^{(2)}(\vartheta) = 0 . \tag{17c}$$

General solution for the homogenous part of differential equations (16) assumes the form

$$\mathbf{y} = \left[\sum_{i=1}^{4} \underset{(2\times2)}{\mathbf{Y}}_i \underset{(2\times2)}{\mathbf{C}}_i \right] \underset{(2\times1)}{\mathbf{e}} \tag{18a}$$

where

$$\mathbf{Y}_1 = \begin{bmatrix} \cos\varphi & 0 \\ \sin\varphi & 0 \end{bmatrix}, \quad \mathbf{Y}_2 = \begin{bmatrix} -\sin\varphi & 0 \\ \cos\varphi & 0 \end{bmatrix}, \quad \mathbf{Y}_3 = \begin{bmatrix} \cos\chi\varphi & \mathcal{M}\varphi \\ \chi\sin\chi\varphi & -1 \end{bmatrix}, \quad \mathbf{Y}_4 = \begin{bmatrix} -\sin\chi\varphi & 1 \\ \chi\cos\chi\varphi & 0 \end{bmatrix} . \tag{18b}$$

Here $\mathbf{C}_i$ is an arbitrary constant matrix, $\mathbf{e}$ is an arbitrary column matrix and

$$\mathcal{M} = \frac{m+1}{m(1+\varepsilon_{o\xi})} . \tag{18c}$$

Solutions to the boundary value problems (16,17a), (16,17b) and (16,17c) are sought in the form

$$\mathbf{y}(\varphi) = \int_a^b \mathbf{G}(\varphi, \psi)\mathbf{r}(\psi)\mathrm{d}\psi, \quad \mathbf{G}(\varphi, \psi) = \begin{bmatrix} G_{11}(\varphi, \psi) & G_{12}(\varphi, \psi) \\ G_{21}(\varphi, \psi) & G_{22}(\varphi, \psi) \end{bmatrix} \tag{19}$$

where $\mathbf{G}(\varphi, \psi)$ is the Green function matrix defined by the following properties [5, 1975]:
(1) the Green function matrix is a continuous function of φ and ψ in each of the triangles $-\vartheta \le \varphi \le \psi \le \vartheta$ and $-\vartheta \le \xi \le \varphi \le \vartheta$. The functions $(G_{11}(\varphi, \psi), G_{12}(\varphi, \psi))$ $[G_{21}(\varphi, \psi), G_{22}(\varphi, \psi)]$ are (2 times) [4 times] differentiable with respect to φ and the derivatives

$$\frac{\partial^\nu \mathbf{G}(\varphi, \psi)}{\partial x^\nu} = \mathbf{G}^{(\nu)}(\varphi, \psi) \quad (\nu = 1, 2) , \quad \frac{\partial^\nu G_{2i}(\varphi, \psi)}{\partial x^\nu} = G_{2i}^{(\nu)}(\varphi, \psi) \quad (\nu = 1, \ldots, 4; \ i = 1, 2)$$

are continuous functions of φ and ψ.
(2) Let ψ be fixed in $[-\vartheta, \vartheta]$. Though the function and the derivatives

$$G_{11}(\varphi, \psi), \ G_{12}^{(1)}(\varphi, \psi), \ G_{21}^{(\nu)}(\varphi, \psi) \ (\nu = 1, 2, 3), \ G_{22}^{(\nu)}(\varphi, \psi) \ (\nu = 1, 2) \tag{20}$$

are continuous everywhere the derivatives $G_{11}^{(1)}(\varphi, \psi)$ and $G_{22}^{(3)}(\varphi, \psi)$ have a jump if $\varphi = \psi$:

$$\lim_{\varepsilon \to 0} \left[G_{11}^{(1)}(\varphi + \varepsilon, \varphi) - G_{11}^{(1)}(\varphi - \varepsilon, \varphi) \right] = 1/\overset{1}{P}_{11}(\varphi), , \quad \lim_{\varepsilon \to 0} \left[G_{22}^{(3)}(\varphi + \varepsilon, \varphi) - G_{22}^{(3)}(\varphi - \varepsilon, \varphi) \right] = 1/\overset{4}{P}_{22}(\varphi) . \tag{21}$$

(3) Let α be an arbitrary, otherwise constant vector. For a fixed $\varphi \in [-\vartheta, \vartheta]$ the vector $\mathbf{G}(\varphi, \psi)\alpha$ as a function of φ ($\varphi \ne \psi$) should satisfy the homogeneous differential equation $\mathbf{K}[\mathbf{G}(\varphi, \psi)\alpha] = 0$.
(4) The vector $\mathbf{G}(\varphi, \psi)\alpha$ as a function of φ should satisfy the boundary conditions (17a), (17b) and (17c) – there belongs one Green function matrix to each of the boundary value problems considered.

If the Green function matrix – defined above for the boundary value problems considered – exists, then vector (19) satisfies differential equation (16) and boundary conditions (17).

Consider the system of differential equations

$$\mathbf{K}[\mathbf{y}] = \lambda\mathbf{y} \tag{22}$$

where $\mathbf{K}[\mathbf{y}]$ is given by (16) and λ is a parameter (the eigenvalue sought). The system of ordinary differential equations (16) is associated with linear homogeneous boundary conditions (17) – they together constitute three eigenvalue problems.

The vectors $\mathbf{u}^T = [u_1|u_2]$ and $\mathbf{v}^T = [v_1|v_2]$ are said to be comparison vectors if they are different from zero, satisfy the boundary conditions and are differentiable as many times as required.

The eigenvalue problems (16), (17) are self adjoint if the product $(\mathbf{u}, \mathbf{v})_M = \int_{-\vartheta}^{\vartheta} \mathbf{u}^T \mathbf{K} \mathbf{v}\, d\varphi$ is commutative, i.e., $(\mathbf{u}, \mathbf{v})_M = (\mathbf{v}, \mathbf{u})_M$ over the set of comparison vectors and it is positive definite if $(\mathbf{u}, \mathbf{u})_M > 0$ for any comparison vector $\mathbf{u}$.

If the eigenvalue problems (16), (17) are self adjoint then the Green function matrices are cross symmetric: $\mathbf{G}(\varphi, \psi) = \mathbf{G}^T(\varphi, \psi)$.

4. NUMERICAL SOLUTION TO THE EIGENVALUE PROBLEMS

With (19) the eigenvalue problems (16), (17) can be replaced by homogeneous integral equation systems of the form

$$\mathbf{y}(\varphi) = \lambda \int_{-\vartheta}^{\vartheta} \mathbf{G}(\varphi, \psi)\mathbf{y}(\psi) d\psi. \tag{23}$$

Numerical solution to any eigenvalue problem determined by (23) can be sought by quadrature methods [9, 1977]. Consider the integral formula

$$J(\phi) = \int_{-\vartheta}^{\vartheta} \phi(\psi)\, d\psi \equiv \sum_{j=0}^{n} w_j \phi(\psi_j) \qquad \psi_j \in [-\vartheta, \vartheta] \tag{24}$$

where $\psi_j(\varphi)$ is a vector and the weights w_j are known. Making use of the latter equation we obtain from (23) that

$$\sum_{j=0}^{n} w_j \mathbf{G}(\varphi, \psi_j)\tilde{\mathbf{y}}(\psi_j) = \tilde{\kappa}\tilde{\mathbf{y}}(\varphi) \qquad \tilde{\kappa} = 1/\tilde{\lambda} \qquad \in [-\vartheta, \vartheta] \tag{25}$$

is the solution which yields an approximate eigenvalue $\tilde{\lambda} = 1/\tilde{\kappa}$ and a corresponding approximate eigenfunction $\tilde{\mathbf{y}}(\varphi)$. After setting φ to ψ_i $(i = 0, 1, 2, \ldots, n)$ we have

$$\sum_{j=0}^{n} w_j \mathbf{G}(\psi_i, \psi_j)\tilde{\mathbf{y}}(\psi_j) = \tilde{\kappa}\tilde{\mathbf{y}}(\psi_i) \qquad \tilde{\kappa} = 1/\tilde{\lambda} \qquad \psi_i, \psi_j \in [-\vartheta, \vartheta] \tag{26}$$

or

$$\mathcal{G}\mathcal{D}\tilde{\mathcal{Y}} = \tilde{\kappa}\tilde{\mathcal{Y}} \tag{27}$$

where $\mathcal{G} = [\mathbf{G}(x_i, x_j)]$ is symmetric if the problem is self adjoint,

$$\mathcal{D} = \mathrm{diag}(\underbrace{w_0, \ldots, w_0}_{l} | \ldots | \underbrace{w_n, \ldots, w_n}_{l})$$

and $\tilde{\mathcal{Y}}^T = [\tilde{\mathbf{y}}^T(x_0)|\tilde{\mathbf{y}}^T(x_1)| \ldots |\tilde{\mathbf{y}}^T(x_n)]$. After solving the generalized algebraic eigenvalue problem (27) we have the approximate eigenvalues $\tilde{\lambda}_r$ and eigenvectors $\mathcal{Y}_r$ while the corresponding eigenfunction is obtained by a substitution into (25):

$$\tilde{\mathbf{y}}_r(\varphi) = \tilde{\lambda}_r \sum_{j=0}^{n} w_j \mathbf{G}(\varphi, \psi_j)\tilde{\mathbf{y}}_r(\psi_j) \qquad r = 0, 1, 2, \ldots, n. \tag{28}$$

Divide the interval $[-\vartheta, \vartheta]$ into equidistant subintervals of length h and apply the integration formula to each subinterval. By repeating the line of thought leading to (28) one can readily show that the algebraic eigenvalue problem obtained is of the same structure as (28).

It is also possible to consider the integral equation (23) as if it were a boundary integral equation and apply isoparametric approximation on the subintervals, i.e., on the elements. If this is the case one can approximate the eigenfunction on the e-th element (the e-th subinterval which is mapped onto the interval $\eta \in [-1, 1]$ and is denoted by $\mathfrak{L}_e$) by

$$\overset{e}{\mathbf{y}} = [\mathbf{N}_1(\eta)|\mathbf{N}_2(\eta)|\mathbf{N}_3(\eta)] \begin{bmatrix} \overset{e}{\mathbf{y}}_1 \\ \overset{e}{\mathbf{y}}_2 \\ \overset{e}{\mathbf{y}}_3 \end{bmatrix} \tag{29}$$

where quadratic local approximation is assumed, $\mathbf{N}_i = \mathrm{diag}(N_i)$, $N_1 = 0.5\eta(\eta - 1)$, $N_2 = 1 - \eta^2$, $N_3 = 0.5\eta(\eta + 1)$, $\overset{e}{\mathbf{y}}_i$ is the value of the eigenfunction $\mathbf{y}(\varphi)$ at the left endpoint, the midpoint and the right endpoint of the element, respectively. Upon substitution of approximation (29) into (23) we have

$$\tilde{\mathbf{y}}(\varphi) = \tilde{\lambda} \sum_{e=1}^{n_{be}} \int_{\mathfrak{L}_e} \mathbf{G}(x, \eta)[\mathbf{N}_1(\eta)|\mathbf{N}_2(\eta)|\mathbf{N}_3(\eta)] d\eta \begin{bmatrix} \overset{e}{\mathbf{y}}_1 \\ \overset{e}{\mathbf{y}}_2 \\ \overset{e}{\mathbf{y}}_3 \end{bmatrix} \tag{30}$$

in which n_{be} is the number of elements (subintervals). Using equation (30) as a point of departure and repeating the line of thought leading to (27) we shall arrive again at an algebraic eigenvalue problem.

5. CALCULATION OF THE GREEN FUNCTION MATRICES

Based on the definition presented in Section 3 we detail the calculation of the Green function for pinned-pinned beams only. The reason is to keep the length of the paper under the limit prescribed. With regards to property (3) – see the definition – the Green function matrix can be given in the form

$$\underbrace{\mathbf{G}(\varphi, \psi)}_{(2 \times 2)} = \sum_{j=1}^{4} \mathbf{Y}_j(\varphi) \left[\mathbf{A}_j(\psi) \pm \mathbf{B}_j(\psi) \right] \tag{31}$$

where (a) the sign is [positive](negative) if $[\varphi \le \psi](\varphi \ge \psi)$; (b) the matrices $\mathbf{A}_j$ and $\mathbf{B}_j$ have the following structure

$$\mathbf{A}_j = \begin{bmatrix} \overset{j}{A}_{11} & \overset{j}{A}_{12} \\ \overset{j}{A}_{21} & \overset{j}{A}_{22} \end{bmatrix} = \begin{bmatrix} \mathbf{A}_{j1} & \mathbf{A}_{j2} \end{bmatrix}, \qquad \mathbf{B}_j = \begin{bmatrix} \overset{j}{B}_{11} & \overset{j}{B}_{12} \\ \overset{j}{B}_{21} & \overset{j}{B}_{22} \end{bmatrix} = \begin{bmatrix} \mathbf{B}_{j1} & \mathbf{B}_{j2} \end{bmatrix} \quad j = 1, \dots, 4; \tag{32}$$

(c) the coefficients in $\mathbf{B}_j$ are independent of the boundary conditions; (d) matrices $\mathbf{Y}_j$ are given by (18b).

For the sake of brevity let us now introduce the following notational conventions

$$a = \overset{1}{B}_{1i}, \; b = \overset{2}{B}_{1i}, \; c = \overset{3}{B}_{1i}, \; d = \overset{3}{B}_{2i}, \; e = \overset{4}{B}_{1i}, \; f = \overset{4}{B}_{2i}$$

We remark that $\overset{1}{B}_{21} = \overset{2}{B}_{21} = \overset{1}{B}_{22} = \overset{2}{B}_{22} = 0$ – we refer back to Section 3. The equation systems for the unknowns $a, \dots, f$ can be set up by fulfilling property (2) of the Green function matrix – i.e. on the basis of equations (20) and (21). If $[i = 1]$ we have

$$\begin{bmatrix} \cos \psi & -\sin \psi & \cos(\chi\psi) & \mathcal{M}\psi & -\sin(\chi\psi) & 1 \\ \sin \psi & \cos \psi & \chi\sin(\chi\psi) & -1 & \chi\cos(\chi\psi) & 0 \\ -\sin \psi & -\cos \psi & -\chi\sin(\chi\psi) & \mathcal{M} & -\chi\cos(\chi\psi) & 0 \\ \cos \psi & -\sin \psi & \chi^2\cos(\chi\psi) & 0 & -\chi^2\sin(\chi\psi) & 0 \\ -\sin \psi & -\cos \psi & -\chi^3\sin(\chi\psi) & 0 & -\chi^3\cos(\chi\psi) & 0 \\ -\cos \psi & \sin \psi & -\chi^4\cos(\chi\psi) & 0 & \chi^4\sin(\chi\psi) & 0 \end{bmatrix} \begin{bmatrix} a \\ b \\ c \\ d \\ e \\ f \end{bmatrix} = \begin{bmatrix} 0 \\ 0 \\ \frac{1}{2m} \\ 0 \\ 0 \\ 0 \end{bmatrix} \tag{33}$$

from where it follows that

$$a = \overset{1}{B}_{11} = \frac{\chi^2}{(1-\chi^2)(1-\mathcal{M})m} \frac{\sin \psi}{2}; \qquad b = \overset{2}{B}_{11} = \frac{\chi^2}{(1-\chi^2)(1-\mathcal{M})m} \frac{\cos \psi}{2},$$

$$c = \overset{3}{B}_{11} = -\frac{\chi^2}{(1-\chi^2)(1-\mathcal{M})m} \frac{\sin \chi\psi}{2\chi^3}; \qquad d = \overset{3}{B}_{21} = -\frac{1}{2(1-\mathcal{M})m}; \tag{34}$$

$$e = \overset{4}{B}_{11} = -\frac{1}{\chi(1-\chi^2)(1-\mathcal{M})m} \frac{\cos \chi\psi}{2}; \qquad f = \overset{4}{B}_{21} = \frac{1}{2}\mathcal{M} \frac{\psi}{m(1-\mathcal{M})}.$$

If $\{i = 2\}$ then

$$\begin{bmatrix} \cos \psi & -\sin \psi & \cos(\chi\psi) & \mathcal{M}\psi & -\sin(\chi\psi) & 1 \\ \sin \psi & \cos \psi & \chi\sin(\chi\psi) & -1 & \chi\cos(\chi\psi) & 0 \\ -\sin \psi & -\cos \psi & -\chi\sin(\chi\psi) & \mathcal{M} & -\chi\cos(\chi\psi) & 0 \\ \cos \psi & -\sin \psi & \chi^2\cos(\chi\psi) & 0 & -\chi^2\sin(\chi\psi) & 0 \\ -\sin \psi & -\cos \psi & -\chi^3\sin(\chi\psi) & 0 & -\chi^3\cos(\chi\psi) & 0 \\ -\cos \psi & \sin \psi & -\chi^4\cos(\chi\psi) & 0 & \chi^4\sin(\chi\psi) & 0 \end{bmatrix} \begin{bmatrix} a \\ b \\ c \\ d \\ e \\ f \end{bmatrix} = \begin{bmatrix} 0 \\ 0 \\ 0 \\ 0 \\ 0 \\ -\frac{1}{2} \end{bmatrix} \tag{35}$$

is the equation system with the solutions

$$a = \overset{1}{B}_{12} = \frac{1}{2} \frac{\cos \psi}{(1-\chi^2)}; \qquad b = \overset{2}{B}_{12} = -\frac{1}{2} \frac{\sin \psi}{(1-\chi^2)}; \qquad c = \overset{3}{B}_{12} = -\frac{1}{2} \frac{\cos \chi\psi}{(1-\chi^2)\chi^2};$$

$$d = \overset{3}{B}_{22} = 0; \qquad e = \overset{4}{B}_{12} = \frac{1}{2} \frac{\sin \chi\psi}{(1-\chi^2)\chi^2}; \qquad f = \overset{4}{B}_{22} = \frac{1}{2\chi^2}. \tag{36}$$

As regards the constants $\mathbf{A}_j$, or which is the same the unknown scalars

$$\overset{1}{A}_{1i}(\psi), \; \overset{2}{A}_{1i}(\psi), \; \overset{3}{A}_{1i}(\psi), \; \overset{3}{A}_{2i}(\psi), \; \overset{4}{A}_{1i}(\psi), \; \overset{4}{A}_{2i}(\psi) \qquad i = 1, 2; \qquad \psi \in [-\vartheta, \vartheta]$$

$(\overset{1}{A}_{21} = \overset{2}{A}_{21} = \overset{1}{A}_{22} = \overset{2}{A}_{22} = 0!)$ property (4) and boundary conditions (17a) yield

$$\begin{bmatrix} \cos\vartheta & \sin\vartheta & \cos(\chi\vartheta) & -\mathcal{M}\vartheta & \sin(\chi\vartheta) & 1 \\ \cos\vartheta & -\sin\vartheta & \cos(\chi\vartheta) & \mathcal{M}\vartheta & -\sin(\chi\vartheta) & 1 \\ -\sin\vartheta & \cos\vartheta & -\chi\sin(\chi\vartheta) & -1 & \chi\cos(\chi\vartheta) & 0 \\ \sin\vartheta & \cos\vartheta & \chi\sin(\chi\vartheta) & -1 & \chi\cos(\chi\vartheta) & 0 \\ \sin\vartheta & -\cos\vartheta & \chi^3\sin(\chi\vartheta) & 0 & -\chi^3\cos(\chi\vartheta) & 0 \\ -\sin\vartheta & -\cos\vartheta & -\chi^3\sin(\chi\vartheta) & 0 & -\chi^3\cos(\chi\vartheta) & 0 \end{bmatrix} \begin{bmatrix} \overset{1}{A}_{1i} \\ \overset{2}{A}_{1i} \\ \overset{3}{A}_{1i} \\ \overset{3}{A}_{2i} \\ \overset{4}{A}_{1i} \\ \overset{4}{A}_{2i} \end{bmatrix} =$$

$$= \begin{bmatrix} -a\cos\vartheta - b\sin\vartheta - c\cos(\chi\vartheta) + d\mathcal{M}\vartheta - e\sin(\chi\vartheta) - f \\ a\cos\vartheta - b\sin\vartheta + c\cos(\chi\vartheta) + d\mathcal{M}\vartheta - e\sin(\chi\vartheta) + f \\ a\sin\vartheta - b\cos\vartheta + c\chi\sin(\chi\vartheta) + d - e\chi\cos(\chi\vartheta) \\ a\sin\vartheta + b\cos\vartheta + c\chi\sin(\chi\vartheta) - d + e\chi\cos(\chi\vartheta) \\ -a\sin\vartheta + b\cos\vartheta - c\chi^3\sin(\chi\vartheta) + e\chi^3\cos(\chi\vartheta) \\ -a\sin\vartheta - b\cos\vartheta - c\chi^3\sin(\chi\vartheta) - e\chi^3\cos(\chi\vartheta) \end{bmatrix} . \quad (37)$$

With the introduction of the constants

$$\mathcal{C}_{11} = (1-\chi^2)\sin\vartheta, \quad \mathcal{C}_{12} = \chi(1-\chi^2)\sin\chi\vartheta ,$$
$$\mathcal{D}_{11} = \cos\vartheta\sin\chi\vartheta - \chi^3\sin\vartheta\cos\chi\vartheta - M\chi\vartheta(1-\chi^2)\cos\vartheta\cos\chi\vartheta$$

we can write the solutions in the following forms

$$\overset{1}{A}_{1i} = \frac{1}{\mathcal{C}_{11}}\left[b(1-\chi^2)\cos\vartheta + d\chi^2\right];$$
$$\overset{2}{A}_{1i} = \frac{1}{\mathcal{D}_{11}}\left[a\chi^3\cos\vartheta\cos\chi\vartheta - a\chi\vartheta(1-\chi^2)\mathcal{M}\sin\vartheta\cos\chi\vartheta + a\sin\vartheta\sin\chi\vartheta + c\chi^3 + \chi^3 f\cos\chi\vartheta\right];$$
$$\overset{3}{A}_{1i} = -\frac{1}{\mathcal{C}_{12}}\left(d - e\chi(1-\chi^2)\cos\chi\vartheta\right);$$
$$\overset{3}{A}_{2i} = -\frac{\chi}{\mathcal{D}_{11}}(1-\chi^2)\chi(a\cos\chi\vartheta + c\cos\vartheta + f\cos\vartheta\cos\chi\vartheta);$$
$$\overset{4}{A}_{1i} = -\frac{1}{\mathcal{D}_{11}}\left(a + c(1-\chi^2)\mathcal{M}\chi\vartheta\cos\vartheta\sin\chi\vartheta + c\left(\chi^3\sin\vartheta\sin\chi\vartheta + \cos\vartheta\cos\chi\vartheta\right) + f\cos\vartheta\right);$$
$$\overset{4}{A}_{2i} = -\frac{1}{\mathcal{C}_{12}\sin\vartheta}\left(b\chi(1-\chi^2)\sin\chi\vartheta - dM\vartheta\chi(1-\chi^2)\sin\vartheta\sin\chi\vartheta + d\chi^3\cos\vartheta\sin\chi\vartheta - \right.$$
$$\left. -d\sin\vartheta\cos\chi\vartheta + e\chi\sin\vartheta - e\chi^3\sin\vartheta\right) .$$

6. THE LOAD-STRAIN RELATIONSHIP AND THE CRITICAL STRAIN

It is vital to be aware of how the loading affects the strain of the centerline. In practise the loading is the known quantity. However our formulation holds the strain as a variable. We can establish the relationship $\varepsilon_{o\xi} = \varepsilon_{o\xi}(\mathcal{P})$ on the basis of equations (7). If the model is a linear one – we shall attack this problem by assuming linearity – the effect the deformations have on the equilibrium conditions are neglected. Under this assumption we have to solve the differential equations

$$\begin{bmatrix} 0 & 0 \\ 0 & 1 \end{bmatrix}\begin{bmatrix} U_o \\ W_o \end{bmatrix}^{(4)} + \begin{bmatrix} -m & 0 \\ 0 & 2 \end{bmatrix}\begin{bmatrix} U_o \\ W_o \end{bmatrix}^{(2)} + \begin{bmatrix} 0 & -m \\ m & 0 \end{bmatrix}\begin{bmatrix} U_o \\ W_o \end{bmatrix}^{(1)} + \begin{bmatrix} 0 & 0 \\ 0 & 1+m \end{bmatrix}\begin{bmatrix} U_o \\ W_o \end{bmatrix} = \begin{bmatrix} 0 \\ 0 \end{bmatrix}$$
$$(39)$$

which follow from (7) by setting $\varepsilon_{o\xi}$ to zero. For a pinned-pinned arch the above equations are associated with the boundary conditions

$$U_o|_{\pm\vartheta} = W_o|_{\pm\vartheta} = \psi_{o\eta}|_{\pm\vartheta} = 0 \quad (40)$$

and the continuity (discontinuity) conditions

$$U_o|_{\varphi=-0} = U_o|_{\varphi=+0} , \qquad W_o|_{\varphi=-0} = W_o|_{\varphi=+0} , \qquad \psi_{o\eta}|_{\varphi=-0} = \psi_{o\eta}|_{\varphi=+0} ,$$
$$N|_{\varphi=-0} = N|_{\varphi=+0} , \qquad M|_{\varphi=-0} = M|_{\varphi=+0} , \qquad T|_{\varphi=-0} - P_\zeta|_{\varphi=0} = T|_{\varphi=+0} \quad (41)$$

prescribed at the crown (more details are presented in thesis [10, 2011] by Kiss). Omitting the long formal transformations we get the axial strain in the form

$$\varepsilon_{o\xi} = -\frac{2\mathcal{P}}{\vartheta}\frac{\cos\vartheta[\vartheta\tan\vartheta + 2(\cos\vartheta - 1)]}{2(m+1)\vartheta\cos^2\vartheta + \vartheta m - 3m\sin\vartheta\cos\vartheta} , \qquad \mathcal{P} = \frac{P_\zeta\rho_o^2\vartheta}{2I_{e\eta}} . \quad (42)$$

For a fixed-fixed beam a similar line of thought results in

$$\varepsilon_{o\xi} = \frac{\mathcal{P}}{\vartheta}\, \frac{(1 - \cos\vartheta)\,(\sin\vartheta - \vartheta)}{\vartheta\,(1 + m)\,[\vartheta + \sin\vartheta\cos\vartheta] - 2m\sin^2\vartheta}\,. \tag{43}$$

We remark that the formula valid for mixed boundary conditions is also available but is not presented here.

We should remark that the presence of $\varepsilon_{o\xi}$ in equation (7) (we assumed that $\varepsilon_{o\xi} = 0$ when we derived (44) and (45)) may seriously influence the relationship $\varepsilon_{o\xi} = \varepsilon_{o\xi}\,(\mathcal{P})$. Investigations concerning the problem wether we can set $\varepsilon_{o\xi}$ to zero in (7) are in progress.

The critical strain (at which the arch loses its stability) can be obtained if we solve the eigenvalue problems defined by equations (12) with the right side set to zero (the arch is in static equilibrium under the action of the force exerted at the crown point – there are no other loads on the arch) and by the corresponding homogenous boundary conditions. The general solutions for the displacement increments are:

$$W_{ob} = -A_2 - A_3\cos\varphi + A_4\sin\varphi - \chi A_5\cos\chi\varphi + \chi A_6\sin\chi\varphi\,, \tag{44a}$$

$$U_{ob} = A_1 + \frac{m+1}{m\,(1 + \varepsilon_{o\xi})}A_2\varphi + A_3\sin\varphi + A_4\cos\varphi + A_5\sin\chi\varphi + A_6\cos\chi\varphi\,, \qquad \chi^2 = m\varepsilon_{o\xi} - 1\,. \tag{44b}$$

For a pinned-pinned arch equations

$$W_{ob}|_{\pm\vartheta} = \psi_{o\eta b}|_{\pm\vartheta} = U_{ob}|_{\pm\vartheta} = 0$$

are the boundary conditions and the lowest critical value of $\chi\vartheta$ is π – the details are omitted again –, consequently

$$\varepsilon_{o\xi\,\text{crit}} = -\frac{1}{m}\left(\chi^2 - 1\right) = -\frac{1}{m}\left[\left(\frac{\pi}{\vartheta}\right)^2 - 1\right] \tag{45}$$

is the critical strain. Similarly for a fixed-fixed arch we have

$$\varepsilon_{o\xi\,\text{crit}} = -\frac{1}{m}\left(\chi^2 - 1\right) = -\frac{1}{m}\left[\left(\frac{g_{21n}}{\vartheta}\right)^2 - 1\right] \tag{46a}$$

in which

$$\chi\vartheta = g_{21n}\,(\vartheta) = 3.689\,334\,516 \times 10^{-2}\vartheta^4 - 0.131\,139\,9068\vartheta^3 + 0.259\,573\,7664\vartheta^2$$
$$- 9.600\,584\,516 \times 10^{-2}\vartheta + 4.506\,225\,066\,. \tag{46b}$$

7. COMPUTATIONAL RESULTS

A program has been developed in Fortran90 for solving the eigenvalue problems governed by the Fredholm integral equations. The numerical results have been compared to those valid for the free vibrations of curved beams with the same geometric and material properties. (For more details about the natural frequencies of planar arches see [5, 1975].) Figure 2 shoes the quotient $\alpha_1/\alpha_{1\,free}$ against the quotient $\varepsilon_{o\xi}/\varepsilon_{o\xi\,crit}$ [Graph (a)] { Graph (b)} shows the results for [the pinned-pinned beam] {the beam fixed at both ends}. It has turned out that the results are independent of both the parameter m and the central angle 2ϑ. As the loading increases (or what is the same the axial strain grows) the frequencies and therefore the quotients decrease. Observe that there is hardly any noticeable difference between the results for the two support arrangements.

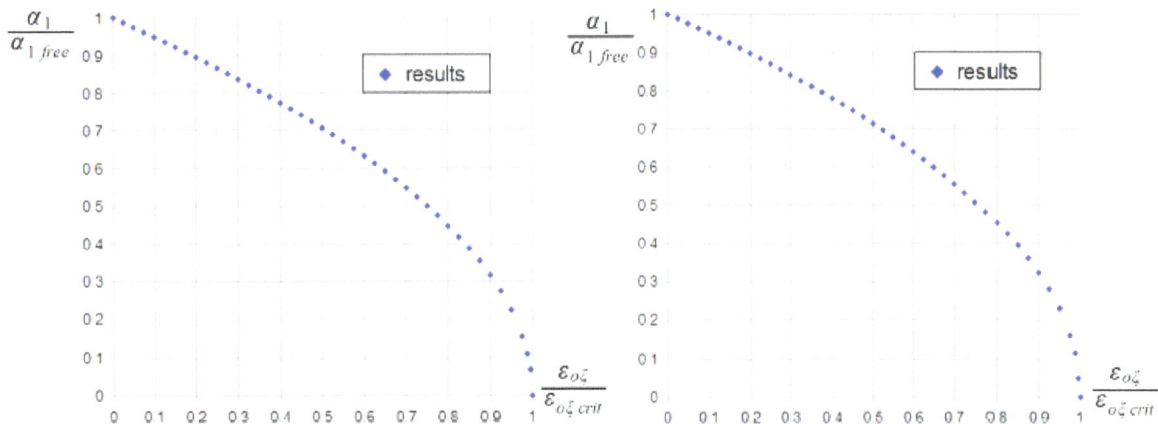

Figure 2. Results for (a) pinned-pinned, and (b) fixed-fixed beams

Not less interesting and illustrative are Figures 3 and 4. Considering a pinned-pinned arch Figure 3 represents the quotients $\alpha_i^2/\alpha_{i\,free}^2$ $(i = 1, 2)$ against the quotient $\varepsilon_{o\xi}/\varepsilon_{o\xi\,crit}$. In addition to being independent of m and ϑ these relationships are linear with a very good accuracy. Equations

$$\frac{\alpha_1^2}{\alpha_{1\,free}^2} = 1.00046 - 1.00038\frac{\varepsilon_{o\xi}}{\varepsilon_{o\xi\,crit}} , \qquad \frac{\alpha_2^2}{\alpha_{2\,free}^2} = 1.00204 - 0.43440\frac{\varepsilon_{o\xi}}{\varepsilon_{o\xi\,crit}} . \tag{47}$$

fit onto the computational results with a very good accuracy. The character of these results is the same as that valid for a compressed straight beam if the latter one also vibrates.

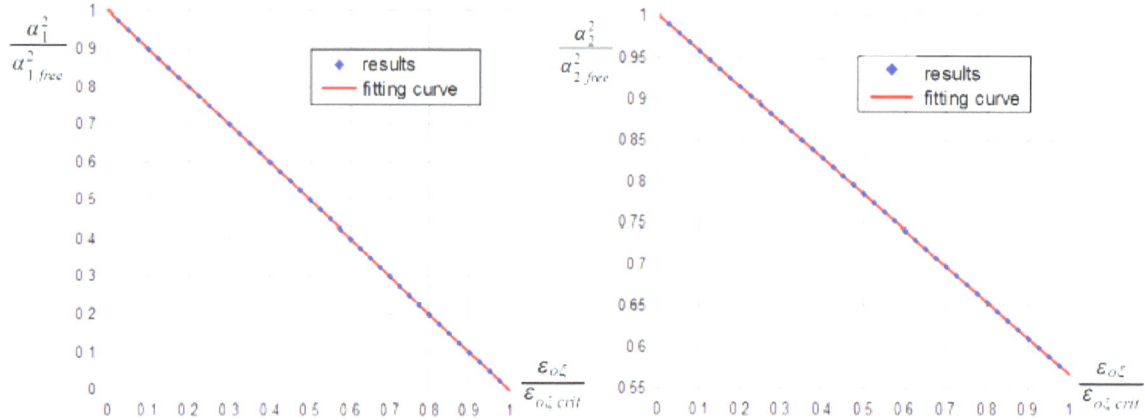

Figure 3. Results valid for pinned-pinned beams

The results are incredibly similar for fixed-fixed beams (Figure 4) however a quadratic term shall be added to the approximative polynomials for a better fit:

$$\frac{\alpha_1^2}{\alpha_{1\,free}^2} = 1.00190 - 0.96824 - \frac{\varepsilon_{o\xi}}{\varepsilon_{o\xi\,crit}} - 0.03280 \left(\frac{\varepsilon_{o\xi}}{\varepsilon_{o\xi\,crit}}\right)^2 , \tag{48}$$

$$\frac{\alpha_2^2}{\alpha_{2\,free}^2} = 1.00175 - 0.43260 - \frac{\varepsilon_{o\xi}}{\varepsilon_{o\xi\,crit}} - 0.00181 \left(\frac{\varepsilon_{o\xi}}{\varepsilon_{o\xi\,crit}}\right)^2 . \tag{49}$$

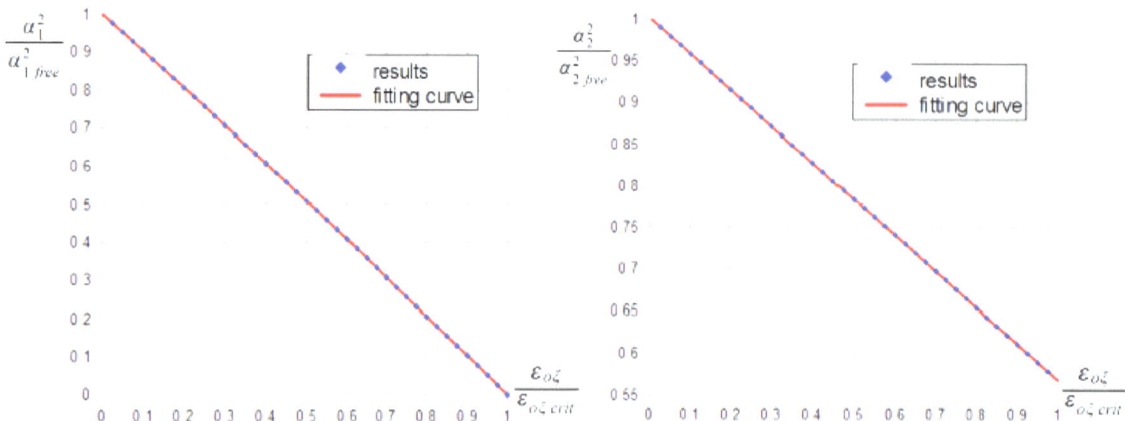

Figure 4. Results valid for fixed-fixed beams

8. CONCLUDING REMARKS

In accordance with our aims we have investigated the vibrations of curved beams with cross sectional heterogeneity subjected to a vertical force at the crown point.

(1) We have derived the governing equations of those boundary value problems which make it possible to clarify how the radial load affects the natural frequencies of the beam.

(2) For pinned-pinned and fixed-fixed beams as well as for beams fixed at one end and pinned at the other end we have determined the Green function matrices assuming that the beams are prestressed by a radial load. Since the length of the paper is limited details were mainly presented for the pinned-pinned beam.

(3) Making use of the Green function matrices we have reduced the eigenvalue problems set up for the frequencies to eigenvalue problems governed by systems of Fredholm integral equations.

(4) Numerical solutions were provided. For the loaded beam considered the square of the natural frequencies depend linearly (pinned-pinned beam) or linearly with a good accuracy (fixed-fixed beam) on the axial strain $\varepsilon_{o\xi}$. With the knowledge of the relationship $\varepsilon_{o\xi} = \varepsilon_{o\xi}(\mathcal{P})$ we can determine that value of $\varepsilon_{o\xi}$ which belongs to a given load and then the natural frequency of the loaded structure. Accuracy of the relations (44) and (45)) is, however, to be investigated. This work is in progress.

Acknowledgement: This research was supported in the framework of TÁMOP 4.2.4. A/2-11-1-2012-0001 „National Excellence Program – Elaborating and operating an inland student and researcher personal support system convergence program" The project was subsidized by the European Union and co-financed by the European Social Fund.

REFERENCES

[1] A.E.H. Love. *Treatese on the mathematical theory of elasticity*. New York, Dower, 1944.

[2] S. Márkus and T. Nánási. Vibration of curved beams. *Shock. Vib. Dig.*, 13(4):3–14, 1981.

[3] P. A. A. Laura and M. J. Maurizi. Recent research on vibrations of arch-type structures. *Shock. Vib. Dig.*, 19(1):6–9, 1987.

[4] P. Chidamparam and A. W. Leissa. Vibrations of planar curved beams, rings and arches. *Applied Mechanis Review, ASME*, 46(9):467–483, 1993.

[5] G. Szeidl. *Effect of change in length on the natural frequencies and stability of circular beams*. Ph.D Thesis, Department of Mechanics, University of Miskolc, Hungary, 1975. (in Hungarian).

[6] Ray Lawther. On the straightness of eigenvalue iterations. *Computational Mechanics*, 37:362–368, 2005.

[7] Gy. Szeidl and L. Kiss:. A Nonlinear Mechanical Model For Heterogeneous Curved Beams. In S. Vlase, editor, *Proceedings of the 4th International Conference on Advanced Composite Materials Enginnering, COMAT*, volume 2, pages 589–596, 18 - 20 October 2012, Braşov, Romania.

[8] L. P. Kiss. Free vibrations of heterogeneous curved beams. *GÉP*, LXIV(5):16–21, 2013. (in Hungarian).

[9] Christopher T. H. Baker. *The Numerical Treatment of Integral Equations – Monographs on Numerical Analysis edited by L. Fox and J. Walsh*. Clarendon Press, Oxford, 1977.

[10] L. Kiss. *Solutions to some problems of heterogenous curved beams*. MSc Thesis, Department of Mechanics, University of Miskolc, 2011. (in Hungarian).

STRUCTURAL SYNTHESIS FOR REDUNDANT INDUSTRIAL ROBOTS WITH MORE THAN 6 AXES

Ionel Staretu

Transilvania University of Braşov, Romania, staretu@unitbv.ro

Abstract: *This paper presents the most possible structures with 7 or 8 axes, which correspond to non-degenerate workspaces of which one can choose other variants than existing ones, to be manufactured.*

The Structures of kinematic chains with more of 8 axes can be obtained from the 6-axis ones, by adding three axes (three monomobile kinematic couplings) of rotation or translation or combined RRR, respectively, TTT, than by adding four axes, five axes or six axes. Some of these structures may have specific functional advantages in certain situations, which will be validated for sure by future practical applications.

The redundant kinematic chains with 7 or 8 axes (7 or 8 monomobile rotation or translation couplings) or more axes, but not more 12 axes, have proved to be useful and some variants are already applied in some industrial robots currently in practice. In this paper is a representation of workspace that will help the designer to choose structures that may have a higher degree of functionality for a given range of applications.

The Kinematic analysis based on the method of homogeneous operators exemplifies the possibility to resolve this issue in the case of these redundant robots too.

Keywords: redundant robot, structural systematization, kinematic chain

1. INTRODUCTION

According to generally accepted principles, a serial industrial robot kinematic chain must have 6 axes (3 axes corresponding to the positioning kinematic chain and 3 axes corresponding to the orientation kinematic chain), any additional number of axes (monomobile kinematic couplings), leading to a certain redundancy.

First, it was considered that redundant kinematic chains are not desirable because they increase the complexity unnecessarily, leading to more time to calculate trajectories and to higher costs.

In recent years, there have been attempts to promote and even make industrial robots with redundant serial kinematic chains, especially those with 7 axes, but also with 8 axes [1], [2], [3], [4], [5], [6], [7], [8], [9]. They prove to have larger workspaces and more handling possibilities within these areas.

The early disadvantages mentioned have reduced their importance significantly, by increasing the calculation speed of trajectories due to existing processors' calculation power boost and lower design and manufacturing costs. Under these new conditions, a 7-axis structure, or structure with more 7 axes relatively more complicated, has become a very important functional advantage for a robot. To provide a useful tool, including for the promotion of new redundant structures, except existing ones, which are applied, in this paper there is a synthesis of kinematic chain structures with 7 axes, 8 axes and more.

2. STRUCTURAL SYNTHESIS OF SERIAL KINEMATIC CHAINS WITH 7 AXES FOR ROBOTS

According to [10], there are some serial kinematic chain structures with 6 axes for robots, with non-degenerated workspaces (reduced to a line or an area) which are usable. They are obtained identifying possible combinations between 3 rotation axes (monomobile couplings), respectively translation (RRR-TTT), considering only situations when two successive couplings are perpendicular or parallel, for positioning kinematic chains, considering possible eccentricities that equals or not zero and of 3 axes (monomobile couplings) of rotation (RRR), corresponding to the orientation kinematic chain. We consider only structures corresponding to non-degenerated workspaces. They are structures whose relative positions of successive couplings are perpendicular or parallel because these structures are more frequently applied to industrial robots and they are validated in practice as the most functional.

The structures with 7 axes (monomobile kinematic chains) are obtained from 6-axis ones adding a monomobile kinematic rotation coupling (R) or translation coupling (T), also in a relative position, perpendicular or parallel with the anterior or posterior coupling. Obviously, for each structure, the corresponding workspace is generated and we only consider those corresponding to non-degenerated workspaces. Basic versions are obtained if the eccentricity between two successive couplings equals zero, and derived versions if the eccentricity is not zero, developed along an axis, two axes, or three axes. Thus, according to [10], there are 20 different structures of tri-mobile positioning kinematic chains, generating volume-non-degenerated workspaces and 8 tri-mobile orientation kinematic chain structures. In Fig. 1 there are 2 examples of positioning kinematic chains highlighting significant kinematic parameters: a_1, a_2, d, r, S_0, S and M-is the characteristic point of the robot extremity [10].

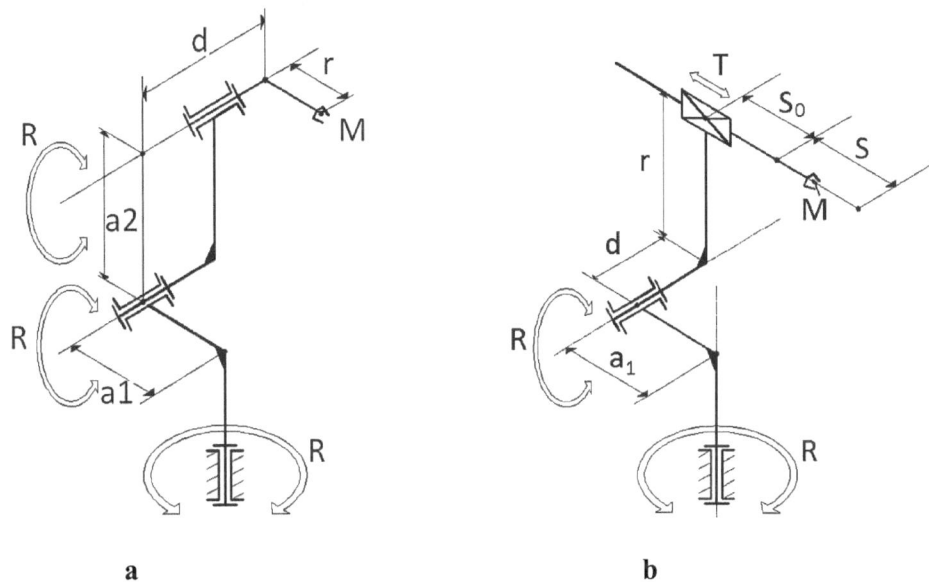

Figure 1: Two examples of tri-mobile positioning kinematic chains: $R\perp R\|R(a)$, $R\perp R\perp T(b)$

In Fig. 2 there are two examples of orientation kinematic chains (made of 3 rotation couplings -$R\perp R\perp R$; a,b – geometric parameters; l,n – the gripper position, longitudinal or perpendicular to the last axis of the orientation mechanism; for version b : a=0) [10].

Connecting in series a positioning kinematic chain with an orientation kinematic chain, we obtain the guiding kinematic chain, in this case having six axes (six monomobile rotation or translation kinematic couplings), where two successive axis-couplings are perpendicular ($\perp$) or parallel ($\|$).

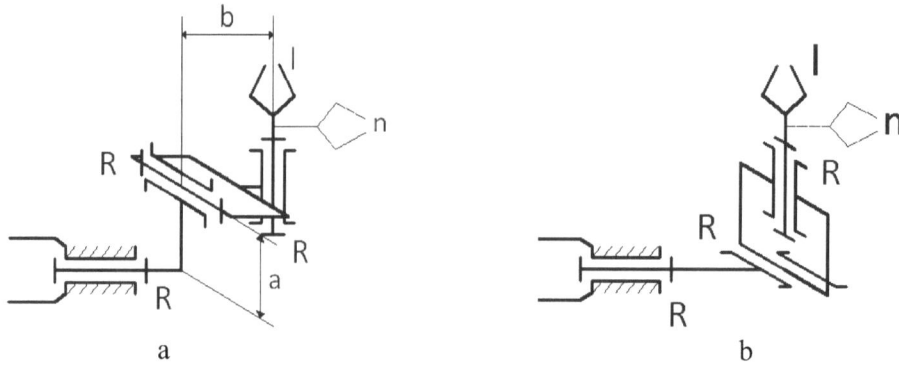

Figure2: Two examples of tri-mobile orientation kinematic chains: R⊥R⊥R

Guiding kinematic chains with seven axes (seven monomobile kinematic couplings of rotation or translation, positioned relatively perpendicular or parallel) we obtain out of those of six axes adding one monomobile kinematic coupling (an additional axis) of rotation (R) or translation (T). This coupling can be added at the beginning, at the end of the guiding kinematic chain or between the two component kinematic chains.

To obtain all the possible combinations, without losing the positioning kinematic chain and the orientation kinematic chain functionality, we consider the positioning kinematic chain a different kinematic module called positioning kinematic module (MP) and the orientation kinematic chain called orientation kinematic module (MO).

The relative position of the seventh coupling (axis) to the first or last coupling of one of the two modules can be perpendicular (⊥) or parallel (∥). We obtain 12 possible combinations for each structure with 6 axes(see e.g. Fig.3): R⊥(MP)(MO)- a1,a2, R∥(MP) (MO)-b, T⊥(MP)(MO)-c1,c2, T∥(MP)(MO)-d, (MP)⊥R(MO)-e, (MP)∥R(MO)-f, (MP)⊥T(MO)-g, (MP)∥T(MO)-h, (MP)(MO)⊥R-i, (MP)(MO)∥R-j. (MP)(MO)⊥T-k, (MP)(MO)∥T-l.

There are altogether, for the 20 structures with 6 axes, 240 structures with 7 axes.

In Fig. 3 there are the 12 kinematic versions with seven axes, representing in detail the positioning module, type: R⊥R⊥T and the compact version of the orientation module (R⊥R⊥R). Notations represent:

T,R- monomobile translation couplings, respectively rotation couplings; a_0, a_1, a_2, b_1, b_2, b_3, c, c_1, c_2, c_3, d-significant kinematic parameters, and M is the characteristic point of the robot. We note that in the case of the perpendicular position of the axis, there are, in general, two versions: the first perpendicular along the direction between the two axes (as in the case of Fig. 3a1, and the second collinear with the direction between the two axes,see Fig. 3a2, versions that will be detailed in following papers).

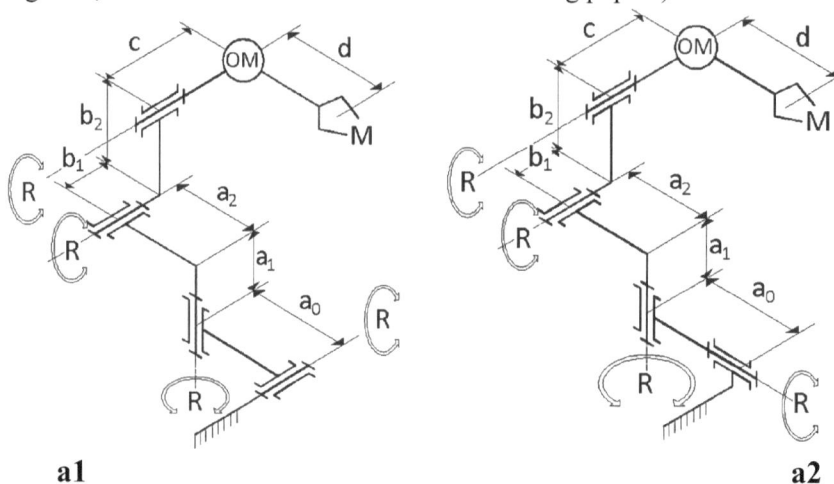

a1   a2

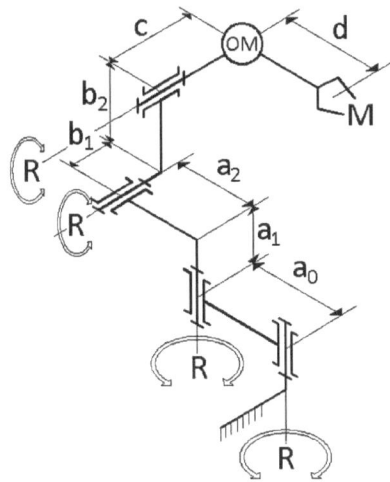

b

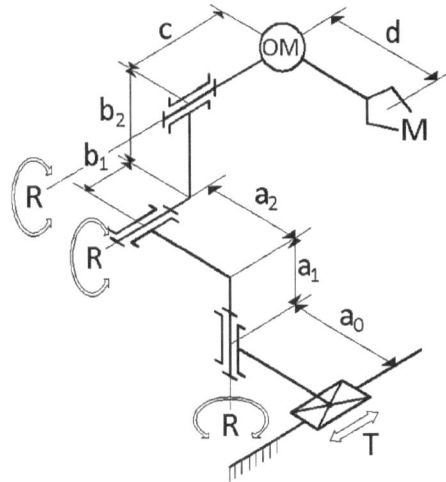

c1

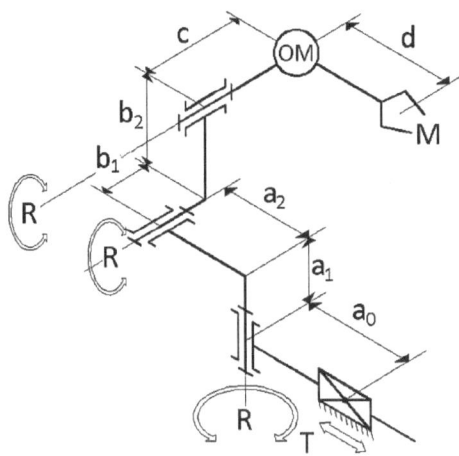

c2

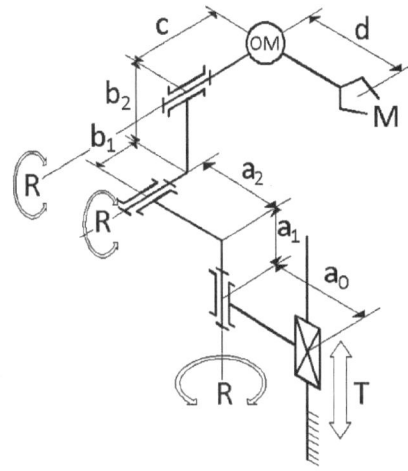

d

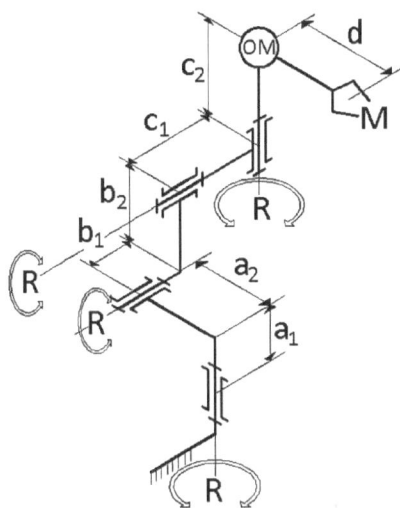

e

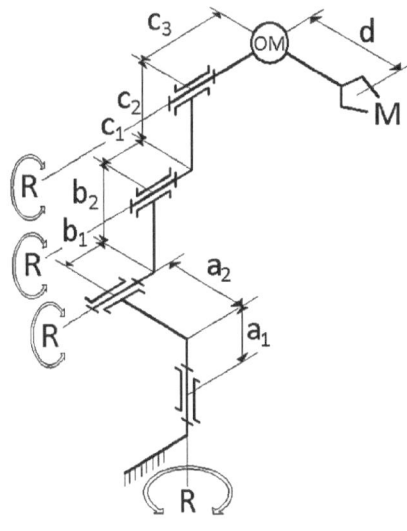

f

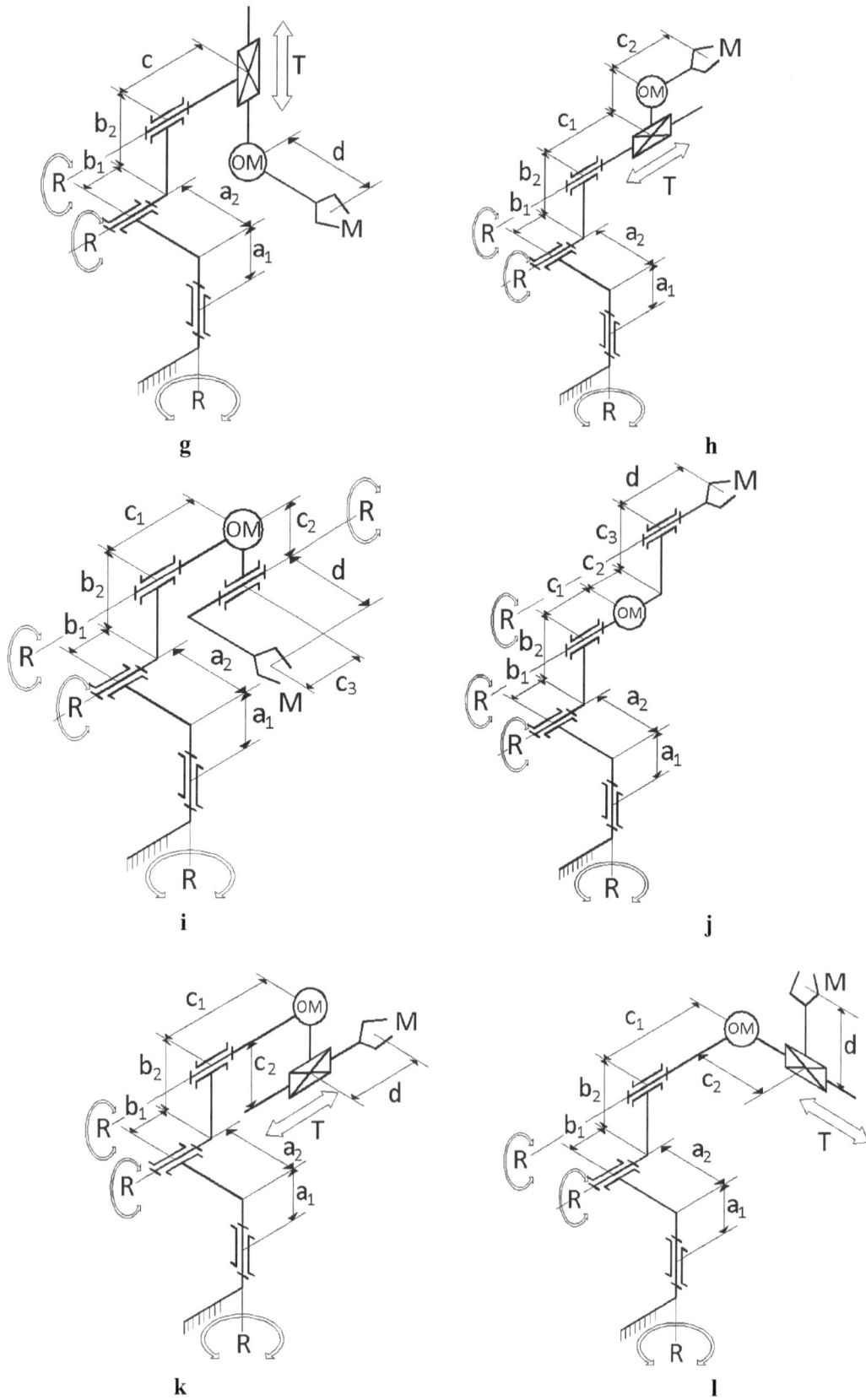

Figure 3: Guiding kinematic chains with seven axes

3. STRUCTURAL SYSTEMATIZATION OF KINEMATIC CHAINS WITH 8 AXES FOR ROBOTS

The structures of kinematic chains with 8 axes can be obtained from 6-axis ones, by adding two axes (two monomobile kinematic couplings) of rotation or translation or combined RT, respectively, TR. They can be both at one end of the 6-axis kinematic chain, or one at one end and the other at the other end of the said kinematic chain, or by adding R or T axes to 7-axis structures. This additional axis is added similarly to the case of obtaining structures with 7 axes out of 6-axis ones. Therefore, we obtain a number of distinct structures corresponding to nondegenerate workspaces. Thus by adding a rotation (R) or translation (T) coupling, perpendicular ($\perp$) or parallel ($\parallel$) to the first four structures $(R/T)(\perp/\parallel)(PM)(OM)$ of the seven-axis kinematic chain, we obtain 16 8-axis kinematic chain structures: $R\perp R\perp$ (PM)(OM),...,$T\parallel T\parallel(PM)(OM)$.Similarly, we obtain structural variants with eight axes based on 7-axis structures such as: $(PM)(\perp/\parallel)(R/T)(OM)$ and $(PM)(OM)(\perp/\parallel)(R/T)$. There are 96 final versions, 16 for each combination: $(R/T)(\perp/\parallel)(R/T)(\perp/\parallel)(PM)(OM)$, $(R/T)(\perp/\parallel)(PM)(\perp/\parallel)(R/T)(OM)$, $(R/T)(\perp/\parallel)(PM)(OM)(\perp/\parallel)(R/T)$, $(PM)(\perp/\parallel)(R/T)(OM)(\perp/\parallel)(R/T)$, $(PM)(\perp/\parallel)(R/T)(\perp/\parallel)(R/T)(OM)$ and $(PM)(OM)(\perp/\parallel)(R/T)(\perp/\parallel)(R/T)$.

In Fig. 4 are two kinematic structures with 8 axes with representation in detail of the positioning module structure and brief representation of the orientation module. Corresponding structural diagrams can be represented for all the 96 different possible combinations.

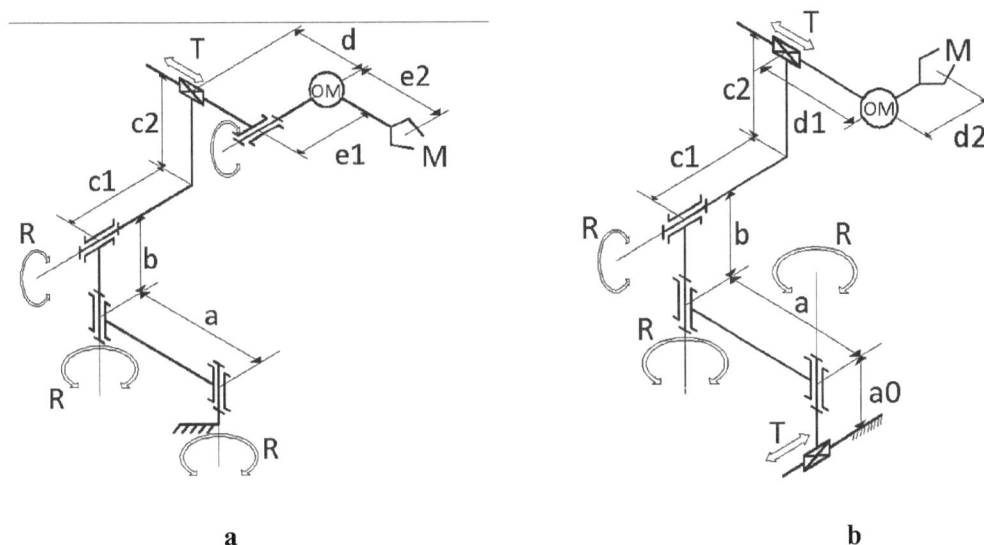

a   b

Figure 4: Structural diagrams of two types of kinematic chains with 8 axes

4. STRUCTURAL SYSTEMATIZATION OF KINEMATIC CHAINS WITH MORE OF 8 AXES FOR ROBOTS

It is noted that in a similar way you can obtain redundant structures that can be used in industrial robots with more than 8 axes: 9, 10, 11 or even 12 axes.

The structures of kinematic chains with more of 8 axes can be obtained from 6-axis ones, by adding three axes (three monomobile kinematic couplings) of rotation or translation or combined from RRR, respectively, till TTT(RRR, RRT, RTR, TRR, RTT, TRT, TTR, TTT), than by adding four axes(RRRR, RRRT, RRTR,..., TTTR, TTTT), five axes(RRRRR, RRRRT, RRRTT,...,TTTTT) or six axes(RRRRRR, RRRRRT, RRRTRR, TTTTTT). Some of these structures may have specific functional advantages in certain situations, which will be validated for sure by future practical applications.

A number higher than 12 axes already approaches the robotic trunk like structure formed by linking in a row several identical or similar constructive modules [10]. So robots with more twelve axes(six plus six axes) formed vertebrate or robots type trunk.

5. CONCLUSIONS

Redundant kinematic chains with 7 or 8 axes (7 or 8 monomobile rotation or translation couplings) have proved to be useful and some variants are already applied in some industrial robots already existing in practice.

This paper illustrates a method to obtain 12 possible structures with 7 axes, for each structure with 6 axes, with the axes in perpendicular or parallel relative positions, corresponding to non-degenerate workspaces, out of which other versions can be chosen, different from existing ones, to be manufactured. Then the paper shows the method to obtain the structures with 8 axes and more than 8 axes.

The workspaces representation helps the designer to choose the structure, which can have maximum functionality degree for a given range of applications. This aspect will be the topic in the next papers.

The structural synthesis method and the method of redundant serial kinematic chains with 7 axes representation can be extrapolated to structures with 8 axes, 9, 10, 11, even 12 axes.

6. ACKNOWLEDGMENTS

We express our gratitude to the company CLOOS from Germany and its representative in Romania, Timișoara ROBCON company, to support research whose results are presented in part in this paper.

REFERENCES

[1]   Brell-Cokcan S., Reis, M., Schmiedhofer, H. and Braumann, J., *Digital Design to Digital Production: Flank Milling with a 7-Axis Robot and Parametric Design*. In: Computation: The New Realm of Architectural Design – Proceedings of the 27th eCAADe Conference, Istanbul, 2009,p. 323-330.

[2]   Qi, R., Lam, T. L., Qian, H., Xu, Y., Arc tracking on an eight-axis robot system. In: ROBIO 2011, p. 678-683.

[3]   Wang, J., Li, Y. and Zhao, X., Inverse Kinematics and Control of a 7-DOF Redundant Manipulator Based on the Closed-Loop Algorithm. In: International Journal of Advanced Robotic Systems, Vol. 7, No. 4 (2010), p. 1-9.

[4]   www1:www.foodengineeringmag.com/.../seven-ax.

[5]   www2: www.motoman.co.

[6]   www3:www.yaskawa.co.jp/en/.../robotics/01.html.

[7]   www4:www.ragroup.com.au/robots/.../sda_sia_seri....

[8]   www5:www.cloos.de/QIROX/.../roboter/.../index.p...

[9]   www6:www.densorobotics.com/news/34.

[10].   Dutita, Fl.,a.a., Linkages mechanisms (in Romanian). Bucharest, Tehnica Press, Romania, 1987.

A DUAL VECTORS BASED FORMALISM FOR PARAMETRIZATION OF RIGID BODY DISPLACEMENT AND MOTION

Daniel Condurache[1], Adrian Burlacu[2]

[1] Gheorghe Asachi Technical University of Iasi, Iasi, ROMANIA, daniel.condurache @gmail.com
[2] Gheorghe Asachi Technical University of Iasi, Iasi, ROMANIA, aburlacu@ac.tuiasi.ro

Abstract: *This paper reveals a set of dual vectors based methods for rigid body displacement and motion parameterization. When parameterization methods are designed, a very important objective is to obtain a reduced number of algebraic equations and fewer variables for a more compact notation. This feature is achieved by acknowledging that parameterization of the rigid body motion is a problem strongly connected with the definition and properties of proper orthogonal dual tensors. Tensor analysis expresses the invariance of the laws of physics with respect to the change of basis and change of frame operations. First, we propose a method for computing orthogonal dual tensors based on the dual vectors derived from the motion laws of both points and lines attached to the rigid body. The rigid body motion parameterization using dual vectors gives the possibility of constructing new computational methods for the screw axis (SA) and instantaneous screw axis (ISA) motion parameters. These methods are based on two results: the computation of SA is equivalent to the computation of the logarithm of an orthogonal dual tensor, the computation of ISA is equivalent with the computation of an algebraic entity entitled "the velocity dual tensor".*
Keywords: *dual vector, rigid body, motion.*

1. INTRODUCTION

A rigid body can be characterized through different types of features, among them being points and lines. Starting with classical manipulator robot kinematics and dynamics description and finishing with the new results obtained in robotics, machine vision, astrodynamics or neuroscience, the range of applications involving points or lines transformation is very large [1-3].

If points are considered then any coordinates transformation can be parametrized using homogeneous transformations. For line features, parametrization techniques were developed using the dual numbers theory [4-7]. The combination of dual numbers, dual vectors or dual matrices calculus with elements of screw theory generates different techniques for rigid body motion modeling [8-10]. Orthogonal dual tensors are a complete tool for computing rigid body displacement and motion parameters. A reduced number of algebraic equations and a more compact notation with fewer variables are two of the advantages of orthogonal dual tensor based parametrization methods.

The first goal of our research is to give a more compact algebraic description regarding rigid body motion parametrization using tensors and to discuss the advantages over the methods involving dual matrices [8-14]. Our tensorial parametrization method is generated by the properties of the dyadic product between dual vectors. The second contribution represents a set of new computational methods for the screw axis (SA) and instantaneous screw axis (ISA) motion parameters.

The mathematical preliminaries and the notations are presented in section 2. Different algebraic sets were used to construct the parametrization methods proposed in this paper, their most important properties being detailed in the appendixes.

The construction of the dual tensors module using the dyadic product between a basis of dual vectors and its reciprocal is discussed in section 3.

Section 4 focuses on a new rigid body motion parameterization using bases of dual vectors. The construction of orthogonal dual tensors and screw and instantaneous screw parameters computation techniques are detailed. Also, a short and constructive proof of the famous *Mozzi - Chasles* theorem using dual tensors is presented.

Section 5 contains the conclusions and future work.

2. MATHEMATICAL PRELIMINARES AND NOTATIONS

This section outlines briefly the notations used in this paper and the algebraic properties of dual numbers, dual vectors and dual tensors. Regarding notation, in order to avoid name clashes, the following are considered: $\underline{x}$ denotes a dual number, $\underline{\mathbf{x}}$ a dual vector, ε represents the imaginary entity which fulfills $\varepsilon^2 = 0$. Details over the dual numbers, dual vectors and dual tensors sets can be found in [7], [14], [15].

• *Dual numbers*

Let the set of real dual numbers be denoted by:

$$\underline{\mathbb{R}} = \mathbb{R} + \varepsilon\mathbb{R} = \{\underline{a} = a + \varepsilon a_0 \mid a, a_0 \in \mathbb{R}, \varepsilon^2 = 0\}. \tag{1}$$

where $a = Re(\underline{a})$ is the real part of $\underline{a}$ and $a_0 = Du(\underline{a})$ the dual part.

Any differentiable function of a dual number variable $\underline{x} = x + \varepsilon x_0$ can be decomposed as:

$$f(\underline{x}) = f(x) + \varepsilon x_0 f'(x). \tag{2}$$

The inverse of $\underline{a} \in \underline{\mathbb{R}}$, denoted by $\underline{a}^{-1} \in \underline{\mathbb{R}}$, exists if and only if $Re(\underline{a}) \neq 0$ and is computed using $\underline{a}^{-1} = \dfrac{1}{\underline{a}} = \dfrac{1}{a} - \varepsilon\dfrac{a_0}{a^2}$. Also, $\underline{a} \in \underline{\mathbb{R}}$ is a zero divisor if and only if $Re(\underline{a}) \neq 0$.

• *Dual vectors*

In the Euclidean space, the linear space of free vectors with dimension 3 will be denoted by V_3. The ensemble of dual vectors is defined as

$$\underline{V_3} = V_3 + \varepsilon V_3 = \{\underline{\mathbf{a}} = \mathbf{a} + \varepsilon\mathbf{a}_0; \mathbf{a}, \mathbf{a}_0 \in V_3, \varepsilon^2 = 0\}, \tag{3}$$

where $\mathbf{a} = Re(\underline{\mathbf{a}})$ is the real part of $\underline{\mathbf{a}}$ and $\mathbf{a}_0 = Du(\underline{\mathbf{a}})$ the dual part. For dual vectors, three products will be considered: scalar product (denoted by $\underline{\mathbf{a}} \cdot \underline{\mathbf{b}}$), cross product (denoted by $\underline{\mathbf{a}} \times \underline{\mathbf{b}}$) and triple scalar product (denoted by $<\underline{\mathbf{a}}, \underline{\mathbf{b}}, \underline{\mathbf{c}}>$).

The magnitude of $\underline{\mathbf{a}}$, denoted by $|\underline{\mathbf{a}}|$, is the dual number which fulfills $|\underline{\mathbf{a}}| \cdot |\underline{\mathbf{a}}| = \underline{\mathbf{a}} \cdot \underline{\mathbf{a}}$ and can be computed using

$$|\underline{\mathbf{a}}| = \begin{cases} \|\mathbf{a}\| + \varepsilon\dfrac{\mathbf{a}_0 \cdot \mathbf{a}}{\|\mathbf{a}\|}, & Re(\underline{\mathbf{a}}) \neq 0 \\ \varepsilon\|\mathbf{a}_0\|, & Re(\underline{\mathbf{a}}) = 0 \end{cases}, \tag{4}$$

where $\|.\|$ is the Euclidean norm. If $|\underline{\mathbf{a}}| = 1$ then $\underline{\mathbf{a}}$ is called unit dual vector.

Thus, based on these properties results that $(\underline{\mathbb{R}}, +, \cdot)$ is a commutative and unitary ring and any element $\underline{a} \in \underline{\mathbb{R}}$ is either invertible or zero divisor, while $(\underline{V_3}, +, \cdot_{\underline{\mathbb{R}}})$ is a free $\underline{\mathbb{R}}$-module.

• *Dual tensors*

An $\underline{\mathbb{R}}$-linear application of $\underline{V_3}$ into $\underline{V_3}$ is called an Euclidean dual tensor:

$$T(\underline{\lambda}_1\underline{\mathbf{v}}_1 + \underline{\lambda}_2\underline{\mathbf{v}}_2) = \underline{\lambda}_1 T(\underline{\mathbf{v}}_1) + \underline{\lambda}_2 T(\underline{\mathbf{v}}_2),$$
$$\forall\underline{\lambda}_1, \underline{\lambda}_2 \in \underline{\mathbb{R}}, \forall\underline{\mathbf{v}}_1, \underline{\mathbf{v}}_2 \in \underline{\mathbf{V}}_3. \tag{5}$$

From now on, the Euclidean dual tensor T will be shortly called dual tensor and $L(\underline{V}_3, \underline{V}_3)$ will denote the free $\mathbb{R}$-module of dual tensors. To the authors knowledge, the properties of $L(\underline{V}_3, \underline{V}_3)$ can be found only in a few articles like.

3. DUAL TENSOR CONSTRUCTION USING DUAL VECTORS

The rigid body displacement and motion parameterization methods proposed in section 4 are based on the properties of dual tensors. Thus, the present section the design of the dual tensor set is discussed. The key of the chosen design is the combination between dual bases and the dyadic product of dual vectors [16, 17]. In order to set-up the base of the dual tensor construction technique, we first uncover some algebraic results for dual bases.

Theorem 1 *If $\underline{a} \in \underline{V}_3$ then a dual number $\underline{\lambda} \in \mathbb{R}$ and a unit dual vector $\underline{u} \in \underline{V}_3$ exist in order to have $\underline{a} = \underline{\lambda}\,\underline{u}$. Also, if $Re(\underline{a}) \neq \mathbf{0}$ then $\underline{\lambda}$ and $\underline{u}$ are unique up to a sign change.*

Proof. If $\|.\|$ is the Euclidean norm then $\underline{\lambda} = \|\mathbf{a}\| + \varepsilon\dfrac{\mathbf{a}_0 \cdot \mathbf{a}}{\|\mathbf{a}\|}$ and $\underline{u} = \dfrac{\mathbf{a}}{\|\mathbf{a}\|} + \varepsilon\dfrac{\mathbf{a} \times (\mathbf{a}_0 \times \mathbf{a})}{\|\mathbf{a}\|^3}$ proves the theorem when

$Re(\underline{a}) \neq \mathbf{0}$. If $Re(\underline{a}) = \mathbf{0}$ then $\underline{\lambda} = \varepsilon\|\mathbf{a}_0\|$ and $\underline{u} = \dfrac{\mathbf{a}_0}{\|\mathbf{a}_0\|} + \varepsilon\underline{v} \times \dfrac{\mathbf{a}_0}{\|\mathbf{a}_0\|}$, $\forall \underline{v} \in \underline{V}_3$.

The geometrical interpretation of Theorem 1 is that any dual vector $\underline{a}$ from $\underline{V}_3$, with $Re(\underline{a}) \neq \mathbf{0}$, can be associated with a *labeled* line in the Euclidean three dimensional space. The elements of the dual vector $\underline{u} = \mathbf{u} + \varepsilon\mathbf{u}_0$ give the direction of the line parametrized as Plucker coordinates [2,3], while the dual number

$\underline{\lambda} = |\underline{a}| = \|\mathbf{a}\| + \varepsilon\dfrac{\mathbf{a}_0 \cdot \mathbf{a}}{\|\mathbf{a}\|}$ represents the label. If $Re(\underline{a}) = \mathbf{0}$ then the geometrical interpretation is a set of parallel

lines described by $\mathbf{a}_0$ and labeled with $\underline{\lambda} = |\underline{a}| = \varepsilon\|\mathbf{a}_0\|$.

Definition 1 *A set of three dual vectors $\mathbf{B} = \{\underline{e}_1, \underline{e}_2, \underline{e}_3\}$ will be called dual basis if the dual vectors are $\mathbb{R}$ linear independent and also represent a span set for $\underline{V}_3$.*

Proposition 1 *If any three dual vectors $\underline{e}_k \in \underline{V}_3, k = \overline{1,3}$, fulfill $Re(<\underline{e}_1, \underline{e}_2, \underline{e}_3>) \neq 0$ then there are uniquely determined $\{\underline{e}^1, \underline{e}^2, \underline{e}^3\}$ using the conditions $\underline{e}_i \cdot \underline{e}^j = \delta_i^j, i, j = \overline{1,3}$, where δ_i^j is the Kronecker symbol.*

Proof. Let $\{\underline{e}_1, \underline{e}_2, \underline{e}_3\}$ be a set constructed by the following rules:

$$\underline{e}^1 = \frac{\underline{e}_2 \times \underline{e}_3}{<\underline{e}_1, \underline{e}_2, \underline{e}_3>}, \quad \underline{e}^2 = \frac{\underline{e}_3 \times \underline{e}_1}{<\underline{e}_1, \underline{e}_2, \underline{e}_3>}, \quad \underline{e}^3 = \frac{\underline{e}_1 \times \underline{e}_2}{<\underline{e}_1, \underline{e}_2, \underline{e}_3>} \tag{6}$$

Using (6) the conditions $\underline{e}_i \cdot \underline{e}^j = \delta_i^j, i, j = \overline{1,3}$ are fulfilled.

Remark 1 *For a dual basis $\mathbf{B} = \{\underline{e}_1, \underline{e}_2, \underline{e}_3\}$, the set $\mathbf{B}^* = \{\underline{e}^1, \underline{e}^2, \underline{e}^3\}$ represents its reciprocal dual basis. The dual basis $\mathbf{B}$ coincides with $\mathbf{B}^*$ if and only if $\mathbf{B}$ is an orthonormal basis (aka $\underline{e}_i \cdot \underline{e}_j = \delta_{ij}$).*

Given two dual vectors $\underline{a}$ and $\underline{b} \in \underline{V}_3$, $\underline{a} \otimes \underline{b}$ denotes a dual tensor called ***tensor (dyadic) product*** and is defined by:

$$\underline{a} \otimes \underline{b} : \underline{V}_3 \times \underline{V}_3 \to \underline{V}_3, \quad (\underline{a} \otimes \underline{b})\underline{v} = (\underline{v} \cdot \underline{b})\underline{a}, \forall \underline{v} \in \underline{V}_3 \tag{7}$$

An important property of (7) is that $(\underline{a} \otimes \underline{b})(\underline{c} \otimes \underline{d}) = (\underline{b} \cdot \underline{c})\underline{a} \otimes \underline{d}$. From this point on we uncover how the dyadic product can be used to construct a dual tensor.

Theorem 2 *The following statements are true:*

1. A dual tensor $T : \underline{V}_3 \rightarrow \underline{V}_3$ is uniquely determined by the values obtained after T is applied to the elements of the dual basis $\mathbf{B} = \{\underline{\mathbf{e}}_1, \underline{\mathbf{e}}_2, \underline{\mathbf{e}}_3\}$:

$$T = (T\underline{\mathbf{e}}_i) \otimes \underline{\mathbf{e}}^i. \tag{8}$$

2. The ensemble $\mathbf{L}(\underline{V}_3, \underline{V}_3)$ is a free $\mathbb{R}$ - module of rank equal to 9.

Proof. Starting with (8) the Einstein's rule for mute indexes summation, when i varies from 1 to 3, will be used. Let $\underline{\mathbf{v}} \in \underline{V}_3$ be an arbitrary vector that has the following expression in the basis $\mathbf{B} = \{\underline{\mathbf{e}}_1, \underline{\mathbf{e}}_2, \underline{\mathbf{e}}_3\}$:

$$\underline{\mathbf{v}} = (\underline{\mathbf{v}} \cdot \underline{\mathbf{e}}^j)\underline{\mathbf{e}}_j. \tag{9}$$

Using (5) it results that

$$T\underline{\mathbf{v}} = T[((\underline{\mathbf{v}} \cdot \underline{\mathbf{e}}^j)\underline{\mathbf{e}}_j)] = (\underline{\mathbf{v}} \cdot \underline{\mathbf{e}}^j)(T\underline{\mathbf{e}}_j) = [(T\underline{\mathbf{e}}_j) \otimes \underline{\mathbf{e}}^j]\underline{\mathbf{v}} \tag{10}$$

which proves the first part of the theorem.

If T is a dual tensor then the dual vectors $T\underline{\mathbf{e}}_j, j = \overline{1,3}$ can be written as

$$T\underline{\mathbf{e}}_j = [\underline{\mathbf{e}}^i \cdot (T\underline{\mathbf{e}}_j)]\underline{\mathbf{e}}_i, i = \overline{1,3}. \tag{11}$$

Denoting with $T^i_j = \underline{\mathbf{e}}^i \cdot (T\underline{\mathbf{e}}_j)$, $T^i_j \in \mathbb{R}$ and combining (14) with (11) generates

$$T = T^i_j \underline{\mathbf{e}}_i \otimes \underline{\mathbf{e}}^j, \tag{12}$$

which represents a linear combination of tensors $\{\underline{\mathbf{e}}_i \otimes \underline{\mathbf{e}}^j\}_{i,j=\overline{1,3}}$ that is equivalent with a spanning set. The previous result, together with the remark (which can be easily proven) that $\{\underline{\mathbf{e}}_i \otimes \underline{\mathbf{e}}^j\}_{i,j=\overline{1,3}}$ are $\mathbb{R}$ linearly independent in $\mathbf{L}$, imply that $\{\underline{\mathbf{e}}_i \otimes \underline{\mathbf{e}}^j\}_{i,j=\overline{1,3}}$ is a basis in $\mathbf{L}(\underline{V}_3, \underline{V}_3)$ and $rank_{\mathbb{R}} \mathbf{L}(\underline{V}_3, \underline{V}_3) = 9$.

For any dual vector $\underline{\mathbf{a}} \in \underline{V}_3$ the associated skew-symmetric dual tensor will be denoted by $\tilde{\underline{\mathbf{a}}}$ and defined by:

$$\tilde{\underline{\mathbf{a}}}\underline{\mathbf{b}} = \underline{\mathbf{a}} \times \underline{\mathbf{b}}, \forall \underline{\mathbf{b}} \in \underline{V}_3. \tag{13}$$

The set of skew-symmetric dual tensors is structured as a free $\mathbb{R}$-module of dimension 3, module which is isomorph with $\underline{V}_3$. The following notation are considered $\underline{\mathbf{a}} = \text{vect} \tilde{\underline{\mathbf{a}}}$, $\tilde{\underline{\mathbf{a}}} = \text{spin} \underline{\mathbf{a}}$ [2].

For an arbitrary dual tensor T the following entities can be computed

$$\text{sym} T = \frac{1}{2}[T + T^\mathsf{T}], \quad \text{skew} T = \frac{1}{2}[T - T^\mathsf{T}], \tag{14}$$

where "sym" is the symmetric part of the dual tensor and "skew" is its skew-symmetric part. Also, the axial dual vector and the trace of tensor T are given by:

$$\text{vect } T = \text{vect} \frac{1}{2}[T - T^\mathsf{T}], \quad \text{trace } T = \frac{<T\underline{\mathbf{e}}_1, \underline{\mathbf{e}}_2, \underline{\mathbf{e}}_3> + <\underline{\mathbf{e}}_1, T\underline{\mathbf{e}}_2, \underline{\mathbf{e}}_3> + <\underline{\mathbf{e}}_1, \underline{\mathbf{e}}_2, T\underline{\mathbf{e}}_3>}{<\underline{\mathbf{e}}_1, \underline{\mathbf{e}}_2, \underline{\mathbf{e}}_3>}. \tag{15}$$

Both $\text{vect} T$ and $\text{trace} T$ have the $\mathbb{R}$ -linearity property: $\forall \underline{\lambda}_1, \underline{\lambda}_2 \in \mathbb{R}, \forall T_1, T_2 \in \mathbf{L}(\underline{V}_3, \underline{V}_3)$

$$\begin{aligned} \text{vect}(\underline{\lambda}_1 T_1 + \underline{\lambda}_2 T_2) &= \underline{\lambda}_1 \text{vect} T_1 + \underline{\lambda}_2 \text{vect} T_2 \\ \text{trace}(\underline{\lambda}_1 T_1 + \underline{\lambda}_2 T_2) &= \underline{\lambda}_1 \text{trace} T_1 + \underline{\lambda}_2 \text{trace} T_2 \end{aligned} \tag{16}$$

If the dual tensor defined by (7) is analyzed, the following results emerge: $(\underline{\mathbf{a}} \otimes \underline{\mathbf{b}})^\mathsf{T} = \underline{\mathbf{b}} \otimes \underline{\mathbf{a}}$, $\text{vect}(\underline{\mathbf{a}} \otimes \underline{\mathbf{b}}) = \frac{1}{2}(\underline{\mathbf{b}} \times \underline{\mathbf{a}})$ and $\text{trace}(\underline{\mathbf{a}} \otimes \underline{\mathbf{b}}) = \underline{\mathbf{a}} \cdot \underline{\mathbf{b}}$. These results combined with (16), when T is given by (11), lead to:

$$T^\mathsf{T} = \underline{\mathbf{e}}^i \otimes (T\underline{\mathbf{e}}_i), \tag{17}$$

$$\text{vect} T = \frac{1}{2}(T\underline{\mathbf{e}}_i) \times \underline{\mathbf{e}}^i, \tag{18}$$

$$\text{trace} T = (T\underline{\mathbf{e}}_i) \cdot \underline{\mathbf{e}}^i. \tag{19}$$

In (31), (32), (33) the Einstein's rule for mute indexes summation has been used, where i varies from 1 to 3.

4. RIGID BODY MOTION AND DISPLACEMENT PARAMETERIZATION

In order to have a more intuitive view of the equations that will be used in this subsection, the following notations must be considered:

$$\{f : \mathbb{R} \to V_3\} = V_3^{\mathbb{R}}, \{f : \mathbb{R} \to SO_3\} = SO_3^{\mathbb{R}}. \tag{20}$$

In (20), SO_3 denotes the special orthogonal group of tensors [2]. The method proposed by the authors emerges from the remark that any rigid body motion can be modeled using elements from the set of proper orthogonal dual tensors denoted by

$$\underline{SO}_3 = \{R \in \mathbf{L}(\underline{\mathbf{V}}_3, \underline{\mathbf{V}}_3) \mid RR^T = I, \det R = 1\} \tag{21}$$

and time depending functions, which can be grouped in a set denoted by $\underline{SO}_3^{\mathbb{R}}$:

$$\{f : \mathbb{R} \to \underline{SO}_3\} = \underline{SO}_3^{\mathbb{R}} \tag{22}$$

The internal structure of any dual tensor $R \in \underline{SO}_3$ is illustrated by the following result:

Theorem 4 *For any* $R \in \underline{SO}_3^{\mathbb{R}}$*, an unique decomposition is viable*

$$R = Q + \varepsilon \rho Q, \tag{23}$$

where $Q = Q(t) \in SO_3^{\mathbb{R}}$ and $\rho = \rho(t) \in V_3^{\mathbb{R}}$.

Based on Theorem 4, a representation of any dual tensor from $\underline{SO}_3$ can be given:

Theorem 5 *For any orthogonal dual tensor* R *defined as in (37), a dual number* $\underline{\alpha} = \alpha + \varepsilon d$ *and a dual unit vector* $\underline{\mathbf{u}} = \mathbf{u} + \varepsilon \mathbf{u}_0$ *exists in order to have the following expression*

$$R = I + \sin \underline{\alpha}\,\underline{\mathbf{u}} + (1 - \cos \underline{\alpha})\underline{\mathbf{u}}^2, \tag{24}$$

where $\mathbf{u}$ and α are recovered from the linear invariants of Q, while $d = \rho \cdot \mathbf{u}$ and

$$\mathbf{u}_0 = \frac{1}{2}\rho \times \mathbf{u} + \frac{1}{2}\cot\frac{\alpha}{2}\mathbf{u} \times (\rho \times \mathbf{u}).$$

Remark 3 *For every choice of an orthogonal dual tensor* $R \in \underline{SO}_3$ *a unit dual vector* $\underline{\mathbf{u}} \in \underline{V}_3$ *exists so that* $R\underline{\mathbf{u}} = \underline{\mathbf{u}}$.

The fundamental Mozzi-Chasles [2] theorem states that: "any rigid displacement may be represented by a planar rotation about a suitable axis passing through that point, followed by a translation along that axis". Theorem 5 and Remark 3 are in fact the steps of a very short, elegant and constructive proof of this famous theorem. A screw axis is characterized by an unitary dual vector $\underline{\mathbf{u}}$ and the screw parameters (angle of rotation about the screw and the translation along the screw axis) structured an a dual angle $\underline{\alpha}$. For the following results, lets recall that the $\underline{SO}_3$ is a Lie group and its Lie algebra can be identified by the skew-symmetric dual tensors set $\underline{so}_3$.

Theorem 6 *If the skew-symmetric dual tensors set is denoted* $\underline{so}_3 = \{\underline{\alpha} \in \mathbf{L}(\underline{\mathbf{V}}_3, \underline{\mathbf{V}}_3) \mid \underline{\alpha} = -\underline{\alpha}^T\}$ *then the mapping*

$$\exp : \underline{so}_3 \to \underline{SO}_3, \exp(\underline{\alpha}) = e^{\underline{\alpha}} = \sum_{k=0}^{\infty} \frac{\underline{\alpha}^k}{k!} \tag{25}$$

is well defined and surjective.

The screw parameters computation are linked with the problem of finding the logarithm of an orthogonal dual tensor R, which is defined by

$$\log : \underline{SO}_3 \rightarrow \underline{so}_3, \log R = \left\{ \underline{\psi} \in \underline{so}_3 \mid exp(\underline{\psi}) = R \right\} \tag{26}$$

and is the inverse of (25). If the dual vector $\underline{\psi} = \psi + \varepsilon \psi_0$ is computed as $\underline{\psi} = vect(\underline{\psi})$ then using Theorem 1 results that $\underline{\psi} = \underline{\alpha} \cdot \underline{u}$, where $\underline{\alpha} = \parallel \psi \parallel + \varepsilon \dfrac{\psi \times \psi_0}{\parallel \psi \parallel}$ and $\underline{u} = \dfrac{\psi}{\parallel \psi \parallel} + \varepsilon \dfrac{\psi \times (\psi_0 \times \psi)}{\parallel \psi \parallel^3}$. This result implies that $\underline{\psi}$ can be used to parametrize any type of rigid motion. Before computing the logarithm of an orthogonal dual tensor, we need to analyze the behavior and influence of the tensor's natural invariants.

Remark 4 *A direct result of Theorems 5 and 6 is that the logarithm of a dual tensor is the product $\underline{\alpha}\underline{u}$. The parameters $\underline{\alpha}$ and $\underline{u}$ are called the natural invariants of **R** and can be recovered from the linear invariants [2] using (24):*

$$\underline{u}\sin\underline{\alpha} = vectR, \tag{27}$$

$$1 + 2\cos\underline{\alpha} = traceR. \tag{28}$$

The above equations can be transformed into

$$\underline{u}\sin\underline{\alpha} = vectR$$
$$\cos\underline{\alpha} = \frac{1}{2}[traceR - 1] \tag{29}$$

and the parameters $\underline{\alpha}, \underline{u}$. The unit dual vector $\underline{u}$ gives the Plucker representation of the Mozzi-Chalses axis, while the dual angle $\underline{\alpha} = \alpha + \varepsilon d$ contains the rotation angle α and the translation distance d. The computational formulas for $\underline{\alpha}, \underline{u}$ are extracted from (29):

$$\underline{u} = \begin{cases} \pm\dfrac{vectR}{|vectR|}, & Re(vectR) \neq 0 \\[2ex] \dfrac{Qv + v}{\parallel Qv + v \parallel} + \varepsilon\dfrac{1}{2}\rho \times \dfrac{Qv + v}{\parallel Qv + v \parallel}, & \forall v \in V_3, \ Re(vectR) = 0 \ and \ traceQ = -1 \\[2ex] \dfrac{\rho}{\parallel \rho \parallel}, & Re(vectR) = 0 \ and \ traceQ = 3 \end{cases} \tag{30}$$

$$\underline{\alpha} = atan2(\pm|vectR|, \frac{1}{2}[traceR - 1]). \tag{31}$$

The line containing the points of a rigid body undergoing minimum-magnitude velocities is called the instant screw axis (ISA) of the body under a given motion. The instantaneous motion of the body is equivalent to that of the bolt of a screw of ISA and is called instantaneous screw. An instantaneous screw axis, which will be defined as a dual vector denoted by $\underline{\omega}$, is characterized by an unit dual vector $\underline{u}$, a dual number $|\underline{\omega}|$ called magnitude and a number p called the pitch.

Let $\underline{h}_0$ embed the Plucker coordinates of a line at $t = t_0$ then:

$$\underline{h}(t) = R(t)\underline{h}_0. \tag{32}$$

Theorem 7 *In a general rigid motion, described by an orthogonal dual tensor, the **velocity dual tensor** Φ defined as*

$$\dot{\underline{h}} = \Phi\underline{h} \tag{33}$$

is expressed by:

$$\Phi = \dot{R}R^T. \tag{34}$$

The form of the velocity dual tensor described by (34) can be taken a step further if R is decomposed as in (23). This implies

$$\dot{R} = \dot{Q} + \varepsilon(\dot{\rho}Q + \rho\dot{Q}).$$ (35)

Because $\Phi \in \underline{so}_3^{\mathbb{R}}$ results that we can consider $\Phi = \underline{\omega}$, which gives:

$$\underline{\omega} = \dot{Q}Q^{\mathrm{T}} + \varepsilon(\dot{\rho} - \dot{Q}Q^{\mathrm{T}}\rho).$$ (36)

Let $\omega = \dot{Q}Q^{\mathrm{T}}$ and $\mathbf{v} = \dot{\rho} - \dot{Q}Q^{\mathrm{T}}\rho$, then

$$\underline{\omega} = \omega + \varepsilon\mathbf{v}.$$ (37)

This equation leads to the internal structure of the dual vector $\underline{\omega}$:

$$\underline{\omega} = \omega + \varepsilon\mathbf{v}$$ (38)

where ω is the angular velocity and $\mathbf{v}$ represents the linear velocity of the point of the body that coincides instantaneously with the origin.

The dual vector $\underline{\omega}$ completely characterize, at a certain time, the velocity field of an rigid body in motion.

Based on Theorem 1, for $\|\omega\| \neq 0$ the instantaneous screw axis unit dual vector is $\underline{\mathbf{u}} = \dfrac{\omega}{\|\omega\|} + \varepsilon\dfrac{\omega \times (\mathbf{v} \times \omega)}{\|\omega\|^3}$ and $|\underline{\omega}| = \|\omega\| + \varepsilon\dfrac{\mathbf{v} \times \omega}{\|\omega\|}$. If $p = \dfrac{\mathbf{v} \times \omega}{\|\omega\|^2}$ denotes the pitch of the screw axis [9] then $|\underline{\omega}| = \|\omega\|(1 + \varepsilon p)$. For $\|\omega\| = 0$ we have an intantaneous pure translation.

5. CONCLUSIONS

The research presented in this paper is focused on developing a new rigid body motion parametrization method using dual vectors. Our studies showed that, in the dual tensors free module, the dual bases of dual vectors can completely characterize the rigid body motion from the Euclidean three-dimensional space.

The proposed parametrization method was used to generate the orthogonal dual tensor that can model the motion of a rigid body. Screw and instantaneous screw parameters computational algorithms were also developed using dual bases of dual vectors.

For the fundamental Mozzi-Chasles theorem a very short, elegant and constructive proof was provided.

As future research goals the authors will analyze higher order kinematic properties using the free module of dual tensors. From the applicative point of view, the dual tensors algebra can be a solution for direct and inverse kinematics problems, multi-body problems, dual Euler-Rodrigues parameters and dual quaternions computation from direct measurements

REFERENCES

[1]   B. Siciliano, O. Khatib (Eds.), Springer Handbook of Robotics, Springer, Berlin, Heidelberg, 2008.

[2]   J. Angeles, Fundamentals of Robotic Mechanical Systems Theory, Methods, and Algorithms Third Edition, Springer, 2007.

[3]   J. Davidson, K. Hunt, Robots and Screw Theory applications of kinematics and statistics to robotics, Oxford, 2004

[4]   Y.-L. Gu, J. Luh, Dual-number transformation and its applications to robotics, IEEE Journal of Robotics and Automation 3 (1987) 615–623.

[5]   J. McCarthy, Dual orthogonal matrices in manipulator kinematics, The International Journal of Robotics Research
5 (1986) 45–51.

[6]   D. P. Chevallier, Lie algebras, modules, dual quaternions and algebraic methods in kinematics, Mechanism and Machine Theory 26 (1991) 613–627.

[7]   I. Fischer, Dual-Number Methods in Kinematics, Statics and Dynamics, CRC Press, 1999.

[8]   J. Angeles, Automatic computation of the screw parameters of rigid-body motions. Part I: Finitely-separated positions, ASME J. of Dynamic Systems, Measurement, and Control 108 (1986) 32–38.

[9]   J. Angeles, Automatic computation of the screw parameters of rigid-body motions. part II: Infinitesimally separated positions, ASME J. of Dynamic Systems, Measurement, and Control 108 (1986) 39–43.

[10]   M. Keler, On the theory of screws and the dual method, proc. of A Symposium Commemorating the Legacy, Works and Life of Sir Robert Stawell Ball Upon the 100th Anniversary of A Treatise on the Theory of Screws (2000) 1–12.

[11]   D. Condurache, M. Matcovschi, Algebraic computation of the twist of a rigid body through direct measurements, Comput. Methods Appl. Mech. Engineering 190 (2001) 5357–5376.

[12]   D. Condurache, M. Matcovschi, Computation of angular velocity and acceleration tensors by direct measurements, Acta Mechanica 153 (2002) 147–167.

[13]   R. Vertechy, V. Parenti-Castelli, Accurate and fast body pose estimation by three point position data, Mechanism and Machine Theory 42 (2007) 1170–1183.

[14]   E. Pennestri, P. P. Valentini, Linear dual algebra algorithms and their application to kinematics, Multibody Dynamics: Computational Methods and Applications 12 (2009) 207–229.

[15]   O. Bauchau, L. Li, Tensorial parameterization of rotation and motion, Journal of Computation and Nonlinear Dynamics 6 (2011) 031007.1–031007.8

[16]   D. Condurache, A. Burlacu, Rigid Body Pose Estimation using Dual Quaternions Computed from Direct Measurements, 43rd International Symposium on Robotics, Taipei Taiwan (2012).

[17]   D.Condurache, A. Burlacu, On Six D.O.F Relative Orbital Motion Parametrization using Rigid Bases of Dual Vectors, proc. of AAS/AIAA Astrodynamics Specialist Conference, Hilton Head South Carolina USA (2013).

ON THE EFFECTIVE MODULI OF SONIC COMPOSITE

Ligia Munteanu[1], Veturia Chiroiu[1], Ştefania Donescu[2], Ruxandra Ilie[2], Valerica Moşneguţu[1]

[1]Institute of Solid Mechanics, Ctin Mille 15, Bucharest 010141, ligia_munteanu@hotmail.com, valeriam732000@yahoo.com

[2]Technical University of Civil Engineering, Bd. Lacul Tei nr.122-124, Bucharest 020396
stefania.donescu@yahoo.com, rux_i@yahoo.com

To the memory of Prof. Petre P. Teodorescu (1929-2013).
The Professor's name would forever be engraved in our hearts.

Abstract: *The auxetic behavior is interpreted in the light of Cosserat elasticity which admits degrees of freedom not present in classical elasticity, i.e. the rotation of points in the material, and a couple per unit area or the couple stress.*
The Young' modulus evaluation for a sonic composite designed in order to provide suppression of unwanted noise for jet engines, with emphases on the nacelle of turbofan engines for commercial aircraft, is presented in this paper.
Key-Words: *Auxetic material, Cosserat elasticity, Young's modulus, sonic composite.*

1. INTRODUCTION

The sonic composites we discuss in this paper consist of an array of acoustic scatterers having the shape of spherical shells and made of the auxetic material (negative Poisson coefficient), embedded into the epoxy matrix. Let us begin with the auxetic materials, which have a negative Poisson ratio. The term *auxetic* comes from the Greek word *auxetos*, meaning *that which may be increase*. Instead of getting thinner like an elongated elastic band, the auxetic material grows fatter, expanding laterally when stretched. All the major classes of materials (polymers, composites, metals, ceramics honeycomb structures, reticulated metal foams, re-entrant structures, the skin covering a cow's teats, certain rocks and minerals, living bone tissue) can exist in the auxetic form [1-6].

The idea is to transform a non-auxetic material into auxetic forms as foams or cellular materials, or to employ new techniques for architecture new auxetic materials.

The simulation of a sonic composite based on the auxetic materials was studied in [7] and [8]. We briefly describe in the following the principal results reported in [7] and [8] because the present paper is devoted to evaluate the Young'modulus for the sonic composite described in these papers.

These sonic composites are characterized by the existence of a large sound attenuation band which is done by the superposition of multiple reflected waves within the array according to the Bragg's theory. The band-gaps which characterises a sonic composite correspond to the Bragg reflections that occur at different frequencies inverse proportional to the central distance between two scaterers [7-9]. If the band-gaps are not wide enough, their frequency ranges do not overlap. These band-gaps can overlap due to reflections on the surface of the scatterers, as well as due to wave propagation inside them. Then, any wave is reflected completely from this periodic array in the frequency range where all the band-gaps for the different periodical directions overlap.

This is the fundamental mechanism for the formation of a full band-gap which is required for sonic composites. The complete reflection on the boundaries of scatterers is due to the full band-gap property itself, independent of the incident angle. This makes sharp bends of the wave-guide in the sonic composite. The evanescent waves distribute across the boundary of the waveguide into the surrounding composite by several times the lattice constant.

The geometry of the scatterers and the material of which they are made, are not assumed a priori. A new technique for choosing the geometry and the material for the acoustic scatterers is proposed. This technique is performed in two stages. In the first stage, the acoustic scatterer is the traditional one, i.e. a sphere filled with conventional foam with positive Poisson ratio. In the second stage, a new geometry and a new material are looking for the scatterer so that the efficiency of the sonic line to be high. The achieving of this stage is made as simple and inexpensive as the first one.

The secret is the property of the equations that describe the behavior of the conventional sonic line. These equations can be reduced to Helmholtz equations which are invariant under geometric transformations. In other words, by choosing an appropriate geometric transformation, the conventional foam-filled sphere can be changed into a new scatterer with different geometry filled with a new material obtained from the initial one by spatial compression.

We must specify that the auxetic foam manufacturing relies on the compression of the conventional one. The conventional foam has pores with an average diameter of around 900, with an isotropic distribution of the major axis of the cell in different directions. The manufacturing of auxetic materials is based on the cell size reduction through radial compression molds.

Therefore, by a careful handling of the geometric transformation we can lead to a simple, cheap and efficient simulation of the manufacturing process of the material required for a high reduction / removal of the noise in the sonic line. The proposed technique is based on the Cosserat theory combined with cnoidal method and a genetic algorithm, respectively.

The standard continuum models cannot describe the phenomenon of band- gaps (the regions in the frequency-wavenumber space where the energy does not propagate) for heterogeneous materials that appear to be homogeneous at the meso-scale. The Cosserat or micropolar theory is an alternative continuum model that incorporate a length scale.

The conventional foams are materials with microstructure which exhibit chiral effects. These effects cannot be expressed within the classical elasticity since the modulus tensor, which is of the fourth rank, is unchanged under inversion

$$C_{ijkl} = x_{m,i} x_{n,j} x_{o,k} x_{p,l} C_{mnop} = (-1)\delta_{im}(-1)\delta_{jn}(-1)\delta_{ok}(-1)\delta_{pl}C_{mnop} = (-1)^4 C_{ijkl} = C_{ijkl} .$$ (1)

This provides a motivation towards developing enriched continuum models endowed with intrinsic length scales that synthesize the key features of the sub-scale material architecture. This is the reason we have chose to describe the chiral properties to Cosserat elasticity [10], [11].

2. THEORY

Let us consider a liner consisted of a porous or perforated facing sheet forming the interior duct wall, bonded on a composite layer and is terminated with a rigid back wall (Figure 1) [8]. The composite layer is a thin plate consisting of an array of acoustic scatterers embedded in an epoxy matrix. The acoustic scatterers are spheres made from conventional foam and the matrix is made from an epoxy resin (Figure 2).

The plate consists of 144 local spherical resonators of diameter a. The length of the plate is L, its width is d, while the diameter of the sphere is a and its thickness is $e > a$. The purpose of this sonic liner is to suppress the noise generated by the fan before it radiates out of the fan inlet and the fan exhaust ducts and in some instances, to reduce the combustion and turbine noise in the exhaust duct of the core engine.

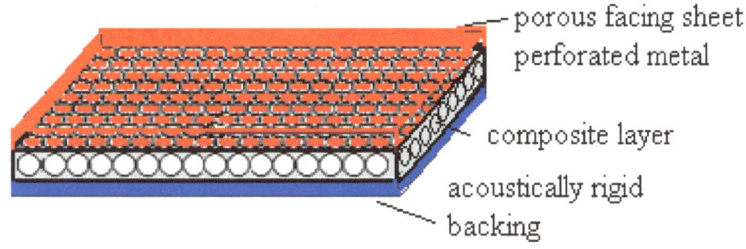

Figure 1: The plate with spherical resonators [8]

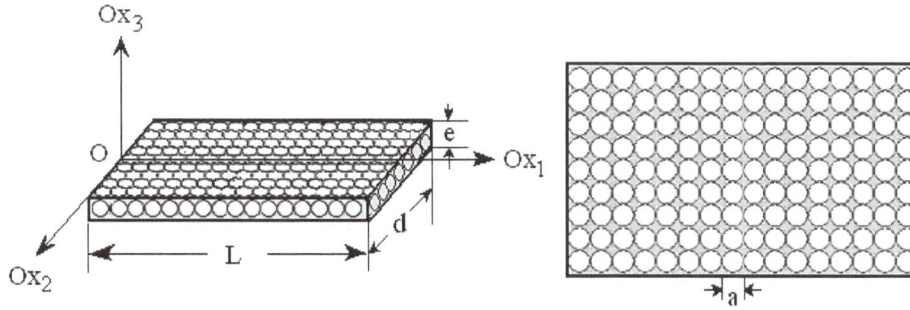

Figure 2: Sketch of the composite layer with spherical scatterers [8]

Consider a chiral Cosserat medium, in a Cartesian coordinates system $x \equiv (x, y, z)$. The equations of motion for the case without body forces and body couples are [12]-[15]

$$\sigma_{kl,k} - \rho \ddot{u}_l = 0, \ m_{rk,r} + \varepsilon_{klr}\sigma_{lr} - \rho j \ddot{\varphi}_k = 0. \tag{2}$$

In (2), σ_{kl} is the stress tensor, m_{kl} is the couple stress tensor, u is the displacement vector, φ_k is the microrotation vector which in Cosserat elasticity is cinematically distinct from the macrorotation vector $r_k = 1/2\varepsilon_{klm}u_{m,l}$, and ε_{klm} is the permutation symbol. The quantity φ_k refers to the rotation of points themselves, while r_k refers to the rotation associated with movement of nearby points. In (2) ρ is the mass density and j the microinertia. The constitutive equations are

$$\sigma_{kl} = \lambda e_{rr}\delta_{kl} + (2\mu + \kappa)e_{kl} + \kappa\varepsilon_{klm}(r_m - \varphi_m) + C_1\varphi_{r,r}\delta_{kl} + C_2\varphi_{k,l} + C_3\varphi_{l,k}, \tag{3}$$

$$m_{kl} = \alpha\varphi_{r,r}\delta_{kl} + \beta\varphi_{k,l} + \gamma\varphi_{l,k} + C_1 e_{rr}\delta_{kl} + (C_2 + C_3)e_{kl} + (C_3 - C_2)\varepsilon_{klm}(r_m - \varphi_m), \tag{4}$$

where $e_{kl} = 1/2(u_{k,l} + u_{l,k})$ is the macrostrain vector. λ, and μ are Lamé elastic constants, κ is the Cosserat rotation modulus, α, β, γ, the Cosserat rotation gradient moduli, and C_i, $i = 1, 2, 3$ are the chiral elastic constants associated with noncentrosymmetry. For $C_i = 0$ the equations of isotropic micropolar elasticity are recovered.

For $\alpha = \beta = \gamma = \kappa = 0$, (1) reduces to the constitutive equations of classical isotropic linear elasticity theory {16]-[18].

The initial conditions are

$$u_i(x, 0) = u_i^0(x), \ \varphi_i(x, 0) = 0, \tag{5}$$

$$m_{ij}(x, 0) = 0, \ \sigma_{ij}(x, 0) = 0. \tag{6}$$

The touchstone of obtaining auxetic materials is that the governing equations (2) of the non-auxetic foams are invariant under geometric transformations. The equations (2) are reducing to the Helmholtz equations

$$\nabla \cdot S : \nabla U + \omega^2 U = 0 , \tag{7}$$

where S is the fourth-order material tensor, ω is the wave angular frequency, $U(x,t) = U(x)\exp(-i\omega t)$, $U = (u_i, \phi_i)$, $i = 1,2,3$, $U(x,t) = U(x)\exp(-i\omega t)$.

The sonic composite layer finally consists from an array of spherical shells scatterers embedded in an epoxy matrix. The spherical shells occupy the region $R_1 < r < R_2$, $r = \sqrt{x^2 + y^2 + z^2}$, which is filled with auxetic material, and the matrix is made from an epoxy resin (Figure 3). In Figure 3, ω is the frequency of the incident sound.

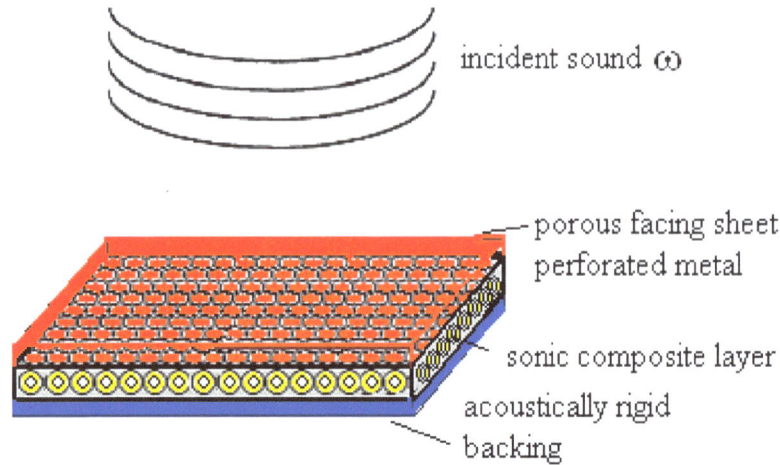

Figure 3: The sonic composite with spherical shells resonators [8].

3. EVALUATION OF THE YOUNG'S MODULUS

The overlapping of all pseudo gaps obtained from reflections on the scatterers as well as due to wave propagation in the scatterers, generates the full band-gap. Any wave is reflected completely in the frequency range where all the pseudo band-gaps for the different directions overlap.

This is the fundamental mechanism for the formation of a full band-gap. Figure 4 shows the first, the second and the 210th pseudo gaps (red, blue and green zones) delimited by the attenuation peaks lines calculated along four symmetric directions, and the full band-gap (grey zone), with respect to the compressive strain. For a given compressive strain $1 - 9$, this plot predicts the full band-gap.

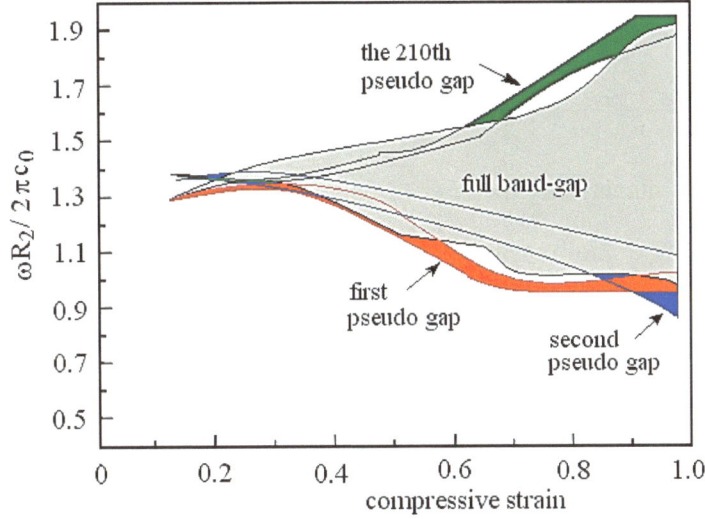

Figure 4: The full-band gap structure [8]

We are interested in the following in the knowing the influence of ☐ the Cosserat ☐ ☐ rotation modulus κ', the Cosserat ☐ ☐ rotation gradient moduli α', β', γ', and the chiral elastic constants C_i', $i = 1,2,3$, on the effective Young' modulus value of the sonic composite. The material new constants $\tilde{C} = \{\lambda', \mu', \kappa', \alpha', \beta', \gamma', C_1', C_2', C_3', \lambda_{ep}', \mu_{ep}', \kappa_{ep}'\}$ of the sonic composite consisted from an array of spherical shells scatterers embedded in an epoxy matrix, are done by the geometric transformations $dx = J_{xx'} dx'$, $J_{xx'} = \dfrac{\partial(x)}{\partial(x')}$ [8], as

$$\lambda' = \frac{\lambda}{\det(J_{xx'})} , \; \mu' = \frac{\mu}{\det(J_{xx'})} , \; \kappa' = \frac{\kappa}{\det(J_{xx'})} , \; \alpha' = \frac{\alpha}{\det(J_{xx'})} , \; \beta' = \frac{\beta}{\det(J_{xx'})} ,$$

$$\gamma' = \frac{\gamma}{\det(J_{xx'})} , \; \rho' = \frac{\rho}{\det(J_{xx'})} , \; \kappa' = \frac{\kappa}{\det(J_{xx'})} , \; \rho' = \frac{\rho}{\det(J_{xx'})} , \; C_i' = \frac{C_i}{\det(J_{xx'})} , \; i = 1,2,3 . \tag{8}$$

$$\lambda_{ep}' = \frac{\lambda_{ep}}{\det(J_{xx'})} , \; \mu_{ep}' = \frac{\mu_{ep}}{\det(J_{xx'})} , \; \kappa_{ep}' = \frac{\kappa_{ep}}{\det(J_{xx'})}$$

where the index *ep* is denoting the epoxy constants. The constants have a periodical character

$$\tilde{C}(r + a) = \tilde{C}(r) \tag{9}$$

where a is the diameter of the spherical resonator. A new length scale $\eta = \dfrac{r}{\varepsilon}$ is added, where $\varepsilon > 0$ is a parameter, $\dfrac{\partial}{\partial x} = \dfrac{\partial}{\partial x} + \dfrac{1}{\varepsilon} \dfrac{\partial}{\partial \eta}$. The problem of Bécus homogenization via multiple scale expansion consists in studying (7) as $\varepsilon \to 0$ [6, 19].

As $\varepsilon \to 0$, the periodic variations of $\tilde{C}$ become frequent, so that the study of equations will bring information on the solutions for different values a. We find

$$E = F(\tilde{C}) + \frac{1}{3}\gamma + \frac{1}{2}\tilde{p}^2 + \frac{3}{4}\sqrt{\delta}\tilde{p} , \tag{10}$$

where

$$\gamma = \frac{(2\mu' + \kappa')(3\lambda' + 2\mu' + \kappa_{aux})}{(2\lambda' + 2\mu' + \kappa_{aux})}, \quad \delta = \frac{(2\mu' + \kappa')(3\alpha' + 2\beta' + \gamma')}{(2C_1' + 2C_2' + C_3')}, \quad \tilde{p}^2 = \frac{2\kappa_{aux}}{\zeta},$$

$$\zeta = \frac{(C_{1aux} + 3C_{2aux} + C_{3aux})(3\lambda' + 2\mu' + \kappa_{aux})}{(\lambda_{aux} + 2\mu_{aux} + \kappa_{aux})(\alpha_{aux} + 2\beta_{aux} + \gamma_{aux})} \tag{11}$$

The function $F(\tilde{C})$ is numerically determined only. The most important physical parameter which dominates the negative Poisson's ratio transformation is the compression ratio $\vartheta = \frac{(R_2'^3 - R_1'^3)}{R_2^3}$, where prime denotes the final parameters. This parameter is directly related to the capacity of damping of the sonic composite. The qiuantity $1 - \vartheta$ represents the compressive strain.

Figure 5 represents the variation of the homogenized Young's modulus with respect to the compressive strain $1 - \vartheta$, and the Poisson's ratio ν of the auxetic material. In the simulation, the spherical specimen has $R_2 = 15$mm the initial radius, and $\lambda = 2.59$GPa, $\mu = 0.77$GPa, $\kappa = 0.0144$GPa, $\alpha = 1.77 \times 10^4$ N, $\beta = 3.37 \times 10^4$ N, $\gamma = 0.33 \times 10^4$ N, $C_1 = -0.5 \times 10^4$ N/m, $C_2 = -2.9 \times 10^4$ N/m, $C_3 = -6.8 \times 10^4$ N/m.

The radius R_1 depends on ϑ. So, for $\vartheta = 0.25$, $R_1 = 13.63$ mm, for $\vartheta = 0.3$, $R_1 = 13.32$ mm, for $\vartheta = 0.35$, $R_1 = 12.99$ mm, and for $\vartheta = 0.4$, $R_1 = 12.65$ mm, respectively.

We observe that Young's modulus is increasing and decreasing in a complex way. The Poisson's ratio significanntly depends on the compressive strain.

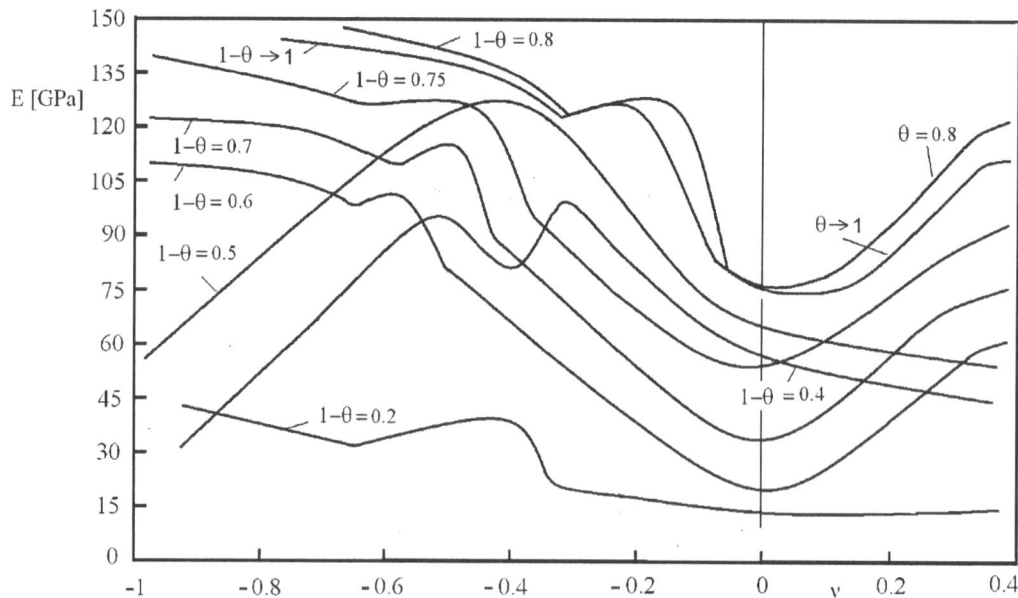

Figure5: The homogenized Young's modulus variation with respect to Poisson's ratio of the auxetic material and the compressive strain

ACKNOWLEDGEMENT

The authors gratefully acknowledge the financial support of the National Authority for Scientific Research ANCS/UEFISCDI through the project PN-II-ID-PCE-2012-4-0023.

REFERENCES

[1]   Donescu, Şt., Chiroiu, V., Munteanu, L., *On the Young's modulus of a auxetic composite structure*, Mechanics Research Communications, 36, 294-301, 2009.

[2]   Munteanu, L., Chiroiu, V., Dumitriu, D., Beldiman, M., *On the characterization of auxetic composites*, Proceedings of the Romanian Academy, Series A: Mathematics, Physics, Technical Sciences, Information   Science, 9(1), 33-40, 2008.

[3]   Lakes, R.S., *Experimental micro mechanics methods for conventional and negative Poisson's ratio cellular solids as Cosserat continua*, J. Engineering Materials and Technology, 113, 148–155, 1991.

[4]   Lakes, R.S., *Foam structures with a negative Poisson's ratio*, Science, 235, 1038–1040, 1987.

[5]   Lakes, R.S., *Experimental microelasticity of two porous solids*, Int. J. Solids, Structures, 22, 1986, pp.55–63, 1986.

[6]   Munteanu, L., Dumitriu, D., Donescu, Şt., Chiroiu, V., *On the complexity of the auxetic systems*, Proc.of the European Computing Conference, Lecture Notes in Electrical Engineering 2(28), 1543-1549, Springer-Verlag (eds. N.Mastorakis, V.Mladenov), 2009.

[7]   Munteanu, L., Chiroiu, V., *On the dynamics of locally resonant sonic composites*, European Journal of Mechanics-A/Solids, 29(5), 871–878, 2010.

[8]   Munteanu L., Chiroiu V., *On the response of a sonic liner under severe acoustic loads*, European Journal of Mechanics-A/Solids, 2013 (in press).
[9] Hirsekorn, M., Delsanto, P.P., Batra, N.K., Matic, P., *Modelling and simulation of acoustic wave propagation in locally resonant sonic materials*, Ultrasonics, 42, 231–235, 2004.

[10]   Munteanu L., *Nanocomposites*, Editura Academiei, 2012.

[11]   Cosserat, E., and F., *Theorie des Corps Deformables*, Hermann et Fils, Paris, 1909.

[12]   Eringen, A.C., *Linear Theory of Micropolar Elasticity*, J. Math. & Mech., 15, 909–924, 1966.

[13]   Eringen, A.C., *Theory of micropolar elasticity*, in Fracture (ed. R.Liebowitz), Academic Press, 2, 621–729, 1968.

[14] Mindlin, R.D., *Microstructure in linear elasticity*, Arch. Rat. Mech. Anal., 16, 51–78, 1964.

[15] Mindlin, R.D., *Stress functions for a Cosserat continuum*, Int J. Solids Structures, **1**, 265–271, 1965.

[16] Gauthier, R.D., *Experimental investigations on micropolar media*, 395–463, in: Mechanics of Micropolar Media, World scientific, 1982.

[17] Teodorescu, P.P., Munteanu, L., Chiroiu, V., *On the wave propagation in chiral media*, New Trends in Continuum Mechanics, Ed. Thetha Foundation, Bucharest, 303–310, 2005.

[18] Teodorescu, P.P., Badea, T., Munteanu, L., Onişoru, J., *On the wave propagation in composite materials with a negative stiffness phase*, New Trends in Continuum Mechanics, Ed. Thetha Foundation, Bucharest, 295–302, 2005.

[19] Bécus, G.A., *Homogenization and random evolutions: Applications to the mechanics of composite materials*, Quarterly of Applied Mathematics, XXXVII (3), 209–217, 1979.

PREDICTION OF ELASTIC PROPERTIES OF SOME SHEET MOLDING COMPOUNDS

H. Teodorescu-Draghicescu[1], S. Vlase[1], M.V. Munteanu[1]
[1] Transilvania University of Brasov, Brasov, ROMANIA,
e-mail draghicescu.teodorescu@unitbv.ro , svlase@unitbv.ro, v.munteanu@unitbv.ro

Abstract: *This paper presents an original approach to compute the longitudinal tensile break stress of multiphase composite materials with short fibers as reinforcement. The model is seen as consisting of three phase compounds: resin, filler and fibers, model that is reduced to two phase compounds: substitute matrix and fibers. The upper and lower limits of the homogenized coefficients for a 27% fibers volume fraction Sheet Molding Compound (SMC) are computed. It is presented a comparison between the upper and lower limits of the homogenized elastic coefficients of a SMC material and the experimental data.*
Keywords: *multiphase materials, Sheet Molding Compounds, homogenization, homogenized coefficients*

1. INTRODUCTION

A typical Sheet Molding Compound (SMC) material is composed of the following chemical compounds: calcium carbonate; chopped glass fibers rovings; unsaturated polyester resin; low-shrink additive; styrene; different additives; pigmented paste; release agent; magnesium oxide paste; organic peroxide; inhibitors.

The matrix (resin) system play a significant role within a SMC, acting as compounds binder and being "embedded material" for the reinforcement. To decrease the shrinkage during the cure of a SMC prepreg, filler (calcium carbonate) have to be added in order to improve the flow capabilities and the uniform fibers transport in the mold.

For the materials that contain many compounds, an authentic, general method of dimensioning is hard to find. In a succession of hypotheses, some authors tried to describe the elastic properties of SMCs based on ply models and on material compounds.

The glass fibers represent the basic element of SMC prepreg reinforcement. The quantity and rovings' orientation determine, in a decisive manner, the subsequent profile of the SMC structure's properties.

There are different grades of SMC prepregs: R-SMC (with randomly oriented reinforcement), D-SMC (with unidirectional orientation of the chopped fibers), C-SMC (with unidirectional oriented continuous fibers) and a combination between R-SMC and C-SMC, known as C/R-SMC.

The following informations are essential for the development of any model to describe the composite materials behaviour: the thermo-elastic properties of every single compound and the volume fraction concentration of each compound.

Theoretical researches regarding the behaviour of heterogeneous materials lead to the elaboration of some homogenization methods that try to replace a heterogeneous material with a homogeneous one. The goal is to obtain a computing model which takes into account the microstructure or the local heterogeneity of a material.

The homogenization theory is a computing method to study the differential operators' convergence with periodic coefficients. This method is indicated in the study of media with periodic structure like SMCs. The matrix and fillers elastic coefficients are very different but periodical in spatial variables. This periodicity or frequency is suitable to apply the homogenization theory to the study of heterogeneous materials [1-11].

2. TENSILE BEHAVIOR MODEL FOR A SHEET MOLDING COMPOUND MATERIAL

A SMC material can be regarded as a system of three basic compounds: resin, filler and reinforcement (fibers). We can consider the resin–filler system as a distinct phase compound called substitute matrix, so a SMC can be regarded as a two phase compound material (fig. 1). This substitute matrix presents the virtual volume fractions V_r' for resin and V_f' for filler. These virtual volume fractions are connected to the real volume fractions V_r and V_f, through the relations:

$$V_r' = \frac{V_r}{V_r + V_f} \; ; \quad V_f' = \frac{V_f}{V_r + V_f},$$

(1)

such that $V_r' + V_f' = 1$.

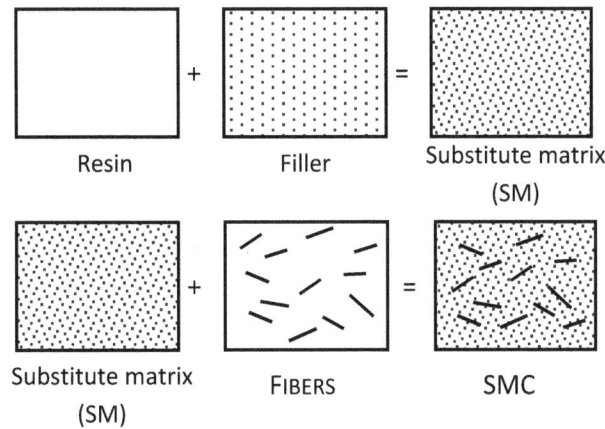

Figure 1: Schematic representation of resin-filler system

It is known that during the manufacturing process of a SMC, there is dependence between the production line speed and the fibers plane orientation on its advance direction. Therefore, this material can be assumed to have the fibers oriented almost parallel to the production line of the SMC.

Due to the longitudinal tensile loading, the SMC strain (ε_C) is identical with the substitute matrix strain (ε_{SM}) and fibers strain (ε_F), see figure 2.

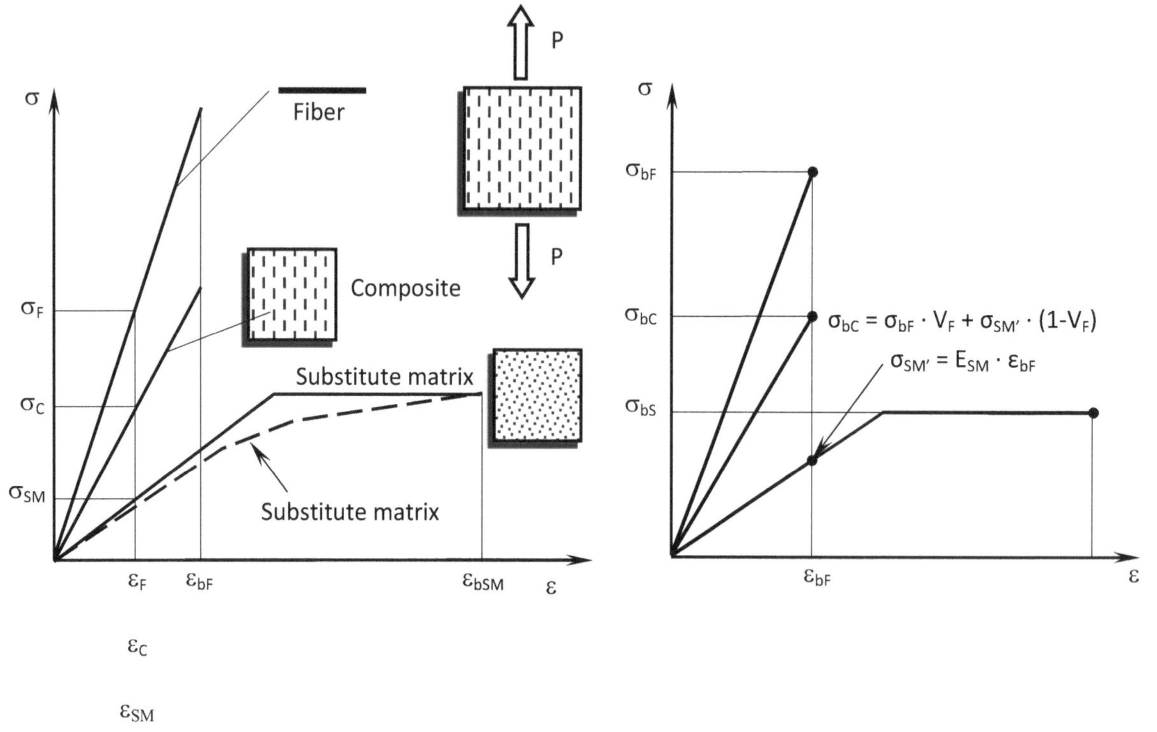

Figure 2: Schematic representation of stress-strain behaviour of a SMC material

Assuming the fact that both fibers and substitute matrix present an elastic linear behaviour, the respective longitudinal stresses are:

$$\sigma_F = E_F \cdot \varepsilon_F = E_F \cdot \varepsilon_C, \tag{2}$$

$$\sigma_{SM} = E_{SM} \cdot \varepsilon_{SM} = E_{SM} \cdot \varepsilon_C. \tag{3}$$

The tensile force applied to the entire composite is taken over by both fibers and substitute matrix [12]:

$$P = P_F + P_{SM} \tag{4}$$

or:

$$\sigma_C \cdot A_C = \sigma_F \cdot A_F + \sigma_{SM} \cdot A_{SM}, \quad \sigma_C = \sigma_F \cdot \frac{A_F}{A_C} + \sigma_{SM} \cdot \frac{A_{SM}}{A_C}, \tag{5}$$

where σ_C is the medium tensile stress in the composite, A_F is the net area of the fibers transverse surface, A_{SM} represents the net area of the substitute matrix transverse surface and $A_C = A_F + A_{SM}$. The ratio: $\frac{A_F}{A_C} = V_F$ is the fibers volume fraction and $\frac{A_{SM}}{A_C} = V_{SM} = 1 - V_F$ represents the substitute matrix volume fraction, such that (5) becomes:

$$\sigma_C = \sigma_F \cdot V_F + \sigma_{SM} \cdot (1 - V_F). \tag{6}$$

Taking into account (2) and (3) and dividing both terms of (6) through ε_C, the longitudinal elasticity modulus for the composite is:

$$E_C = E_F \cdot V_F + E_{SM} \cdot (1 - V_F). \tag{7}$$

The equation (7) shows that the value of the longitudinal elasticity modulus of the composite is situated between the values of the fibers and substitute matrix longitudinal elasticity moduli.

In general, the fibers break strain is lower than the matrix break strain, so assuming that all fibers present the same strength, their break lead inevitable to the composite break. According to equation (6), the break strength at longitudinal tensile loads of a SMC material, is:

$$\sigma_{bC} = \sigma_{bF} \cdot V_F + \sigma_{SM'} \cdot (1 - V_F), \tag{8}$$

where σ_{bF} is the fibers break strength and $\sigma_{SM'}$ represents the substitute matrix stress at the moment when its strain reaches the fibers break strain ($\varepsilon_{SM} = \varepsilon_{bF}$).

Assuming that the stress-strain behaviour of the substitute matrix is linear at the fibers break strain, (8) becomes:

$$\sigma_{bC} = \sigma_{bF} \cdot V_F + E_{SM} \cdot \varepsilon_{bF} \cdot (1 - V_F). \tag{9}$$

The estimation of the substitute matrix longitudinal elasticity modulus in case of a heterogeneous material like SMC, obtained by mixing some materials with well defined properties, depends both on the basic elastic properties of the isotropic compounds and the volume fraction of each compound.

If we note down E_r the basic elastic property of the resin, E_f the basic elastic property of the filler, V_r the resin volume fraction and V_f the filler volume fraction, the substitute matrix longitudinal elasticity modulus can be estimated computing the harmonic media of the basic elastic properties of the isotropic compounds, as follows:

$$E_{SM} = \frac{2}{\dfrac{1}{E_r \cdot V_r} + \dfrac{1}{E_f \cdot V_f}}. \tag{10}$$

3. ELASTIC PROPERTIES OF SMC-R27 COMPOSITE MATERIAL

In the case of a SMC composite material which behaves macroscopically as a homogeneous elastic environment, is important the knowledge of the elastic coefficients. Unfortunately, a precise calculus of the homogenized coefficients can be achieved only in two cases: the one-dimensional case and the case in which the matrix- and inclusion coefficients are functions of only one variable.

For a SMC material is preferable to estimate these homogenized coefficients between an upper and a lower limit. Since the fibers volume fraction of common SMCs is 27%, to lighten the calculus, an ellipsoidal inclusion of area 0.27 situated in a square of side 1 is considered. The plane problem will be considered and the homogenized coefficients will be 1 in matrix and 10 in the ellipsoidal inclusion.

In figure 3 the structure's periodicity cell of a SMC composite material is presented, where the fibers bundle is seen as an ellipsoidal inclusion.

Let us consider the function $f(x_1, x_2) = 10$ in inclusion and 1 in matrix. To determine the upper and the lower limit of the homogenized coefficients, first the arithmetic mean as a function of x_2-axis followed by the harmonic mean as a function of x_1-axis must be computed. The lower limit is obtained computing first the harmonic mean as a function of x_1-axis and then the arithmetic mean as a function of x_2-axis. If we denote $\varphi(x_1)$ the arithmetic mean against x_2-axis of the function $f(x_1, x_2)$, it follows:

$$\phi(x_1) = \int_{-0,5}^{0,5} f(x_1, x_2) dx_2 = 1, for \quad x_1 \in \left(-0,5; \ -0,45\right) \cup \left(0,45; \ 0,5\right), \tag{11}$$

$$\phi(x_1) = \int_{-0,5}^{0,5} f(x_1, x_2) dx_2 = 1 + 9,45\sqrt{0,2025 - x_1^2}, for \quad x_1 \in \left(-0,45; \ 0,45\right) \tag{12}$$

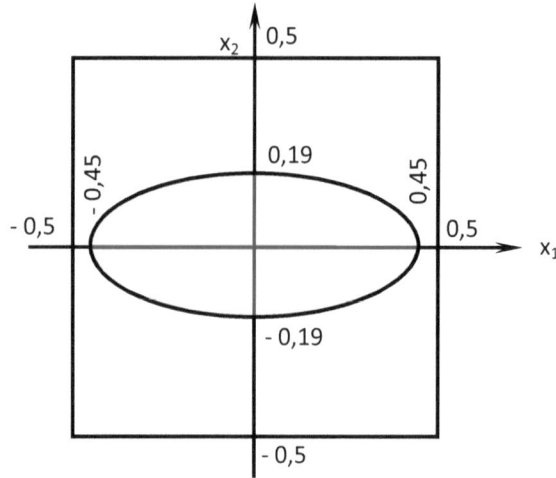

Figure 3: Structure's periodicity cell of a SMC material with 27% fibers volume fraction

The upper limit is obtained computing the harmonic mean of the function $\varphi(x_1)$:

$$a^+ = \frac{1}{\displaystyle\int_{-0,5}^{0,5} \frac{1}{\phi(x_1)}dx_1} = \frac{1}{\displaystyle\int_{-0,5}^{-0,45} dx_1 + \int_{-0,45}^{0,45} \frac{dx_1}{1+9,45\sqrt{0,2025-x_1^2}} + \int_{0,45}^{0,5} dx_1}. \quad (13)$$

To compute the lower limit, we consider $\psi(x_2)$ the harmonic mean of the function $f(x_1, x_2)$ against x_1:

$$\psi(x_2) = \frac{1}{\displaystyle\int_{-0,5}^{0,5} \frac{1}{f(x_1,x_2)}dx_1} = 1, \text{ for } x_2 \in (-0,5; \ -0,19) \cup (0,19; \ 0,5) \quad (14)$$

$$\psi(x_2) = \frac{1}{\displaystyle\int_{-0,5}^{0,5} \frac{1}{f(x_1,x_2)}dx_1} = \frac{1}{1-3,42\sqrt{0,0361-x_2^2}}, \text{ for } x_2 \in (-0,19; \ 0,19) \quad (15)$$

The lower limit will be given by the arithmetic mean of the function $\psi(x_2)$:

$$a_- = \int_{-0,5}^{0,5} \psi(x_2)dx_2 = \int_{-0,5}^{-0,19} dx_2 + \int_{-0,19}^{0,19} \frac{dx_2}{1-3,42\sqrt{0,0361-x_2^2}} + \int_{0,19}^{0,5} dx_2. \quad (16)$$

Since the ellipsoidal inclusion of the SMC structure may vary angular against the axes' centre, the upper and lower limits of the homogenized coefficients will vary as a function of the intersection points coordinates of the ellipses, with the axes x_1 and x_2 of the periodicity cell.

4. RESULTS

Typical elasticity properties of the SMC isotropic compounds and the composite structural features are: E_M=3.52 GPa; E_F=73 GPa; E_f=47.8 GPa; G_M=1.38 GPa; G_F=27.8 GPa; G_f=18.1 GPa; φ_M=30%; φ_F=27%; φ_f=43%.

According to equations (10) and (7), the longitudinal elasticity moduli E_{SM} (for the substitute matrix) and E_C (for the entire composite) can be computed.

A comparison between these moduli and experimental data is presented in figure 4. In practice, due to technical reasons, the fraction of each isotropic compound is expressed as percent of weight, so that the dependence between volume- and weight fraction can be determined:

$$\phi = \frac{1}{1 + \frac{1-\psi}{\psi} \cdot \frac{\rho_F}{\rho_{SM}}}, \qquad (17)$$

where φ and ψ are the volume respective the weight fraction, ρ_F as well as ρ_{SM} are the fibers- respective the substitute matrix density.

From figure 4, it can be noticed that the Young modulus for the entire composite is closer to the experimental value unlike the Young modulus for the substitute matrix. This means that the rule of mixture used in equation (7) give better results than the inverse rule of mixture presented in equation (10), in which the basic elastic property of the filler and the filler volume fraction can be replaced with fibers Young modulus and fibers volume fraction, appropriate for a good comparison.

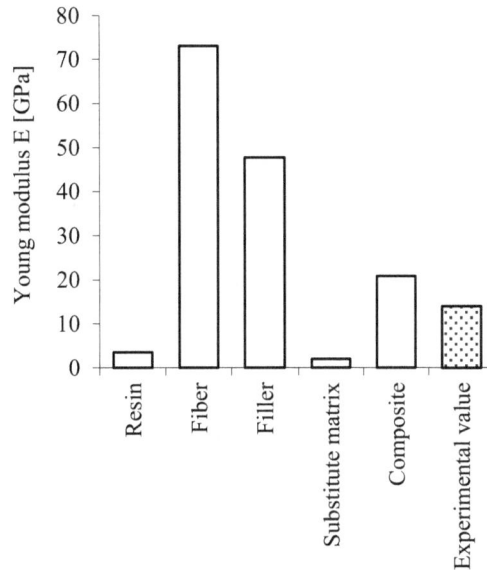

Figure 4: Young's moduli E_{SM} and E_C for a 27% fibers volume fraction SMC material

The material's coefficients estimation depends both on the basic elasticity properties of the isotropic compounds and the volume fraction of each compound. If we write P_M, the basic elasticity property of the matrix, P_F and P_f the basic elasticity property of the fibers respective of the filler, φ_M the matrix volume fraction, φ_F and φ_f the fibers- respective the filler volume fraction, then the upper limit of the homogenized coefficients can be estimated computing the arithmetic mean of these basic elasticity properties taking into account the volume fractions of the compounds:

$$A^+ = \frac{P_M \cdot \phi_M + P_F \cdot \phi_F + P_f \cdot \phi_f}{3}. \qquad (18)$$

The lower limit of the homogenized elastic coefficients can be estimated computing the harmonic mean of the basic elasticity properties of the isotropic compounds:

$$A_- = \frac{3}{\frac{1}{P_M \cdot \phi_M} + \frac{1}{P_F \cdot \phi_F} + \frac{1}{P_f \cdot \phi_f}}, \qquad (19)$$

where P and A can be the Young modulus respective the shear modulus. Figure 5 shows the Young's moduli of the isotropic SMC compounds as well as the upper and lower limits of the homogenized elastic coefficients.

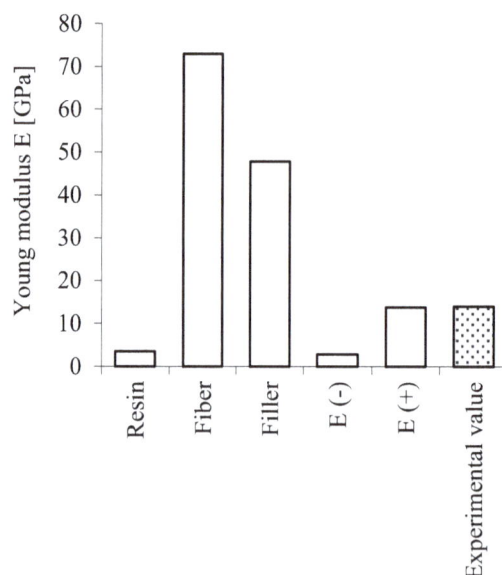

Figure 5: Young's moduli of the SMC compounds and the upper/lower limits of the homogenized coefficients

5. CONCLUSION

For the same fibers length (e.g. l_F = 4.75 mm) but with a shear stress 10 times greater at the fiber-matrix interface, it results an increase with 18% of the longitudinal break strength of the composite. Therefore, improving the bond between fibers and matrix by using a technology that increases the fibers adhesion to matrix, an increase of composite longitudinal break strength will be achieved. The computing model regarding the longitudinal tensile behaviour of multiphase composite materials like SMCs shows that the composite's Young modulus computed by help of rule of mixture is closer to experimental data than the inverse rule of mixture.

6. ACKNOWLEDGEMENT

Instrumente Structurale
2007 – 2013

UNIUNEA EUROPEANĂ

GUVERNUL ROMÂNIEI

PROGRAMUL OPERAŢIONAL SECTORIAL
„CREŞTEREA COMPETITIVITĂŢII ECONOMICE”
„Investiţii pentru viitorul dumneavoastră"
Proiect: Dezvoltarea de componente din materiale compozite avansate cu aplicatii in industria auto civilă şi militară.

Proiect cofinanţat de Uniunea Europeană prin Fondul European de Dezvoltare Regională

REFERENCES

[1] Schürmann H., Konstruiren mit Faser-Kunststoff-Verbunden, Springer, 2005.

[2] Teodorescu-Draghicescu H., Vlase S., Scutaru M.L., Serbina L., Calin M.R., Hysteresis Effect in a Three-Phase Polymer Matrix Composite Subjected to Static Cyclic Loadings, Optoelectron. Adv. Mater. – Rapid Comm. 5(3), 273, 2011.

[3] Vlase S., Teodorescu-Draghicescu H., Motoc D.L., Scutaru M.L., Serbina L., Calin M.R., Behavior of Multiphase Fiber-Reinforced Polymers Under Short Time Cyclic Loading, Optoelectron. Adv. Mater. – Rapid Comm. 5(4), 419, 2011.

[4] Vlase S., Teodorescu-Draghicescu H., Calin M.R., Serbina L., Simulation of the Elastic Properties of Some Fibre-Reinforced Composite Laminates Under Off-Axis Loading System, Optoelectron. Adv. Mater. – Rapid Comm. 5(4), 424, 2011.

[5] Teodorescu-Draghicescu H., Stanciu A., Vlase S., Scutaru M.L., Calin M.R., Serbina L., Finite Element Method Analysis Of Some Fibre-Reinforced Composite Laminates, Optoelectron. Adv. Mater. – Rapid Comm. 5(7), 782, 2011.

[6] Teodorescu-Draghicescu H., Vlase S., Homogenization and Averaging Methods to Predict Elastic Properties of Pre-Impregnated Composite Materials, Computational Materials Science, 50(4), 1310, Elsevier, Feb. 2011.

[7] Stanciu A., Teodorescu-Draghicescu H., Vlase S., Scutaru M.L., Calin M.R., Mechanical Behavior of CSM450 and RT800 Laminates Subjected to Four-Point Bend Tests, Optoelectron. Adv. Mater. – Rapid Comm. 6(3-4), 495, 2012.

[8] Vlase S., Teodorescu-Draghicescu H., Calin M.R., Scutaru M.L., Advanced Polylite composite laminate material behavior to tensile stress on weft direction, J. Optoelectron. Adv. Mater., 14(7-8), 658, 2012.

[9] Teodorescu-Draghicescu H., Scutaru M.L., Rosu D., Calin M.R., Grigore P., New Advanced Sandwich Composite with twill weave carbon and EPS, J. Optoelectron. Adv. Mater., 15(3-4), 199, 2013.

[10] Modrea A., Vlase S., Teodorescu-Draghicescu H., Mihalcica M., Calin M.R., Astalos C., Properties of Advanced New Materials Used in Automotive Engineering, Optoelectron. Adv. Mater. – Rapid Comm. 7(5-6), 452, 2013.

[11] Vlase S., Purcarea R., Teodorescu-Draghicescu H., Calin M.R., Szava I., Mihalcica M., Behavior of a new Heliopol/Stratimat300 composite laminate, Optoelectron. Adv. Mater. – Rapid Comm. 7(7-8), 569, 2013.

[12] Mallik, P.K., Fibre Reinforced Composite materials. Manufacturing and Design, Marcel Dekker Inc., 1993.

BUS BODYWORK SECTION MATHCAD REPRESENTATIONS

Mihai ULEA, Mihai C. TOFAN

Transilvania University of Braşov, ROMANIA, e-mail ulea@unitbv.ro

Abstract: The roll over test of a bus bodywork section is one of the most important safety condition according to the EU rules. During the test some areas became plastic joints. The paper presents a MathCAD model of this test, which it is used to calculate the energy and speeds developed.
Keywords: mechanical structure MathCAD, energy, plastic joint, energy, bus bodywork section

1. INTRODUCTION

The roll over test of a bus bodywork section is one of the most important safety condition according to the EU rules as Directive 2001/85/EC [2].

There are some test conditions as:
- The bodywork section shall represent a section of the unladen vehicle.
- The geometry of the bodywork section, the axis of rotation and the position of the centre of gravity in the vertical and lateral directions shall be representative of the complete vehicle.
- The impact area shall consist of concrete or other rigid material.
- The difference between the height of the horizontal starting plane and the horizontal lower plane on which impact takes place shall be not less than 800 mm.

In the book [1] and paper [3] the authors presented the Integrated Force Method and developed MathCAD programs for some structure applications.

The paper [4] developed the method for the generation of fundamental contours of structure graph and the paper [6] presents MathCAD design for fundamental contours of a bus bodywork section graph using addressing matrix for sequence chain of going over it.

The simulation of the roll over test for a plane model of a Romanian bus bodywork section was developed in Math CAD [5].

2. THE CONTOUR BASE GENERATION OF THE BUS BODYWORK SECTION MODEL

For the study of the bodywork section in elastic field, a MathCAD plane model was developed. In figure 1 is presented the first node numbering according to the *t* vector, and in matrix A the corresponding node coordinates in meters.

Considering from the 16 nodes only the nodes of the *t0* vector, a new node numbering is made and the new node coordinates matrix is B, as in figure 2.

The new model has 14 nodes and 23 bar elements.

The node number 0 is the cylindrical joint which axis is the roll over axis of the bus bodywork section.

The node number 13 is the resulting gravity centre G.

$A =$

	0	1	2	3	4	5	6	7	8	9	10	11	12	13	14	15	16	17
0	0	0	0	0	0	0	0	0	0	0	0	0	0	0	0	0	0	0
1	0	-1.1	1.16	1.24	1.24	1.16	1.16	1.24	1.24	1.16	1.1	0	1.23	0	0	0	0.62	0.62
2	3.13	3.08	2.84	1.5	0.8	0	0	0.8	1.5	2.84	3.08	3.13	0.38	1.16	0	0.8	3.13	3.13

$$t := (0 \ 1 \ 2 \ 3 \ 4 \ 13 \ 7 \ 4 \ 5 \ 6 \ 12 \ 14 \ 6 \ 7 \ 8 \ 9 \ 10 \ 0)^T$$

$$DIM(t)^T = (18 \ 1)$$

$$t_{17} = 0$$

$$t0 := (12 \ 6 \ 7 \ 9 \ 10 \ 17 \ 16 \ 1 \ 2 \ 4 \ 5 \ 14 \ 15 \ 13)^T$$

$$it := 0 .. 13$$

$$ja := 0 .. 12$$

$$B^{\langle it \rangle} := A^{\langle t0_{it} \rangle}$$

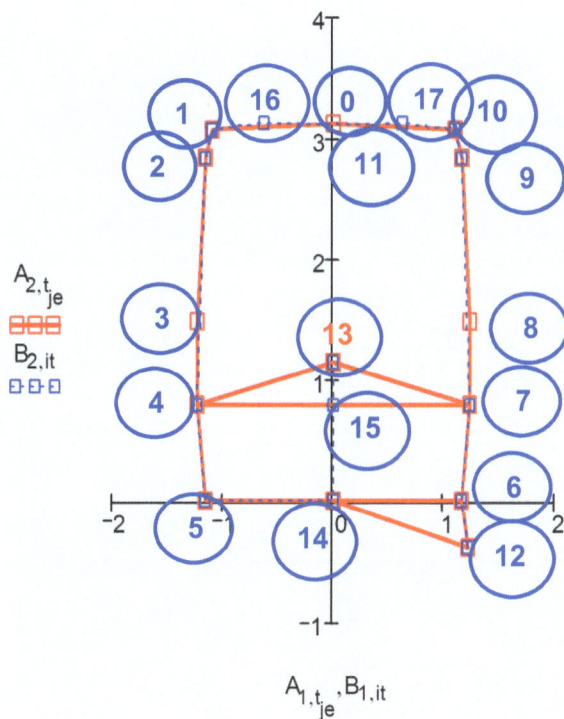

Figure 1: First node numbering

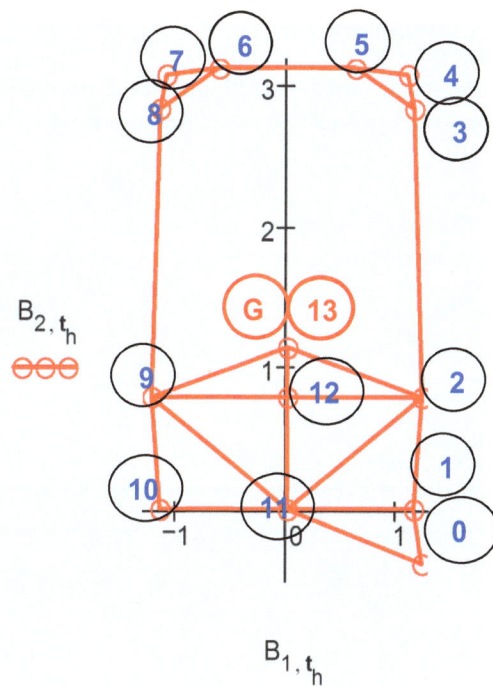

Figure 2: Tree nodes

The 23 elements are presented in figure 3.

The element number 0 starts from the cylindrical joint and the last elements are the two reinforcements of the bus roof.

The tree contour is shown in figure 4.

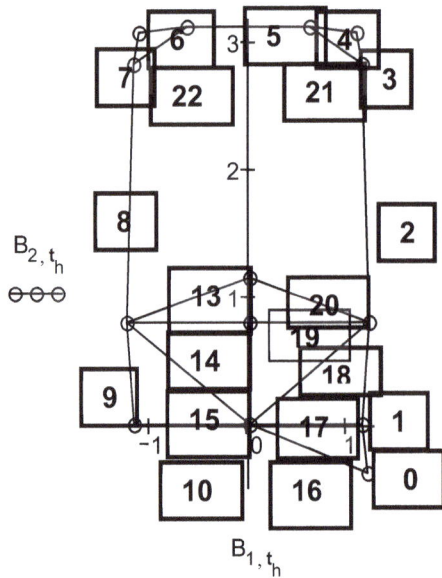

Figure 3: Elements numbering   **Figure 4**: Tree Contour Elements

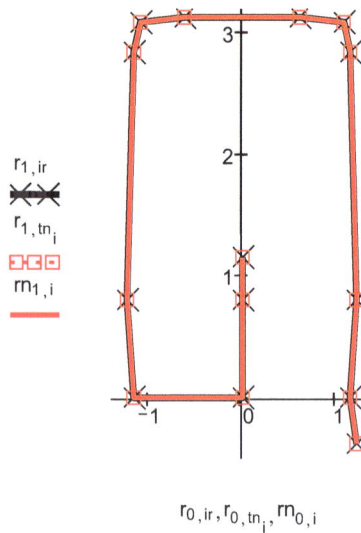

The matrix I of the tree connecting elements, the matrix II of the junctions and the nodal junction elements matrix **I** of bus bodywork section is presented in table 1.

Table 1: Nodal junctions' matrix

I =		0	1	2	3	4	5	6	7	8	9	10	11	12
	0	0	1	2	3	4	5	6	7	8	9	10	11	12
	1	1	2	3	4	5	6	7	8	9	10	11	12	13

$$II := \begin{pmatrix} 9 & 9 & 9 & 0 & 1 & 2 & 2 & 13 & 3 & 6 \\ 13 & 12 & 11 & 11 & 11 & 11 & 12 & 2 & 5 & 8 \end{pmatrix} \quad I := EXT(I, II) \quad DIM(I)^T = (2\ \ 23)$$

I =		0	1	2	3	4	5	6	7	8	9	10	11	12	13	14	15	16	17	18	19	20	21	22
	0	0	1	2	3	4	5	6	7	8	9	10	11	12	9	9	9	0	1	2	2	13	3	6
	1	1	2	3	4	5	6	7	8	9	10	11	12	13	13	12	11	11	11	11	12	2	5	8

According to [6] one can notice that the fundamental contours can be grouped in the bus roof contours, as in figure 5, and in the bus floor contours shown in figure 6.

$$j\sigma := 0 .. 7 \qquad\qquad jl := 0 .. 16$$

$$r_{1,tS_{j\sigma,1}}$$

$$r_{0,tS_{j\sigma,1}}$$

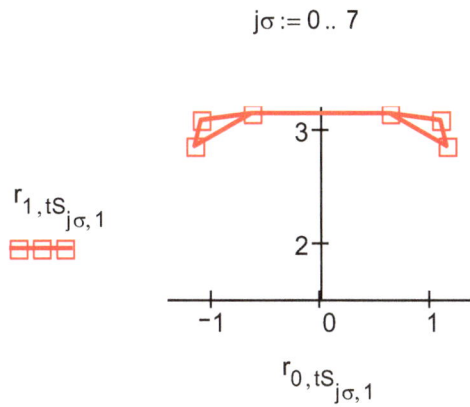

Figure 5: Roof fundamental contours

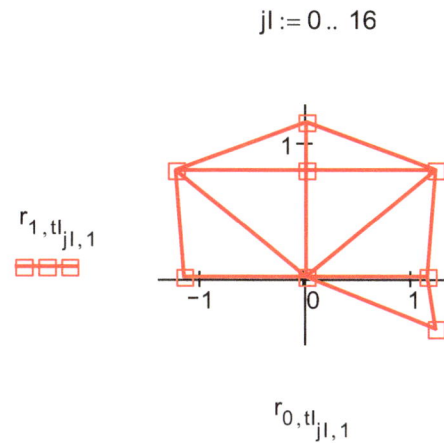

$$r_{1,tl_{jl,1}}$$

$$r_{0,tl_{jl,1}}$$

Figure 6: Floor fundamental contours

3. THE ROLLOVER TEST MODEL

The m vector contains the mass in kg of the own bodywork section and the mass expressed as a percentage of the unladen mass of the vehicle in running order. It is useful to consider this mass of the section.

The Z vector contains the vertical coordinates of the corresponding gravity centers,

ZG is the vertical coordinate of the resulting gravity centre,

RC is the rotation radius to the axis of tilting for the resulting gravity centre, and

EP is the potential energy in J.

The matrix **v** describes the horizontal and vertical components for the impact speed of the resulting gravity center and for the node which strike first the horizontal lower plane [m/s].

$$m := \begin{pmatrix} 392 & 1203 \end{pmatrix}^T$$

$$g := 9.807$$

$$mt := \begin{pmatrix} 1 & 1 \end{pmatrix} \cdot m \,; \quad mt := 1595$$

$$Z := \begin{pmatrix} 1.563 & 1.436 \end{pmatrix}^T$$

$$ZG := mt^{-1} \cdot \left[\sum_u (Z_u - 0.28) \cdot m_u \right] \,; \quad ZG = 1.187$$

$$RC := \left| rG^{\langle 10 \rangle} - rA \right|$$

$$EP := mt \cdot g \cdot RC \cdot \left(1 - c(\theta_0)\right)$$

$$RC = 1.965$$

$$vc := \sqrt{2 \cdot g \cdot \left[RC \cdot \left(1 - c(\theta_0)\right) \right]}$$

$$\omega = \frac{vc}{RC}$$

$$EP = 1.78 \times 10^4$$

$$\begin{pmatrix} vc & \omega \end{pmatrix} = \begin{pmatrix} 5.855 & 2.673 \end{pmatrix}$$

$$rv := EXT\left(rG^{\langle 10 \rangle}, r^{\langle 16 \rangle} \right)$$

$$sv := 0.2 \,; \quad iv := 0..3 \,; \quad k_{iv,u} := u \cdot 4 + iv$$

$$k^T = \begin{pmatrix} 0 & 1 & 2 & 3 \\ 4 & 5 & 6 & 7 \end{pmatrix}$$

$$rV^{\langle k_{1,u} \rangle} := rv^{\langle u \rangle} - rA$$

$$V^{\langle u \rangle} := -\omega \cdot E^{\langle 0 \rangle} \times rV^{\langle k_{1,u} \rangle}$$

$$V = \begin{pmatrix} 1.988 & -1.683 \\ -4.286 & -8.161 \end{pmatrix}$$

$$V_u := \left(\left\| V^{\langle u \rangle} \right\| \right)$$

The impact gravity center speed is v = 4,724 m/s as in figure 5. This is perpendicular to the gravity center radius. The impact speed of the contact node to the rigid horizontal lower plane is v_1 = 8,333 m/s.

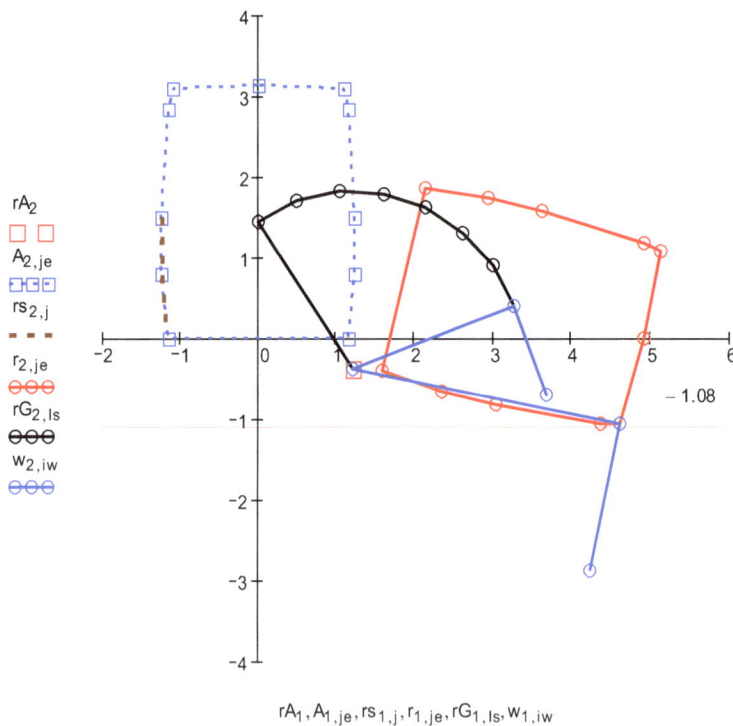

rA_2
$A_{2,je}$
$rs_{2,j}$
$r_{2,je}$
$rG_{2,ls}$
$w_{2,iw}$

$rA_1, A_{1,je}, rs_{1,j}, r_{1,je}, rG_{1,ls}, w_{1,iw}$

Figure 5: The speeds

4. CONCLUSIONS

From the computation it results an impact energy EP = 1,78 $.10^4$ J.
MathCAD models are usefully to calculate the effects of this energy and simulate the plastic joints.

REFERENCES

[1]   Tofan, M.C., Goia, I., Ţierean, M., Ulea, M. - *Deformatele Structurilor*, Editura Lux Libris, Braşov, 1995,   ISBN 973-96854-2-0

[2]   Directive 2001/85/EC of the European Parliament and of the Council of 20 November 2001

[3] Tofan, M.C., Ulea, M., Goia, I., Burcă, I. – *On Structure Deformations by Integrated Force Method* – in Proceedings of the 2nd International Conference "Computational Mechanics and Virtual Engineering" COMEC 2007, October 11-13th, Braşov, ISBN 978-973-598-117-4, pg. 513-519

[4] Tofan, M.C., Ulea, M., Vlase, S. - *MathCAD Application for Structures Fundamental Contours Base Generation* - in Proceedings of the 2nd International Conference "Computational Mechanics and Virtual Engineering" COMEC 2007, October 11-13th, Braşov, ISBN 978-973-598-117-4, pg. 521-526

[5] Ulea, M., Tofan, M.C.: - *MathCAD Model for Roll Over Test of a Bus Bodywork Section* - in Proceedings of the 2nd International Conference "Advanced Composite Materials Engineering " COMAT 2008, 9 – 11 October 2008, Braşov, Romania, vol. 1B, ISSN 1844-9336, pg. 593-596

[6] Tofan, M., Ulea, M. - *Contour Base Generation for Bus Bodywork Section*- in Proceedings of the 3rd International Conference "Advanced Composite Materials Engineering and International Conference Research & Innovation in Engineering " COMAT 2010, 27 – 29 October 2010, Braşov, Romania, vol. 3, ISSN 1844-9336, pg. 258-263

[7] Ulea, M., Itu, C. - *FEM Analysis for Staticaly Roll Over Test of a Bus Bodywork Section* - in Proceedings of the 3rd International Conference "Advanced Composite Materials Engineering and International Conference Research & Innovation in Engineering " COMAT 2010, 27 – 29 October 2010, Braşov, Romania, vol. 3, ISSN 1844-9336, pg. 264 -267

EVALUATION OF STRESS AND STRAIN STATES BY FINITE ELEMENT METHOD OF PANORAMIC STRUCTURE MADE OF BARS AND PLATES

Mariana D. Stanciu, Dragos Apostol, Ioan Curtu,

Transilvania University of Braşov, Braşov, ROMANIA, mariana.stanciu@unitbv.ro [1]

Abstract: *This paper presents modal analysis of a panoramic structure made up of bars and plates in order to determine its own frequency. The panoramic structures are constructions by tourism purposes for which reason in our country they may be in a niche area of the economy in terms of existing and future travel demands. In this study, the structure has been designed to meet both the aesthetic, functional and endurance. The structure consists of rods and plates, was analyzed in terms of stresses and strains, as well as their frequencies by finite element method, varying the material and plate thickness. The most effective and optimum combinations were determined.*

Keywords: *strain, stress, plates, panoramic*

1. INTRODUCTION

The panoramic views of natural formations, protected archaeological areas are accessible to the public by creating access and visitation areas, which are known as "panoramic structures". In Romania, "National Strategy for Regional Development (SNDR), developed based on Regional Development Plans and the National Strategic Reference Framework 2007-2013 identified tourism development as a priority of regional development given the existing tourism potential in all regions. This justifies the potential financial support to infrastructure rehabilitation and enhancement of tourist areas of natural, historical and cultural, for inclusion in the tourist circuit and their promotion to attract tourists". In this sense the easiest and least costly option would be the location of stations in places where there are already panoramic tourist access (cable car lift, lift, paved roads or forest trails). The idea is inspired by the experience of other countries such as Austria, Germany, China, USA, Switzerland, Sweden, Norway, Peru, Mexico which have highlighted the natural beauty, history by creating access and visitation areas seemingly inaccessible places (Figure 1).

a) „Heaven's Gate" – China

b) „Grand Canyon Skywalk - USA

c) „5 fingers" – Austria   d) „Top of Tyrol" -Austria

Figure 1: Types of panoramic structures [7]

In this paper is presented the panoramic ARDDOR which carried out on a metal structure and floor with glass or acrylic being suspended from the cliff of the mountain Tampa - Postăvaru, about 400m above the city, offering panoramic views over the historical center of Braşov and surroundings – Bârsa (Figure 2). To choose the optimal structure and materials, the static and dynamic analysis of plates with different degrees of reinforcement used as platform in the panoramic structure were performed (Figure 3).

Figure 2: Panoramic views of Braşov city

The basic panoramic variant is made up of three main parts: the assembly (1) is the glass platform; subassembly (2) is a system of vertical and horizontal beams with annular section for supporting the structure; subassembly (3) is composed of several panels of transparent material (glass or acrylic glass, polyvinyl) and serves to protect tourists (Figure 3). In this paper, the stress and strain states of panoramic structure are analyzed with finite element method.

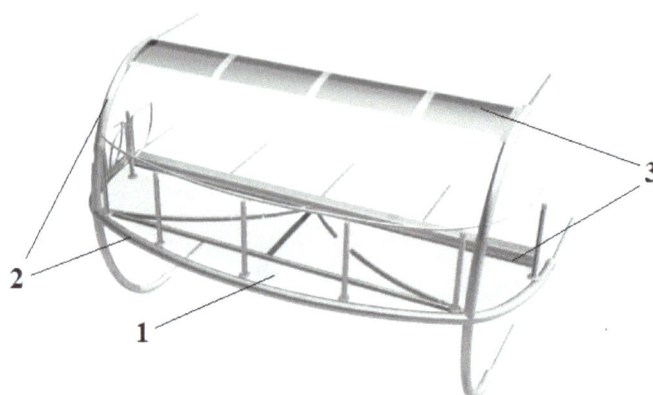

Figure 3. Design of ARDDOR Panoramic

2. EVALUATION OF STRESS AND STRAIN STATES BY FINITE ELEMENT METHOD

The geometry of the entire structure was completed in Catia and imported into Abaqus CAE [1, 2]. The structure with dimensions (length x width) 6500 mm x1700 mm was embedded in the anchoring area placed on the end of vertical and horizontal bars [3, 4 .5, 6]. The loading was distributed over the board surface as shown in Figure 4 and was calculated to safely hold the weight of up to 15 tourists, with medium mass of 80 kg. The structure was discretized using hexahedron elements.

The material used for bars was aluminum with elastic modulus $E = 69000$ MPa and Poisson's ratio $\nu = 0.33$, and for plate structure several simulations were performed using different valuse of elastic modulus ($E = 70000$ MPa and Poisson's ratio $\nu = 0.33$, $E_L = 13000$ MPa, $E_T = 4000$ MPa and Poisson's ratio $\nu = 0.2$, $E = 210000$ MPa and Poisson's ratio $\nu = 0.3$). Besides static analysis of the structure, dynamics analysis was performed and determining the eigenvalues and frequencies response.

Figure 4: Equivalent loading and boundary conditions for the panoramic structure

Firstly, the stress-strain states of plates with different stiffening systems were obtained. The simulation results are presented in Figure 5. Stress distribution, shown in Figure 5 highlights high values in anchorage areas and in the joints between the bars and the horizontal bar half.

a)   b)

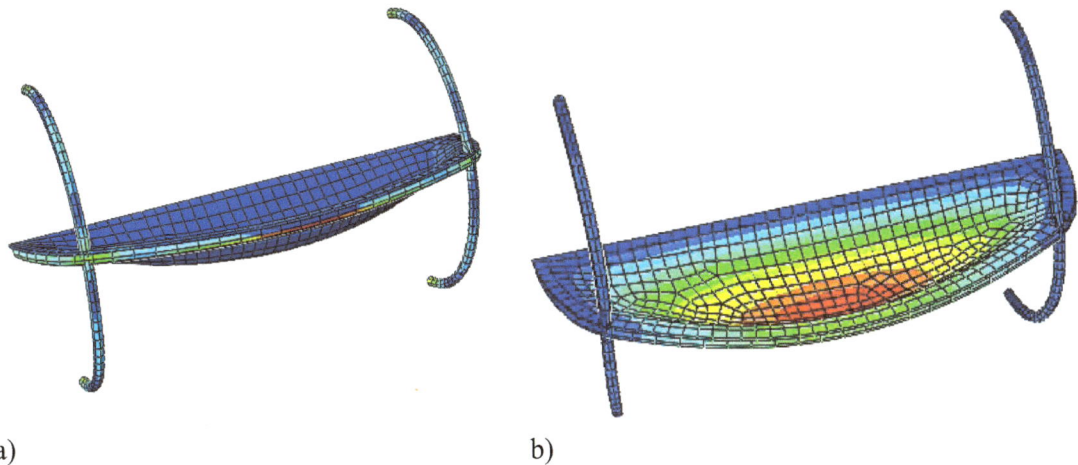

Figure 5: Stress and strain states of analyzed structure: a) displacements; b) stress

Following the results of the FEM will be considered as the design and implementation of optimal structure to use procedures to ensure the stability and sustainability of the structure. Also the structures of the shelf plates are provided with reinforcement to minimize deformations in the center of the plate. Thus, it is proposed a system of stiffening of three bars and three radial bars.

In Table 1 are selected the most representative normal modes and modal shapes of the entire structure. Modal shapes shown in Table 1 shows how the bearing of the panoramic structure vibrates at its own frequency and harmonics. It is important to know the frequencies and modes to avoid resonance that may occur when developing forced vibration due to tourist traffic and wind / seismic

Table 1: Natural frequencies and modal shape of panoramic structure

Mode/ f[Hz]	(0,0) 10,211	(0,1) 22,536	(0,2) 32,28
Mode/ f[Hz]	(0,3) 46,472	(1,0) 27,221	(1,1) 41,176

3. RESULTS and DISCUSSION

The results obtained in the analysis by FEM of the whole structure (denoted Str Int) were compared with those obtained for panoramic platform modeled as simple rectangular plate (denoted PS) or as curved contour plates (denoted PSC). In Figure 6 it can be noticed that with increasing of complexity of structure, the stress value increasing. With increasing of elasticity modulus, the stresses decrease. Figure 7 presents the variation of displacement with respect to material and shape parameters studied. The higher value of displacement is recorded in case of complex structure with wood plate – around 28 mm, the smaller one recording in case of a high rigidity plate (E=210000 MPa). From these results one can choose from the point of view of resistance and the aesthetic, gauge and costs. Thus, the optimum structure of panoramic ARDDOR consist of plate with three radial bars, with thickness h = 40 mm and made of acrylic glass (E = 70000 MPa).

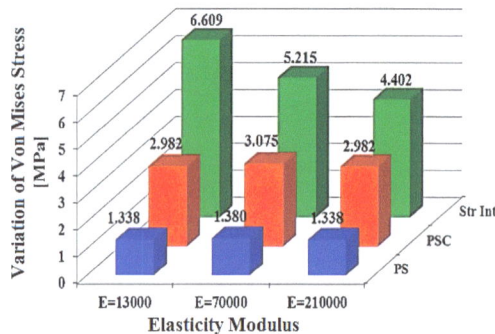

Figure 6: Variation of Von Mises stress reported to elasticity modulus and complexity of structure

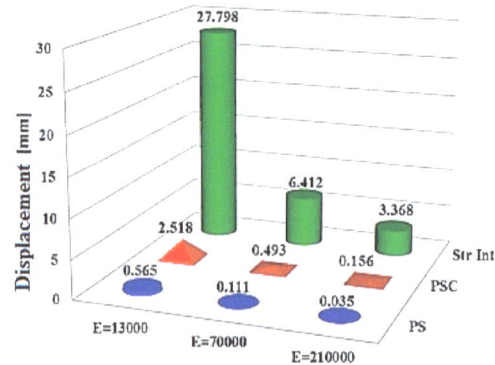

Figure 7: Variation of displacement reported to elasticity modulus and complexity of structure

4. CONCLUSSION

Panoramic structure ARDDOR consists of two basic elements: bars and plates. These elements are complex in their geometry, the method of attachment, the load on the subject and type of solidarity between them. The design and analysis of plates within the panoramic ARDDOR aimed at determining, through numerical simulations (FEM) and analytical optimal alternative that meets all the criteria proposed for this structure: aesthetic, functional, strength, security, environmental and economic.

5. REFERENCES

[1]   Dacko M, Nowak J.: Numerical simulation of stress and strain state induced by shrinkage of concrete in large-size plate, Journal of KONES Powertrain and Transport, Vol. 17, No. 1 2010;

[2]   Stanciu Mariana, Curtu Ioan, Rosca I. Calin, Application of Modal Analysis on the Behaviour Diagnosis of Lignocelluloses Plates, in Proceedings of 14th International Research/Experts Conference „ Trends in the Development of Machinery and Associated Technology" TMT2010, 11-18 September 2010 – Mediterranean Cruise, ISSN 1840-4944, pp. 281-284.

[3]   Năstăsescu, V., Bârsan, Gh., *Metoda SPH (Smoothed Particle Hydrodynamics)*, Sibiu, Editura Academiei Fortelor Terestre "Nicolae Balcescu", 2012, ISBN 978-973-153-130-4

[4]   Curtu, I., Stanciu, M.D., Motoc, D.L., "Diagnosis of dynamic behaviour of ligno-cellulose composite plates", in *Proceedings of 14th European Conference on Composite Materials*, Budapets, 2010, p. 75, ISBN 978-963-313-008-7

[5]   Cozzens, R., *CATIA V5 Workbook, Release 3*, Editată de Schroff Development Corporation, 2009, ISBN: 978-1-58503-544-1

[6]   Mocibob D., Belis J., Crisinel M., Lebet Jp. : Stress distribution at the load introduction point of glass plates subjected to compression, in *Proceedings of the International Association for Shell and Spatial Structures (IASS) Symposium 2009, Valencia Evolution and Trends in Design, Analysis and Construction of Shell and Spatial Structures* 28 September – 2 October 2009, Universidad Politecnica de Valencia, Spain

[7] *** http://china-attractions.info, http://grand-canyon.com, http://www.choices.co.uk, http://en.wikipedia.org/wiki/5_Fingers_(Austria)

ON THE USE OF CHARPY TRANSITION TEMPERATURE AS REFERENCE TEMPERATURE FOR THE CHOICE OF A PIPE STEEL

A. Coseru, J. Capelle, G. Pluvinage
LABPS, ENIM, Route d'Ars Laquenexy, Metz France

Abstract: *Transition temperature is not intrinsic to material but depends on specimens and mode of loading used for tests. Here, the linear dependance of transition temperature with constraint is shown. Constraint is evaluated by the effective T stress, which is the value of the stress difference distribution for the effective distance provided by the volumetric method.*

The application of this approach is a better choice of the reference transition temperature and its degree of conservatism when comparing with transition temperature of the studied structure.

Key words: *Transition temperature, constraint, effective T stress, reference temperature*

1. INTRODUCTION

The concept of brittle-ductile transition temperature was developed during the Second World War, because of the rupture of liberty ships at sea. The ductile-brittle transition temperature (DBTT), nil ductility temperature (NDT), or nil ductility transition temperature (NDTT) of a metal represents the point at which the fracture energy passes below a pre-determined value.

Design against brittle fracture considers that the material exhibits at service temperature, a sufficient ductility to prevent cleavage initiation and sudden fracture with an important elastic energy release. Concretely, this means that service temperature T_s is higher than transition temperature T_t:

$$T_s \geq T_t \tag{1}$$

The service temperature is conventionally defined by codes or laws according to the country where the structure or the component is built or installed. For examples, in France, a law published in July1974 indicates that service temperature in France is -20°C.

However, despite the introduction during the 1960's of Fracture Mechanics tests to measure fracture resistance of materials, the practice of the Charpy impact test remains. It always gives a simple and inexpensive method to classify materials by their resistance to brittle fracture. The current trend is also to use these tests to measure fracture toughness and ductile tearing strength.

The comparison of the two methods requires taking into account two major differences:

- Charpy test uses a notched sample, and fracture mechanics tests use a pre-cracked specimen (but pre-cracked Charpy specimens may also be used).
- Charpy tests are dynamic tests, although the conventional fracture mechanics tests are static ones.

Different Charpy specimens are used in standard. The most widely used is Charpy V specimens (V notch, notch radius $\square$ = 0.25 mm , notch depth a=2 mm). Other specimens like Charpy U (U notch, notch radius $\square$ =1 mm , notch depth a= (5 mm) are also used in standards [1].

The increase of notch acuity of Charpy specimen shift transition temperature to higher value and increase scatter in transition temperature [2] [3].

Several definition of transition temperature are used in Charpy test:

- The temperature at a conventional level of Charpy energy (generally 27 joules) and called T_{K27},
- The temperature corresponding to half also at half the jump between brittle and ductile plateau ($T_{K1/2}$)
- The temperature corresponding to 50% of fracture cristallinity T_{K50}

A Fracture Mechanics based design ensures that the design stress intensity factor is lower than admissible fracture toughness and fracture toughness is greater than 100 MPa√m (i.e. the reglementary service temperature defined is above the reference temperature). This additional criterion introduces the concept of reference temperature RT and is expressed by:

$$T_s \geq RT_t + \Delta T \tag{2}$$

where ΔT is the uncertainty on reference temperature (8°C for ASME API 579 code) [4]. This reference temperature RT_i varies according to codes (RT_{NDT} : Nil ductility transition reference tempeature or RT_{T0} : reference temperature for a conventional value of 100 MPa√m):

$$RT_{NDT} = T_{NDT}$$
$$RT_{T0} = T_0 + 19.4 \text{ °C} \tag{3}$$

Generally, in codes the choice of the reference temperature is under the responsibility of the designer.

Due to the fact that different fracture tests give different transition temperatures, the choice of the most adequate tests to provide a value close to the "structure or component" transition temperature T_{struct} is an open question. Thus, it is necessary to know the degree of conservatism of the designer approach.

It has been seen that transition temperature is sensitive to constraint [5]. The transition temperature decreases when effective T stress decreases. Therefore, the choice of the reference temperature can be made on the basis of a specimen providing a constraint value close to the structure to minimise conservatism or to increase safety factor. One notes that choosing a specimen providing high constraint like Charpy V test is conservative.

In this paper, a selected pipeline steel API 5L X65 is controlled by three different instrumented Charpy impact using three types of specimens (Charpy V, Charpy U and a modified Charpy U). Then, the transition temperatures are expressed versus effective T stress computed by the finite element method. A discussion on the effect of loading rate and comparison with T_0 transition temperature is proposed.

2. MATERIAL

The typical chemical composition is given in Table 1; mechanical properties at room temperature are given in Table 2, and the microstructure in Figure 2.

Table 1: Typical chemical composition of pipe steel API 5L X65 (wt %)

	C	Si	Mn	P	S	Mo	Ni	Al	Cu	V	Nb
min.	0.05	0.15	1.00	-	-	-	-	0.01	-	-	-
max.	0.14	0.35	1.50	0.020	0.005	0.25	0.25	0.04	0.080	0.080	0.040

Yield stress R_e (MPa)	Ultimate strength R_m(MPa)	Elongation at failure A %	Charpy Energy K_{CV} (J)	Fracture Toughness K_{Jc} (MP√am)	Hardness HV
465.5	558.6	10.94	285.2	280	205

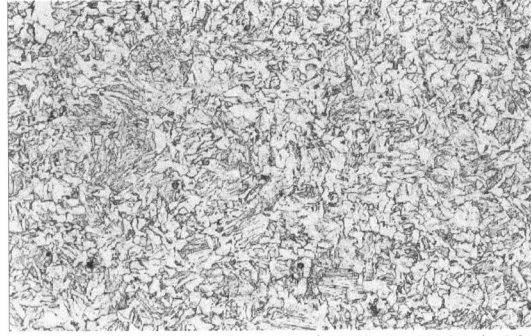

Figure 2: Microstructure of pipeline steel API 5L X65 (x100, nital etching)

Tensile tests at very low temperature exhibits brittle fracture and ductile failure at high temperature. At very low temperature, the fracture always occurs at yield stress. This phenomenon was proven by compressive tests where no failure occurs, but yield stress is easily determined. When test temperature reaches the transition temperature, failure occurs with plasticity at ultimate stress. Plasticity is a thermal activated process and yield stress decreases exponentially with temperature according to the following relationship:

$$R_e = R_e^{\mu} + AExp(-BT)$$
(4)

where R_e^{μ} is a threshold, A and B are constants and T is temperature in Kelvin.

Similarly, the ultimate strength decreases to temperature according to:

$$R_m = R_m^{\mu} + CExp(-DT)$$
(5)

where R_m^{μ} is a threshold, C and D are constants.

Tensile tests have been performed on standard specimens in a temperature range [120 - 293 K] with a strain rate of about $10^{-3}s^{-1}$. Stress-strain diagrams have been recorded and the (static) yield stress and ultimate strength determined. The values of yield stress R_e, and ultimate strength R_m are reported on Figure 3. Data are fitted with equation (4) and (5). The values of R_e^{μ}, R_m^{μ} and constants A, B, C, D are reported in Table 3.

The yield stress value at 0K is independent of loading rate and equal to 2320 MPa. This value is generally considered as equal to cleavage stress.

Table 3: Values of constants of equation (3) and (4) for API 5L X65 pipeline steel for static loading.

$R_e^{\square}$ (MPa)	A(MPa)	B (T⁻¹)	$R_m^{\square}$ (MPa)	C(MPa)	D (T⁻¹)
434	1910	-0.01405	507	843	-0.0094

3. THE USE OF INSTRUMENTED CHARPY IMPACT TEST TO DETERMINED DYNAMIC YIELD STRESS AND TRNSITION TEMPERATURE

The test campaign was conducted with an instrumented Charpy pendulum with initial energy of 300 Joules and an impact rate of 5.5 m/s. The frictions of the hammer were determined by vacuous load tests before each test campaign. The corresponding loss energy is 1.2 J.

The acquisition of the results takes the form of voltage-time data $V_F = f(t)$. The treatment algorithm presented in Figure 16, allows to draw a dynamic behaviour law of the material in force-displacement $F_N = f(\square)$

With: $F_V = f(t)$ is the voltage versus time recorded during the test; $F_N = C F_V$ is a calibration force / voltage, $v(t)$ is the instantaneous hammer velocity, $\delta(t)$ is the displacement of the hammer, $F_N = f(\delta)$ is the function force-displacement m is the hammer mass, v_0 the initial velocity of impact, t_0 time at the beginning of deformation and t later time.

These curves show more or less light oscillations, consequently of the impact exciting system vibration characterized by loss of contact (hammer-specimen and/or anvil-specimen) so the peaks of inertia [6].

For ductile failure, the load increases up to reach the point (F_{GY}) denoting the beginning of plastic flow of the ligament and the end of the strip loading at point (F_{max}) the load reaches its peak load. Crack initiation occurs between load at general yielding and maximum load at critical load (F_c), figure 3.1. For brittle fracture, after several oscillations, failure occurs at maximum load (F_{max}), figure 3.2.

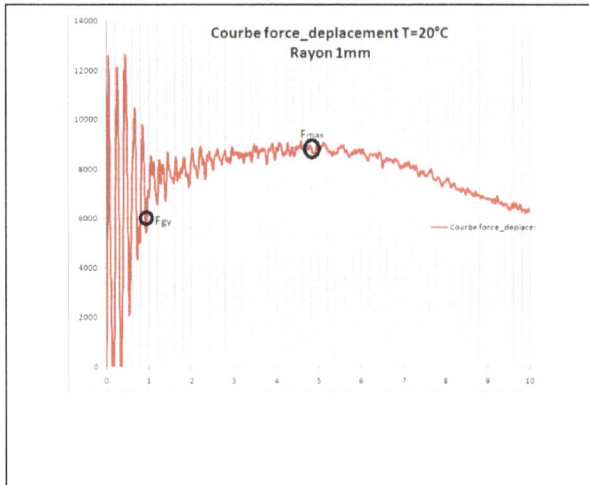

Figure 3.1: Instrumented Charpy impact test load-displacement curve for ductile failure, notch type U₁.	Figure 3.2: Instrumented Charpy impact test load-displacement curve for brittle fracture, notch type U₁.

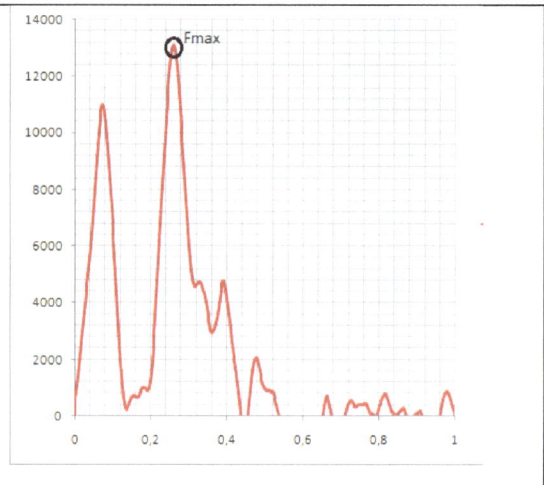

According to Chaoudi et Puzzolante [7] the critical load is given by the folllowing relation:

$$F_c = (F_{max} - F_{gy})/2 \qquad (6)$$

The dynamic yield stress (strain rate 10^2 s^{-1}) is determined by the method of instrumented Charpy impact test. The load versus time diagram is recorded and the load at general yielding P_{GY} is evaluated (see Figure 3.1). Dynamic yield stress is then obtained using the Green and Hundy solution [8]:

$$R_{e,d} = \frac{4 \cdot W \cdot F_{gy}}{B \cdot L \cdot (W - a)^2} \qquad (7)$$

where W is specimen's width, B thickness, a is notch depth, and L is the constraint factor with a value of L=1.31 for Charpy V specimen [4].

Data for Charpy V specimen are reported in figure 4 and have been fitted using equation (4). The values of the coefficients $R_{e,d}$, A$_d$ and B$_d$ are reported also in this figure.

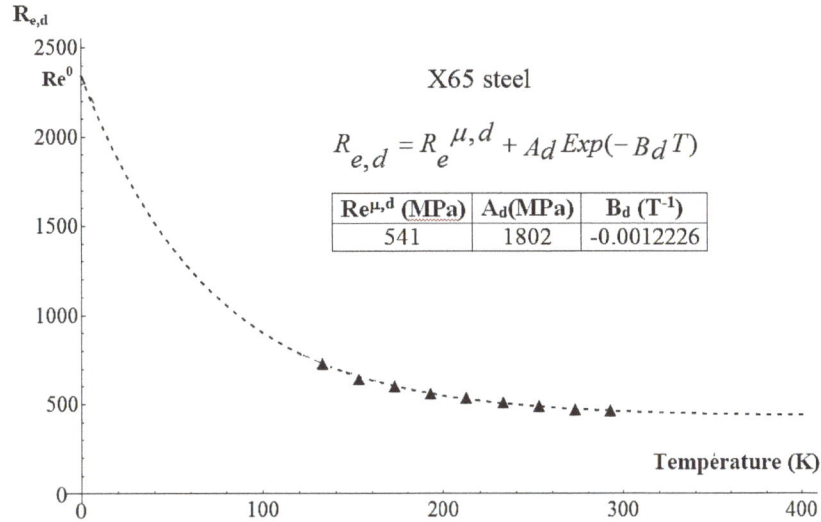

Figure 4 : Evolution of dynamic yield stress with temperature for X65 steel.

The dynamic yield stress can be also evaluated using U Charpy specimen but the values of constraint factor are different. Assuming that the yield stress is independent of notch geometry, one can find corresponding values of the constraint factor.

Table 4: Values of constraint factor for specimens with notch type U_1, U_{05} and V.

specimen	U_1	U_{05}	V
L	1.13	1.25	1.38

The values of the dynamic yield stress will be used for a loading rate correction of transition temperature later.

4. TRANSITION TEMPERATURE

Plotting the fracture energy Kc_V (J) temperature and fitting data according to equation (1), one gets a S shape curve (figure 5a, 5b):

$$K_{CV} = A_{CV} + B_{CV} \cdot \tanh\left[\frac{T - Dcv}{Ccv}\right]$$ (8)

where A_{CV}, B_{CV}, C_{CV}, and D_{CV} are constants. A_{CV} represents Charpy energy at transition temperature D_{cv}, B_{CV} is the energy jump between brittle and ductile plateaus and $2C_{CV}$ is the temperature range of the Charpy energy transition.

The transition temperature has been determined at conventional level of 27 joules and called T_{K27} and also at half the jump between brittle and ductile plateau ($T_{K50} = D_{CV}$).

The Charpy impact tests have been performed on API 5L X65 pipe steel with V, U_1 and U_{05} Charpy specimens at temperature range [-196°C up to 20 °C].

The Charpy energy and fracture aspects reveal the two failure modes below and above the transition temperature. For U_1 and U_{05} Charpy U notch, one notes a bimodal fracture mode and not for Charpy V. In the temperature range [187 - 195 K] for U_{05} and [150 - 190 K] for U_1, the two failure modes coexist. The values of the four constants A_{CV}, B_{CV}, C_{CV} and D_{CV} are reported in Table 5 Transition temperature T_{K27} and $T_{K1/2}$ for each specimen type are reported in table 6. Due to this bimodality, the transition temperature $T_{K1/2}$ at half jump between ductile and brittle plateau has been considered and corresponds to D_{CV} values.

Table 5: Values of constants A_{CV}, B_{CV}, C_{CV} et D_{CV} for curves related to specimen types U_1, U_{O5} et V

	$A_{cv}[J]$	$B_{cv}[J]$	$C_{cv}[K]$	$D_{cv}[K]$
U 1	50.70	59.01	22.88	149.47
U 0,5	61.67	59.67	8.22	187.48
V	141.3	135.6	4.43	179.2

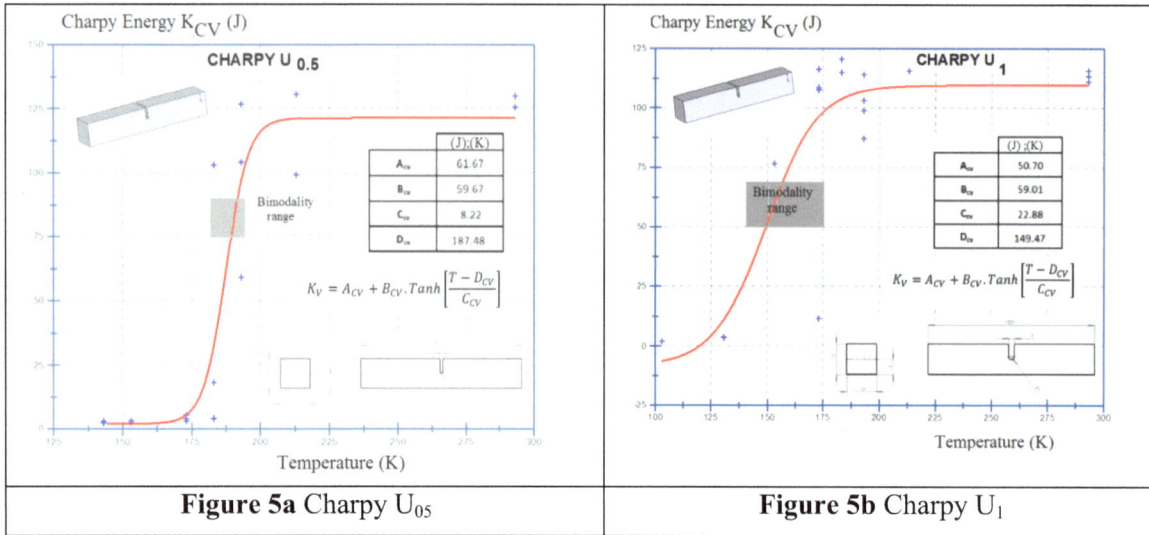

Figure 5a Charpy U_{05}	Figure 5b Charpy U_1

Figure 5 Charpy energy versus temperature curve for API 5L X65 pipe steel values of parameters of equation (4).

The fact that bimodality temperature range increases with decreasing notch radius has been note previously (3).

Table 6: Transition temperature T_{K27} and $T_{K1/2}$ for specimen types U_1, U_{05} and V

Specimen	U_1	U_{O5}	V
T_{K27} (K)	141	185	178
$T_{K1/2}$ (K)	150	187	179

5. EFECTIVE T STRESS FOR A NOTCH TIP STRESS DISTRIBUTION

The material strength like all mechanical properties is sensitive to geometric parameters such as size, specimen geometry, thickness loading mode, etc. The influence of these parameters is related to the plastic constraint. This is the consequence of the Poisson effect limitation due to the material elasticity nearby the localized plastic zone. Then a greater load is needed to get the same level of deformation.

Several parameters have been proposed to describe this phenomenon: constraint factor L [10], stress triaxiality β [11], Q parameter [12] and stress difference [13] $((\sigma_{xx} - \sigma_{yy}) = \sigma_{yy} (\nu_{ap}-1))$ where ν_{ap} is the apparent Poisson's ratio indicating how lateral contraction is hampered. This stress difference is now widely used to translate plastic constraint

In the case of a singular stress distribution at crack tip, this stress difference is identical to T stress [13]. Several methods have been proposed in the literature to determine T stress for a cracked specimen (Chao et al. [15], Ayatollahi et al. [16] and Wang [17]). Here, the difference in stress method (SDM) proposed by

Yang and Chatel [18] is used and evaluated from the stress distribution calculated by the Finite Elements method.

Physically, T is the stress acting parallel to the crack line in direction xx to the extension of crack with the amplitude proportional to the gross stress. The term non-singular T may be positive (tensile) or negative (compression). A positive T stress leads to an increase in constraint, a negative T constraint to a loss In the crack tip singular distribution, for a tensile smooth specimen, the strain difference is rater used and the parameter has the same dimension as T.

$$u = E(\varepsilon_{xx} - \varepsilon_{yy}) \tag{9}$$

In the case of a distribution reflecting a stress concentration the stress difference $(\sigma_{xx} - \sigma_{yy})$ is not constant along ligament and increases slowly after a given distance. This stress difference $(\sigma_{xx} - \sigma_{yy})$ is called $T_\square$. It is therefore evaluated for a conventional distance X_{ef} given by the volumetric method [13] and related to the size of the fracture process zone.

The volumetric method is a local failure criterion used for fracture emanating from notch. It is assumed in this method, that fracture process requires a physical volume. This volume is assumed to be quasi-cylindrical and centered at notch tip. The radius of the cylinder is called the "effective distance". By calculating averaging opening stress in this volume, one gets the effective stress. This local failure criterion is therefore based on two parameters, namely, the effective distance X_{ef} and stress the effective σ_{ef}. The distance corresponding to the minimum of relative stress gradient is conventionally regarded as the relevant effective distance. The value of the relative stress gradient is given as follows:

$$\chi(r) = \frac{1}{\sigma_{yy}(r)} \frac{\partial \sigma_{yy}(r)}{\partial r} \tag{10}$$

χ is he stress gradient and σ_{yy} the maximum principal or opening stress respectively. The effective stress is defined as the average of the opening stress weighted within the area of the fracture process:

$$\sigma_{ef} = \frac{1}{X_{ef}} \int_0^{X_d} \sigma_{yy}(r)\Phi(r)dr \tag{11}$$

where σ_{ef}, X_{ef} Φ are the effective stress, effective distance and weight function, respectively.

The Unit and Peterson weight functions are the simplest weight functions. The Unit weight function deals with the mean stress and Peterson weight function gives the stress value at a specific distance.

Figure 6 represents the stress distribution in the case of Charpy specimens U_1. The maximum stress is related to a stress concentration factor $k_t = 1.95$.

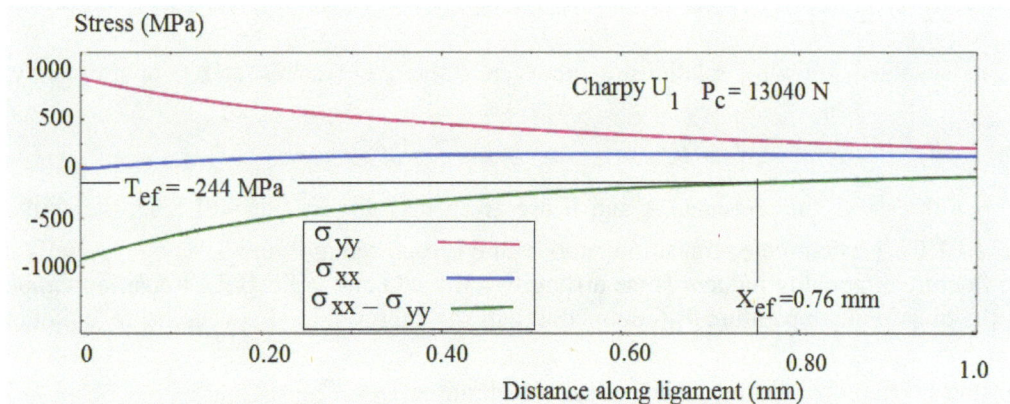

Figure 6: Stress distribution at notch of in the case of Charpy specimens U1. Determination of Tef on stress difference distribution.

The values of T_{ef} have been determined for different ligament ratio (a/W) and for the three notch geometries. The results are reported in Table 7.

Table 7: Values of T_{ef} for different ligament ratio (a/W) 3 notch geometries.

	U_1		U_{05}		V	
a/W	T_{ef}[MPa]	X_{ef}[mm]	T_{ef}[MPa]	X_{ef}[mm]	T_{ef}[MPa]	X_{ef}[mm]
0.2	-	-	-	-	-230.8	0.49
0,3	-318,9	0,74	-242,3	0,56	-221.4	0.51
0,4	-285,2	0,75	-236,8	0,58	-202.5	0.53
0,5	-244,2	0,76	-228,1	0,6	-194.7	0.57
0,6	-228,6	0,77	-225,1	0,62	-189.5	0.6
0,7	-221,9	0,78	-223,8	0,65	-186.3	0.64

One notes that T_{ef} increases when (a/W) ratio increases. The values associated which each notch geometry are reported in table 8.

Table 8: T_{ef} Values associated which each notch geometry.

	U_1	U_{05}	V
a/W	0.5	0.5	0.2
T_{ef}[MPa]	-244,2	-228,1	-230.8

One notes that T_{ef} values are relatively close. The increase of constraint by increase (a/W) ratio is counterbalance by the increase of the notch acuity when comparing U_{05} and V notch.

6. DISCUSSION

The use of Charpy V T_{K27} or $T_{K1/2}$ transition temperature as reference temperature needs to introduce a correction due to the fact that this test is made at a loading rate of about $10\ s^{-1}$ and consequently higher than the loading rate corresponding to a static loading(10^{-3s-1}). An empirical relationship between transition T_t and yield stress Re proposed by Rolfe and Barsom [14] is used.

$$T_t = 0.17\ Re - 125\ (K, MPa) \tag{12}$$

Knowing transition temperature for Charpy Test $T_{t,d,}$ (174K), it is easy to know the equivalent static transition temperature $T_{t,s}$ by reporting dynamic yield stress (661MPa) at this temperature into modify equation (1).

$$\Delta T_t = 0.17\left(Re_d\left(T_{t,d}\right) - Re_{,s}\left(T_{t,s}\right)\right) \tag{13}$$

This equation is solved knowing relationship between static yield stress and temperature by dichotomy method.

$$R_e = R_e^{\mu} + AExp(-BT) \tag{14}$$

R_e^{μ} is a threshold, the constants A and B are given in Table 3. For API X65, the shift in transition temperature is 10°C. This corrected transition temperature is reported in figure 8.

The fracture bimodality induces some difficulty to fit data and defined the transition temperature. For this reason, the transition temperature $T_{t1/2}$ defined at half the energy jump between britle and ductile plateaus is used.

The stress difference method (SDM) is use to determine T or T_ρ

$$T, T_\rho = \sigma_{xx}(\theta = 0) - \sigma_{yy}(\theta = 0) \tag{15}$$

It appears on figure 6 that this difference is not constant but presents a plateau or a small increase at some distance of crack or notch tip. Therefore, the chosen value T_{ef} is defined by a conventional manner. In this paper two way have been used:
- The method proposed by Maleski et al [15],
- The method using the effective distance obtained from Volumetric method.

In the method proposed by Maleski et al, effective T stress T_{ef} is obtained by linear extrapolation to origin of T distribution. This value corresponds to the effective distance on this distribution. One notes that these values are very closed for V notch but far for U notch.

Table 9: Comparison of Methods to determine T_{ef}

Method	Maleski et al [15],	From effective distance
T_{ef} for U_1 specimen	- 410 MPa	-244,2 MPa
T_{ef} for V specimen	-230.8 MPa	-220 MPa

The method based on volumetric method has been chosen in this paper because of difficulty to choose the appropriate linear extrapolation.
The transition temperatures have been determined for specimens with U_1, U_{05} and V notch in the present study together with effective T stress T_{ef}. From a previous study [5], data from tensile tests and fracture toughness on CT specimen are also reported in Table 11.

Table 10: Transition temperature and Effective T stress for different specimens made in X65.

Notch	T_{ef}(MPa)	T_t(K)
U_1	-244,2	150
U_{05}	-228,1	187
V	-230,8	179
CT	-330	156
Tensile	-510	123

The transition temperature for Charpy specimens (U_1, U_{05} and V) are corrected to take into account the strain rate effect and reported in figure 7.

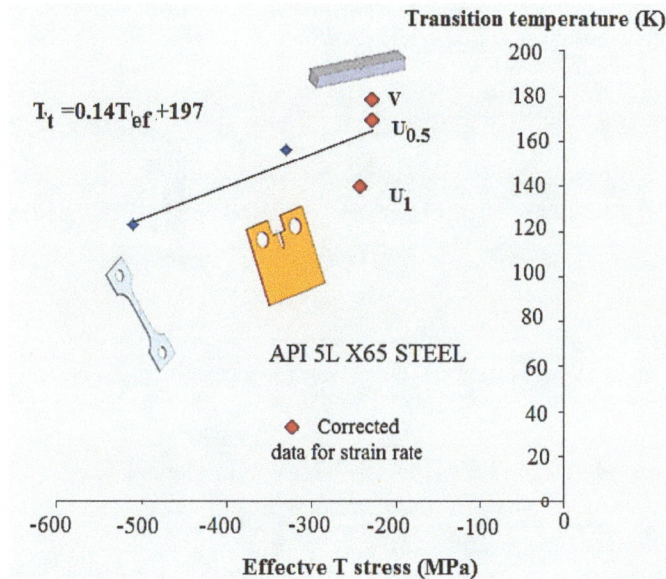

Figure7: Material master curve $T_t = f (T_{ef})$ for pipe steel API 5LX65.

Data are fitted according linear interpolation and relationship between transition temperature and effective T stress is given by:

$$T_t = 0.14T_{ef} + 197 \tag{16}$$

This equation represents the material master curve $T_t = f (T_{ef})$ which is the key to determine the appropriate reference transition temperature by comparison with structure transition temperature.

The values of T_{ef} are close for Charpy V, U_1 and U_{05} like values of transition temperature relative to the same specimen notch geometry. They are higher than CT specimen which exhibits a lower plastic constraint than 3PB specimen. CT specimens loaded both by bending and tension have a transition temperature intermediate with those of tensile and Charpy specimens.

The transition temperature relative to the investigated component is obtained using two material master curves: transition temperature master curve $T_t = f (T_{ef})$ and material master curve $K_c = f (T_{ef})$, where K_c is fracture toughness.

The effective T stress for a component $T_{ef, comp}$ is obtained through a procedure described in [13]. The material master curve at transition temperature has been drawn using the fracture toughness at transition temperature of CT specimen (156K, 100 MPa√m) and Charpy V (156 K, 92 MPa√m).

Transition temperature of component has then been determined for pipe steel made in API 5L X65 with 355mm diameter and 19 mm thickness. This pipe exhibits a surface notch with a notch angle $\square\square= 0°$, a notch radius $\square\square= 0.25$ mm and a notch depth (a) to thickness (t) ratio equal to a/t = 0.5. Loading curve $K_{ap} = f (T)$ has been computed by finite element assuming elastic behaviour, the steel is considered as brittle at transition temperature. This loading curve $K_{ap} = f (T)$ intercept the material master curve at point (T^*_{ef}, K_c) figure 8.

The obtained value of T^*_{ef} is -495MPa. In [16], the determination of the material master curve is more precise because 4 specimen types has been used (SENT, CT, TR and DCB), each specimen with different a/W ratio. In this procedure, we have used available data in order to have a cheaper and faster procedure. To take into account this uncertainty, it is better to estimate the constraint range -450 MPa and -550 MPa.

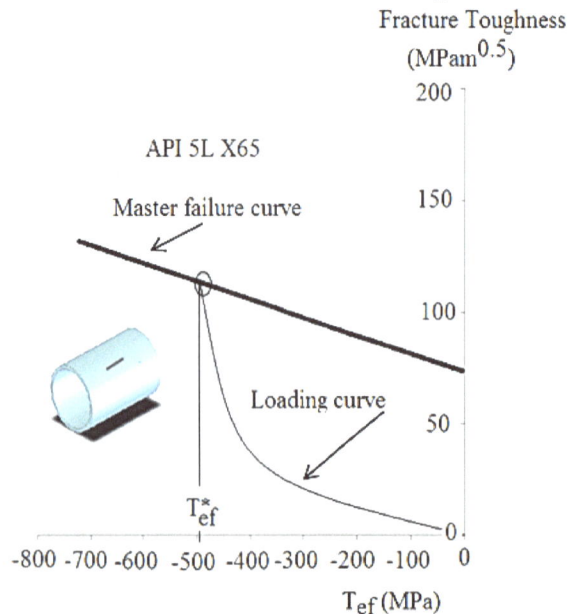

Figure 8 : Material Failure curve for API 5L X65 steel and loading curve $K_{ap} = f (T)$ for a pipe exhibiting a surface notch

The conservative range induces by the choice of the reference transition temperature RT is determined from the transition temperature of the component. $T_{t, comp}$ is obtained by reporting T^*_{ef} value on transition temperature master curve $T_t = f (T_{ef})$ and equal to $T_{t, comp} = 150$ K. The critical exposure temperature + margin of 8°C [4] is equal to185 K. Conservative range $\square T^*$ according to RT choice a given in table 11.

Table 11: Conservative range $\Box T^*$ according to RT

Transition temperature	$T_{t,comp}$	T_0	$T_{K1/2}$
$\Box T^*$ (K)	35 ± 10	29	6

Due to uncertainties on material master curve $K_c = f\,(T_{ef})$, a error of $\pm 10°C$ on component transition temperature has been estimated. One notes that the transition temperature T_0 is close to component one For both CT specimen and pipe, values of plastic constrain are very close. However the choice of Charpy V transition temperature as reference temperature is justified. A sufficient conservative range of $11°C$ is obtain and offers the possibility to enlarge material choice.

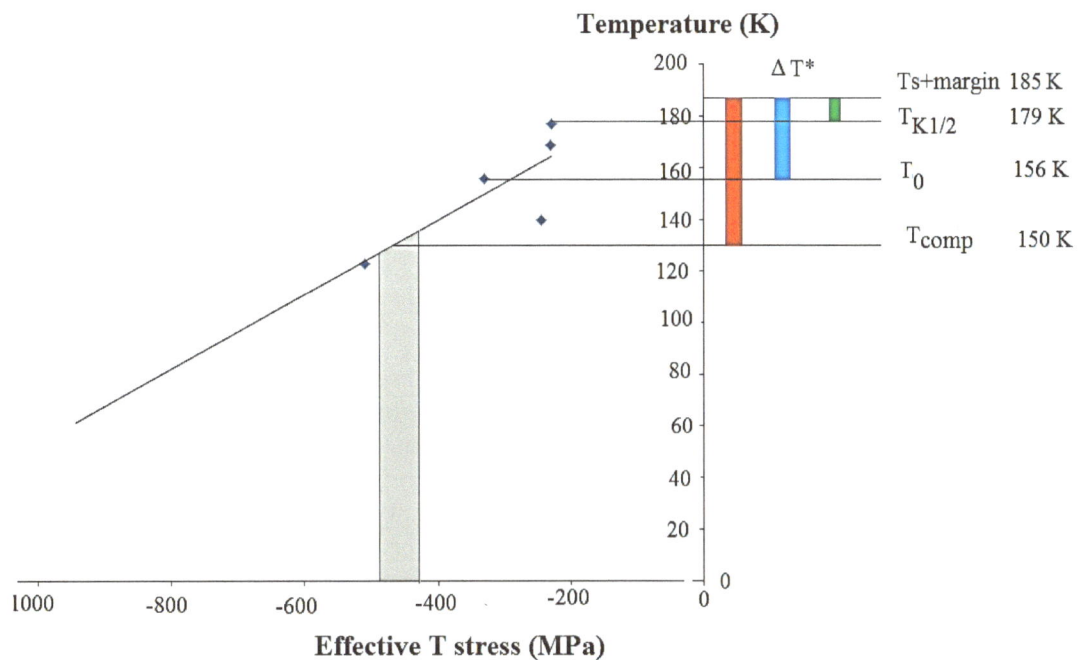

Figure 9: Estimation of conservative range $\Box T^*$ according to RT from material failure curve.

7. SUMMARY

The choice of the referece temperature is, according to codes like API 579-1 ASME FFS-1 is open and is under responsibility of designer.

In this paper, we propose a method to estimate the degree of conservatism induces by a choice of a reference transition temperature. This method is based on the relationship between transition temperature and constraint and a transition temperature associated with component.

This one is obtain by intersection of material master curve $K_c = f\,(T_{ef})$ and loading curve $K_{ap} = f\,(T)$.

Reducing the conservatism range, offer the possibility to enlarge material choice according to availability in large quantities, price and time for delivering.

However, this procedure is time consuming and costly because it need to determine several transition temperature with different specimens and finite element computing.

At least, selection based on Charpy V transition temperature $T_{K1/2}$ offer the most conservative approach. This is due that plastic constraint induced by bending is lower than in tension or mixture of bending and tension like for CT specimen.

REFERENCE

[1]   AFNORstandard: NF EN ISO 14556 (2001). *Acier. Essai de flexion par choc sur éprouvette Charpy à entaille en V. Méthode d'essai instrumentée*, Association Française de Normalisation, Saint Denis La Pleine.

[2]   Toth. L. "Rissbildungs und Ausbreitungsarbeit hinsichtlich der kerb-geometrie von kerbschlag biegeversuchen".Publications of the Technical University for Heavy Industry. Series C, Machinery, Vol 34, pp31-47, (1978).

[3]   G. Pluvinage et F. Montariol. " Contribution à l'étude des transitions de résilience dans le cas d'un acier doux" Rev. Met,. LXV, No 4, pp. 297-308, Avril, (1968).

[4]   API 579-1 ASME FFS-1 June 5 (2007)

[5]   Capelle, J. Furtado, Z. Azari. S. Jallais And G. Pluvinage, "Design based on ductile-brittle transition temperature for API 5l X65 steel used for dense CO_2 transport" , to appear to appear in Engineering Fracture Mechanics

[6]   G. Fearnehoug et C. Hoy, "Mechanism of deformation and fracture in the Charpy test as revealed by dynamic recording of impact loads", J. of the Iron and Steel Institut, 1964, pp. 912-920.

[7]   R. Chaouadi, J.L. Puzzolante, "Loading rate effect on ductile crack resistance of steels using precracked Charpy Specimens", International Journal of Pressure Vessels and Piping 85 (2008) 752–761

[8]   Green, A. P., and Hundy, B. B. (1956). Initial plastic yielding in notch bend tests, *Journal of Mechanics and Physics of Solids*, **4**, 128-144.

[9]   ASTM E1921-11a, (2003). *Standard Test Method for Determination of Reference Temperature, T_0 for Ferritic Steels in the Transition Range*, American Society for Testing and Materials, Philadelphia.

[10]   M. Mouwakeh, G. Pluvinage, S. Masri . « Failure of water pipes Containing Surface Cracks Using Limit Analysis Notions". Res. J. of Aleppo Univ.   Engineering Science Series   No.63, (2011).

[11]   Henry, B. S., Luxmore, A. R. (1997). The stress triaxiality constraint and the Q-value as a ductile fracture parameter, *Engineering Fracture Mechanics*, **57**, pp. 375-390.

[12]   Ruggieri, C., Gao, X., and Dodds, R. H., (2000). Transferability of elastic-plastic fracture toughness using the Weibull stress approach: significance of parameter calibration; *Engineering Fracture Mechanics*, **67**, 101-117.

[13]   Hadj Meliani, M., Matvienko, Y. G., Pluvinage, G. (2011). Two-parameter fracture criterion ($K_{\square\square\square}$-$T_{ef,c}$) based on notch fracture mechanics, *International Journal of Fracture*, **167**, 173-182.

[14]   S.T Rolfe andJ.M Barsom, Fracture and Fatigue control in structures. Prentice-Hal (1977)

[15]   M.J. Maleski, M.S. Kirugulige and H.V. Tippur. A Method for Measuring Mode I Crack Tip Constraint Under Static and Dynamic Loading Conditions. Society for Experimental Mechanics. Vol. 44, No. 5, October 2004.

[16]   M. Hadj Meliani, Z. Azari, G. Pluvinage, J. Capelle Gouge assessment for pipes and associated transferability problem, *Engineering Failure Analysis*, Volume 17, Issue 5, July 2010, Pages 1117-1126

NUMERICAL MODEL FOR NITI WIRES DESIGNED AS A TOOL FOR ACTUATOR APPLICATIONS

V. Gheorghita[1,3], A. Chiru[3], J. Strittmatter [1,2]

[1] University of Applied Sciences Konstanz, Brauneggerstrasse 55, 78467 Konstanz, Germany,
[2] WITg Institut für Werkstoffsystemtechnik Thurgau an der Hochschule Konstanz, Konstanzer Strasse 19, 8274 Tägerwilen, Switzerland,
[3] University Transilvania Braşov, Str.Politehnicii no. 1, 500024 Braşov, Romania.

Abstract: Shape Memory Alloys (SMA) have unique properties which do not exist in many materials traditionally used in engineering applications. Most important of these extraordinary characterisitcs are the pseudo-elasticity and the shape memory effect. This paper offers a new mathematical model for the shape memory effect when electrically activated. Many engineering applications start with the question concerning the required electrical energy, the prevenient displacement and the needed space for the assembly. The behaviour of the NiTi-SMA actuator wire can be mathematical described. Numerical simulations to show qualitatively the ability of the model to capture the behavior of the shape memory alloys are also presented. The background of this paper is the development of a mathematical model which is able to give informations about electrical tension, current intensity and deformation, when concrete values for diameter, length and/or mechanical tension of the SMA actuator wires are given. These informations are very useful as an engineering tool in order to design an actuator.

Keywords: shape memory alloy, NiTi wires, mathematical model, numerical model, SMA model

1. INTRODUCTION

The shape memory alloys have unique properties which do not exist in many materials traditionally used in engineering applications. The NiTi alloys are known as the most important shape memory alloys (SMAs) because of their multitude of applications based on the shape memory effect (SME) and pseudo-elasticity (PE) [6], [8-9]. Those properties are related to the martensite austenite (and reverse) transformation. This comes from the fact that NiTi alloys have superior properties in ductility, fatigue, corrosion resistance, biocompatibility and recoverable strain [6] [9]. When they are used as actuatorsis it is important to know the displacement vs. time at constant loading and the electrical energy vs. time. In order to save time and costs a numerical model is realized as an easy way to describe the functionality of SMA when the memory effect is electrical activated. Several researchers have already realized different models for SMA using the Joule principle, that transforms electrical energy into heat.

During this process the martensite structure is changing to austenite and the sample will execute a change of its shape. It is known from other models [1-3], [7], [11-18] that it is possible to fulfill the shape change when the sample is electrically activated. But in actuator industry this deformation is a function of time and stress. If the stress applied to the specimen is higher, then the phase change temperatures will linearly increase (see Fig. 1). This behaviour is the commonly known stress dependency of the SME and has to be considered in each model.

These two assumptions were the background for some numerical models. The models are divided in three groups: microscopic, mesoscopic and macroscopic. Each one considers the memory effect like a thermo-mechanical effect [1-3], [7-18].

The microcscopic [7] models study the atoms and the internal energy which is able to dislocate them. On the other side the macroscopic [1] models analyse the external dimension using also the thermodynamic principles. In the actuator area, where one of the most important things is the reduction of the assembly space, it therefor makes sense to use one of the macroscopic models to find the best combination (SME – space) [17]. The technical applications take care of the thermo-mechanical aspects from SMA and also the space of assembly. These models are mesoscopic. The last examples of such mesoscopic models are the Lexcellent and Lagoudas model[8-9].

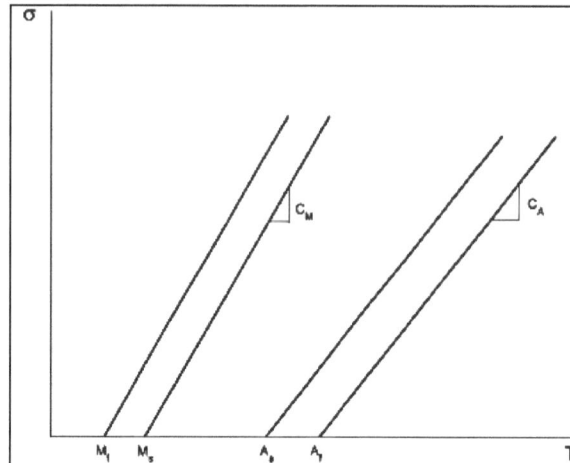

Figure 1: Relation between temperature and stress [8]

Also, in the realm of the so-called non-equilibrium (or irreversible) thermodynamics, among others models [3], [7], [10-18] the proposed models based on the use of a set of thermomechanical equations describing the kinetics of thematensitic transformations.

The constitutive equations are developed in a non - linear manner on the basis of a free energy driving force and the laws of thermodynamics. Boyd and Lagoudas [8] have extended the theory of the thermomechanical approach by Ortin and Planes [12-13] and Raniecki and Lexcellent [9] in order to be valid for more general loading conditions such as non-proportional loading and combined isotropic and kinematic hardening [15]. The models are inconvenient for large-scale computations. The main reason is the difficulty in establishing the constant parameters. This aspect is very important and therefor it is necessary for the model accuracy to determinate the constants for the SMA experimental. One way to determinate some constant parameters (phase change temperatures, elastic module, heat coefficient, etc.) is using the DSC device [19].

These informations are very important as an engineering tool in order to design an actuator. The focus of the numerical model presented in this work is related to the automotive safety system [5]. But this tool is also useful for other shape memory actuators that find their applications in other fields, e.g. a thermally activated shape memory tube that represents the driving element in an intramedullary nail for bone lengthening [20].

However the SMAs are characterized by solid to solid displacive transformation between the austenite and the martensite fraction, in response to mechanical loading, thermal loading and electrical loading.

2. DESCRIPTION OF THE MODEL

The current model starts from the Ortin and Planes approach. The issues are to determinate the elongation when SMAs are electrically activated and mechanically loaded.

The backgrounds of this model are the first and second thermodynamic laws and also the Clausius – Duhem inequality [4].

The model was divided in four small systems:
- Mechanical system;

- Thermal system;
- Hysteresis system;
- Energy balance system.

It can be written that the electrical energy should be equal with the output heat, as shown in the equation 1.

$$P_{el} = \dot{Q}_{out} \tag{1}$$

The next function should be mathematically described:

$$R = f(T, \xi) \tag{2}$$
$$\xi = f(T, \sigma) \tag{3}$$
$$T = f(R, \xi) \tag{4}$$

where:

- R - the electrical resistance of the NiTi wire;
- T – temperature value (can be A_s, A_f, M_s or M_f);
- σ – mechniacal tension;
- ξ – martensite fraction.

After solving the energy balance (considering mechanical energy, electrical energy and also heat losses) it can be designed in MathLab-Simulink a new numerical model for SMA, which is divided in the mentioned four small system models: mechanical, hysteresis, thermal and energy balance.

The main equation for structure changing is in connection with temperature.

For the Martensite – Austenite changing the heating process will be determined according to equation 5:

$$\xi(T_C) = \frac{\xi_{Start}}{\pi} \cdot \tan^{-1}\left[\pi \cdot \left(\frac{A_s - T_C}{A_f - A_s} + \frac{1}{2}\right)\right] + \frac{\xi_{Start}}{2} \tag{5}$$

Also for the Austenite-Martensite changing, the cooling process can be calculated according to equation6

$$\xi(T_C) = \frac{1 - \xi_{Start}}{\pi} \cdot \tan^{-1}\left[\pi \cdot \left(\frac{M_s - T_C}{M_f - M_s} + \frac{1}{2}\right)\right] + \frac{1 + \xi_{Start}}{2} \tag{6}$$

where:

- A_s and A_f are temperature points for the austenite start and finish structure;
- M_s and M_f the martensite start and finish temperature;
- T_c – room temperature;
- ξ – Martensite fraction.
-

The MathLab program is a close loop using equation 1 till 6 and is presented in the figure no. 2. First application is to see how much electrical energy is needed to obtain at least 4% elongation.

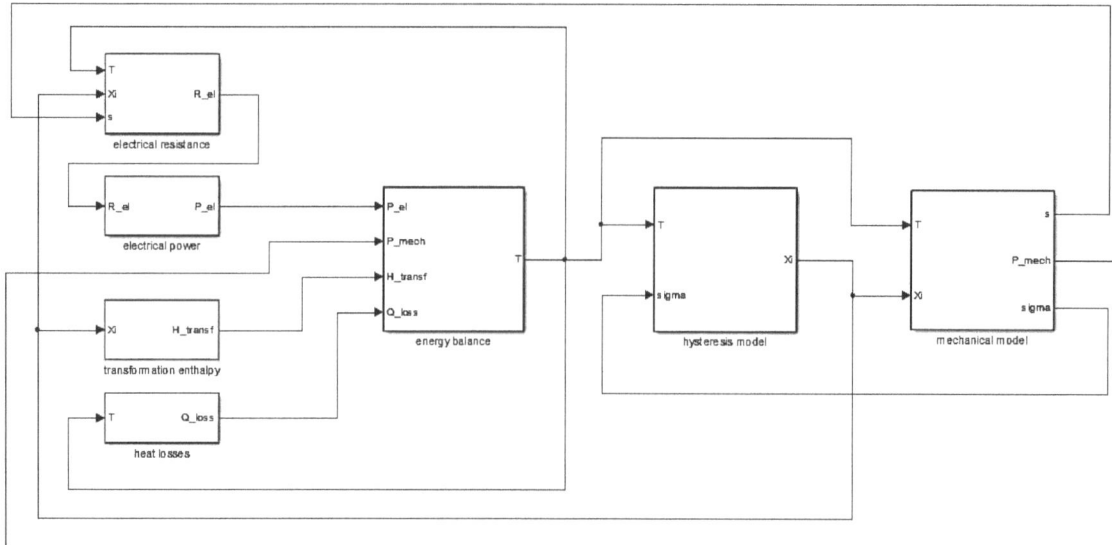

Figure 2: Internal loops for the SMA numerical model.

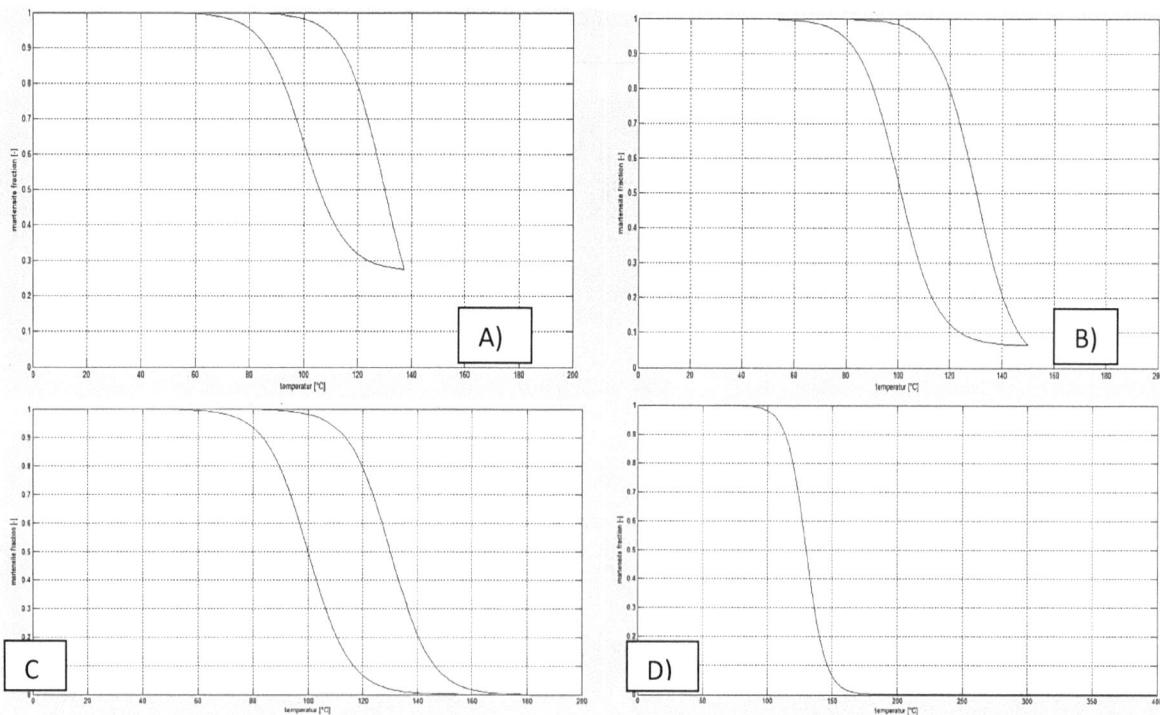

Figure 3: Martensite fraction vs. temperature at U=12V and a) I=3A, b) I=3.5A, c) I=4A, d) I=7A

The most important graphic for this simulation is the martensite fraction vs. temperature diagram. To record the real values for the electrical tension and current intensity, when the sample is loaded with 400N/mm^2 and elongated at least 4% in 1s, actuator samples with 0.5mm diameter and active length 250mm were used.

If the full martensite is 1, in figure 3a, it can be observed that at a current value of 3A the structure doesn't change in full austenite. The SMA is heated till 135°C, but is not full austenite. If the current intensity is increased till 4A the structure changes into the full austenite. When the current value is greater than 4A the SMA is overheated and in the worst case damaged.

3. RESULTS

The results from the model are compared with experimental test. For this comparison tests were performed with NiTi actuator wires between diameters of d=0.1-0.5mm. In the table 1 are presented the results for a mechanical tension of σ=250N/mm^2; more tests at different mechanical tension values were performed in the material testing laboratory. It can be observed that the calculated values of the numerical value correspond very well in comparison to the measured values in most of the cases. The biggest difference was found for the wire with 0.3mm diameter, where the elongation value of the model differs 9.2% from the measured value. It has to be mentioned that the real tests were performed at different points in time and the room temperature was not recorded. This can be a reason for the slight differences of the measured and modelled values, as the model is referring to a room temperature of 20°C in this case. In spite of these slight differences it can be stated, that in the actuator field this model is very useful[5].

Table 1: Comparison between measured and model results

		Measured			Results from Modell		
d [mm]	Settings at Source	U [V]	I [A]	ε [%]	U [V]	I [A]	ε [%]
0,5	12 V; 4,5 A	9,60	4,45	4,33	9,60	4,45	4,03
0,4	12 V; 2,8 A	10,68	2,81	4,24	10,68	2,81	4,13
0,3	12 V; 1,5 A	10,19	1,51	4,01	10,19	1,51	3,64
0,2	12 V; 0,75 A	11,06	0,76	4,37	11,06	0,76	4,33
0,1	16 V; 0,25 A	15,43	0,29	4,34	15,43	0,29	4,37

4. CONCLUSION

The results prove that the NiTi wires are able to be activated using electrical energy in order to show the desired function. This activation can be described with a numerical model that only shows little differences in comparison to the measured values obtained in real tests. For the next future more real tests have to be carried out at precise conditions (e.g. recording the room temperature) and the model has to be slightly modified with some additional parameters in order to minimize the differences of the calculated values. This model is very useful for engineering applications in order to design a device with NiTi actuators.

In the automotive industry the electrical energy is limitated, but the actuators require a specific energy. Before starting the design of a new actuator for this industrial field, the presented model can be very useful to define the SMA wires, e.g.: number, diameter and length of wires.

ACKNOWLEDGEMENT

Gratitudes are given to the HTWG Konstanz for the financial support in order to realize this work within the framework of a "Small Research Project".

REFERENCES

[1]   F. Auricchio, Shape memory alloys: applications, micromechanics, macromodeling and   numerical simulations, Ph.D. Dissertation, University of California at Berkeley, Berkeley, CA,   USA, 1995.
[2]   J.C. Boyd, D.C. Lagoudas, A thermomechanical constitutive model for shape memory materials. Part I. The monolithic shape memory alloy, Int. J. Plasticity 12, 805, 1996.

[3]   L.C. Brinson, M.S. Huang, Simplifications and comparisons of shape memory alloy constitutive models, J. Intell. Matl.Syst. &Struct, 7 108, 1996.

[4]   M. Fremond, The clausius-Duhem inequality - an Interesting and Productive Inequality, Advance in Mechanics and Mathematics, 12, 107-118, 2006

[5]   V. Gheorghita, P. Gümpel, J. Strittmatter, A. Chiru, T. Heitz and M. Senn, Using Shape Memory Alloys in automotive safety systems, Proceedings of the FISITA 2012 World Automotive Congres, Lecture Notes in Electrical Engineering Volume 195, 909-917, 2013

[6]   P. Gümpel, Formgedächtnislegierungen – Einsatzmöglichkeiten in Maschinenbau, Medizintechnik und Aktuatorik, Expert-Verlag, Renningen, 156 p., 2004

[7]   A. Kelly, K. Bhattacharya, R.M. Murray, A constitutive Relation of Shape Memory Alloys, California Institut of Technology, 180 p., 2009

[8]   D.C. Lagoudas, Shape Memory Alloys – Modeling and Engineering Applications, New York, Springer, 350p, 2008

[9]   C. Lexcellent, Shape memory Alloys Handbook, Wiley, London, 390 p., 2013

[10]   C. Liang, C.A. Rogers, One-dimensional thermomechanical constitutive relations for shape memory materials, J. Intell. Mater.Systems Struct. 1, 207, 1990

[11]   H. Maier and A. Czechowicz, Computer –Aided Development and Simulation for Shape Memory Actuators, Metallurgical and Materials Transaction, VOL 43A, 2882-2890, 2012

[12]   J. Ortin, A. Planas, Thermodynamic analysis of thermal measurements in thermoelastic   martensitic transformations, Acta Metall. Mater.36, 1873, 1988

[13]   J. Ortin, A. Planas, Thermodynamics of thermoelastic martensitic transformations, Acta Metall. Mater. 37, 143, 1989

[14]   K. Otsuka, and C.M. Wayman, Shape Memory Materials, Press Syndicate of the University of Cambridge, Cambridge, 2002.

[15]   V.P. Panoskaltsis, S. Bahuguna, D. Soldatos, On the thermomechanical modeling of shape memory alloys, International Journal of Non-Linear Mechanics 39, 709-722, 2004.

[16]   B. Raniecki, C. Lexcellent, RL-models of pseudoelasticity and their specification for some shape memory solids, Eur. J. Mech. A 13, 21, 1994.

[17]   F. Schiedeck, Entwicklung eines Modells für Formgedächtnisaktoren im geregelten dynamischen Betrieb. PhD, Leibniz, 150 p. 2009

[18]   K. Tanaka and R. Iwasaki, A phenomenological theory of transformation superelasticity, Eng. Fract. Mech. 21, 709, 1985.

[19]   G. Laino, R. De Santis, A. Gloria, T. Russo, D. S. Quintanilla, A. Laino, R. Martina, L. Nicolais and L. Ambrosio, Calorimetric and Thermomechanical Properties of Titanium-Based Orthodontic Wires: DSC–DMA Relationship to Predict the Elastic Modulus, Journal of Bimaterials Applications, 26, 829-844, 2012

[20]   T. M. Boes, P. Guempel, R. Storz Irion and J. Strittmatter, Implantatvorrichtung zur Gewebe- und/oder Knochendistrak- tion sowie Verfahren zum Betreiben einer solchen, Patent no. EP2173267B1, February 2009.

EX-IN VITRO TESTING OF TOTAL KNEE REPLACEMENTS – FIRST PART

C. Drugă[1], M. Mihai[2]

[1] Transilvania University/Product Design, Mechatronic and Environmental Department, Braşov, ROMANIA,
druga@unitbv.ro

[2] Transilvania University/ Product Design, Mechatronic and Environmental Department, Braşov, ROMANIA,

Abstract: This paper presents a part of a study being conducted in the Laboratory of Biomechanics and Biomechatronics (DPMM Department) of Transilvania University of Brasov. This study aims to test the total knee prostheses using an experimental test stand that simulates all movements and loading of the knee joint. In the first part of the paper presents some aspects of anatomy, biomechanics and prosthetic knee joint.

Keywords: experimental stand, knee prosthesis, movements, loading

1. INTRODUCTION

Total Knee Arthroplasty fundamentally changed the prognosis of knee degenerative disease, enabling a postoperative knee close to normal functionality.

The knee musculoskeletal mobile segment is considered one of the most used joints of the human body. In 1968, Mr. R. Merryweather used for the first time the total knee prosthesis and it was the beginning of discussion about total knee replacement. Since then the joint replacement has become a successful treatment, so that the moment a significant number of implanted knee prosthesis [1].

The knee is the intermediate joint of the lower limb; it is composed of the distal femur and proximal tibia. It is the largest and most complex joint in the body. The knee joint is composed of the tibio-femoral articulation and the patella-femoral articulation [1].

The shape of the articular surfaces of the proximal tibia and distal femur must fulfill the requirement that they move in contact with one another. The profile of the femoral condyles varies with the condyle examined (Figure 1 and Table 1).

A condyle is the rounded prominence at the end of a bone, often at an articulation joint. The tibial plateau widths are greater than the corresponding widths of the femoral condyles. However, the tibial plateau depths are less than those of the femoral condyle distances. The medial condyle of the tibia is concave superiorly (the center of curvature lies above the tibial surface) with a radius of curvature of 80 mm [4].

The lateral condyle is convex superiorly (the center of curvature lies below the tibial surface) with a radius of curvature of 70 mm [4].

The shape of the femoral surfaces is complementary to the shape of the tibial plateaus. The shape of the posterior femoral condyles may be approximated by spherical surfaces (Table 1).

The mechanism of movement between the femur and tibia is a combination of rolling and gliding.

The backward movement of the femur on the tibia during flexion has long been observed in the human knee.

The magnitude of the rolling and gliding changes through the range of flexion. The tibial–femoral contact point has been shown to move posteriorly as the knee is flexed, reflecting the coupling of anterior/posterior motion with flexion/extension [1].

During flexion, the weight-bearing surfaces move backward on the tibial plateaus and become progressively smaller.

The tibio-femoral joint is mainly a joint with two degrees of freedom. The first degree of freedom allows movements of flexion and extension in the sagittal plane. The axis of rotation lies perpendicular to the sagittal plane and intersects the femoral condyles. The symmetric optimal axis is constrained such that the axis is the same for both the right and left knee [1], [2], [4]. The screw axis may sometimes coincide with the optimal axis but not always, depending upon the motions of the knee joint.

The second degree of freedom is the axial rotation around the long axis of the tibia. Rotation of the leg around its long axis can only be performed with the knee flexed. There is also an automatic axial rotation which is involuntarily linked to flexion and extension. When the knee is flexed, the tibia internally rotates. Conversely, when the knee is extended, the tibia externally rotates. During knee flexion, the patella makes a rolling/gliding motion along the femoral articulating surface.

Throughout the entire flexion range, the gliding motion is clockwise. In contrast, the direction of the rolling motion is counter-clockwise between 0° and 90° and clockwise between 90° and 120°.

The mean amount of patellar gliding for all knees is approximately 6.5 mm per 10° of flexion between 0° and 80° and 4.5 mm per 10° of flexion between 80° and 120°. Between 80° and 90° of knee flexion, the rolling motion of the articulating surface comes to a standstill and then changes direction.

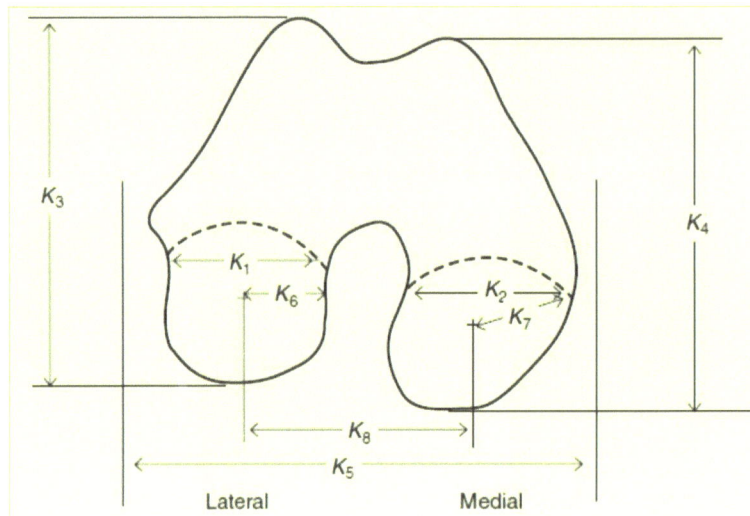

Figure 1: Geometry of distal femur [1], [2]

Table 1: Geometry of Distal Femur [2]

Parameter		Condyle				
		Lateral		Medial		Overall
	Symbol	Distance (mm)	Symbol	Distance (mm)	Symbol	Distance (mm)
Medial/lateral distance	K_1	31 ± 2.3 (male) 28 ± 1.8 (female)	K_2	32 ± 31 (male) 27 ± 3.1 (female)		
Anterior/posterior distance	K_3	72 ± 4.0 (male) 65 ± 3.7 (female)	K_4	70 ± 4.3 (male) 63 ± 4.5 (female)		
Posterior femoral condyle spherical radii	K_6	19.2 ± 1.7	K_7	20.8 ± 2.4		
Epicondylar width					K_5	90 ± 6 (male) 80 ± 6 (female)
Medial/lateral spacing of center of spherical surfaces					K_8	45.9 ± 3.4

2. BIOMECHANICS OF KNEE JOINT

The muscle control of the knee is mostly through the quadriceps and hamstring muscles. The quadriceps attach to the quadriceps tendon, which attaches to the kneecap, which attaches to the patellar tendon, which attaches to the tibia.

The forces on the static lower leg loaded with an ankle weight are shown in Figure 2a, 2b [6],[5]. The forces shown are due to this added weight, the weight of the lower leg, the quadriceps muscle force transmitted by the patellar tendon M (of magnitude M), and the joint reaction force R (of magnitude R), while the angle between the horizontal and the leg is β. In equilibrium, the muscle force and the x and y components of the joint force are

$$M = \frac{(bW_1 + cW_0)\cos\beta}{a\sin\theta} \tag{1}$$

$$R_x = M\cos(\theta + \beta) \tag{2}$$

$$R_y = M\sin(\theta + \beta) - W_0 - W_1 \tag{3}$$

Example: Let $a = 12cm$, $b = 22cm$, $c = 50cm$, $W_1 = 150\ N$, $W_0 = 100\ N$, $\theta = 15°$ and $\beta = 45°$. We see that the muscle force $M = 1,381N$ and the joint force $R = 1,171\ N$.

One function of the patella is to increase the moment arm. We can analyze the equilibrium of the kneecap at the patella-femoral joint between the reaction force on the kneecap from the anterior end of the femoral condyles, the patellar tendon and the quadriceps tendon; this is shown in Figure 3.

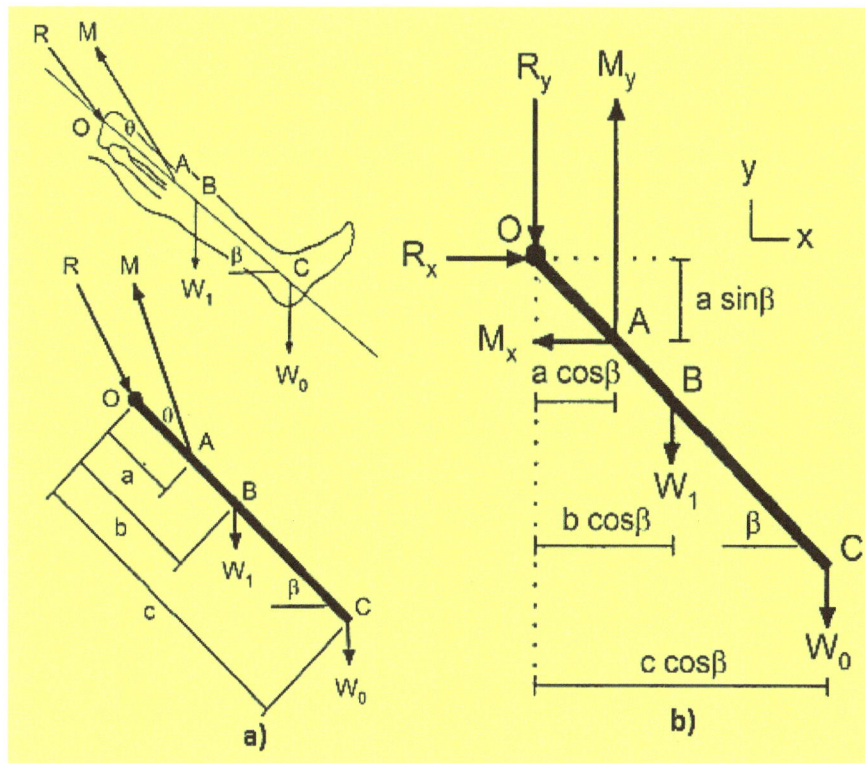

Figure 2: a)-Forces on the lower leg, while exercising the muscle around the knee, b)- Resolution of the forces [6]

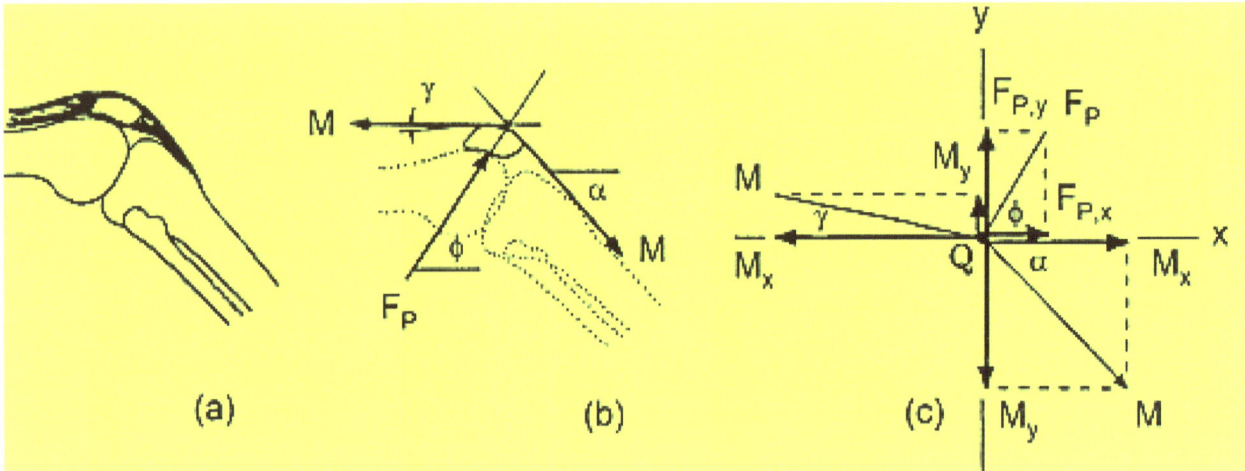

Figure 3: Force diagram of the patella in equilibrium.[6]

The compressive force applied on the kneecap is [6]:

$$F_P = \frac{\cos\gamma - \cos\alpha}{\cos\phi} M \tag{4}$$

At an angle

$$\phi = \arctan\left(\frac{\sin\alpha - \sin\gamma}{\cos\gamma - \cos\alpha}\right) \tag{5}$$

3. TOTAL KNEE JOINT REPLACEMENT

The tibio-femoral joint is a complex joint which can be affected many types of diseases. If the joint degradation is severe, it comes to the replacement of the natural joint with an artificial joint. Nowadays there are sufficient data to demonstrate that knee prosthesis is feasible for treatment of various diseases of the knee joint. The knee prosthesis is characterized by diversity their variability about the hinge [7]. Thus we can classify the prosthetic knees after anatomical parts or joints they replace, the degree of mobility the degree of constraint, the attachment or when implanted.

The rrostheses, produced from biologically compatible metal and plastic materials of a high strength, are used for the total knee joint replacement. Cobalt, chromium, and molybdenum alloys are the metals used most frequently.

The plastic materials are made from a high molecular polyethylene (UHMWPE).

The total implants have been used for around 30 years and the body's tolerance to them has been very good.

High requirements are imposed on the components' production, their surface must have identical properties all the time and it must be smooth and glossy [8]. Only damaged areas of the joint, not the entire knee, are replaced in a total knee replacement in modern medicine. In principle, the surgery lies in the replacement of the joint surface and joint cartilage only. Just a small part of the bone is removed; original ligaments, tendons and muscles are retained and re-fixed [8]. The metal femoral component is the same size and shape as the femur end. The tibial component placed on the apex of the tibia has a metal base but the upper surface is always made from polyethylene. Part of the patella surface may be cut off and also be covered by polyethylene. Several configuration of the joint replacement is shown in Figure 4.

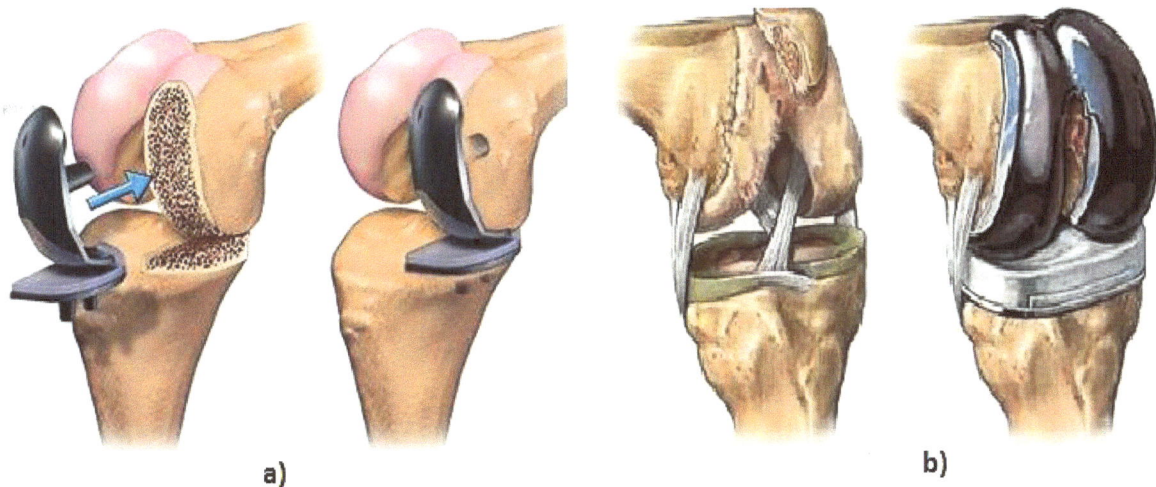

Figure 4: a)- Partial knee prosthesis and b) total knee prosthesis. [7]

The components are frequently fixed to the bone by a special substance (polymethacrylate), a so-called "bone cement". Alternatively some components have a porous surface into which the bone can grow [8].

There is a wide range of models produced in different sizes for all prostheses types. The bone shape, the weight, the physical activity of the patient and the surgeon's experience and philosophy determine the selection of the prostheses [8].

REFERENCES

[1]   Peterson R.D.,Bronzino D.J., Biomechanics - Principles and Applications, CRC Press. Taylor & Francisc Group LLC, New York, 2008, pp.35.

[2]   Yoshioka Y, Siu D, Cooke TDV. 1987. The anatomy of functional axes of the femur, J. Bone Joint Surg. 69A(6):873–880.

[3]   Kurosawa H., Walker PS., Abe S., Garg A., Hunter T., Geometry and motion of the knee for implant and orthotic design, J Biomech 18(7):487, 1985.

[4]   Kapandji I.A., The Physiology of the Joints, Vol 2, Lower Limb, Edinburgh, Churchill Livingstone, 1987.

[5]   Radu C.,Rosca I.,Druga C., Cismaru M.,Biomecanica şi Biomecatronica sistemelor biomecanice, Ed. Univ. Transilvania din Braşov, 2009, pp.22-23.

[6]   Irving P.Herman., Physics of the Human Body, Springer, 2008, pp.62-64.7

[7]   http://ortopediaonline.ro/content/view/41/44/#1

[8]   Horáček M., Charvát O.,Pavelka T.,Sedlák J., Madaj M., Nejedlý J., Medical implants by using RP and investment casting technologies, The 69th WFC Paper, February 2011.

MECHANICAL TESTING OF THE COMPOSITE MATERIALS BASED ON POLYPROPYLENE AND ITS APPLICATION IN AUTOMOTIVE PARTS

Camelia Cerbu[1]

[1]Transilvania University, Braşov, ROMANIA, e-mail cerbu@unitbv.ro

Abstract: The paper shows some advantages of the composite materials based on polypropylene for applications in the automotive field. Then, tensile test is applied to some specimens made of such a composite material. The stress-strain $(\sigma - \varepsilon)$ curves are shown. After statistically processing of the experimental results, these were used for analysis and simulation of the mechanical behavior of a front bumper under the impact with a stone. It is known that the bumper is only a component part of the bumper system of an automotive and it is located in front of the energy absorber and bumper beam.

Keywords: polypropylene, composite, tensile test, simulation

1. INTRODUCTION

The bumper considered in this study is only a part of the front bumper system whose components are: bumper; energy absorber; bumper beam; bumper mounting bracket.

It is known that the front bumper system is designed to protect the vehicle's safety systems in low-speed impacts by absorbing kinetic energy during a collision. On the other hand, the role is to reduce pedestrian injuries in very light accidents.

In the last years, there was a lot of material combination used to manufacture the automotive bumper: a combination of polycarbonate and acrylonitrile butadiene styrene called PC/ABS [1,8]; glass fibre reinforced polypropylene [5]; glass fibre reinforced plastics [2]; thermoplastic olefin elastomers denoted with TPO [4].

There are some works [5,6] that focused on material selection for the bumper taking into account the required properties of this material. Another paper [7] focuses on the analysis of a bumper with capacity of energy release.

The bumper analyzed in this paper is made of thermoplastic olefin elastomers (TPO). Here, are presented some mechanical properties of the material used to manufacture the bumper of a car. These properties were determinated in the tensile test. Then, the tensile properties together with the stress-strain $(\sigma - \varepsilon)$ curve are used to describe the material behavior in numerical modeling of the bumper.

2. WORKING METHOD

2.1. Mechanical testing of the material

Firstly, the tensile specimens were cut from a flat surface of a front automotive bumper. A number of ten specimens were prepared according to the European standard [9] for tensile testing of plastics or reinforced plastics. The dimensions of the cross-section of each specimen were accurately measured before mechanical testing. Then, the dimensions were considered as input data in the software program of the machine.

The testing equipment used for tensile test consists of a hydraulic power supply. The maximum force capacity is ± 15 kN. During the tensile tests, the speed of loading was 1.5 mm/min. according with [9].

The testing equipment allowed us to record pairs of values (force F and elongation Δl of the specimen) in form of files having 200-300 lines.

Therefore, the average values of the following quantities could be computed: Young's modulus E in tensile test; tensile stress σ_{max} at maximum load; tensile strain ε at maximum load; maximum tensile strain ε_{max}.

The experimental results recorded during tensile tests of the specimens, may be graphically drawn by using $\sigma - \varepsilon$ curves (Figure 1). Herein, only five curves recorded in case of 5 specimens are shown. It may be noted that Young's modulus E was computed for data points located on the linear portion of the $\sigma - \varepsilon$ curve in case of each specimen tested. In Figure 1 it may be observed the nonlinear behavior of the material for values of strains greater than 0.01 to 0.015.

Finally, the experimental data were statistically processed and the average values of the tensile properties are shown in the Table 1.

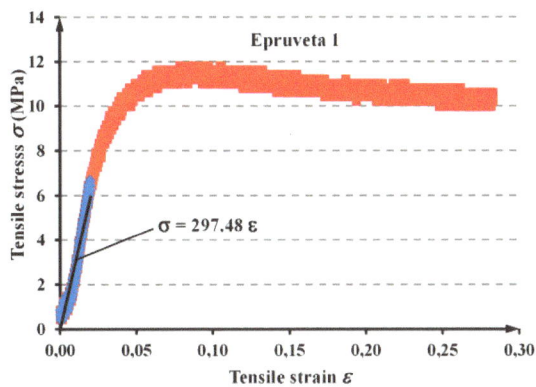

a.

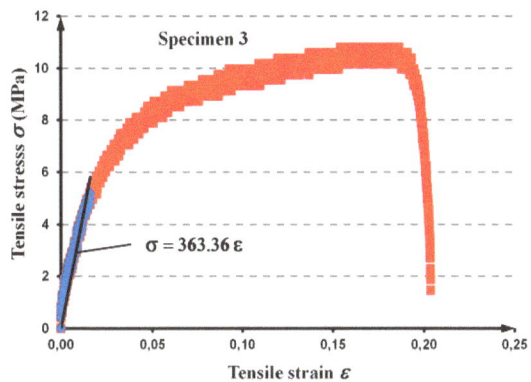

b.

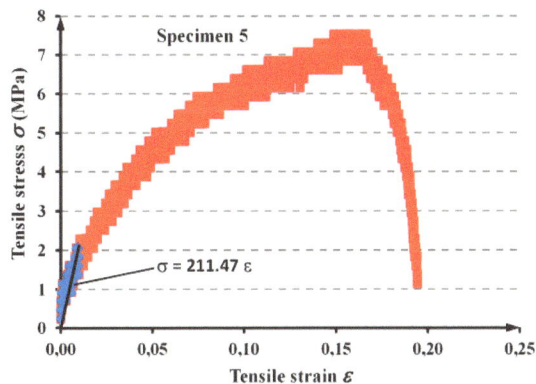

c.

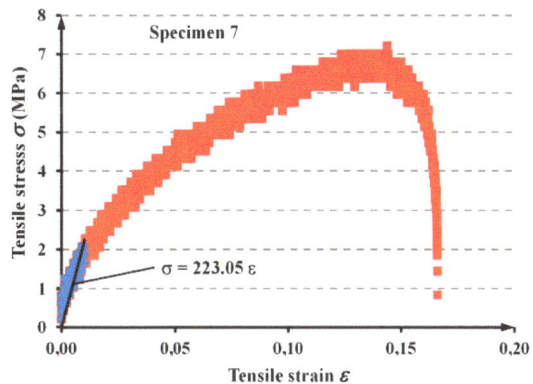

d.

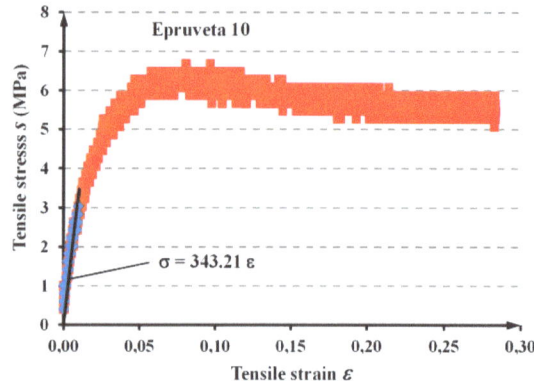

e.

Figure 1: Some of stress-strain $(\sigma - \varepsilon)$ curves recorded in tensile test in case of the material tested

Table 1: Mechanical properties determinate in tensile test in case of the material of the bumper fascia

Specimen	Young's modulus E (MPa)	Max. Force F (N)	Tensile stress σ at max. load (MPa)	Tensile strain ε at max. load [%]	Max.tensile strain ε_{max} [%]
Average value of the mechanical property	287.714	359.08	8.8	13.98	22.62

2.2. Numerical simulation of the mechanical behavior of the bumper

In the second part of the paper, the experimental results obtained in tensile test are applied to the numerical model of the front bumper to describe the mechanical behavior of the material.

The material of the bumper is TPO that is homogeneous and isotropic from macroscopically point of view. Taking into account the mechanical behavior in tensile test, the elastic portion of the $\sigma - \varepsilon$ curve will be described by the following properties: Young's modulus E=287.7 MPa; Poison's ratio $v = 0.26$. The plastic domain of stress-strain $(\sigma - \varepsilon)$ curve is described by using the coordinates of the points corresponding to this portion. For modeling was used the stress-strain $(\sigma - \varepsilon)$ curve showed in Figure 1, a, obtained in the case of the first tensile specimen tested.

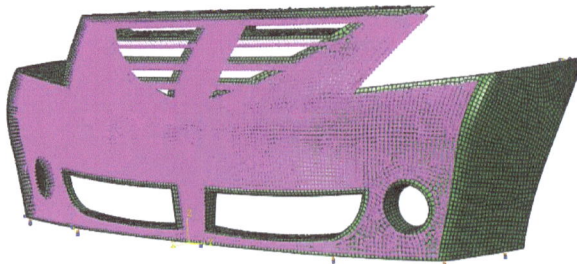

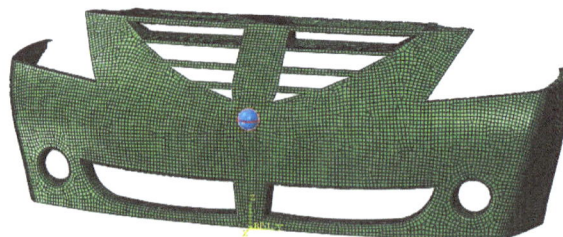

Figure 2: Numerical model for the 1st Scheme of loading (uniform pressure)

Figure 3: Numerical model for the 2nd Scheme of loading (impact with a steel ball)

Two schemes of loading are considered to analyse the mechanical behavior of the bumper.

In the first scheme of loading (Figure 2), the bumper subjected to uniformly pressure $p = 10^{-3} N / mm$ and it is fixed at the holes used for mounting on both bumper beam and car body.

In the second scheme of loading (Figure 3) the impact with a steel ball is analyzed. The diameter of the steel ball is equal to 25 mm. The last scheme of loading is analogous with the impact with a stone. It was chosen this way because the defining of the steel is easier. The low-speed impact was considered, the speed of the steel ball being equal to 2 m/s. It was considered that the steel ball hints perpendicular on the front bumper, parallel to Ox axis.

3. RESULTS

The figures 4 and 5 show some of the results obtained by analysis with finite elements in the case of the first scheme of loading: equivalent normal stress σ (Misses) in the bumper due to the action of the uniform pressure (Figure 4); total displacement in the bumper (Figure 4).

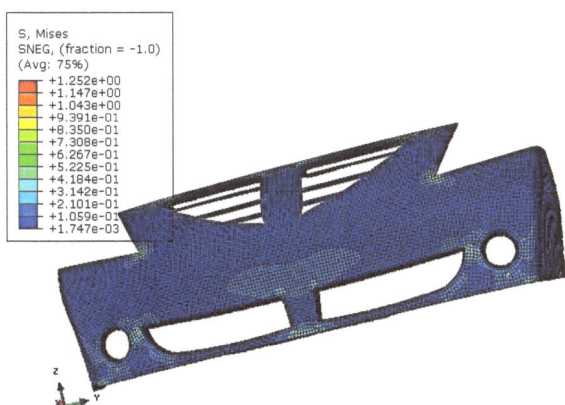

Figure 4: The distribution of the equivalent stress σ in the bumper due to the action of the uniform pressure

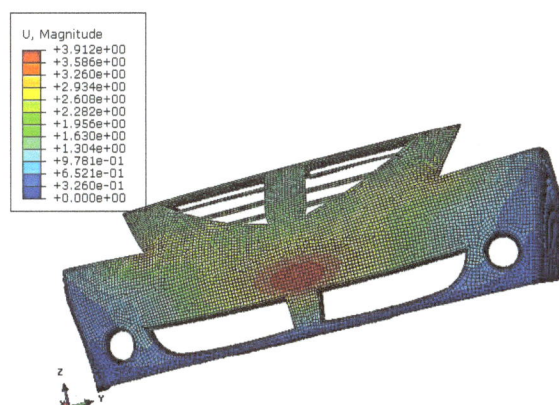

Figure 5: The distribution of the total displacement in the bumper due to the action of the uniform pressure

We must remark that the maximum value of the resultant normal stress σ (Misses) is 1.252 MPa, which does not exceed the average value $\sigma_{max} = 8.8 \, MPa$ of the stress recorded at maximum load (see the forth column of the Table 1). On the other hand, the maximum value of the total displacement is equal to 3.912 mm corresponding to the plastic portion of $\sigma - \varepsilon$ curve.

Figures 6 and 7 show the results obtained in the case of the second scheme of loading that involved a dynamically analysis of the bumper in impact loading.

In this case, the maximum value of the resultant normal stress σ (Misses) is 2.246 MPa which does not exceed again the average value $\sigma_{max} = 8.8 \, MPa$ recorded in tensile tests.

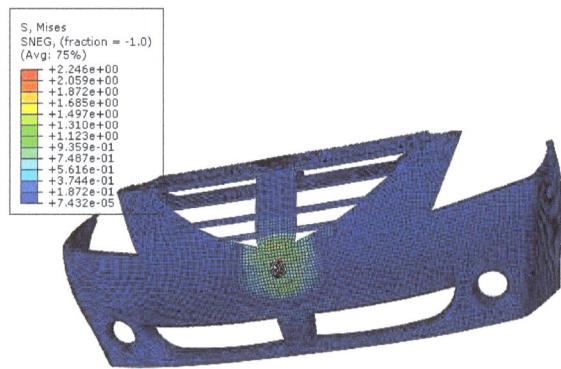

Figure 6: Distribution of the equivalent stress σ in the bumper due to the impact with a steel ball

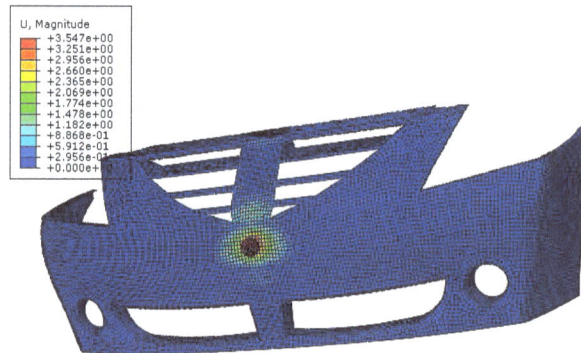

Figure 7: Distribution of the total displacement in the bumper due to the impact with a steel ball

4. CONCLUSIONS

In this paper, it was highlighted the behavior in tensile test concerning to the material thermoplastic olefin elastomers (TPO). The usage of the real stress-strain $(\sigma - \varepsilon)$ curve to model the front bumper lead to the conclusion that the maximum resultant normal stress σ (Misses) does not exceed the average limit stress 8.8MPa allowed for the material tested (Table 1).

These are only preliminary results because in reality the bumper is included in the bumper system that contains the energy absorber part. To be closer to the real behavior, in the futer the entire bumper system should be considered.

REFERENCES

[1]   Dick Mann M. Inst.M., Jan C Van den Bos, Arthur Way, Automotive Plastics & Composites. Worldwide Markets and Trends to 2007, Second edition, cap. 2. Plastics andReinforcements usedin Automobile Construction, Published by Elsevier Advanced Technology, 1999, pages: 26 – 60;

[2]   Mohan R., SRIM composites for automotive structural applications [C]// Proceedings of the Third Annual Conference on Advanced Composites. Detroit, Michigan: ASM Int, 1987: 57–62.

[3]   Clark G L, Bals C K, Layson M A., Effects of fibre and property orientation on 'C' shaped cross sections [R]. SAE Technical Paper, 910049, 1991.

[4]   Luda M. P., Brunella V., Guaratto D., Characterisation of Used PP-Based Car Bumpers and Their Recycling Properties, ISRN Materials Science, Volume 2013, Hindawi Publishing Corporation, http://dx.doi.org/10.1155/2013/531093;

[5]   Hambali A., Sapuan S. M., Ismail N., Nukman Y., Material selection of polymeric composite automotive bumper beam using analytical hierarchy process, J. Cent. South Univ. Technol. 17(2010), pp: 244–256;

[6]   Lucrare de licenta: Aspecte privind starile de tensiuni si deformatii in bara fata din constructia automobilului, solicitata mecanic. Cornel Dima, coordinator: Camelia Cerbu, 2011;

[7]   Mortazavi Moghaddam A.R., Ahmadian M. T., Design and analysis of an automobile bumper with the capacity of energy release using GMT materials, World Academy of Science, Engineering and Technology, 52, 2011;

[8]   ***http://www.plastics-car.com/bumpers;

[9]   ***EN ISO 527-2: 2000, Determination of the tensile properties of the plastics - Part 2: Test conditions for moulding and extrusion plastics, European Committee for Standardization, Bruxelles, 2000.

THE DISCRETIZATION OF THE LIMIT OF A BOUNDARY VALUE DIFFUSION PROBLEM IN A PERFORATED DOMAIN

Camelia Gheldiu[1], Mihaela Dumitrache[2]

[1] University of Piteşti, Piteşti, ROMANIA, camelia.gheldiu@upit.ro

[2] University of Piteşti, Piteşti, ROMANIA, mihaela_dumitrache_1@yahoo.com

Abstract: *This paper presents the approximation of a boundary value problem of diffusion in a perforated domain with very small holes. For the beginning, we homogenize the problem using two-scale convergence and than, the limit problem is discretized applying the finite element method and the operator-splitting method.*

Keywords: *perforated domain with small holes, homogenization, operator splitting, finite element method.*

1. INTRODUCTION

In this paper we resume the idea from the article [3], the different is that the domain is perforated with holes with a diameter much smaller than the period which are distributed in the domain. In the second, third and fourth sections of the article we discuss the homogenization of the non-stationary diffusion problem with Robin conditions. The limit problem obtained is a non-stationary convection-diffusion problem considered on the cylindrical domain $Q = \Omega \times (0,T)$, where Ω is the initial fixed domain without holes.

The novelty of the present article is the approximation of the limit problem on $Q = \Omega \times (0,T)$. The fifth section presents the spatial discretization of the locale problems which were obtained after the homogenization process from the fourth section, using the finite element method. In the sixth section we approximated the limit problem of the section four, combining the operator splitting – the Glowinski's scheme of the fractional step for the temporal discrimination (decomposition of the convection-diffusion operator) with the finite element method for the spatial discretization. The last section presents the result of the convergence.

2. THE PERFORATED DOMAIN

We consider the open and bounded domain $\Omega \subset R^n$, with the Lipschitz border $\partial\Omega$, the reference cell $Y = (o,l_1) \times (o,l_2) \times \cdots \times (o,l_n)$ and let be an open domain $S \subset Y$ such that $\overline{S} \subset Y$ with the smooth border ∂S. Let $r_\varepsilon << \varepsilon$ such that $\lim_{\varepsilon\to 0} \dfrac{r_\varepsilon}{\varepsilon} = 0$ and $\lim_{\varepsilon\to 0} \dfrac{\varepsilon^n}{r_\varepsilon^{n-2}} = 0$.

We consider $\Im(r_\varepsilon S)$ the translated of $r_\varepsilon \overline{S}$ with the form $\left(\varepsilon kl + r_\varepsilon \overline{S}\right)$, where $k \in Z^n$, $kl = (k_1 l_1, k_2 l_2, ..., k_n l_n)$, these translated representing the micro holes from R^n.

We denote $S_\varepsilon = \bigcup_{k \in K_\varepsilon} \left(\varepsilon kl + r_\varepsilon \overline{S}\right)$, where $K_\varepsilon = \left\{ k \in Z^n \middle| \left(\varepsilon kl + r_\varepsilon \overline{S}\right) \cap \Omega \neq \Phi \right\}$, S_ε represents the finite reunion of the holes from Ω, which can intersect $\partial\Omega$.

We're defining now the perforated domain $\Omega_\varepsilon = \Omega \setminus \overline{S}_\varepsilon$ where the holes are distributed with the period ε and the diameter r_ε is much smaller than ε.

3. THE STATE OF THE PROBLEM

We consider the following non-stationary diffusion problem in the perforated domain Ω_ε.

$$\begin{cases} \dfrac{\partial u_\varepsilon}{\partial t} - div(A_\varepsilon \nabla u_\varepsilon) + \mu_\varepsilon u_\varepsilon = f_\varepsilon & \text{in} \quad \Omega_\varepsilon \times (0,T) \\[2mm] (A_\varepsilon \nabla u_\varepsilon)\nu_\varepsilon + \alpha_\varepsilon u_\varepsilon = \dfrac{\varepsilon^n}{r_\varepsilon^{n-1}} g_\varepsilon & \text{on} \quad \Sigma_\varepsilon \times (0,T) \\[2mm] u_\varepsilon = 0 & \text{on} \quad \partial\Omega \times (0,T) \\[2mm] u_\varepsilon(0) = u_\varepsilon^0 & \text{on} \quad \Omega_\varepsilon \end{cases} \qquad (1)$$

under the following conditions:

i. $f_\varepsilon \in L^2(\Omega_\varepsilon \times (0,T))$, $g_\varepsilon \in L^2(\Sigma_\varepsilon \times (0,T))$, where $\Sigma_\varepsilon = \partial S_\varepsilon$ represents the border of the holes from the domain Ω_ε, $\partial S = \Sigma$. The estimation $\|f_\varepsilon\|_{L^2(\Omega_\varepsilon)} + \sqrt{\varepsilon}\|g_\varepsilon\|_{L^2(\Sigma_\varepsilon)} \leq c$ is true, where c is a positive constant, independent of ε.

ii. $A \in \left(L_{per}^\infty(Y)\right)^{n\times n}$, $m|\xi|^2 \leq A_{ij}(y)\xi_i\xi_j \leq \beta|\xi|^2$, $\forall \xi \in R^n$ a.e. $y \in Y$.

iii. $\mu \in L_{per}^\infty(Y)$, $\int_Y \mu(y)dy \geq \mu_0 > 0$; $\alpha \in L_{per}^2(\Sigma)$ so that $\int_\Sigma \alpha(y)d\sigma(y) = 0$, $u_\varepsilon^0 \in L^2(\Omega_\varepsilon)$.

We denote by $A_\varepsilon(x) = A\left(\dfrac{x}{\varepsilon}\right)$, $\mu_\varepsilon(x) = \mu\left(\dfrac{x}{\varepsilon}\right)$, $\alpha_\varepsilon(x) = \alpha\left(\dfrac{x}{r_\varepsilon}\right)$, $g_\varepsilon(x) = g\left(x, \dfrac{x}{r_\varepsilon}\right)$ with

$g \in L^2(\Omega \times \Sigma)$ and ν_ε is the external normal to Σ_ε.

4. THE HOMOGENIZATION

Using the homogenization method of the multiple scales like in the paper [3] and proving the convergence of the homogenization process with two-scale convergence method like in [1], where we are taking into account the convergences

$$\frac{\varepsilon^n}{r_\varepsilon^{n-1}}\chi_{\Omega_\varepsilon}u_\varepsilon \xrightarrow{2s} u(x,t), \quad \frac{\varepsilon^n}{r_\varepsilon^{n-1}}\chi_{\Omega_\varepsilon}\nabla u_\varepsilon \xrightarrow{2s} \nabla_x u(x,t) + \nabla_y U(x,y,t),$$

respectively [7]

$$\lim_{\varepsilon \to 0} \int_{\Sigma_\varepsilon} \frac{\varepsilon^n}{r_\varepsilon^{n-1}} g_\varepsilon \varphi d\sigma^\varepsilon(x) = \frac{1}{|Y|}\int_\Omega \left[\int_\Sigma g(x,y,t)d\sigma(y)\right]\varphi(x)dx, \quad \forall \varphi \in H_0^1(\Omega)$$

we obtain the homogenized problem with convection

$$\begin{cases} \dfrac{\partial u}{\partial t} - div(A^{eff}\nabla u(x,t)) + B\nabla u(x,t) + \lambda u(x,t) = F(x,t) & \text{in} \quad \Omega \times (0,T) \\[2mm] u = 0 & \text{on} \quad \partial\Omega \times (0,T) \\[2mm] u(0) = u_0 & \text{on} \quad \Omega \end{cases} \qquad (2)$$

where $\chi_{\Omega_\varepsilon} u_\varepsilon^0 \xrightarrow{2s} u_0$, and the other constants are:

$$a_{ik}^{eff} = \frac{1}{|Y|} \int_Y a_{ij}(y) \frac{\partial(\chi_k(y) + y_k)}{\partial y_j} dy \tag{3}$$

where the correctors χ_k satisfies the local problem

$$\begin{cases} -div_y \left[A(y) \nabla_y (y_j - \chi_j(y)) \right] = 0 \quad \text{in} \quad Y \\ \chi_j \text{ is } Y\text{-periodically}, \quad j = \overline{1,n}. \end{cases} \tag{4}$$

$$b_i = -\frac{1}{|Y|} \int_{Y^*} a_{ij}(y) \frac{\partial \gamma}{\partial y_j}(y) dy + \frac{1}{|Y|} \int_\Sigma \alpha(y) \chi_i(y) d\sigma(y), \quad i = \overline{1,n}, \tag{5}$$

where $B = (b_i)_{1 \le i \le n}$ is the convection vector.

$$\lambda = \overline{\mu} + \int_\Sigma \alpha(y) \gamma(y) d\sigma(y), \tag{6}$$

where $\overline{\mu} = \frac{1}{|Y|} \int_Y \mu(y) dy$, and γ satisfies the local problem:

$$\begin{cases} -div_y \left[A(y) \nabla \gamma(y) \right] = 0 \quad \text{in} \quad Y^*, \\ \left[A(y) \nabla \gamma(y) \right] \nu = -\alpha(y) \quad \text{on} \quad \Sigma, \\ \gamma \text{ is } Y\text{-periodically}, \end{cases} \tag{7}$$

$$F(x,t) = \frac{1}{|Y|} \left[\int_Y f(x,y,t) dy + \int_\Sigma g(x,y,t) d\sigma(y) \right]. \tag{8}$$

5. THE SPATIAL DISCRETIZATION

The spatial discretization of the local problems is made with the finite element method. Because the local problems (4) and (7) are considered on two different domains Y, respectively Y^*, and these two problems are independent of each other, we will consider two different triangulations: $\Im_{h/2}$ on Y and respectively $\Im_{h/4}$ on Y^*, with $h > 0$. In the following figures we present these two triangulations.

About the discretizated coefficients $a_{ik,h}^{eff}$, $b_{i,h}$ and λ_h we apply a quadrature scheme.

We consider the bidimensional case and we choose $Y = [0,1]^2$, $S = \left(\frac{3}{8}, \frac{5}{8} \right)^2$.

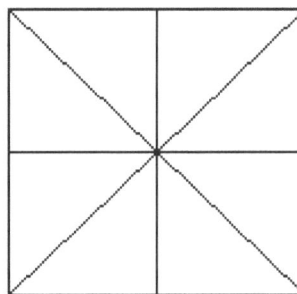

Figure 1: $\Im_{h/2}$

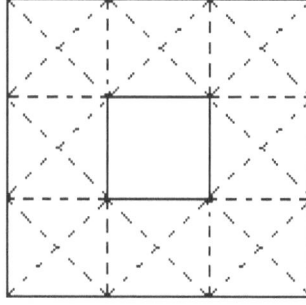

Figure 2: $\mathfrak{I}_{h/4}$

We consider the finite dimensional spaces

$$V_{h/2} = \left\{ v_h \middle| v_h \in C^0\left(\overline{Y}\right), v_{h|K} \in P_{11}, \forall K \in \mathfrak{I}_{h/2} \right\},$$

$$P_{11} = \left\{ p(x_1, x_2) = \sum_{0 \le i, j \le 1} c_{ij} x_1^i x_2^j, c_{ij} \in R \right\},$$

$$W_{h/4} = \left\{ w_h \middle| w_h \in C^0\left(Y^*\right), w_{h|K} \in P_{11}, \forall K \in \mathfrak{I}_{h/4} \right\},$$

and the subspations

$$V_{per,h/2} = \left\{ v_h \in V_h \middle| v_h(0, y_2) = v_h(1, y_2), v_h(y_1, 0) = v_h(y_1, 1), \forall y_1, y_2 \in [0,1] \right\},$$

$$W_{per,h/4} = \left\{ w_h \in W_h \middle| w_h \text{ is periodically} \right\}.$$

In this case, the spatial discretisation of the local problems (4) and (7) is to find $\chi_{j,h} \in V_{per,h/2}$ such that:

$$\int_Y A_{h/2}(y) \nabla_y \chi_{j,h}(y) \cdot \nabla_y v_h(y) dy = \int_Y A_h(y) \cdot e_j \cdot \nabla_y v_h(y) dy, \forall v_h \in V_{per,h/2} \tag{9}$$

Respectively, to find $\gamma_h \in W_{per,h/4}$ such that:

$$\int_{Y^*} A_{h/4}(y) \nabla_y \gamma_h(y) \cdot \nabla_y w_h(y) dy = \int_\Sigma \alpha_h(y) w_h(y) d\sigma(y), \forall w_h \in W_{per,h/4} \tag{10}$$

where $A_{h/2}$, $A_{h/4}$ represent the approximations of the matrix $A(y)$ relative to $\mathfrak{I}_{h/2}$, respectively $\mathfrak{I}_{h/4}$.

The relations (9) and (10) represent linear algebraic systems.

The discretized coefficients are obtained from the equations (3), (5) and (6):

$$a_{ik,h}^{eff} = \sum_{K \in \mathfrak{I}_{h/2}} \int_K a_{ij,h}(y) \frac{\partial\left(\chi_{k,h}(y) + y_k\right)}{\partial y_j} dy,$$

$$b_{i,h} = -\sum_{K \in \mathfrak{I}_{h/4}} \int_K a_{ij,h}(y) \frac{\partial \gamma_h}{\partial y_j} dy + \sum_{\substack{K \in \mathfrak{I}_{h/4} \\ K \cap \Sigma \ne \Phi}} \int_{\partial K \cap \Sigma} \alpha_h(y) \chi_{i,h}(y) d\sigma(y), \tag{11}$$

$$\lambda_h = \sum_{K \in \mathfrak{I}_{h/2}} \int_K \mu_h(y) dy + \sum_{\substack{K \in \mathfrak{I}_{h/4} \\ K \cap \Sigma \ne \Phi}} \int_{\partial K \cap \Sigma} \alpha_h(y) \gamma_h(y) d\sigma(y),$$

where α_h and μ_h are the approximations of the functions α (to $\mathfrak{I}_{h/4}$) and μ (to $\mathfrak{I}_{h/2}$). We apply the quadrature schemes to integrals.

Regarding the free term $F(x,t)$, the two integrals of relation (8) are calculated using the quadrature scheme.

6. THE DISCRETIZATION

The discretization of the problem (2) using the operator splitting method and the finite element method.

We discretize the global problem (2) combining the operator splitting – the Glowinski's scheme of the fractional step with the finite element method. We consider Ω a bounded poligonal domain from R^2. Let be $\mathfrak{I}_h$ a triangulation of Ω.

We introduce the spaces:

$$W_h = \left\{ v_h \in C^0(\bar{\Omega}) \big| v_{h/T} \in P_{11}, \forall T \in \mathfrak{I}_h \right\},$$

where P_{11} is the space of polynomials in two variables with degree at most one;

$$W_{0h} = \left\{ v_h \in W_h \big| v_h = 0 \ \ \text{on} \ \ \partial\Omega \right\}.$$

We consider the partition of the interval $[0,T]$:

$$0 = t^0 < t^1 = \Delta t < \cdots < t^n = n\Delta t < t^{n+1} = (n+1)\Delta t < \cdots < t^N = N\Delta t = T,$$

$$\Delta t = \frac{T}{N} \ \text{and} \ t^{n+\theta} = (n+\theta)\Delta t, \ \theta \in \left\{0, \frac{1}{3}, \frac{2}{3}, 1\right\}.$$

We have the following scheme: we denote by $u^0 = u_0$ and let $u_h^0 = (u_0)_h$ an approximation of u_0. We assume that u_h^n is known and we consider the following discrete variational problems:

Let's find $u_h^{n+\frac{1}{3}}$, $u_h^{n+\frac{2}{3}} \in W_h$ and $u_h^{n+1} \in W_{0h}$ such that

$$3\int_\Omega \frac{u_h^{n+\frac{1}{3}} - u_h^n}{\Delta t} v_h dx + \int_\Omega A_h^{eff} \nabla u_h^{n+\frac{1}{3}} \nabla v_h dx = \int_\Omega \left(F_h^n - B_h \nabla u_h^n - \lambda_h u_h^n\right) v_h dx, \forall v_h \in W_h,$$

$$3\int_\Omega \frac{u_h^{n+\frac{2}{3}} - u_h^{n+\frac{1}{3}}}{\Delta t} v_h dx + \int_\Omega B_h \nabla u_h^{n+\frac{2}{3}} v_h dx + \int_\Omega \lambda_h u_h^{n+\frac{2}{3}} v_h dx =$$

$$= \int_\Omega \left(F_h^{n+\frac{1}{3}} v_h - A_h^{eff} \nabla u_h^{n+\frac{1}{3}} \nabla v_h \right) dx, \forall v_h \in W_h, \tag{12}$$

$$3\int_\Omega \frac{u_h^{n+1} - u_h^{n+\frac{2}{3}}}{\Delta t} v_h dx + \int_\Omega A_h^{eff} \nabla u_h^{n+1} \nabla v_h dx =$$

$$= \int_\Omega F_h^{n+\frac{2}{3}} v_h dx - \int_\Omega B_h \nabla u_h^{n+\frac{2}{3}} v_h dx - \int_\Omega \lambda_h u_h^{n+\frac{2}{3}} v_h dx, \forall v_h \in W_{0h},$$

where

$$F_h^n = F_h(t^n) \ \text{in} \ \Omega, \ F_h^{n+\frac{1}{3}} = F_h\left(t^{n+\frac{1}{3}}\right), \ F_h^{n+\frac{2}{3}} = F_h\left(t^{n+\frac{2}{3}}\right),$$

and F_h is an approximation of F relative to T_h and $t^{n+\theta} = (n+\theta)\Delta t$, $\theta \in \left\{0, \frac{1}{3}, \frac{2}{3}, 1\right\}$.

Therefore, the switching from u_h^n to u_h^{n+1} is made passing through $u_h^{n+\frac{1}{3}}$ and $u_h^{n+\frac{2}{3}}$, practically breaking the interval $\left(t^n, t^{n+1}\right)$ with the intermediary points $t^{n+\frac{1}{3}}$ and $t^{n+\frac{2}{3}}$, where $t^{n+\theta} = (n+\theta)\Delta t$, $\theta \in \left\{0, \frac{1}{3}, \frac{2}{3}, 1\right\}$, $n \in \{0, 1, ..., N-1\}$, where we have the partition of the interval $[0, T]$:

$$0 = t^0 < t^1 < \cdots < t^n < t^{n+1} < \cdots < t^N = T, \quad t^n = n\Delta t, \quad t^N = N\Delta t = N\frac{T}{N} = T$$

and also braking the convection-diffusion operator, we denote by $\mathscr{L}_1, \mathscr{L}_2$ the operators:

$$\begin{cases} \mathscr{L}_1 = -div\left(A^{eff}\nabla\right) \\ \mathscr{L}_2 = B\nabla + \lambda. \end{cases}$$

This is the decomposition of the next operator

$$\mathscr{L} = -div\left(A^{eff}\nabla\right) + B\nabla + \lambda.$$

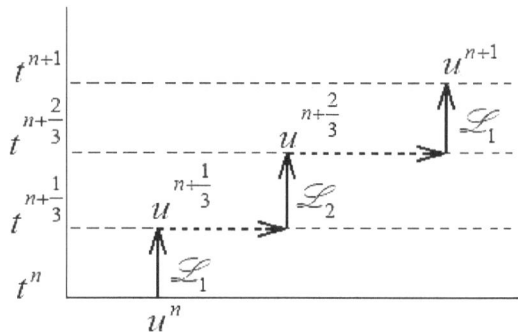

Figure 3:

7. CONCLUSION

In section 6 we combined the finite element method with the operator splitting for the convection-diffusion operator and for the breaking of the interval $\left(t^n, t^{n+1}\right)$, but before we partitioned the time interval $(0, T)$ such that

$$0 = t^0 < t^1 = \Delta t < \cdots < t^n = n\Delta t < t^{n+1} = (n+1)\Delta t < \cdots < t^N = N\Delta t = T, \quad \Delta t = \frac{T}{N}.$$

On the other hand, from the ellipticity of the coefficients and the Schwars's inequality we find the estimation

$$\sum_{n=0}^{N-1} \left\| u_h^{n+\theta} \right\|_h^2 \leq \text{constant}, \text{ for } \theta \in \left\{0, \frac{1}{3}, \frac{2}{3}, 1\right\}, \text{ and } \|\cdot\|_h \text{ is the norm on } W_{0h} - \text{the space which is the}$$

approximation of $H_0^1(\Omega)$.

Also, we obtain the convergence

$$u_h^{n+\theta} \xrightarrow[h \to 0]{} u\left(\cdot, t^{n+\theta}\right) \text{ strong in } L^2(\Omega), \quad \theta \in \left\{0, \frac{1}{3}, \frac{2}{3}, 1\right\}.$$

REFERENCES

[1]   Ainouz A., Two-scale homogenization of a Robin problem in perforated media, Applied Mathematical Sciences, Vol. I, No. 36, 1789-1802, 2007.

[2]   Dean, E. J., Glowinski, R., Operator-splitting methods for the simulation of Bingham visco-plastic flow, Chinese Anals of Mathematics, Vol. 23, No.2, Ser. B, 187-205, 2002.

[3]   Dumitrache, M., Gheldiu, C., Georgescu, R., Homogenization of nonstationary diffusion problem with boundary values in a perforated domain, Bul. Stiintific – UPIT, Seria Mat. si Inf., Nr. 18, 2012.

[4]   Glowinski, R., Numerical methods for nonlinear variational problems, Springer-Verlag, New-York, NY, 1984.

[5]   Banks, H. T., Bokil, V. A., Cioranescu, D., Gibson, N. L., Griso, G., Miara, B., Homogenization of periodically varying coefficients in electromagnetic materials, Journal of Scientific Computing, Vol. 28, No. 2/3, 191-221, 2006.

[6]   Temam, R., Navier-Stokes equations. Theory and numerical analysis, North-Holland, Amsterdam, 1979.

[7]   Conca, C. and Donato, P., Non-homogeneous Neuman problems in domains with small holes, Modelisation Mathematique et Analyse Numerique, 22(4), 561-608.

THE CHECKING OF THE SEMI-PRECAST R.C. FLOORS

Cristina Chilibaru–Opriţescu[1], Amalia Ţîrdea[2], Corneliu Bob[3]

[1, 2] The Research Institute for Construction Equipment and Technology - ICECON S.A. Bucureşti, Research Department of Timişoara, P. Râmneanţu str No.2, Romania, opritescucristina@yahoo.com, amalia_tirdea@yahoo.com

[3]University "Politehnica" of Timişoara, Department CCI, T. Lalescu str. No 2, Romania, cbob@mail.dnttm.ro

Abstract: *The paper deals with the checking of the semi–precast R.C. floors for the technological phases: transport and mounting. A dynamic analysis is presented instead of the manual calculation. The influence of dynamic effect (F=1.5G) is proposed and the most of stresses are smaller than characteristic strength and it presents a safety calculation.*

Keywords: *semi-precast R.C. floors, transport and mounting phases, dynamic analysis, manual calculation, concrete class, concrete strength*

1. INTRODUCTION

The Semi-precast slabs are part of the mixed floors partly monolithic, partly precast. These are composed of a lower layer made of reinforced concrete plates precast with thickness from 30 to 100mm and an upper layer of reinforced concrete monolith with thickness from 80 to 200mm or more. The precast plate contains the reinforcement from the lower part of the plate; the layer of monolithic reinforced concrete contains the upper reinforcement for take up the negative moments in the supporting areas. The precast plate is the formwork and it must be designed for take up the efforts during all phases. [1]

The connection between the prefabricated plate and the monolith reinforced concrete is done by both adherences of the two layers of concrete, as well by some special links of reinforcement.

Lattice truss are made like a plane truss or spatial (triangular) truss, made of reinforced concrete or thin profiles. These are dimensioned such that will ensure the efforts of transport, mounting and weight of monolithic concrete layer after casting. In the same time the stiffened truss links the two layers of monolithic and prefabricated concrete. The stiffened truss can be placed near the wire that constituting the reinforcement of prefabricated plates.

Checks in mould release phases, transport, and mounting on static schemes, depending on suspension system; the calculation loads for these checks are shown in [2], [3]:

- The dead load of partial prefabricated slab add uniformly distributed load for to defeat the adherence of formwork $(1 – 1.5kN/m^2)$ to demoulding;
- The dead load of partial prefabricated slab increased with dynamic coefficient of 1.5 in the transport and mounting phases.

The utilization of cracked semi-precast slab is not allowed for building structures

2. ANALYSIS OF SEMI-PRECAST R.C. FLOORS

The behavior of a partial prefabricated slab for the transportation and mounting is always taking into account by engineers. For usual design, there is used a static analysis for which the self weight of the slab is multiplied by a dynamic coefficient $\eta=1.5$.

For the checking of the state of stress in this design, a dynamic simulation was performed by using a computer program ANSYS LS-Dyna. [4]

The semi-precast slab has taken into account with the next characteristics: geometrical dimensions 4x2x0.1m; density 2400kg/m3; Poisson ration 0.2 and mechanical characteristics of material are presented in Table 1.

Table 1: The mechanical concrete characteristics

Concrete classes	Tensile concrete strength [MPa]			E [$\cdot 10^4$ MPa]	G [$\cdot 10^4$ MPa]
	f_{tm}	f_{tk}	f_{td}		
C8/10	1.41	0.92	0.6	2.1	0.84
C10/15	1.71	1.19	0.8	2.4	0.96
C15/20	1.98	1.43	0.95	2.7	1.08
C20/25	2.24	1.65	1.1	3	1.2
C25/30	2.71	1.86	1.25	3.25	1.3
C30/35	2.94	2.03	1.35	3.45	1.38

A static analysis of the semi-precast slab was performed on the scheme given in Fig.1a with the static loads F=G and F=1.5G, where G is the weight of the slab.

For comparison of the calculated data, a dynamic simulation was made by finite elements using ANSYS LS-Dyna, Fig.1b. This dynamic analysis was performed for the forces with the intensity of G, 1.25G, 1.5G and 2G and with proper strength ftk for each class.

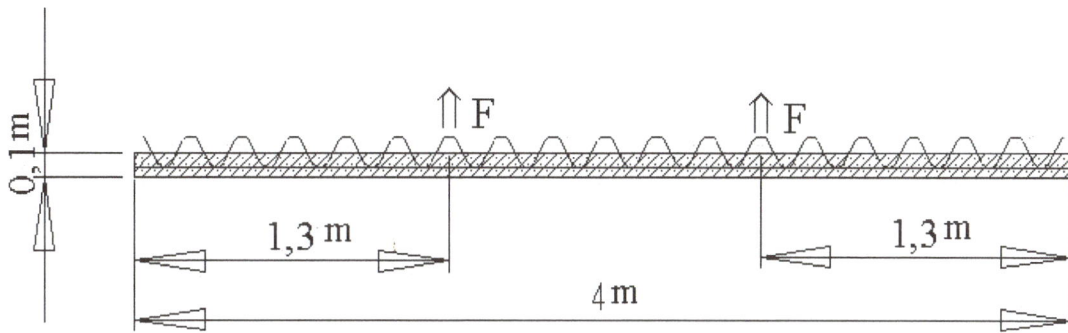

a

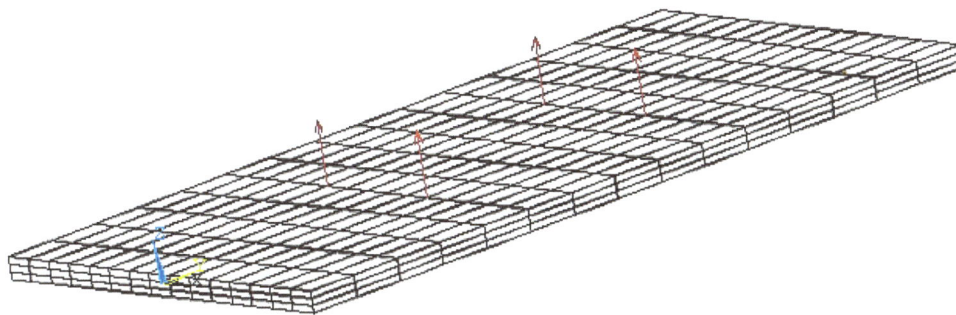

b

Figure 1: a-The scheme for mounting and transport of a semi-precast R.C. floor;
b-The modeling by finite elements of a semi-precast R.C. floor

Table 2: Value of stresses from static and dynamic calculation

Concrete classes	Stress [MPa]					
	static analysis for the force:		dynamic simulation for the force:			
	F=G	F=1.5G	F=G	F=1.25G	F=1.5G	F=2G
C8/10	1.27	2	0.79	0.94	1.06	1.29
C10/15			0.85	1.06	1.22	1.47
C15/20			0.85	1.07	1.28	1.55
C20/25			0.93	1.16	1.39	1.78
C25/30			0.95	1.19	1.43	1.89
C30/35			0.95	1.18	1.42	1.9

From the analysis presented in Figure.2 and Table 2 it can be pointed:
- The safety zone, where $f_t<f_{td}$, is valuable only for the action without dynamic effect F=G and for the concrete class greater than C10/15 as well as for F=1.25G and concrete class C25/30 and C30/35. For the case of static calculation the safety zone will include only the concrete class C30/35 and F=G.
- The zone where $f_{td}<f_t<f_{tm}$ was defined as un-safety due to the stress values witch are greater than designed tension stress of concrete. Most of calculated values are included in this zone, together with the characteristic strength of concrete.
- The cracked zone where $f_t>f_{tm}$ is proper only for static calculation and for concrete classes C8/10 and C10/15.

From the data presented, the dynamic simulation for the action of F=1.5G, the values of stresses are closer to designed stress, f_{td}: for higher concrete class C30/35 – 5%, and for C8/10 – 77%. Such approach is much favorable than manual calculation which is with no security values. Taking into account the purpose of such calculation with the effect of dynamic action for transportation and mounting, the value with F=1.5G is good enough most of stresses obtained with this dynamic coefficient are smaller than characteristic strength f_{tk} which represent quite a safety calculation. A smaller value of dynamic coefficient is not indicated for design and checking.

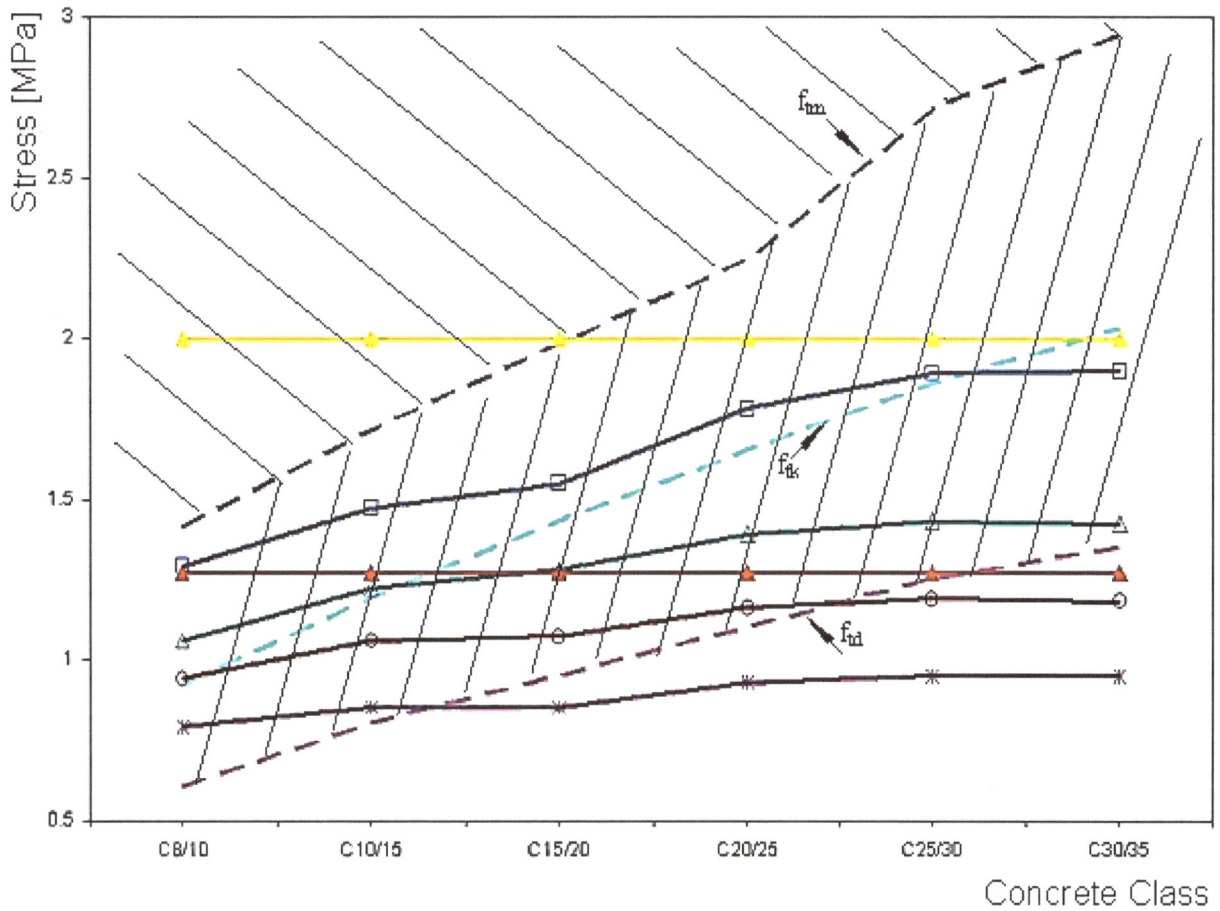

Figure 2: The stresses for concrete class from static and dynamic analysis

3. CONCLUSION

Some practical conclusion, which emerge from the presented analysis are:
- For the check in the transport and mounting phase a dynamic analysis there is necessary
- The manual calculation, with a dynamic coefficient of 1.5 is not satisfactory for concrete of inferior classes
- Dynamic analysis gives good results, even the data are situated in the un-safety zone, but for such technological, short-time, stressed there are not probability of having cracked zone. On the other hand the stresses are smaller than characteristic strength which means a safety calculation.

REFERENCES

[1] Ovidiu MÎRŞU, Corneliu BOB: *Reinforced Concrete Sructures*, vo.I (in Romanian) IPT Facultatea de construcţii, Timişoara, 1990.
[2] EN 1992–*Design of Concrete structures*.
[3] Daniel BOGDĂNESCU, Corneliu BOB, Liana BOB: *Some Considerations for the Assessment of Existing RC Framed Structure*, North Atlantic University Union-Recent Advances in Engineering, Paris, 2-4 Decembrie, 2012.
[4] Cristina OPRIŢESCU, Amalia ŢÎRDEA, Corneliu BOB, Mihai Ilie TOADER: *Numerical simulation of real contact between two bodies*, The 10th International Conference "Acustică. Vibraţii.", 29-30 septembrie 2011 Petroşani, CD

SOME PROBLEMS OF FEM MODELLING OF SANDWICH STRUCTURES

Ctirad Novotný[1], Karel Doubrava[2]
[1] Czech Technical University in Prague, Prague, CZECH REPUBLIC, ctirad.novotny@fs.cvut.cz
[2] Czech Technical University in Prague, Prague, CZECH REPUBLIC, karel.doubrava@fs.cvut.cz

Abstract: *Sandwiches are a form of composites providing high bending stiffness at a reasonable weight. This paper illustrates the possibility of calculating the stiffness of the sandwich structure (tram roof) with glued handles of roof equipment using FEM. Sandwiches can be modelled using different finite element types.*

The use of shell and solid elements is demonstrated on the example of three-point bending of sandwich beam. Another problem is the modelling of adhesive joints of individual sandwich parts or other connected structures. One possibility to model adhesive joints using FEM is the use of special cohesive elements. For a description of the mechanical behaviour of a relatively thin layer of adhesive is suitable definition traction – separation of interfaces. The above was used to calculate the stiffness of composite roof structure loaded through the handles of roof equipment.

Keywords: *sandwich, FE model, adhesive layer, cohesive element, sandwich tram roof*

1. INTRODUCTION

The use of composite materials in industry (especially in transport applications) has in recent years expanded. Composite structures include the sandwich structures. They are a form of composites characterized by a combination of different materials bonded together so that the resulting structures provide high bending stiffness at a reasonable weight.

Sandwiches are usually composed of three main parts: the two outer relatively thin and stiff skins and thicker softer inner core. The skins are connected to the core to allow transfer of loads between the parts of structure.

The unit is designed so that the core transmits primarily shear forces and the faces bending load. High flexural stiffness is result of the distance of stiff skins and low weight is achieved by using a lightweight core.

As a material of faces can be applied orthotropic or quasi-orthotropic laminates with carbon or glass fibres. In the case of massive structure there are used metal faces. The core is usually made of lightweight materials such as aluminium or polymer foam or honeycomb.

This paper illustrates the possibility of calculating the stiffness of the sandwich structure (tram roof) with glued handles of roof equipment using finite element method (FEM). In all cases there was used software Abaqus/ Standard v6.11 (Dassault Systèmes).

2. FEM MODEL OF SANDWICH STRUCTURE

The sandwich structures can be analyzed by finite element method (FEM). With regard to the geometry of the structure, the relative size of the individual components, the accuracy of calculating stress and strain and finally the available processing capacity sandwiches can be modelled in different way.

Thin-walled structures can be modelled using shell elements or a combination of shell and solid elements. For the model of thick-walled structures it is often necessary to use solid elements.

When using shell elements only the reference surface (location of nodes) is modelled. As a reference surface it is usually chosen midsurface of shell, in some cases it is preferable to model reference surface shifted to top or bottom surface. The thickness of the shell is given only as a parameter of element.

115

Using solid elements there is modelled whole geometry of sandwich layer or the entire structure. If it is not appropriate to model the entire structure using shell elements, the thick-walled core can be modelled using solid elements and skins using shell elements. Between the adjacent surfaces of the sandwich structure there is modelled rigid coupling. The most common type of coupling is a *tie*. Element nodes lying on the bound (slave) surface have their degrees of freedom tied to the control (master) surface.

2.1. Different approaches to modelling of sandwiches

As test sample for comparative analysis it was chosen the sandwich beam consisting of the top skin, the core, and the bottom skin.

The comparison was carried out against a three-point bending. The sandwich beam was placed on two supports (Figure 1).

The dimensions of the beam sample are: support distance L = 600 mm, width b = 90 mm, thickness of top skin t_1 = 3.4 mm, thickness of core c = 42.3 mm, thickness of bottom skin t_2 = 4,2 mm.

The skins are made from a laminate with glass fibre mat; the core is made of stiff PVC foam. For the purpose of FEM analysis it was assumed for each sandwich part homogeneous isotropic material with elastic properties.

The following material properties obtained from experiments have been used:
- top skin – tensile modulus E_1 = 22 139 MPa, Poisson ratio v_1 = 0.2
- core – compressive modulus E_c = 32,5 MPa, shear modulus G_c = 12,5 MPa
- bottom skin – tensile modulus E_2 = 19 097 MPa, Poisson ratio v_2 = 0.2

The loading force F is applied to the top skin in the middle of the beam and is evenly distributed throughout the width of the beam. Its size varied from 0 - 1 kN. This range was chosen with regard to the condition of elastic behaviour of materials. However, the analysis considered large strain, rotations and displacements.

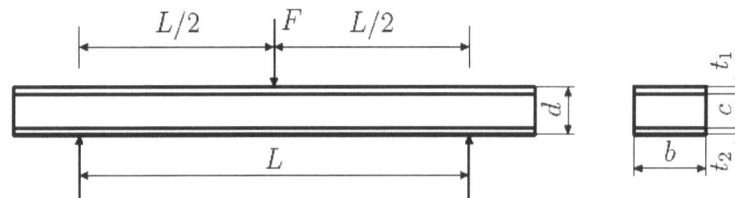

Figure 1: Geometry of beam specimen and loading scheme

The individual FEM modelling methods were assessed for compliance with the experiments on sandwich beams with the same geometry and material. The created FE models consist of the following elements from element library of Abaqus software:
- Shell elements:
  - S4R – 4-nodes linear general-purpose element with reduce integration
  - S8R – 8-nodes quadratic thick element with reduce integration
- Solid elements:
  - C3D8R – 8-nodes linear element with reduce integration
  - C3D20R – 20-nodes quadratic element with reduce integration

The shell elements have the possibility to define multiple layers with different material and thickness. If the core or the face was modelled using solid elements, three elements per thickness were used.

2.2. Comparison of FE sandwich models

There were compared deflections at the location of the loading force. Values of deflection under load of 1 kN are shown in Table 1. Furthermore, a comparison of bending strain courses along the beam thickness was made (Figure 2). The strain path was located on halfway between the left support and the location of force at a load of 1 kN. The models using layered shells (model A, B) are consistent with the assumptions of the linear bending strain course (shear strain is constant). The models with core consist of solid elements (model C, D, E, F) have a non-linear course of bending strain. The strain-inconsistency at the interface skin – core can be explained by influence of interpolation of strain values from the element integration points to the intersection of strain path and the element face.

Table 1: Comparison of models

Model	Elements for Skins	Core	Deflection [mm]
experiment	-	-	1.22 ± 0.16
model A	S4R (layered)		1.28
model B	S8R (layered)		1.29
model C	S4R	C3D8R	1.22
model D	S8R	C3D20R	1.23
model E	C3D8R	C3D8R	1.24
model F	C3D20R	C3D20R	1.23

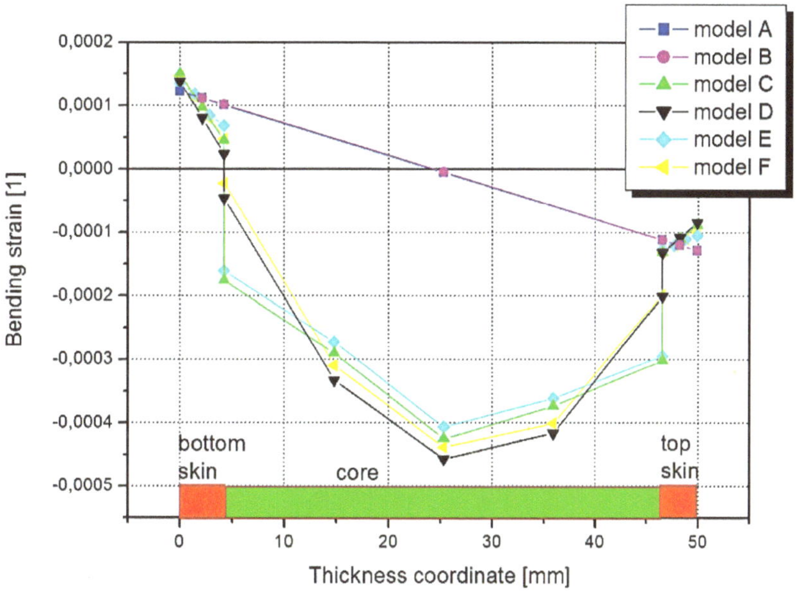

Figure 2: Course of bending strain

2.2. Recommendations for further modelling of sandwich structures

All the above models have good agreement in deflection and thus exhibit a comparable flexural stiffness. From the course of bending deformation follows that more realistic results provide models with core meshed with several solid elements per thickness. Using shell elements on the skins is appropriate for the size of the model and easy change of parameters (thickness, material composition). For further modelling of sandwich structures there was used approach combining shell elements on the skins and solid elements in the core.

117

3. ADHESIVE JOINTS IN SANDWICH STRUCTURES

Another problem is the modelling of joints of individual sandwich parts and other connected structures. One possibility to model adhesive joints using FEM is the use of special cohesive elements that are part of the element library of Abaqus software. For modelling of adhesive it should be only one element per thickness of the adhesive layer. The mechanical response of these elements is described by the dependence traction – separation of interfaces. This approach is suitable for modelling of relatively thin adhesive layer.

The use of cohesive elements is illustrated by the example of the adhesive joint of roof equipment handle and sandwich roof of the tram.

3.1. Cohesive elements – identification of material parameters

The cohesive elements allow modelling of initial loading, damage initiation and damage evolution leading to eventual fracture. The behaviour of the adhesive joint can only be described as linear elastic with stiffness, which is penalized as the material degrades under tensile and shear loads. The pressure load doesn't affect stiffness. In the case of a three-dimensional problem model considers three components of the interfaces separation (the normal component to the interface, two component parallel to the interface) and three corresponding components of stress in the material section. The cohesive elements operate with nominal stress and strain. The nominal stress is defined as the ratio of the force component and the initial cross-section at each element integration point. The nominal strain is defined as the ratio of separation of the bonded surfaces and the initial thickness of the adhesive at each element integration point. The elastic behaviour is expressed by the elastic constitutive matrix that relates the nominal stress to the nominal strain. The elastic behaviour can be written as

$$\begin{bmatrix} t_n \\ t_s \\ t_t \end{bmatrix} = \begin{bmatrix} K_{nn} & K_{ns} & K_{nt} \\ K_{ns} & K_{ss} & K_{st} \\ K_{nt} & K_{st} & K_{tt} \end{bmatrix} \begin{bmatrix} \varepsilon_n \\ \varepsilon_s \\ \varepsilon_t \end{bmatrix} \tag{1}$$

The vector of nominal stress consist of normal stress t_n and two shear stresses t_s, t_t. The vector of the nominal strain consists of normal strain ε_n and two shear strain ε_s, ε_t. In the event that the different modes of loading (normal and shear) don't mutually affect or this can be neglected off-diagonal elements of constitutive matrix are zero. It is possible to model adhesive layer between the parts of sandwich or sandwich skin and another structure. To identify the basic parameters of ductile adhesives for modelling using cohesive elements with response traction – separation there were carried out two basic tests. They are a tension and shear test [1]. The both types of test samples consist of two liners connected by investigated adhesive layer with a thickness of 3 mm. The interfaces in the case of both types of tests are the same - 25 mm^2. The scheme of tensile and shear tests are shown in Figure 3.

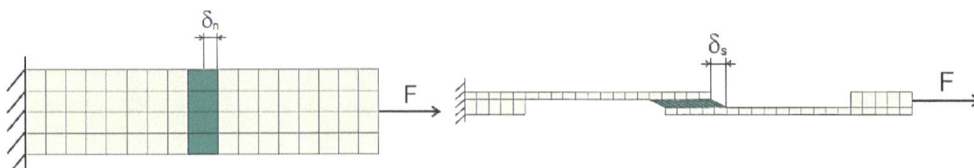

Figure 3: Tensile and shear test – measured separations of interfaces

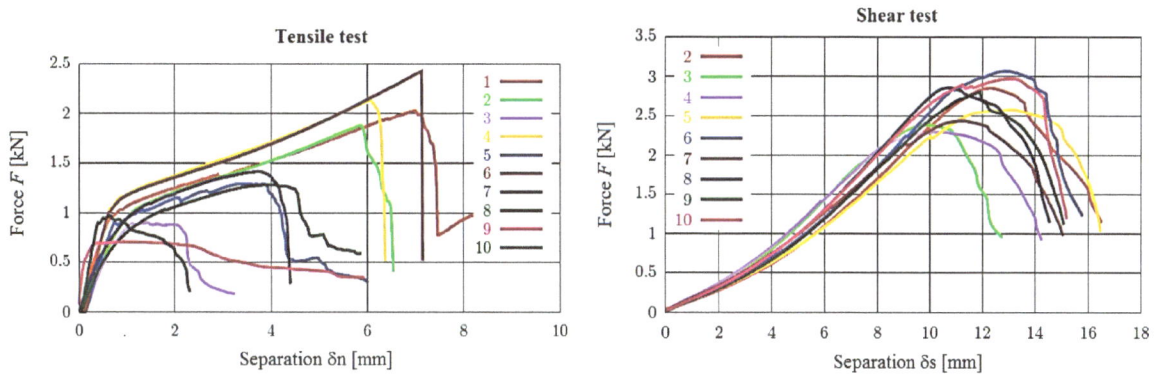

Figure 4: Tensile and shear test – measured separation of interfaces

The dependence force – separation of interfaces for tensile tests and shear tests (Figure 4) can be replaced by bilinear dependence. Force respectively separation values were converted to the corresponding nominal stress respectively nominal strain values. Due to the absence of combined tensile/ shear tests there wasn't taken into account interaction of normal and shear modes. Furthermore, it was considered that the characteristics of the first and second shear direction are the same. From an initial linear dependence in accordance with [1] there were determined diagonal elements of constitutive matrix in equation (1) $K_{nn} = 6.70$ MPa, $K_{ss} = K_{tt} = 0.77$ MPa.

3.2. Verification of traction – separation approach

Using cohesive elements was firstly verified by experiment on a smaller sample. With regard to the possibility of loading in uniaxial tensile testing machine there was roof handle glued to both top and bottom skin of sandwich sample.

The loading force was introduced to the sample through two massive steel arms bolted to the roof handle (Figure 5), so that the adhesive layers were forced to shear and tension/ compression. The sandwich skins are steel plates (thickness 0.5 mm), the core is made of PVC foam (thickness 40.5 mm). The adhesive between skin and core isn't modelled.

The adhesive layer between the skin and the sandwich roof handle has a thickness of 3 mm. The roof handle is made of steel profile.

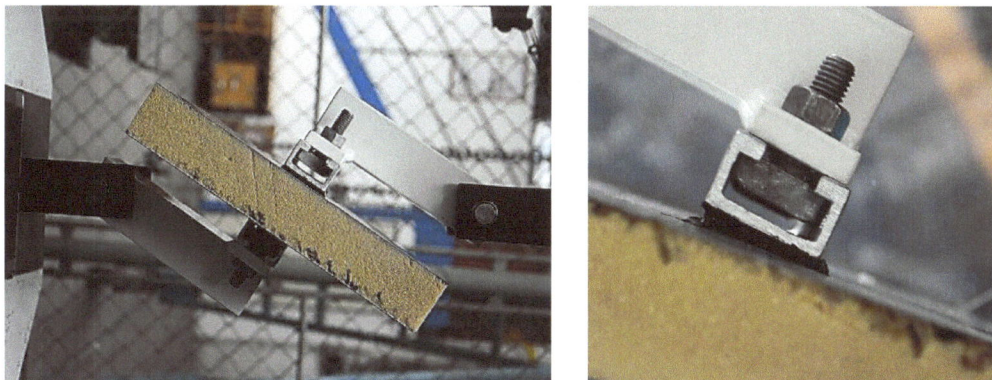

Figure 5: Experimental loading set and detail of loaded adhesive joint

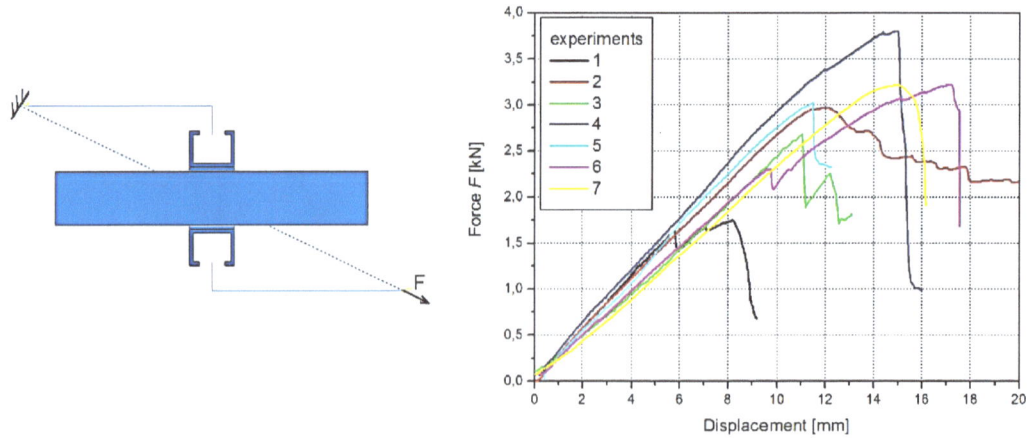

Figure 6: Scheme of loading and experimental dependence force – displacement of connecting eyes

The load force acts in the direction of the connecting line of connecting eyes of both arms.

In the FEM model the load force is introduced into the element type *connector* connecting eyes of both arms (Figure 6). The massive arms are idealized rigid beams. Experimentally determined dependence force – relative displacement of connecting eyes is shown in Figure 6.

The comparison of experiments and FEM simulation was carried out for load 1 kN. The value of mutual displacement of the two connecting eyes of both arms is 3.82 ± 0.42 mm (experiment) and 2.75 mm (FEM simulation). Greater stiffness of FE model can be caused by several factors. The simulation doesn't include compliance of other construction parts in loading chain of experiment. In the linear response of the adhesive model there aren't considered interaction of tensile and shear modes. Nevertheless, this model is suitable for predicting of the initial linear response of adhesive layer.

3.3. Loading of tram roof handle

Thus verified model of adhesive was used to simulate the loading of the roof handle on the larger segment of the real sandwich tram roof (Figure 7).

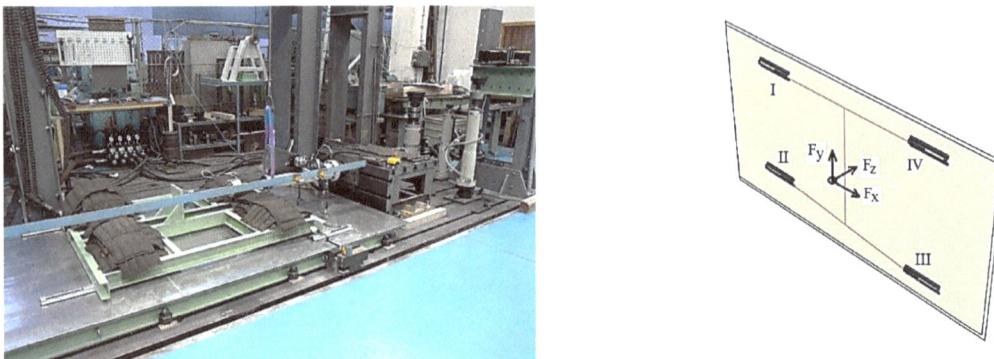

Figure 7: Experimental loading set and detail of loading of roof handles

The kinematic boundary conditions of the roof segment model correspond to placement during the experiment. On the edges of the segment there are prevented displacements.

The loading conditions are given in standard [3]. The force load is introduced to the estimated centre of gravity of proposed roof equipment boxes and through massive frame (in the model rigid beams) distributed into individual handles. Handles were first load vertically ($F_z = 4\,905$ N), which corresponds to the weight of the equipment box. Subsequently they were loaded against the expected drive direction of the tram with considered inertial force $F_x = 14\,715$ N. Then they were unloaded in the x direction and loaded in the side direction (y) with considered inertial force $F_y = 14\,715$ N. There was measured the vertical displacement u_z of mounting points on handles as difference of displacement under load F_x, F_z and F_z only (mode xz) respectively F_y, F_z and F_z only (mode yz). The comparison of calculated and measured values is given in Table 2. The differences in vertical displacement are up to 94% (0.37 mm) - handle IV, yz mode. This may be due to not fully rigid support of the real testing roof segment. This leads to possible further displacements of roof handles which are not affected by the FE model. But, the displacement trends of model are generally corresponding to the experiment. Furthermore, in adhesive layers stress values were controlled. In any adhesive layer there weren't exceeded limit stresses for the validity of proposed characteristic K_{nn}, K_{ss} and K_{tt}.

Table 2: Vertical displacements of mounting points (positive displacement is oriented upwards)

Point on handle	Mode xz: displacement [mm]		Mode yz: displacement [mm]	
	experiment	FEM simulation	experiment	FEM simulation
I	0.49	0.52	-0.73	-0.43
II	0.79	0.72	1.10	0.90
III	-0.45	-0.30	1.06	0.78
IV	-0.41	-0.53	-0.39	-0.76

4. CONCLUSION

This paper describes a string of tests and calculations leading to the verification of the computational approach using FEM for the determination of the stiffness of the flat sandwich structures and the adhesive joints in the global structure. This approach was used for stiffness determination of connection roof handle and sandwich tram roof. The sandwich core was modelled by solid elements, skins using shell elements. The adhesive layer in FE model consists of special cohesive elements with description of mechanical behaviour traction – separation of interfaces.

The uncoupled mode was considered, this means that there wasn't considered the influence of the tensile and the shear loading modes.

This approach has been verified and therefore seems appropriate for examined adhesive. The described calculation methodology can be used to predict the stiffness of sandwich structures.

ACKNOWLEDGMENT

This work had been supported by grant of Ministry of Industry and Trade of the Czech Republic FI-IM5/237.

REFERENCES

[1]   EN 1465:2009 Adhesives - Determination of tensile lap-shear strength of bonded assemblies

[2]   Renton J.; Vinson J. R., On the Behavior of Bonded Joints in Composite Material Structures. Engineering Fracture Mechanics, Vol.7, Issue 1, 1975, pp.41-52.

[3]   EN 12663-1:2010: Railway applications - Structural requirements of railway vehicle bodies - Part 1: Locomotives and passenger rolling stock (and alternative method for freight wagons)

A ROLLOVER TEST OF BUS BODY SECTIONS USING ANSYS

D.A. Micu, 1, M.D. Iozsa2, Gh. Frățilă3
[1] University Politehnica of Bucharest, ROMANIA, dan.alexandru.micu85@gmail.com
[2] University Politehnica of Bucharest, ROMANIA, daniel_iozsa@yahoo.com
[3] University Politehnica of Bucharest, ROMANIA, ghe_fratila@yahoo.com

Abstract: *The homologation of buses depends on its behavior during the rollover tests. This paper presents a computer simulation of rollover test on a body section of a structure constructed in order to be used for new bus models. The pendulum test on a bodywork section from Directive 2001/85/EC is used as a test method for rollover test.*

The geometrical model and the material proprieties are designed through the Ansys 13 software, using a real structure as a reference. The pendulum and the residual space are the other two designed parts for the simulation process, an interconnection are being used just between the structure part and the pendulum part.

The pendulum velocity is calculated using the equivalent mass of the body section, including the loads of elements which are not part of the structure.

The rollover behaviour results should be improved. The next steps will be to verify which properties should be changed to make this improvement.

Keywords: *bus, rollover, Ansys, tests, pendulum*

1. INTRODUCTION

The accident statistics show that bus rollover accidents occur with relative low frequency, taking into account all kinds of bus accidents. Nevertheless, the risk of mortality in the case of rollover is five times greater compared with any other possible bus accident typology [1].

The automotive regulations establish the rules that have to be respected by the bus body designers in order to get the homologations. The vehicle framework should absorb the impact caused by vehicle rollover so that the kinetic energy of the impact is converted into structure deformation work [2].

A very stiff superstructure is needed to keep an intact survival space after deformation, while the respect of the limits on the biomechanical parameters needs a structure with large energy absorption capability [3].

The computer simulation of a rollover test on a complete vehicle is an approval method that can be used. It allows manufacturers to test design and safety features virtually in the crash scenario until they obtain the safest and optimum design, thus saving time and money in developing expensive costly prototypes [4].

For the automotive manufactures the question is which is the best structural geometry or material that should be used in order to satisfy both stiffness and energy absorption capability.

In [2], the simulation tests of the strength of elementary tubular profiles when hitting non-deformable ground surface have revealed that from the point of view of resistance to impacts and energy absorption, the rectangular cross-sections of bus bodies may turn out to be dangerous for bus occupants, while the bus body frameworks with alternative symmetric circular shapes would provide better safety for bus passengers.

The formation process of plastic hinges along the pillars is fundamental for adequate energy absorption, was concluded in [3], in which an optimization program based on the exploration of the response surfaces with fractional factorial plans that allows obtaining a design solution satisfying both deformation and energy absorption requests.

One of conclusions from [1] was that from the point of view of safety it will be necessary to model the vehicle with great detail so simplified finite element models should only be used at preliminary design phases.

The pendulum method according to the EU Directive No. 2001/85/EC of 20 Novembre 2001, entitled "Strength of superstructure" is presented in the next sections. This is the method used to analyze, through computer simulation, the rollover behavior of a body section of a structure constructed in order to be used for new bus models. The geometrical model and the material proprieties are designed through the Ansys 13 software, using a real structure as a reference. The pendulum velocity is calculated using the equivalent mass of the body section, including the loads of elements which are not part of the structure.

2. MODEL DEVELOPMENT

2.1. Pendulum Method

The EU directive 2007/46, establishing a framework for the approval of motor vehicles and their trailers, and of systems, components and separate technical units intended for such vehicles, made two references for strength of superstructure to Directive 2001/85/EC and to UNECE Regulation Number 66. The pendulum method is not anymore a method specified in UNECE Regulation Number 66. It was presented in the old version of the Regulation, but it is still specified in Directive 2001/85/EC.

In pendulum test on a bodywork section, the energy to be transmitted to a particular bodywork section shall be the sum of the energies declared by the manufacturer to be allocated to each of the cross-sectional rings included in that particular bodywork section.

The most important test conditions are:
- a sufficient number of tests shall be carried out for the technical service conducting the test to be satisfied that no displaced part of the vehicle intrudes into the residual space;
- the side of the bodywork section to be impacted shall be at the discretion of the manufacturer;
- high speed photography, deformable templates or other suitable means shall be used;
- the bodywork section to be tested shall be firmly and securely attached to the mounting frame through the cross-bearers or parts which replace these in such a way that no significant energy is absorbed in the support frame and its attachments during the impact;
- the pendulum shall be released from such a height that it strikes the bodywork section at a speed of between 3 and 8 m/s.

The high of the centre of gravity is established by the designer and the fall of the centre of gravity (h) is determined by graphical methods. The total energy (E*) may be taken to be given by the formula:

$$E^* = 0,75 \cdot M \cdot g \cdot h \quad (\text{Nm}) \tag{1}$$

The unladen mass of the vehicle (M) is used to determine the distribution of mass on the tested structure section. The angle between the central longitudinal vertical plane of the bodywork section and the horizontal ground is determined as it is presented in Figure 1.

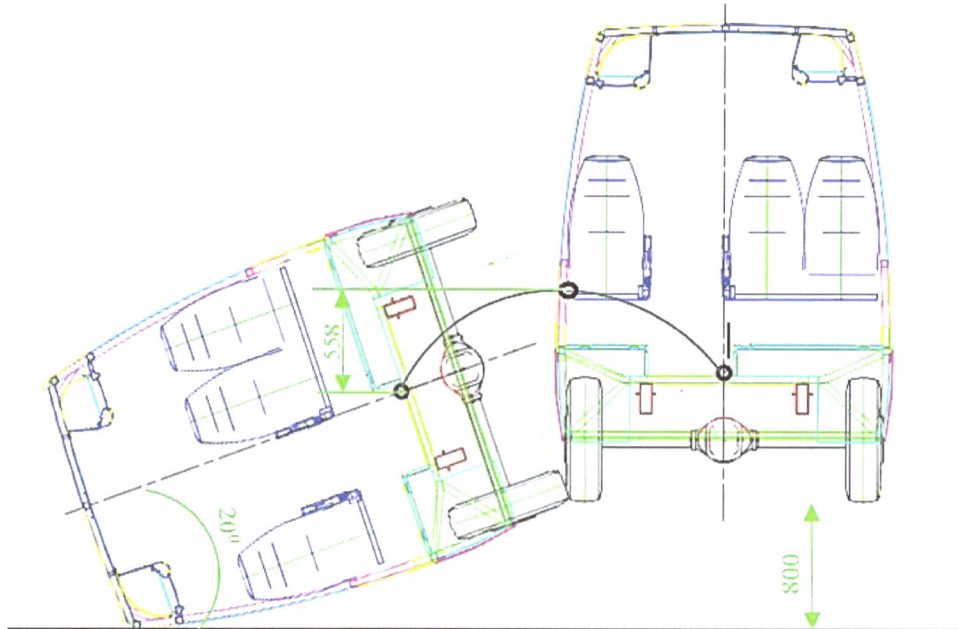

Figure 1: The determination of the fall of the centre of gravity

Using the energy conservation law, by equating the kinetic energy of the block, at the moment of the impact, with the total energy impact, the speed value that needs to be used for the block can be identified. The mass of the pendulum, the value of the initial speed of the pendulum and the absorbed energy of the structure are determined.

2.2. The model
The analyzis started from a real structure realized by a company in order to use it for new versions of buses. It is prepared to be tested it with „Roll-over test on a bodywork section" method. A frontal view of the bodywork is presented in Figure 2.

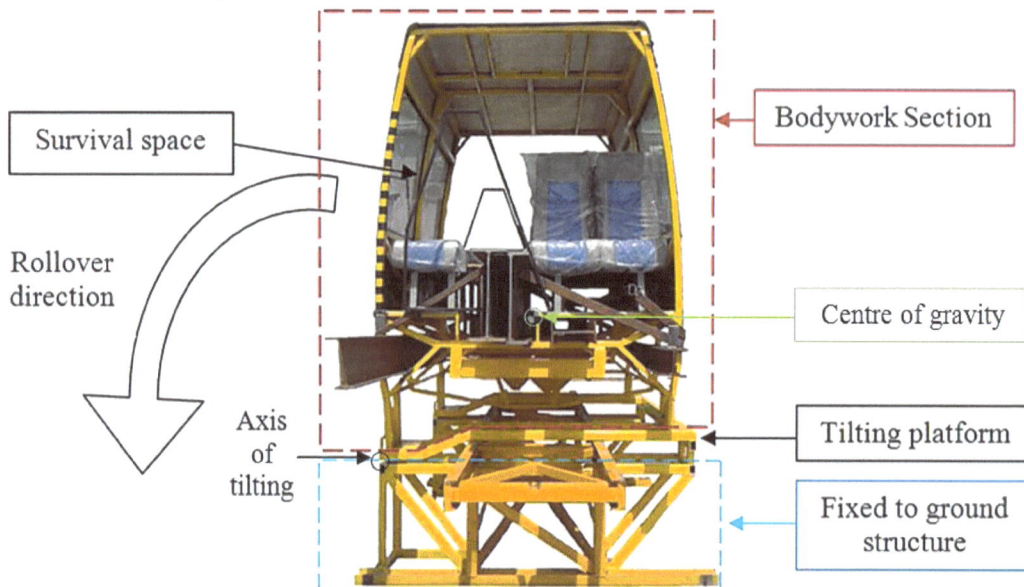

Figure 2: The real structure of the bodywork section model

The bodywork section represents a section of the unladen vehicle. Some additional beams are welded to the structure to simulate the mass of the completed section and to establish the right centre of gravity.

The survival space is defined by two shapes mounted in the front and in the back part of the structure. The bodywork section is installed on the upper part of a tilting platform, which is rotated around the axis of tilting. The fixed to ground structure has a height of 800 mm, like the one specified in Directive 2001/85/EC.

This article presents the verification of strength of the superstructure by calculation, considered it as a preliminary design phase. In order to analyze the structure behaviour during it is subjected to the rollover test, ANSYS was used - computer software specialized in the analysis of structures using the finite element method.

Specific stages are required to make an evaluation: designing the geometrical model, meshing, processing the calculations and results interpretation. The segment presented in Figure 2 was designed. The real model as well as the finite element model contains the space of survival. In this way it will be easier to observe whether the structure enters this space during the rollover process.

The structure is made of rolled bars of a rectangular, square or L-shaped profile. The entire geometrical model was composed of surfaces. The structure is constrained in the bottom part through some fixed support applied on faces of longitudinal and transversal elements. Figure 3 shows the geometrical model developed using ANSYS software.

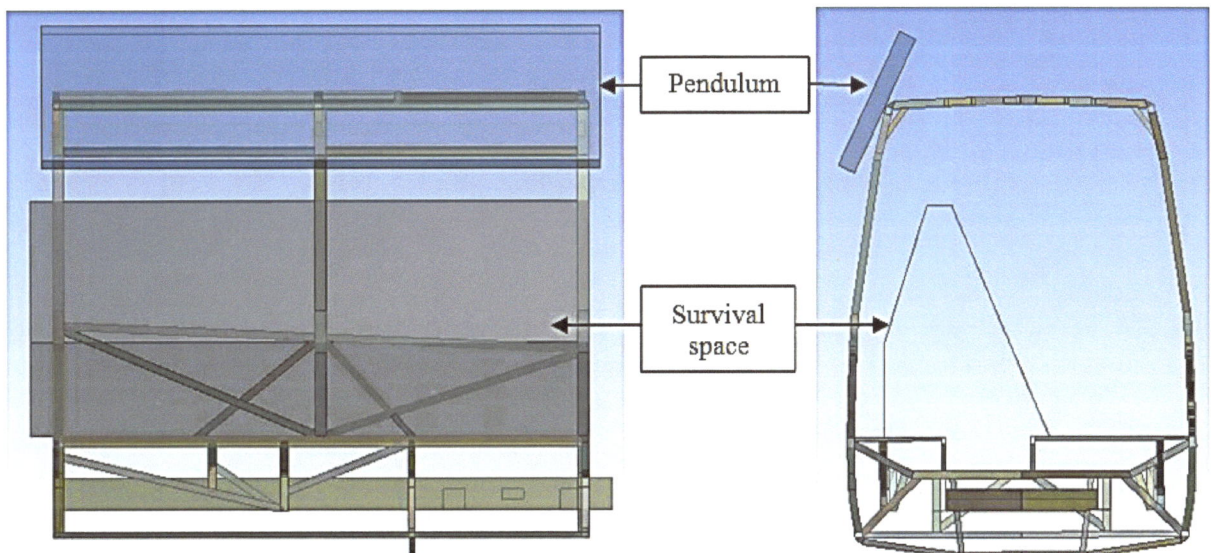

Figure 3: The finite elements section model

2.3. Results

The rollover behaviour using the energy resulted from Directive 2001/85 was studied in the first analyse. The computation has stopped before the pendulum's velocity get to zero. Even so, the structure has intruded into the survival space, as it can be noticed in Figure 4.

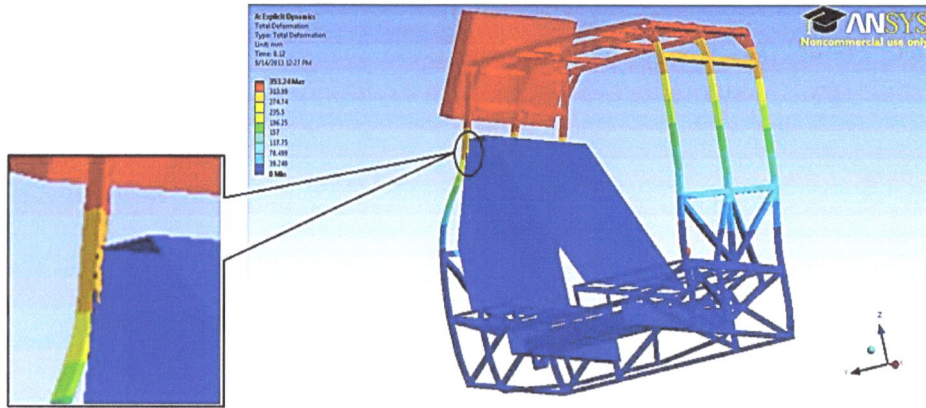

a

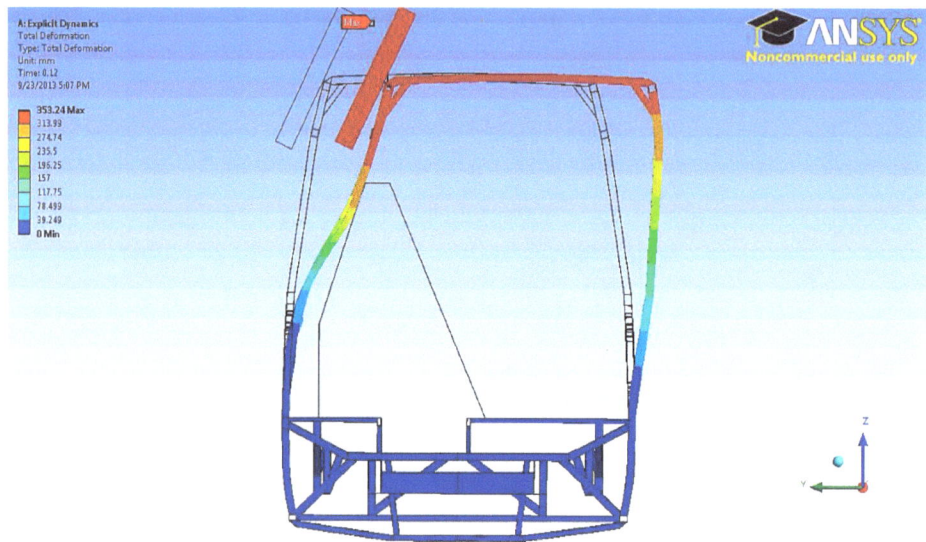

b

Figure 4: The deformations of the finite elements section model

The velocity values of the structure and of the pendulum are presented in Figure 5.a). The structure presents some bigger values due to the vibrations, but the pendulum's velocity decreased. The stress results are smaller than the material limit and it can be seen in Figure 5.b).

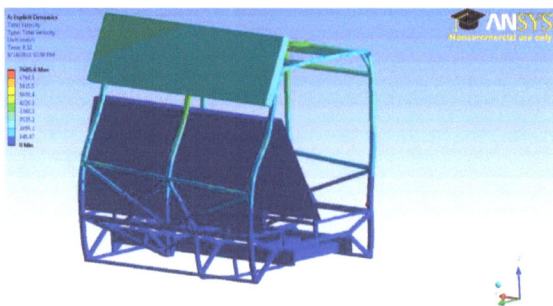

a

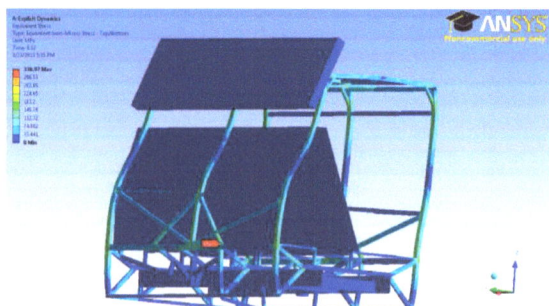

b

Figure 5: The velocities and the stress of the finite elements section model

What's happening when a part of the structure elements are missing was also studied.

Four from the six corner elements used to increase the structure strength were deleted, and the test was repeated. It was noticed that the cross-sectional ring that still have the corner element was deformed more than the others. This difference is presented in Figure 6, where it can be noticed that the maximum displacement is present on a bigger surface of the pole from the cross-sectional ring with the corner element.

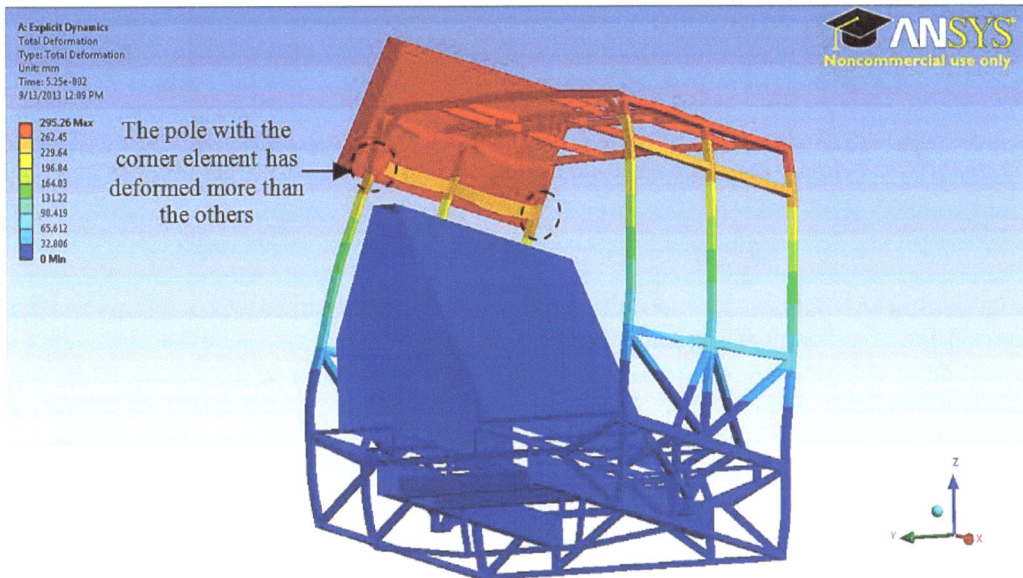

Figure 6: The influence of using an element in one corner of the cross-sectional rings

Another study is referring at a bigger mass of the structure to be considered during the rollover test. This is the mass used in the case the vehicle is fitted with occupant restraints, when Regulation 66 mentions the using of total effective vehicle mass. Using the equivalent energy, the resulted velocity for pendulum was almost doubled.

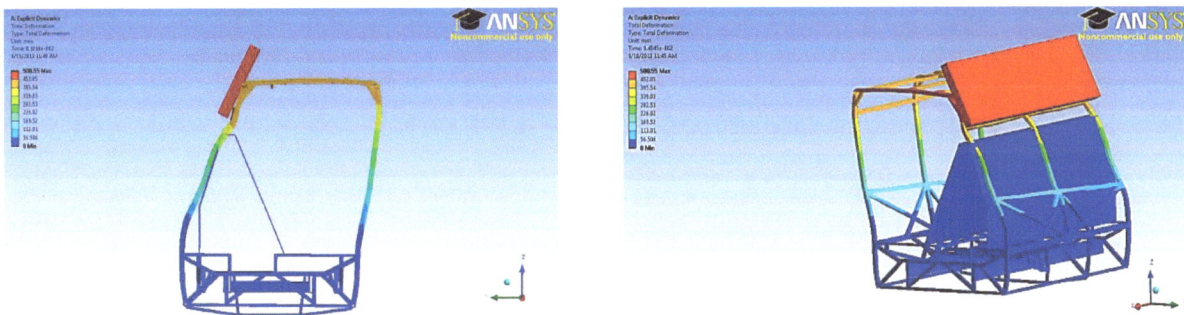

Figure 7: Deformations of the structure when it was considered a bigger mass

In this case the velocity of the pendulum has got to zero as it can be noticed in Figure 8.a. This process happened very fast, the pendulum being stopped in the lost few steps of processing, decreasing from almost half of the maximum speed of the pendulum. The stress values were also under the material limits (Figure 8.b), except one point generated through the computer modelling.

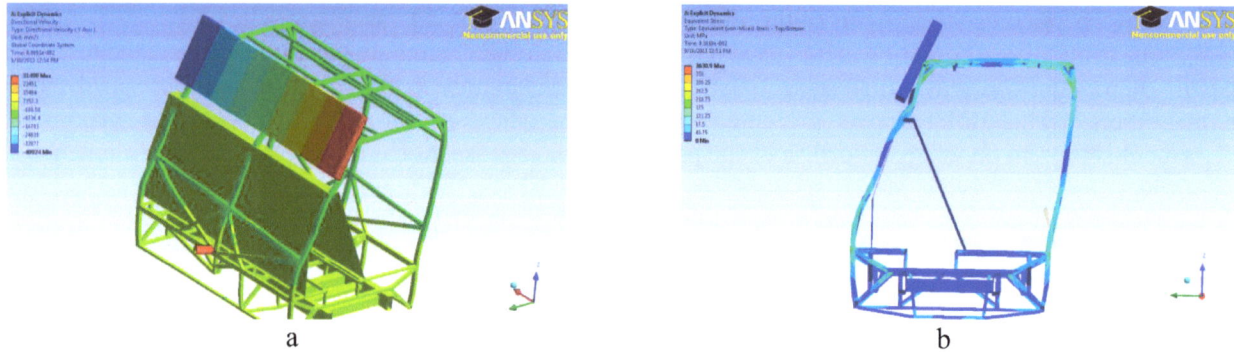

Figure 8: Velocity (a) and stress (b) resulted when it was considered a bigger mass

2.4. Discussion

The model studied in this article doesn't comply with all the specifications and requirements of Directive 2001/85/EC. In all cases the structure has intruded into the survival space, but the energy considered for these computations corresponds to the real model mass of the structure with the additional mass added.

This additional mass is bigger than the one specified in the Directive.

The corner elements used for increasing the strength of the structure can lead to a bigger deformation of the pole of the cross-sectional ring where it is mounted.

The velocity of pendulum decreases very fast in the last deformation steps.

3. CONCLUSION

The verification of strength of a real structure by calculation using the pendulum method presented in Directive 2001/85 EC was studied in this paper. The model was designed and tested, using Ansys 13.0, for a real structure realized by a company in order to use it for new versions of buses.

The model studied in this article doesn't comply with all the specifications and requirements of Directive 2001/85/EC. In all cases the structure has intruded into the survival space.

The corner elements used for increasing the strength of the structure can lead to a bigger deformation of the pole of the cross-sectional ring where it is mounted. The velocity of pendulum decreases very fast in the last deformation steps.

The next steps are to test some modified geometry parts and/or material proprieties, in order to find a better solution.

4. ACKNOWLEDGMENTS

Dan Alexandru Micu gratefully acknowledges the PhD funding from the Sectoral Operational Programme Human Resources Development 2007 - 2013 of the Romanian Ministry of Labour, Family and Social Protection through the Financial Agreement POSDRU/107/1.5/S/76909.

REFERENCES

[1] Valladares D., Miralbes R. ,Castejon L., Development of a Numerical Technique for Bus Rollover Test Simulation by the F.E.M. , Proceedings of the World Congress on Engineering 2010, London, U.K., June 30 - July 2, 2010, Vol II;

[2] MARIAŃSKI M., SZOSLAND A., Research on the Strength of Standard Bus Bodies at Rollover on the Side, Technical University of Lodz, 2011;

[3] Belingardi G., Gastaldin D., Martella P., Peroni L., Multibody Analysis Of M3 Bus Rollover: Structural Behaviour And Passenger Injury Risk, Dipartimento di Meccanica, Politecnico di Torino, 2002;

[4] Micu D.A., Iozsa M.D., Analysis of the Rollover Behaviour of the Bus Bodies, Analele Universității „Eftimie Murgu", Reşiţa, 2011;

[5] Directive 2001/85/EC;

[6] Regulation 66-UNECE.

EFFECTS OF SIMULATED WOOD DRYING SCHEDULES ON DRYING TIME AND ENERGY CONSUMPTION AT AN EXPERIMENTAL KILN

Daniela Şova[1], Bogdan Bedelean[2], Monica A. P. Purcaru[3]

[1] Transilvania University, Faculty of Mechanical Engineering, Braşov, ROMANIA, sova.d@unitbv.ro
[2] Transilvania University, Faculty of Wood Engineering, Braşov, ROMANIA, bedelean@unitbv.ro
[3] Transilvania University, Faculty of Mathematics and Computer Science, Braşov, ROMANIA, mpurcaru@unitbv.ro

Abstract: Wood conventional timber drying is the process in which the heat transfer from the warm air to wood and the mass transfer from wood to air are encountered. It is difficult to obtain exact solutions of the differential heat and mass transfer equations applied to the drying process, due to many process variables, therefore numerical modeling gives important information about the process.

We used in the paper a simulation model, developed for a single board drying, in order to obtain how the drying time and energy consumption are influenced by 64 different drying schedules as combinations of three factors, air temperature (50-80°C), velocity (0.5-0.8m/s) and relative humidity (20-35%), each at four levels.

The relationship between the factor settings and the responses (drying time and energy consumption, respectively) was acquired by means of the multiple quadratic regression models.

It was found that the temperature had the main effect on both, drying time and energy consumption. Instead, the velocity had very little effect on these ones.

Keywords: laboratory drying kiln, spruce, drying simulation, multiple quadratic regression analysis

1. INTRODUCTION

Wood conventional timber drying is the process in which the heat transfer from the warm air to wood and the mass transfer from wood to air are encountered. This process is one of the most important because it enhances wood mechanical and technological properties and ensures the protection of the wood against insect and fungal attack. The drying process aims at a compromise among high quality, short residence time and low energy usage. The quality and the residence time have been usually prioritized before the energy usage [1]. But the wood convective drying efficiency is estimated in all terms: drying time, energy consumption, drying cost and product quality and it depends significantly on the air flow related parameters, like velocity, temperature and relative humidity.

Therefore, one of the main challenges in wood drying is choosing an optimal drying schedule. The optimal drying schedule means the values of temperature, relative humidity and air velocity for which the best quality, shortest drying time and lowest energy consumption are achieved. A numerical simulation can be beneficial in choosing the optimal drying schedule by predicting drying variables, such as final moisture content, very quickly and accurately, without running a kiln. Given sets of predicted output drying variables for different input drying parameters makes it easier to choose the most optimal drying schedule. [2]

There are several reports on different simulation models for conventional timber drying; all based on coupled heat and mass transfer equations [3], [4], [5].

A very practical simulation model for a single board conventional drying was developed by Salin [6] which resulted in a computer program. It can be used for drying schedule optimization and process improvement. The model has been verified against many full scale tests.

130

New or modified drying methods and dryers were also proposed for drying effectiveness improvement in aspects regarding energy saving, drying time shortening and quality of the dried products [7], [8]. Most of the simulation models are validated by experiments on full scale driers. But, some of the experiments are carried out on pilot-scale kilns or different laboratory driers, such as tunnel laboratory driers [9], [10], custom-made laboratory kiln [11], experimental electrical oven [8] and climate chamber [12].o

The statistical techniques are appropriate methods for the evaluation of different properties on the drying outcomes.

On one hand, the multivariate data (principal component regression) analysis is applied to develop a regression model for investigating the wood properties that influence the final moisture content and the measure of this influence [13].

On the other hand, wood characteristics and drying schedule parameters are used as independent variables and a number of drying results (such as defects) as dependent variables in different multivariate regression analyses [8].

The objective of this research was to investigate how temperature, relative humidity and velocity of air in the drying process of spruce, simulated in an experimental kiln, affect the drying time and energy consumption, two out of the three major factors through which the drying efficiency is measured.

2. MATERIALS AND METHODS

2.1. Drying simulation

A wood drying simulation model is based on Fourier's heat transfer equation (1) and Fick's mass transfer equation (2):

$$\frac{\partial T}{\partial t} = \nabla\left(\alpha \nabla T\right) \tag{1}$$

$$\frac{\partial u}{\partial t} = \nabla\left(D \nabla u\right) \tag{2}$$

where: T is temperature, t is time, α is thermal diffusivity, u is wood moisture content, D is diffusion coefficient, ∇ is the gradient operator.

It is difficult to obtain exact solutions of the differential heat and mass transfer equations applied to the wood drying process, due to many process variables (the thermal diffusivity and diffusion coefficient depend on wood species, three principal directions of heat and mass transfer, moisture content, temperature, external boundary conditions etc.), therefore numerical modeling gives important information about the process.

Nowadays models are mostly based on balance equations for energy and moisture [14] that lead to the analogy between heat and mass transfer:

$$\frac{h}{h_m} = \rho c_p \tag{3}$$

where: h and h_m are transfer coefficients for heat and mass, respectively.

But, experimental results indicated a deviation from the analogy; the real mass transfer coefficient is much lower than expected from the analogy [14]. Therefore, a correction factor is included in the drying simulation model, dependent on surface moisture content and temperature [6].

The wood drying simulation model (TORKSIM) we used in the paper [15] has been tested against 28 full scale measurements and the measured and simulated final moisture content resulted in a 1.4% standard deviation [6].

The TORKSIM computer program is based on information regarding wood properties, drying schedule and kiln model and the results consist in drying time, energy consumption and drying costs calculation and quality aspects.

Our simulation input data refer to:

- wood species: spruce (*Picea Abies*)
- board thickness: 20 mm
- wood initial moisture content: 30% (falling rate drying period)
- wood target moisture content: 10%
- drying schedule: the time based drying schedules used in the drying simulation were created on fixed parameters: velocity, dry-bulb temperature and relative humidity of air [16]. The values and the symbol assignments are indicated in Table 1.

Table 1: Drying schedule parameters

Velocity (m/s)	Dry-bulb temperature (°C)	Relative humidity (%)
V_1: 0.5	T_1: 50	F_1: 20
V_2: 0.6	T_2: 60	F_2: 25
V_3: 0.7	T_3: 70	F_3: 30
V_4: 0.8	T_4: 80	F_4: 35

We designed 64 drying schedules by combining the values of the three parameters: velocity, dry-bulb temperature and relative humidity.

- kiln model: the simulation was carried out for a single board drying in a controlled laboratory tunnel, which is the scale model of the full-size industrial drying kiln equipment. The climatic conditions remained constant throughout the drying simulation.

2.2. Statistical analysis

A preliminary evaluation of the simulation results (drying time and energy consumption) was done with univariate analysis. The temperature, relative humidity and velocity, variable factors of the drying schedules, were plotted against drying time and energy consumption, the responses.

The relationships between factor settings and responses were evaluated with the multiple quadratic regression models. All 64 drying schedules were used for drying time and energy consumption evaluation.

The responses were approximated by a quadratic polynomial model (response surface modeling design).

The regression analysis was applied to estimate the regression coefficients in a model that relates the factors (x_1, x_2, x_3), their interaction (x_1x_2, x_1x_3 and x_2x_3) and quadratic terms (x_1^2, x_2^2 and x_3^2) to a response y.

The models for the two responses were calculated individually by using the software MODDE 9.1.1 [17] for construction and analysis of experimental designs and response surface plotting.

The software was also used for statistical evaluation of the models. Quadratic and interaction terms that were not significantly different from zero, with 95% confidence, were excluded from the models.

3. RESULTS AND DISCUSSION

3.1. Univariate analysis

The univariate analysis (Figure 1) shows that both, drying time and energy consumption decrease with temperature increase. The variation of velocity and relative humidity had a minor influence on drying time and energy consumption. A slight decrease of energy consumption and increase of drying time with relative humidity increase can be observed.

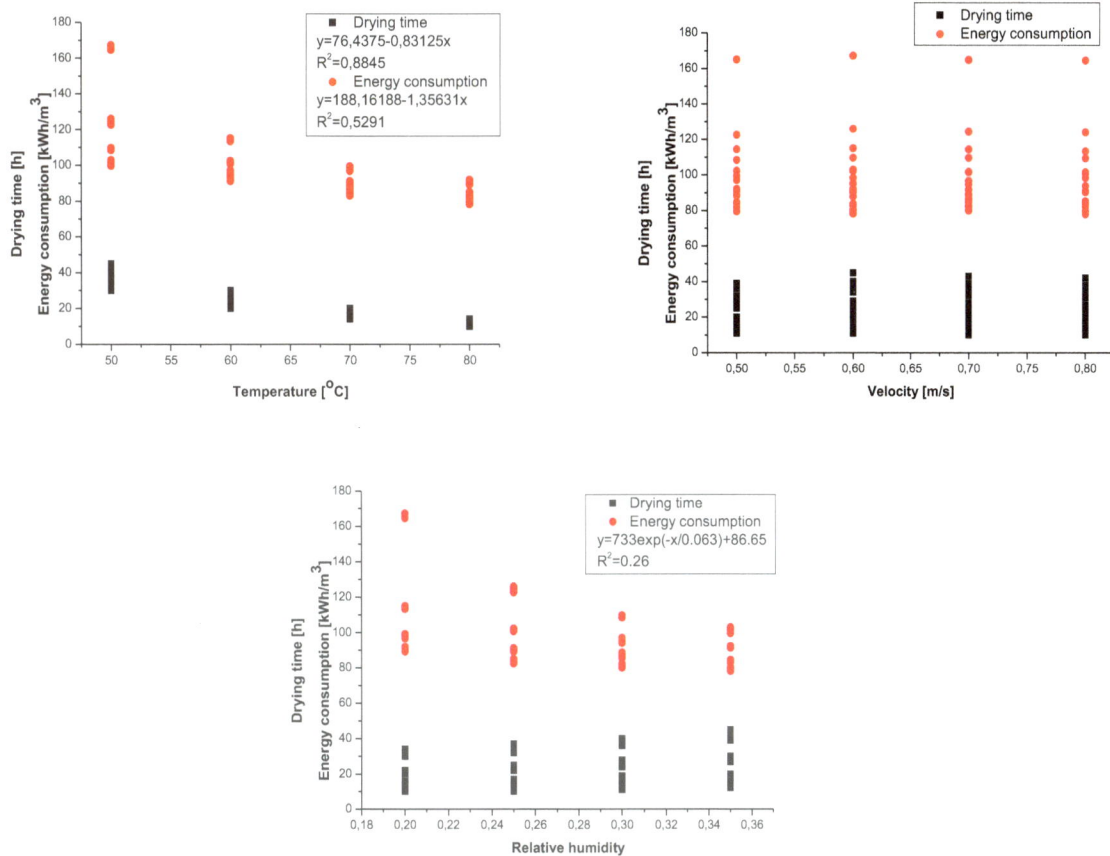

Figure 1: Influence of temperature, velocity and relative humidity on drying time and energy consumption

A correlation between drying time and energy consumption was identified, i.e. the energy consumption increases very much at long drying times, where the points dispersion is notable (Figure 2). For low drying times, the energy consumption is nearly the same.

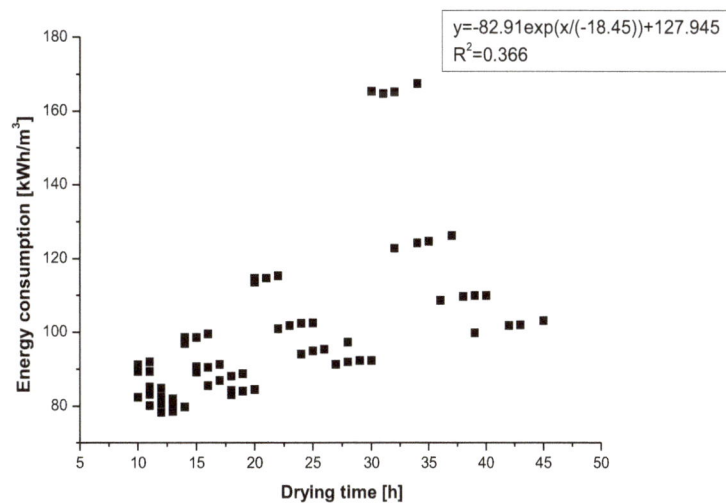

Figure 2: Correlation between drying time and energy consumption

3.2. Multivariate analysis of drying time

Non-significant model terms were deleted and the regression coefficients showed that the temperature and the relative humidity were the two dominating factors. The drying time decreased with increasing temperature and decreasing relative humidity (Figure 3).

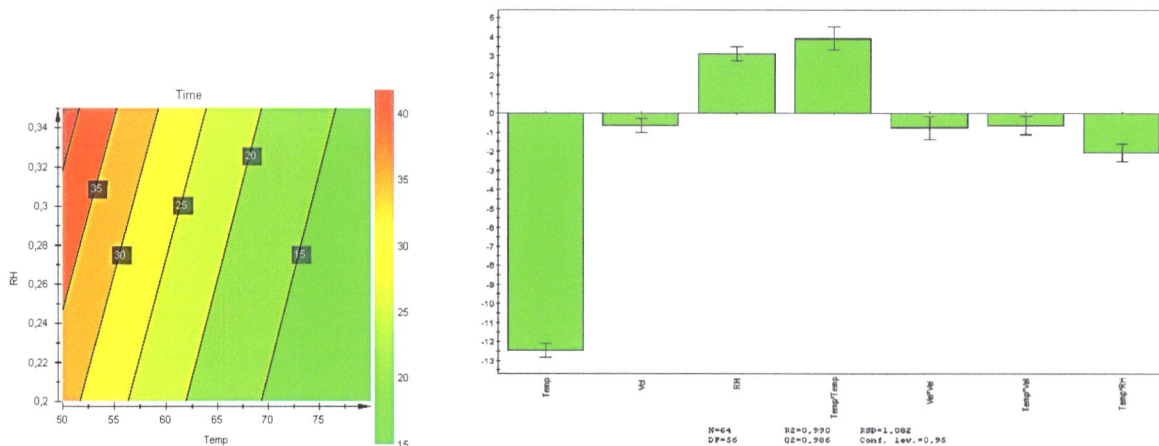

Figure 3: Response surface plot for drying time (boxed figures, hours) at different temperature and relative humidity values at 0.65m/s velocity (left) and regression coefficients with 95% confidence interval (right)

The influence of velocity and factors interaction were significant but of minor importance. The model accuracy was appreciated by the high R^2 and Q^2 values, 0.99 and 0.985, respectively (Table 2).

Table 2: Regression modeling statistics

Response	Constant	Temp	Vel	RH	Temp^2	Vel^2	RH^2	Temp×Vel	Temp×RH	R^2	Q^2
Drying time	20.49	-12.47	-0.65	3.11	3.94	-0.77	ns	-0.64	-2.07	0.99	0.985
Energy consumption	90.54	-20.34	ns	-13.63	10.18	ns	7.27	ns	12.14	0.943	0.903

ns-not significant

3.3. Multivariate analysis of energy consumption

The multiple regression analysis of energy consumption resulted in a similar model to that of drying time, with a good fit according to the regression coefficients (Figure 4). Thus, the energy consumption decreased with increasing temperature and relative humidity. Again, the temperature and relative humidity had important influence. Of importance were also the quadratic terms and the factors interaction.
Both models were validated by high R^2 values and they had good predictive capabilities according to cross validation (high Q^2 values), (Table 2). Most of the responses could be explained, but an unexpected result was the minor influence of velocity. The influence of the velocity was significant only for the drying time, even if its contribution was unimportant (Figs. 3 and 4). A reason for the result could be the low velocity values which were adopted according to the similarity conditions applied to the laboratory kiln. The test section of the

laboratory kiln is larger than the air flow channel from the industrial kiln that was simulated in the wind tunnel.

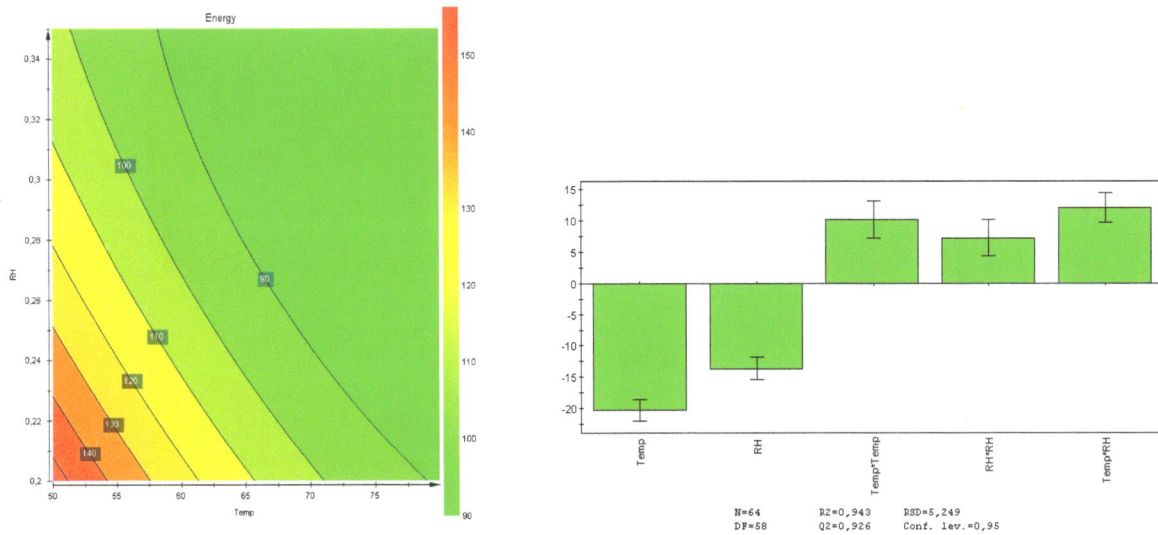

Figure 4: Response surface plot for energy consumption (boxed figures, kW/m^3) at different temperature and relative humidity values at 0.65m/s velocity (left) and regression coefficients with 95% confidence interval (right)

The results shown in Figures 1, 3 and 4 raised an important question: should the drying process be run at high or low relative humidity? Univariate and multivariate analyses show, especially at low temperature values, a different behaviour of the drying time and the energy consumption when the relative humidity increases (the regression coefficients of the relative humidity factor in the two models are positive and negative, respectively). It can be observed that at higher temperatures than 65°C, the drying time and energy consumption are low, regardless the relative humidity values. Since, both drying time and energy consumption decrease with temperature increase, the drying should be run at high temperature and lower relative humidity, considering also the falling rate drying period. In this manner, the wood drying quality would not be negatively influenced.

4. CONCLUSIONS

In this study, within the range of variables we have used, the following conclusions can be drawn:
1. The drying time (10-50 h) and energy consumption (80-170 kWh/m^3) were obtained by wood drying simulation with 64 different drying schedules. The results accuracy is high since the model was validated by many full scale tests.
2. The drying schedule parameters are correlated and their interaction influence drying time and energy consumption.
3. The high temperature is the most important variable to decrease drying time and energy consumption.
4. At high temperature (above 65°C), the relative humidity can take any value, without increasing drying time and energy consumption. But, high temperature and low relative humidity can assure during the falling rate drying period a good wood drying quality.
5. The low velocity values have minor importance on drying time and energy consumption.
6. A correlation was found between drying time and energy consumption.
7. The precision of the quadratic regression models (response surface modeling design) was proved by high R^2 and Q^2 values.

135

The study presented the importance of the main parameters of the drying agent in a limited range. It doesn't cover all possible variations in the wood sample and the process.

In future research, the results obtained on the experimental drying kiln should be compared with those obtained on the actual kiln.

The study can be extended furthermore with a multivariate analysis of the stress development inside the wood board during drying. In this way all aspects of drying efficiency would be covered.

ACKNOWLEDGEMENT

We acknowledge the National University Research Council for its support through the research project PNII-PCE, ID_851: The application of the irreversible processes thermodynamics method to the optimization of the capillary-porous materials drying process.

REFERENCES

[1]   Anderson, J.-O., Improving energy use in sawmills: from drying kilns to national impact, Licentiate thesis, Luleå University of Technology, Sweden, 2012.

[2]   Berberović, A., Numerical Simulation of Wood Drying, Thesis of Master of Science, Oregon State University, 2007.

[3]   Dedic, A. Dj. et al., A three dimensional model for heat and mass transfer in convective wood drying, Drying Technology 1: 1–15, 2003.

[4]   Trcala, M., A 3D transient nonlinear modelling of coupled heat, mass and deformation fields in anisotropic material, International Journal of Heat and Mass Transfer 55: 4588–4596, 2012.

[5]   Truscott, S.L., Turner, I.W., A heterogeneous three-dimensional computational model for wood drying, Applied Mathematical Modelling 29: 381–410, 2005.

[6]   Salin, J.-G., Simulation models; from a scientific challenge to a kiln operator tool, 6th International IUFRO Wood Drying Conference, Stellenbosch, South Africa, 177-185, 1999.

[7]   Kowalski, S.J., Pawłowski, A., Energy consumption and quality aspect by intermittent drying, Chemical Engineering and Processing 50: 384–390, 2011.

[8]   Vansteenkiste, D. et al., High temperature drying of fresh sawn poplar wood in an experimental convective dryer, Holz als Roh- und Wekstoff 55:307-314, 1997.

[9]   Plumb, O. A. et al., Experimental measurements of heat and mass transfer during convective drying of southern pine, Wood. Sci. Technol. 18: 187-204, 1984.

[10]   Straže, A. et al., Impact of various conventional drying conditions on drying rate and on moisture content gradient during early stage of beechwood drying, 'The Future of Quality Control for Wood & Wood Products', The Final Conference of COST Action E53, Edinburgh, 2010.

[11]   Neville, C.J., Vermaas, H.F., Laboratory kiln for the development of low temperature wood drying schedules, Holz als Roh- und Wekstoff 46:269-273, 1988.

[12]   Gatica, Y.A. et al., Modeling conventional one-dimensional drying of Radiata pine based on the effective diffusion coefficient, Latin American Applied Research 41: 183-189, 2011.

[13]   Watanabe, K., Kobayashi, I., Kuroda, N., Investigation of wood properties that influence the final moisture content of air-dried sugi (Cryptomeria japonica) using principal component regression analysis, J Wood Sci. 58:487–492, 2012.

[14]   Salin, J. G., Problems and solutions in wood drying modeling: history and future, Wood Material Science and Engineering 5: 123-134, 2010.

[15]   TRATEK, TORKSIM v.5.0., User manual and instructions for evaluation of calculated results, SP Technical Research Institute of Sweden, 2008.

[16]   Câmpean, M., Marinescu, I., Thermal treatments of wood: a guide for practical works (in Romanian, Tratamente termice ale lemnului: Îndrumar pentru lucrări practice), Transilvania University of Braşov, 2000.

[17]   MODDE 9.1.1, Software for Design of Experiments and Optimization, UMETRICS, Sweden, 2009.

THE APPLYING OF AN AUTOMATIC CONFIGURATION TOOL FOR THE INVESTIGATION OF THE UAV ELECTRICAL NETWORK

Diana CAZANGIU[1], Ileana ROSCA[2], Yves LEMMENS[3]
[1] Transilvania University of Braşov, Braşov, ROMANIA, e-mail: cazangiu.diana@unitbv.ro
[2] Transilvania University of Braşov, Braşov, ROMANIA, e-mail: ilcrosca@unitbv.ro
[3] LMS International, Leuven, BELGIUM, e-mail: yves.lemmens@lmsintl.com

Abstract: These researches were a part of a project which lasted more than a year and were done in collaboration with LMS International from Leuven, Belgium and Katholieke Hogeschool Brugge – Oostende that help us with the technical support.

The main purpose of these researches was to determine the optimum configuration that can offer the best results in a real flight. This was possible by applying the Model – Based System Engineering (MBSE) in the developing and the investigation of an UAV electrical network. Based on the "black box" principle from System Modeling, it was created different models for the main components of the aircraft.

Using specific mathematical formulas and taking account of the basic forces and the moments developed on the surface of the aircraft structure during the real flight, it was computed the functional parameters for the main components (propeller, motor, servo, battery).

Have been developed two different models for the UAV electrical network (electrical and thermal model). Also, using the automatic configuration tool, it was created, for the both models, different configurations.

Keywords: UAV, configured architecture, Model-Based System Engineering, System Simulation.

1. INTRODUCTION

The aim of these researches was to determine the optimum configuration of an UAV electrical network by using an automated configuration tool. According with this goal it established a set of main objectives, as: the designing of two reference architecture for the UAV electrical network (electrical and thermal model), the creation, for each of these, of a set of configurations and the investigation of every configuration. Using the Model – Based System Engineering (MBSE), it was possible to obtain the specific results.

Starting with the 90's interest in the UAVs grew significantly within military and civil areas as well with a large number of producers and covered missions. According to [1], in 2012, it is estimated 76 countries have some sort of UAV.

The UAV (unmanned aerial vehicle) is an aircraft without a human pilot on board. Its flight is controlled either autonomously by computers in the vehicle, or under the remote control of a pilot on the ground or in another vehicle.

The use of drones has grown quickly in recent years because unlike manned aircraft they can fly for many hours (*Zephyr* a British drone under development has just broken the world record by continuous flying for over 82 hours); they are much cheaper than military aircraft and they are flown remotely so there is no danger to the flight crew [2].

The UAVs are well suited for emergency situations. Some typical applications are the following: disaster operations management, transport, search and rescue, firefighting, flood watch, volcano monitoring, hurricane watching, nuclear radiation monitoring etc. Most of these missions require extended time in the air and also real-time video to the ground control situation.

It is expected to see increasing automation built into UAVs. In the near future UAVs will be able to take off, navigate to a destination, return and land without operator intervention [1]. This should improve reliability by reducing the impact of a disruption to radio signals between the UAV and control center.

2. THE MATHEMATICS FUNDAMENTALS IN UAV COMPONENTS MODELING

The strategy of the components modelling is based on an observation and theoretical analyses of the component individually and after that an iterative process is started with first reviewing the AMESim library to find the best matching model available and second, characterizing the component by essential output, input and internal parameters.

The basic principle of the AMESim modelling was taken from the System modelling theory, namely "Black Box" principle that is represented in Figure 1.

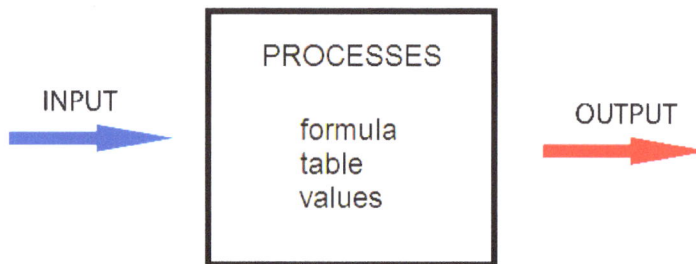

Figure 1: The scheme of Black Box principle

In science and engineering, a "black box" consists in a device, system or object which can be viewed in terms of its input, output and transfer characteristics without any knowledge of its internal workings.

The "black box" principle is considered a mathematical model that consists in a series of mathematical relations between the process variables. These relations link, one with other, the state process variables or more commanded variables of this system.

In the systems modeling theory it specifies that the mathematical models offer high possibilities of study regarding the optimization of a modeled process, because by solving the equations from within the "box" it determines the optimum values of the commanded variables [3].

Based on this concept it created submodels for the main components of the UAV (propeller, motor, battery, servo, ESC).

Data available from the manufacturer, other documentation and/or measurements is used and put into the submodel as parameters, functions or tables [4].

The parameters used for the submodels were computed using the specific formulas and taking account of the forces and environment parameters that are involved in the real flight conditions.

Basing on this principle, it created the models for main components involved in the UAV electrical network that are presented in Figure 2.

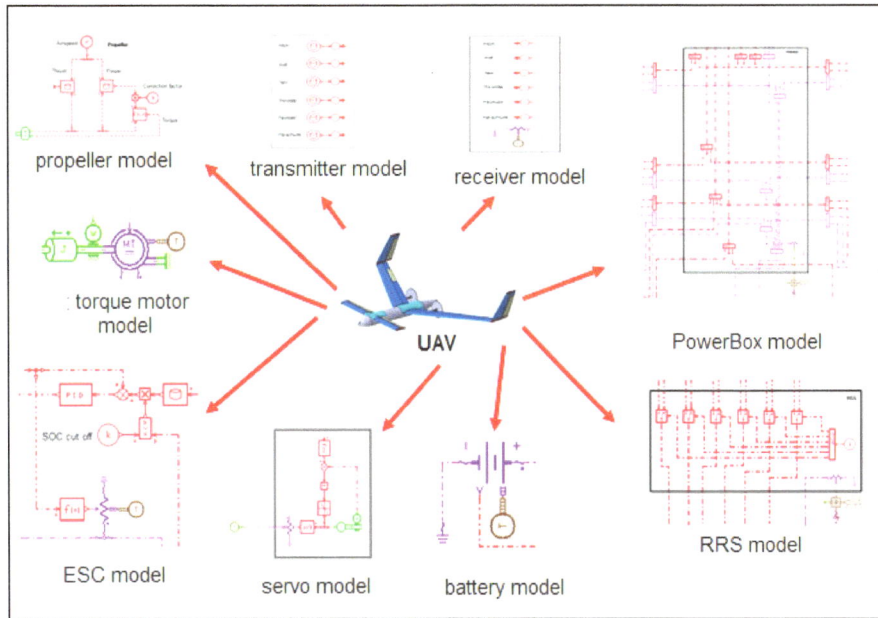

Figure 2: The main components of UAV electrical network

Starting from these UAV models and using AMESim software it created different supercomponents for all the components of the electrical network of UAV (batteries, servos, motors and propellers).

The supercomponents consist of an icon (image) and ports and are associated with the submodels. After creating each supercomponent, it applied the simulation mode for check if the new supercomponent is functionally.

3. THE APPLYING OF MBSE CONCEPT IN THE INVESTIGATION OF UAV ELECTRICAL NETWORK

The Model-based systems engineering (MBSE) is a methodology for designing systems using interconnected computer models. The recent proliferation of MBSE is evidence of its ability to improve the design fidelity and enhance communication among development teams.

One of the main objectives of this research was to investigate of two different electrical models using an automated configuration tool. This could be possible by the system simulation and the evaluation of the output parameters for the both models. For the creating of different simulation scenarios it had to use SysDM and System Synthesis software.

The flow chart from Figure 3 represents the interconnection between the three used software.

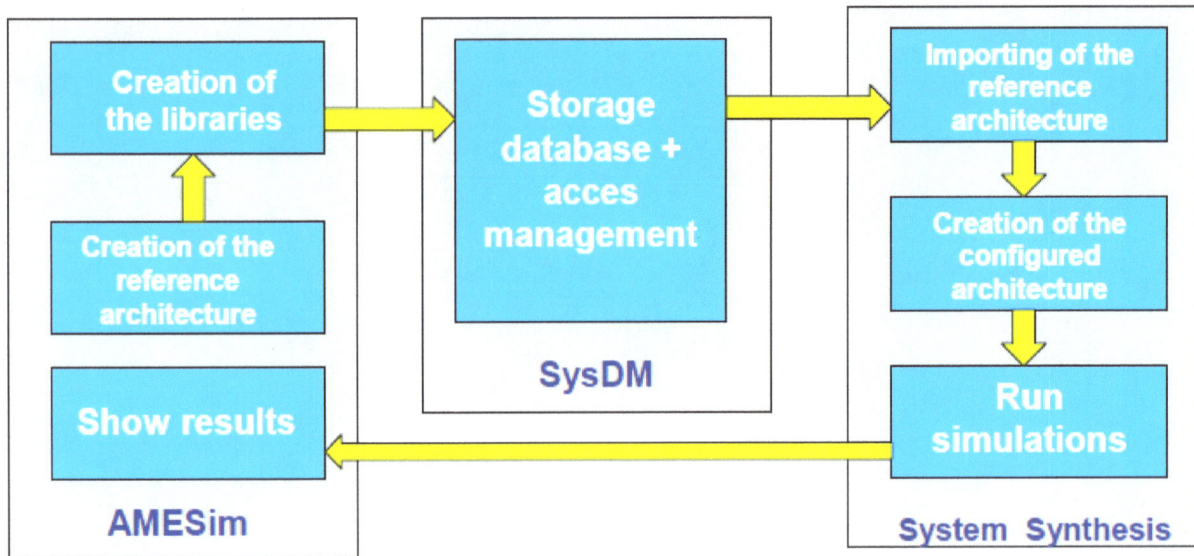

Figure: 3 The interconnection between the three used software

Referred to this scheme, first step is the creation of the reference architecture, using AMESim software, and then it has to create the libraries for the used components.

Next step is to import the libraries of the components, using SysDM software. After that, it needs to import the reference architecture, already created in AMESim software, using System Synthesis software. Base of that it has to create the configured architecture and the following step is to run the simulations.

The final step is to show the results and for that reason it needs to reopen AMESim software.

The original UAV model used in this project was constructed, by KHBO team, in AMESim software, by elements which contain electrical components, which were already presented in Figure 2. The UAV is powered by two electric motors, driving propellers. These motors consume the most electrical energy of all the components installed. This electrical energy is transformed into mechanical energy and transferred via an axle to the load, a propeller.

Starting to this global model, it created two different reference architectures for the UAV electrical network: the electrical model (Fig. 4) and the thermal model (Figure 5).
In the both figures, the lines represent all the connections, red for signals, purple for electrical and green for mechanical connections. The components causing heat production have a thermal port ready for the thermal model.

One of the objectives at the beginning of this research was to investigate to the thermal behavior of the components inside the fuselage. This should indicate if a continuous airstream, whether or not forced, is needed throughout the fuselage. The temperature of the composite structure is limited to 120°C and must be taken into account.

The thermal model makes use of convective heat transfer between components and a medium (air). Their thermal solid properties are characterized in a data file, linked to the solid subcomponent in AMESim and can be copied. Also, the heat flux convection process from the components to a flow of air can easily be copied in the new thermal model.

In the thermal model, subcomponents influencing the surrounding air temperature inside the fuselage were connected to each other.

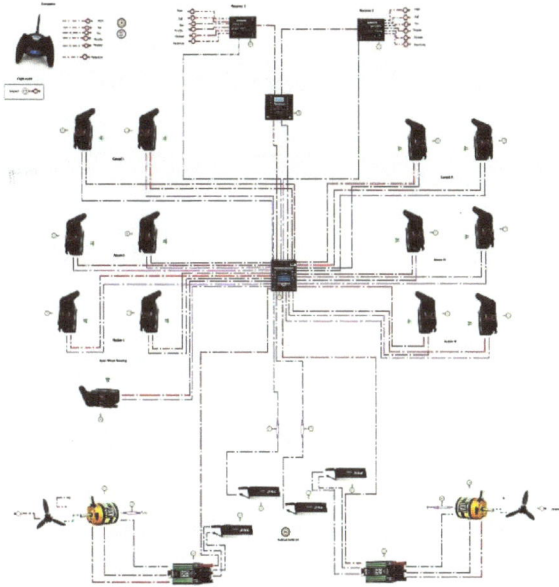

Figure 4: Energy management only reference architecture

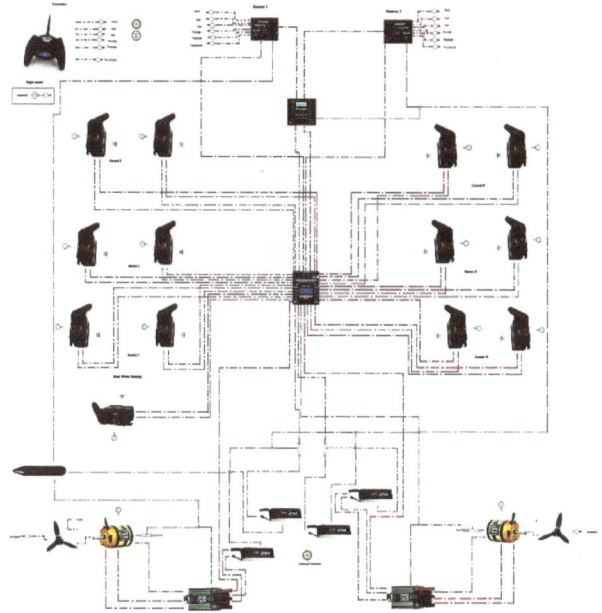

Figure 5: Thermal and Energy management reference architecture

Starting from the reference architecture of UAV electrical model and using System Synthesis software it created 9 configurations with different parameters of some components (motor, propeller). Because it were created 3 models of these components it was easy to choose the configuration that it wanted.

Finally, after that it created all the possible configurations, it started to run simulations in System Synthesis software.

In the same way, for the UAV thermal model, 8 configured architectures were created. Using the 2 models of ESC and 4 models of fuselage it was possible to create the configured architectures, helping by System Synthesis software.

4. SIMULATIONS AND RESULTS

Also, it made simulations on the configurations of UAV electrical model, realized helping by System Synthesis software. It used for these simulations normal flight mission (the maximum time at the value of 3400 s).

The results obtained after simulations were around of the value of 10 % for the state of charge output of the main batteries, but not for all the tested configurations was possible a simulation at 3400 s. It observed that the maximum time of simulation flight was different in function of the motor and the simulations were executed using three sets of parameters for the propellers (AmeProp 17x10, AmeProp 18x10 and AmeProp 19x10). The best results were obtained by using of the propellers with smaller parameters (the diameter).

So, if these parameters (torque motor and the diameter of the propeller) are too big, then, the time of flight simulation decrease because the state of charge for the main batteries reaches more quickly at the value of 10 %.

In the table 1 can be observed the values of output SOC of the main batteries for the 9 configurations of UAV electrical model.

Table 1 The numerical results of the simulation for the configurations of UAV electrical model propeller parameters.

Configuration no.	Value of State of charge output for the main batteries, [%]
1	10.577
2	Failed (the simulation stops after 2100 s)
3	Failed (the simulation stops after 2600 s)
4	Failed (the simulation stops after 2600 s)
5	Failed (the simulation stops after 1600 s)
6	10.5264
7	Failed (the simulation stops after 2400 s)
8	Failed (the simulation stops after 1900 s)
9	37.3146

For the UAV thermal model, the simulations were done at normal flight, which means 3400 s. As output variables were followed the temperature developed in fuselage and in Electronic speed controller. The graphical results for the 8 configurations were plotted.

In the table 2 can be noticed the numerical results for the output temperature for fuselage and ESC at normal flight mission.

Table 2 The numerical results for the output temperature for fuselage and ESC

Configuration no.	Temperature fuselage [°C]	Temperature ESC [°C]
1	98.5151	529.15
2	137.221	536.887
3	137.221	536.887
4	98.5151	529.15
5	72.5153	189.35
6	97.8255	191.214
7	97.8255	191.214
8	72.5153	189.35

The fuselage has a temperature of 98.5151 °C while the electronic speed controllers, mounted onto the fuselage have a staggering of 529.15 °C. This output temperature for the electronic speed controller has obtained because the UAV thermal model has not been calibrated yet and it were not used the realistic parameters of simulations.

In real flight conditions it should be impossible that a controller to resist at this big output temperature; after a temperature of 200 °C this type of controller should be decomposed.

For avoiding a very big increase of the output temperature for the electronic speed controller it used for the last configurations an additional sink of cooling air and the values of the temperature were significantly decreased till the value of 189.35 °C.

142

5. CONCLUSIONS

After the simulations developed on the two models it can identify some conclusions, as:
- the use of the automated configuration tools in system designing significantly decreases the work time for the process of components change;
- the simulation process can be applied of a set of configuration that exists in a list;
- by using the automated configuration tool, it obtained many configurations of the same reference architecture without its modify;
- when a system is investigated, it is always necessary to create a set of configurations. In this way, it can determine, after the simulations were done, the optimum configuration for a specific mission profile. Related to this work the optimum configuration for the electrical model was No. 9 from the Table 1 beacuse after the 3400 s of flight simulation, the level of charge of the main batteries reached at the value of 37,3146. It has to be clear that this imposes an adjustement of the imput parameters for the motor and for the propeller. For the thermal model, it is identified two optimum configurations (No. 5 and No. 8 from Table 2).

As a final conclusion, it can asserts that the cosimulation process between the software tools is, in the current days, a good option for saving time and money.

6. ACKNOWLEDGEMENTS

The authors would like to thank of the KHBO university for their support of the work of this project.

REFERENCES

[1]   Birch M., Lee G., Pierscionek T., *Drones, the physical and psychological implications of a global theatre of war,* Medact 2012, UK
[2]   Austin R., *Unmanned aircraft systems, UAVs design, development and deployment*, Wiley, UK, 2010.
[3]   Lache, S., *System identification used for model updating of the mechanical structures*, în Prooc. of the 6th International Symposium on Automatic Control and Computer Science, SACCS'98, Iasi, 20-21 Noiembrie 1998, Ed. Matrix Rom, ISBN 973-9390-42-0, pp. 25-30.
[4]   VLOC. (2011). *PWO project UAV*. Ostend: KHBO.

OPTIMUM DESIGN OF SPINDLE-BEARING SYSTEMS

Dumitru D. Nicoara

Transilvania University of Brasov, Brasov, ROMANIA, tnicoara@unitbv.ro

Abstract: *In this paper we proposed several optimization models for spindle-bearing systems.*

The goal is to find out the position of the bearings, the diameters of the spindle (different diameters for several segments of the spindle) in order to maximize dynamic stiffness (minimize receptance), i.e. the diminishing of the vibrations. Some constraints are imposed: the distances between bearings, different diameters for several segments of the spindle, etc.

The method is very useful for the design engineers from the very beginning of the design, offering to the designer the optimal values of the parameters.

Keywords: *spindle, bearing, finite element, optimization, dynamic stiffness.*

1. INTRODUCTION

One of the most important parts of machine tool is the spindle-bearing system. The structural properties of the spindle affect the machining productivity and quality of the work pieces. The structural properties of the spindle depend on the dimensions of the shaft, bearings, tool holder, and the design configuration of the spindle systems.

For HMS (high speed machining), the spindle design must be carefully decided by designers. The bearing arrangement, the preload for the bearings, the tool holder, tool interface technologies are important issues for high speed spindles [6], [9], [10].

For design optimization of spindles, Yang [1] conducted static stiffness to optimize a bearing span using two bearings, and described the methods used to solve the multi-bearing spans' optimization method. Taylor et al. [2] developed a program which optimizes the spindle shaft diameters to minimize the static deflection with a constrained shaft mass.

Wang and Chang [3] simulated a spindle-bearing system with a finite element model and compared it to the experimental results. They concluded that the optimum bearing spacing for static stiffness does not guarantee an optimum system dynamic stiffness of the spindle. Hagiu and Gafiteanu [4] demonstrated a system in which the bearing preload of the grinding machine is optimized.

The previous research used only two support bearings, although practical spindles may use more bearings depending on the machining application. In addition, most of them optimize design parameters, such as shaft diameter, bearing span, and bearing preload, to minimize the static deflection.

The machining performance can be raised by improving dynamic stiffness of spindle-bearing system [5].

The dynamic performance of the spindle system are strongly influenced by design parameter such as: distance between bearing, diameter of the different portion of a spindle, bearing preload, bearing spacing etc.

In most papers this influence is studied by varying the parameters and analyzing of its effect on the system.

In this paper we proposed several optimization models for spindle-bearing systems.

The goal is to find out the position of the bearings, the diameters of the spindle (different diameters for several segments of the spindle) in order to maximize dynamic stiffness (minimize receptance), i.e. the diminishing of the vibrations caused by cutting forces, shaft unbalance etc.. Some constraints are imposed: the distances between bearings, different diameters for several segments of the spindle, etc. The method is very useful for the design engineers from the very beginning of the design, offering to the designer the optimal values of the parameters.

To solve the problem we have combined the finite element method with optimization methods. Therefore, the code computer optimization program in MATLAB is obtained by the coupling of the FEM with the non-linear optimization methods with constraints [5], [10]. Spindle, holder, and tool are modeled as multi-segment beams by using Timoshenko beam theory. In the case of the dynamic analysis four degrees of freedom (DOF) per node are considered: two displacements and two slopes. The linearized bearing are commonly modeled as four spring coefficients and four damping coefficients.

Based on the modal analysis we propose an "external" (passive) optimization model for spindle-bearing systems.

2. MATHEMATICAL MODEL

2.1. Finite element model of spindle-bearing systems

The most commonly model for analyzing a spindle systems is shown in Figure 1. In this model are the included tool, tool-holder, spindle shaft, and bearings.

In this study, all components of the spindle–holder–tool assembly are modeled as multi-segment beams. Timoshenko beam model and Euler–Bernoulli beam model is used. The results are compared.

The model consists of a spindle treated as a continuous elastic shaft supported on isotropic or anisotropic elastic bearings. Consider that the dynamic equilibrium configuration of the spindle-bearing system the undeformed shaft is along the x- direction of an inertial x, y, z coordinate system. In the study of the lateral motion of the spindle, the displacement of any point is defined by two translations (v, w) and two rotations (ϕ_y, ϕ_z)

In the following, only axisymmetric spindles are considered. The model could use one of the following two beam finite element types [5]:
- Beam C1 finite element type based on Euler-Bernoulli beam model;
- Beam C1 finite element type based on Timoshenko beam model;

The beam finite element has two nodes. In the case of the dynamic analysis four degrees of freedom (DOF) per node are considered: two displacements and two slopes measured in two perpendicular planes containing the beam. We do a comparative study of the two proposed models and on its basis we adopt the optimal model of the goal. Timoshenko beam model is finally adopted as the beam might be short and therefore the effect of the shear force must be considered. The gyroscopic effect and damping in bearings may be taken into account. The liniarized bearing are commonly modeled as four spring coefficients and four damping coefficients.

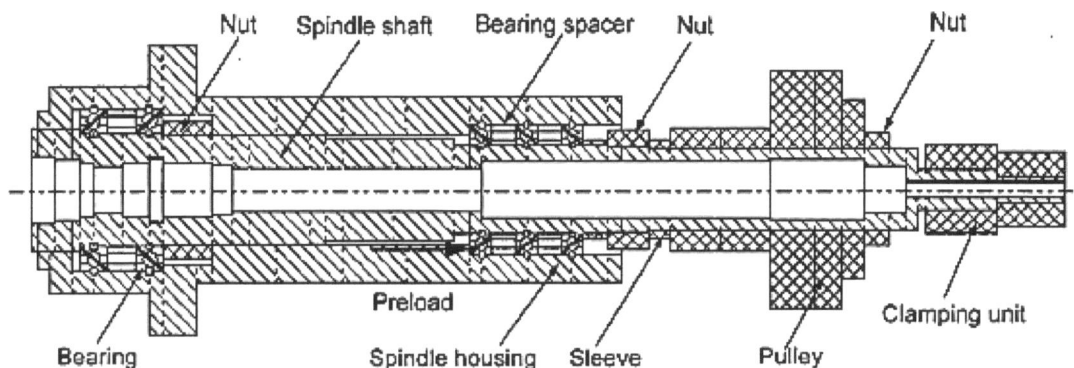

Figure 1: Spindle system

The equation of an anisotropic spindle-bearing systems which consists of a flexible nonuniform shaft and anisotropic bearings may be written as [4], [5]

$$M \ddot{q} + (C + \Omega G) \dot{q} + K q = F \qquad (1)$$

where q is the global displacement vector, whose upper half contains the nodal displacements in the y-x plane, while the lower half contains those in z-y plane, and where the positive definite matrix M is mass (inertia) matrix, the skew symmetric matrix G is gyroscopic matrix, and the nonsymmetric matrices C and K are called the damping and the stiffness matrices, respectively.

The matrices of M, C, G, K, q, and F consist of element matrices given as

$$M = \begin{bmatrix} m & 0 \\ 0 & m \end{bmatrix}_{N \times N} , \quad C = \begin{bmatrix} c_{yy} & c_{yz} \\ c_{zy} & c_{zz} \end{bmatrix}_{N \times N} ,$$

$$G = \begin{bmatrix} 0 & g \\ -g & 0 \end{bmatrix}_{N \times N} , \quad K = \begin{bmatrix} k_{yy} & k_{yz} \\ k_{zy} & k_{zz} \end{bmatrix}_{N \times N} , \qquad (2)$$

$$F = \begin{Bmatrix} f_y(t) \\ f_z(t) \end{Bmatrix}_{N \times 1} , \quad q(t) = \begin{Bmatrix} q_y(t) \\ q_z(t) \end{Bmatrix}_{N \times 1}$$

where $N = 4n$, n is the number of nodes.

2.2 Receptance and dynamic stiffness

The equation of motion (1) can be rewritten in state space form as

$$A \dot{X} + B X = Q \qquad (3)$$

where

$$A = \begin{bmatrix} C + \Omega G & M \\ M & 0 \end{bmatrix}_{2N \times 2N} , \quad B = \begin{bmatrix} K & 0 \\ 0 & -M \end{bmatrix}_{2N \times 2N} , \quad Q = \begin{Bmatrix} F \\ 0 \end{Bmatrix}_{2N \times 1} , \quad X = \begin{Bmatrix} q \\ \dot{q} \end{Bmatrix}_{2N \times 1}$$

The $2N \times 2N$ matrices A and B are real but in general indefinite, nonsymmetric. The resulting system of equations (3) gives nonself-adjoint eigenvalue problem.
In the case of the synchronous excitation

$$F = A_F e^{j\Omega t} , \quad q = A_q e^{j\Omega t} \qquad (4)$$

transforming Eq. (3) into frequency domain, we obtain

$$A_X = R_d A_Q , \quad A_X = \begin{Bmatrix} A_q \\ j\Omega A_q \end{Bmatrix} , \quad A_Q = \begin{Bmatrix} A_q \\ 0 \end{Bmatrix} \qquad (5)$$

where the matrix R_d is receptance matrix

$$R_d = (j\Omega A + B)^{-1} , \quad (j = \sqrt{-1}) \qquad (6)$$

By matrix operational transform the receptance becomes

$$R_d(\Omega) = U(j\Omega a + b)^{-1} V^T \tag{7}$$

where $U = [u_1\ u_2\ ....u_{2N}]$, $V = [v_1\ v_2\ ....v_{2N}]$

are the $2N \times 2N$ matrices of right and left eigenvectors.

Next, let us introduce the dynamic stiffness matrix K_d, defined as the inverse of receptance matrix

$$K_d = R_d^{-1}(\Omega) \tag{8}$$

From the Equation. (5) and (7) we obtain

$$A_q = \sum_{r=1}^{2N} \frac{u_r^* v_r^{*T}}{j\Omega a_r + b_r} A_F \ , v_r^T A u_r = a_r \ , v_r^T B u_r = b_r \tag{9}$$

where u_r^* and v_r^* are the upper halves of the corresponding modal vectors.

3. OPTIMIZATION

3.1 Objectiv functions and design parameters

In this section, based on the modal analaysis, we propose an external (passive) optimization model for spindle-bearing systems. The goal being the diminishing the vibrations by the maximizing of the dynamic stiffness, i.e. by minimizing of the receptance.

To do this we need to find out the design parameters: the position of the bearings, the diameters of the shaft (different diameters for several segment of the shaft).

Therefore, the code computer optimization program in MATLAB is obtained by the coupling of the FEM with the nonlinear optimization methods with constraints [10]. The SQP algorithm is used to optimize the bearing locations. The numerical differentiation and a Newton method are used to calculate the Hessian matrix, and BFGS (Boyden-Fletcher-Goldfarb-Shanno) algorithm is used to update the Hessian matrix.

In the case of synchronous excitation the objective function is the receptance for a given rotating speed, or the average receptance for an interval of rotating speeds. The optimization problem obtained is

$$
\begin{aligned}
&\min_{s_k, d_k} \frac{1}{\Omega_1 - \Omega_2} \int_{\Omega_1}^{\Omega_2} \frac{A_u}{A_F} d\Omega \\
&s_k^i \leq s_k \leq s_k^s \\
&d_k^i \leq d_k \leq d_k^s \\
&\Omega \in (\Omega_1, \Omega_2) \\
&V = const.
\end{aligned}
\tag{10}
$$

The design parameters are the distances s_i between the bearings and the diameters d_i of the different portions of the shaft. We assume that the shaft type Timoshenko with gyroscopic effects is included.

In the above equations A_u is the amplitude of the displacement, A_F is the force amplitude, Ω is the rotor spin speed and ω is the whirl speed. The objective function is a measure of dynamic stiffness defined by relation (8). The authors elaborated several computer codes in MATLAB programming language.

3.2 Numerical example. Optimization of bearing locations

The design variables are bearing spans s1, s2, s3 and s4. In the numerical simulations, the same numerical data set, as in the paper [9], has been used, for compare sake. Figure 3 shows the design variables for the motorized spindle with five bearings. The main spindle specifications of SH-403 are shown in [7].

The proposed system is demonstrated against a commercially existing machine tool (Mori Seiki SH-403) as shown in Figure 2 for comparison. The spindle has a motorized transmission with oil–air type lubrication with four bearings at the front and one at the rear. The maximum spindle speed is 20,000 rpm and the power and torque properties of the spindle motor are set from the data shown in [7].

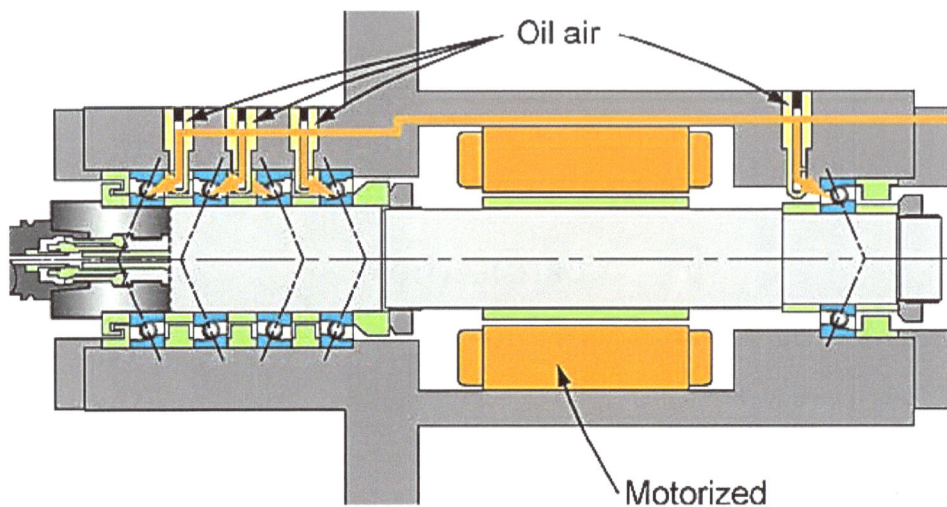

Figure 2: Mori Seiki SH-403 spindle system

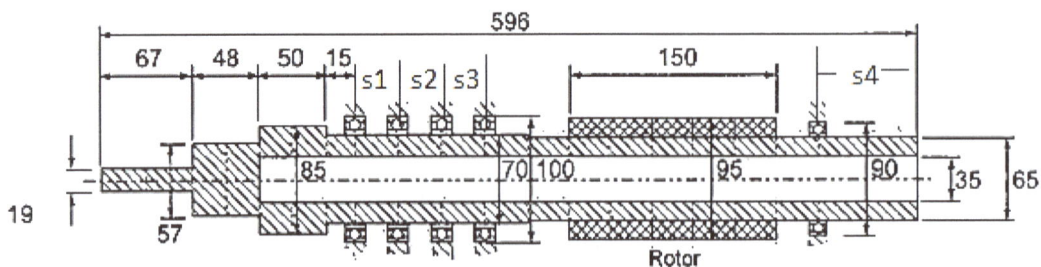

Figure 3: Design variables for the motorized spindle with five bearings

The material parameters: E = 2.07e11; Poisson = 0.3; G = E/(2*(1+Poisson)); rho = 8300

148

Optimization results:

FEM Optimal configuration

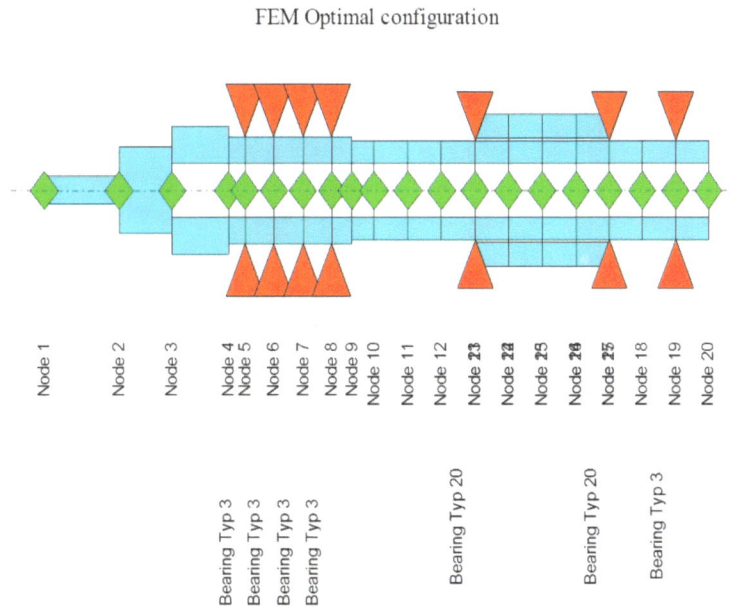

Figure 4: Optimal configuration spindle-holder-tool system

Optimal bearings pozitions: [5 0.180; 6 0.206; 7 0.232; 8 0.258; 19 0.566; 20 0.596; 21 0.386; 23 0.446];

Natural frequencies and response for optimal configuration:

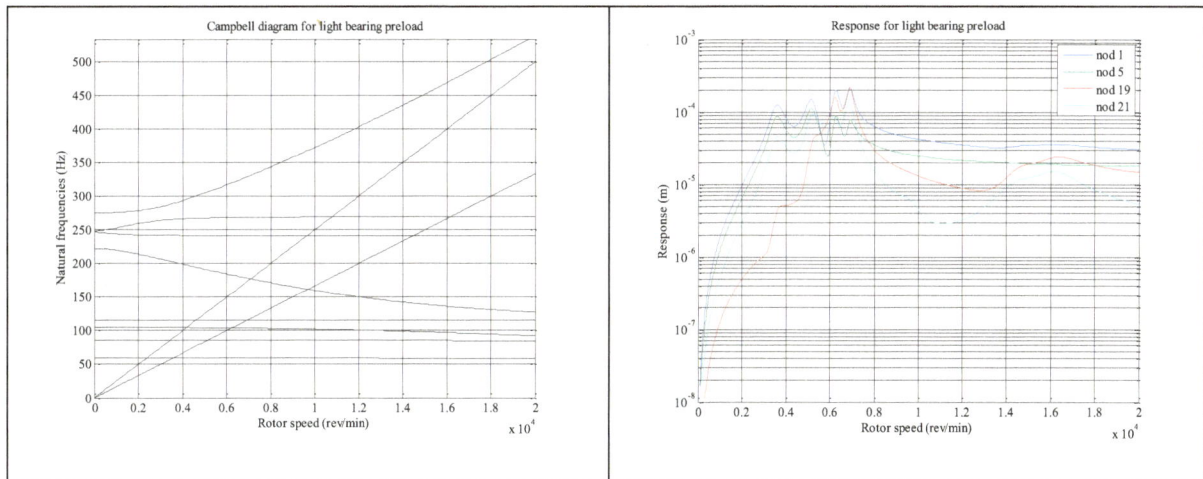

Figure 5: The Campbell diagram and response for bearing leigth preload

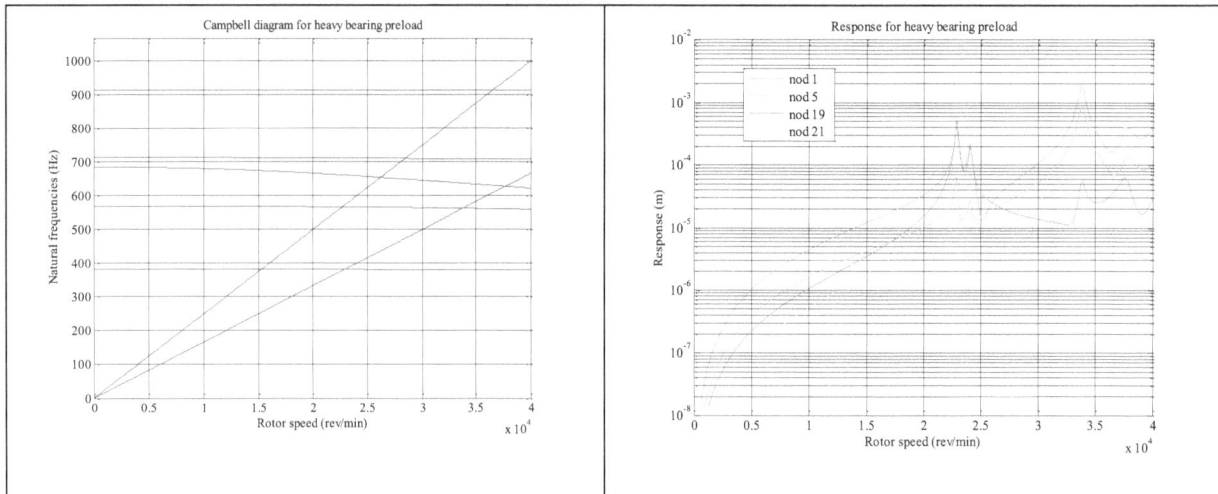

Figure 6: The Campbell diagram and response for bearing heavy preload

3. CONCLUSION

The static and dynamic behavior of machine tools is influenced significantly by the design of the spindle and its bearings. The distance between the bearings and bearing preload has considerable influence on the stiffness of the spindle. It is often important to consider the dynamic behavior of a spindle before establishing an optimum bearing span.

In this paper we propose an external (passive) optimization model for spindle-bearings systems, the goal being the diminishing of the vibrations by maximizing the dynamic stiffness i.e. by minimizing the receptance. The paper proposes a bearing spacing optimization strategy. The spindle is analyzed by a proposed Finite Element Method (FEM) algorithm based on Timoshenko beam elements.

Therefore, the code computer optimization program in MATLAB is obtained by the coupling of the FEM with the nonlinear optimization methods with constraints

The proposed system is demonstrated against a commercially existing machine tool (Mori Seiki SH-403). The spindle has a motorized transmission with oil–air type lubrication with four bearings at the front and one at the rear. The maximum spindle speed is 20,000 rpm and the power and torque properties of the spindle motor are set from the data shown in [7].

REFERENCES

[1]   S. Yang, S., A study of the static stiffness of machine tool spindles, International Journal of Machine Tool Design & Research 21 (1), 23–40, 1981.

[2]   Taylor, S., Khoo, D. B. T., Walton, Microcomputer optimization of machine tool spindle stiffnesses, International Journal of Machine Tools and Manufacture 30 (1), 151–159,1990.

[3]   Wang, W. R., Chang, C. N., Dynamic analysis and design of a machine tool spindle-bearing system, Journal of Vibration and Acoustics, Transactions of the ASME 116 (3), 280–285, 1994.

[4]   Hagiu, G., Gafiteanu, M. D., Preload optimization: a high efficiency design solution for grinding machines main spindles, ASTM Special Technical Publication, No. 1247, 1995.

[5]   Nicoara, D., Optimizarea sistemelor mecanice. Aplicatii la sistemele rotor-lagare, Editura Universitatii Transilvania din Brasov, 2003.

[6]   Y. Altintas, Manufacturing automation-metal cutting mechanics, Machine Tool Vibrations, and CNC Design, Cambridge University Press, Cambridge, 2000.

[7]   SH-403, MORI SEIKI CO., LTD.

[8]   Vanderplaats, S.V., Numerical Optimization Tehniques for Engineering Designe: with Applications, McGraw-Hil, New York, 1984.

[9]   Maeda, O., Cao, Y., Altintas Y., Expert spindle design system, International Journal of Machine Tools & Manufacture 45, 537–548, 2010.

[10]   Ertürk , A., et al., Analytical modeling of spindle–tool dynamics on machine tools using Timoshenko beam model and receptance coupling for the prediction of tool point FRF, International Journal of Machine Tools & Manufacture 46, 1901–1912, 2006.

ROUGH SURFACE CONTACT – APPLICATION TO BEARINGS

Enescu Ioan

Transilvania University Braşov, ROMÂNIA, enescu@unitbv.ro

Abstract: The asperities on the surface of very compliant solids such as soft rubber, if sufficiently small , may be squashed flat elasticity by the contact pressure,so that perfect contact is obtained throughout the nominal contact area. In general, however contact between the solid surfaces is discontinous and the real contact is a small fraction of the nominal contact area. Aplication of these routh asperities on the surfsce is made of rollings bearings
Keyword: surface,contact,pressure,rubber,bearing

1. ELASTIC CONTACT OF ROUGH CURVED SURFACE [2]

The main question posed now is: how the elastic contact stresses and the deformation between curved surfaces in contact influenced by surface roughness.

There are two scales of size in the problem:

(1)   The bulk (nominal) contact dimensions and elastic compression which would be calculated by Hertz theory for the smooth, mean profiles of two surface, and

(2)   The height and spatial distribution of the asperities.

We shall consider axi-symmetric case which we can be simplified to the contact of a smooth sphere of radius R with a nominally flat rough surface having a standard distribution of summit heights σ_s , where R and σ_s are related to the radii and roughness of two surfaces by

$$1/R = 1/R_1 + 1/R_2 \text{ and } \sigma_s^2 = \sigma_{s1}^2 + \sigma_{s2}^2 .$$

Referring to figure 1 a datum is taken at the mean level of the rough surfaces. The profile of the undeformed sfere relative to the datum is given by

$$y = y_0 - r^2/2R$$

At any radius the combined normal displacement of both surfaces is made up of a bulk displacement w_b and a asperity displacement w_a. The separation d between the two surface contain only the bulk deformation

$$d(r) = w_b(r) - y(r) = -y_0 + \frac{r^2}{2R} + w_b (r) \tag{1}$$

The asperities displacement is $w_a = z_s$ - d, where z_s is the height of asperities summit about the datum. The effective pressure at radius found will be

$$p(r) = \left(\frac{4\eta_s E}{3k_s^{\frac{1}{2}}}\right)\int_d^{\infty} [(z_s] - d(r))^{\frac{3}{2}}\phi(Z_s)dz_s \tag{2}$$

For the normal displacement of an axi-symmetric distribution of pressure p(r) can be written

$$w_b (r) = \frac{4}{\pi E}\int_0^a \frac{t}{t + r}p(t)K(k)dt \tag{3}$$

where K is the complete elliptic integral of the first kind with argument

$$k = \frac{2(rt)^{\frac{1}{2}}}{r + t}$$

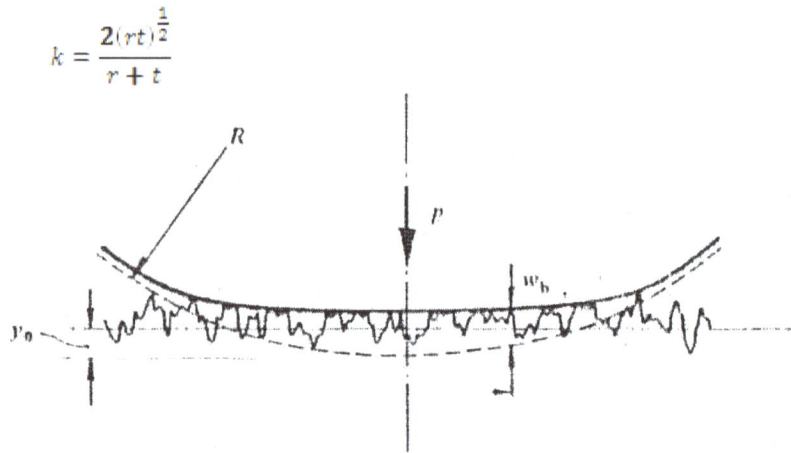

Figure 1

2. BEARING APLICATION [1]

The roughness of the surfaces of the pieces of bearings in contact may be characterized by the following parameters:

- The amplitude parameters
- The distance parameters
- The bastarg parameters

Amplitude parameters (figure 2)

R_x -parameters most frequent used to the general rough

n- number of discrete deviations

R_a -standard deviation of the profile

$$R_a = \frac{1}{L} \int_0^L |y(x)| \, dx \qquad (4)$$

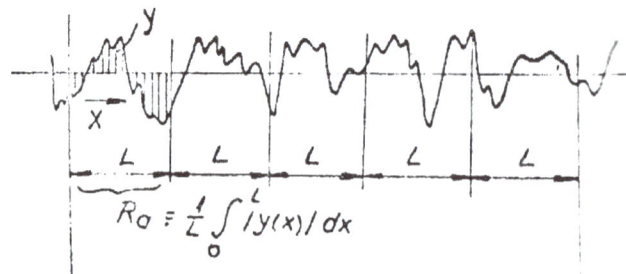

Figure 2

R_{max} -is the distance between the upper- most contact point at the low contact point in the interior point (figure 3)

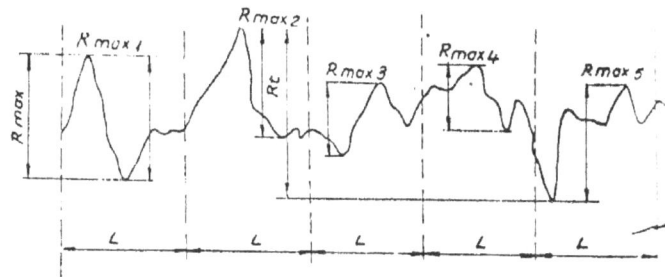

Figure 3

R_t - Distance by the most upper-most contact to the most low contact point in evaluation (figure 4)
R_z - Average of the absolute values of the most 5 upper-most contact points or the most law contact point

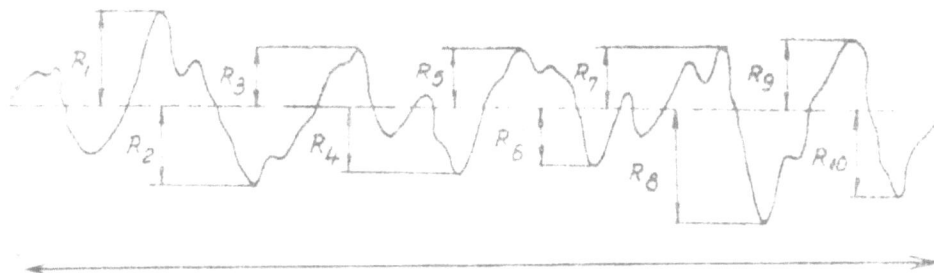

Figure 4

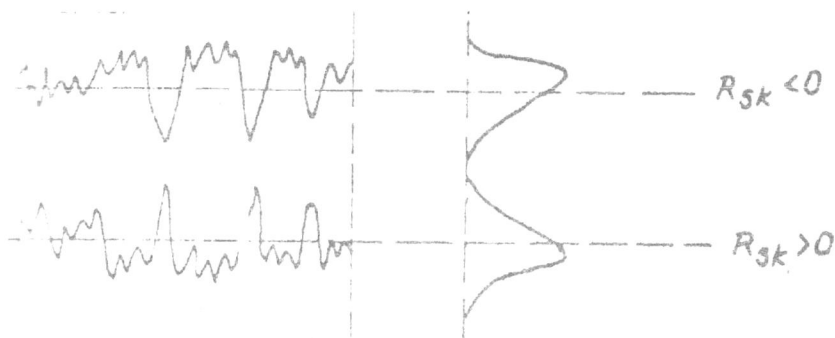

Figure 5

The measure of distribution density of the amplitude of the profile is note S_k where

$$S_k = \frac{1}{Rq^3} \frac{1}{n} \sum_{i=1}^{n} y^3 \tag{5}$$

Distance parameters

Hsc (High Sport Count) – is the number of the upper-most contact completely project upper the median line or by a parallel of the medianline to a preselected distance p, above the reference line.

The counting is making the base length. Another parameter is S_m, the medium pass of the irregularities of the profile

REFERENCES

[1]   I. Enescu, Gh. Ceptureanu, D.Enescu, Rulmenti, Editura Universitatii Transilvania, 2005
[2]   K.L.Johnson, Contact mechanics, Cambridge University Press, 1985

ACTUAL STATUS OF GUSSETED JOINTS OF AEROSPACE WELDED STRUCTURES

G. Dima[1], I. Balcu[2]

[1]Nuarb Aerospace SRL, Braşov, ROMANIA, g.dima@nuarb.ro
[2]Transylvania University, Braşov, ROMANIA, balcu@unitbv.ro

Abstract: *The welded structures are used from early aircraft using a variety of materials, procedures and designs. This paper identifies the gusseted joins of welded structures in a multidisciplinary approach, including aerospace and civil engineering. The weld gusseted joints of thin walled structures are not standardised and there are no methodology for calculation, existing only design reccomandations. The paper identifies the status of different approach of gussets in literature, focusing on the future work to be done in this area.*

Keywords :aircraft structure, latticed beams, tubular joints, gussets, welded structure

1. GENERAL

The welded structures are first used in aerospace structure for Fokker Eindeker (1915) for the fuselage [L02]. Since than, the welded structures passed through a complex process of improvement related to the material employed, topology or technological process. Beacuse of the problemsdiscovered at welding (big dependece of welder skills, poor fatigue behaviour), in time manufacturers searched for alternative designs (especially by riveting), and ending up by the adoption of the welding or semimonocoque structure.

Nowadays weldings are used on aircrafts for the structural applications like fuselage (for light aircrafts), landing gear, equipments or external mounts, seats, engine supports, or non-structural applications like brackets, fairings, tanks or inlets.

2. CLASSIFICATION OF WELDED STRUCTURES ON AIRCRAFT

Heavy loaded structures on aircraft are categorised in primary structure (their failure leads to loose of structure integrity), secondary structure (their failure leads to emergency landing) and tertiary structure (failure allows flight to the first airport).

The welded structures on actual aircrafts are used as follows:
- Primary structures – fuselage (light airplanes and helicopters) (Figure 1a), helicopter tail booms;
- Secondary structures – landing gear, ailerons, elevator, rudder (Figure 1b);
- Tertiary structure and brackets - engine supports (Figure 2a), defence systems, cargo swing (Figure 2b);

The aircraft welded structures have in common some specific features as follows:
- Using of CHS (circula hollow structures) for higher overall buckling strength, higher radius of gyration function of cross sectional area and smaller effective buckling length than that of angle profiles [F02].
- Using of inseted bushing and end memers lugs/ forks for external mounts.
- Using of gussets to decrease stress leve in joins.

There are authors considering circular section similar to nature's optimal response to combined loads as bamboo or bones structure. In [K01] it is shown that, for the same bucling capacity, a CHS column has only 60% of mass of a "I"or "H"profile.

The aerospace structures use CHS (Chircular Hollow Structures) also because of minimum aerodynamic drag, despite geometry of members in connection areas or multiple member nodes. Gussets are employed in structural nodes to improve strength and rigidity.

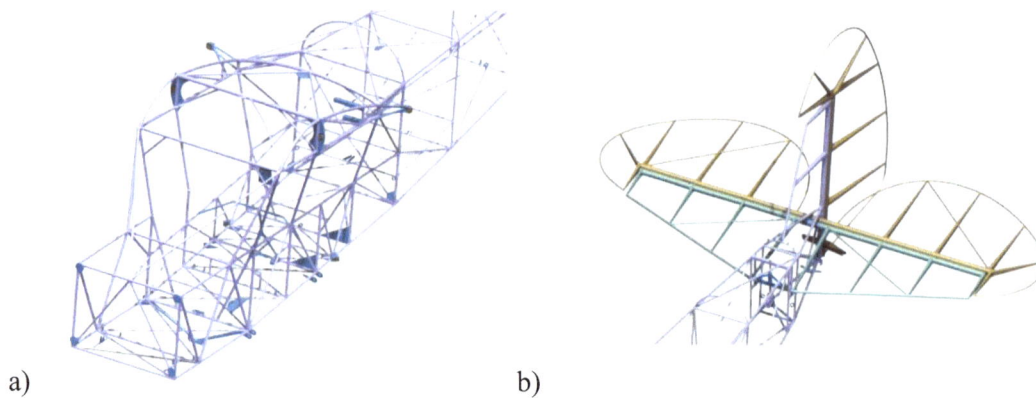

a)   b)

Figure 1: Stol King a) Welded fuselage; b) Empennage

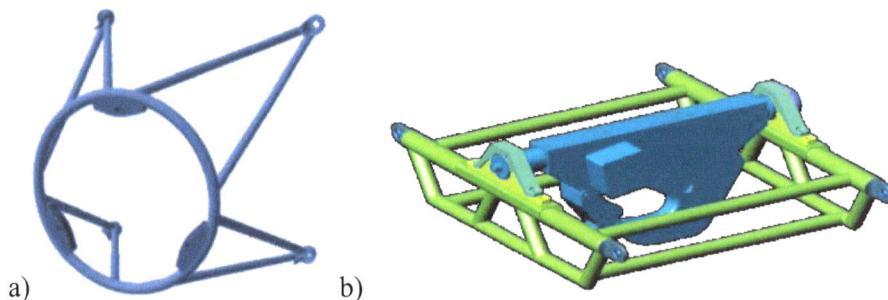

a)   b)

Figure 2: a) Engine support; b) Helicopter cargo swing

3. WELDING VS. OTHER ASSEMBLY TECHNOLOGIES IN AEROSPACE

In aerospace industry are used the processes as riveting, bolting, bonding and machining (for integral structures). The biggest disadvantage of welding is the crack propagation through all the structure, in other built up structures cracks and stopping into the affected member. Another problem is related to the structure nodes with more than three members, because welds overlap it has to be avoided because of altering the characteristics of previous one.

The technological disadvantages of welding are noted in [I01], [N01]:

- The need of positioning and fixture jigs;
- The need of tubes end milling, especially for multiple and non-orthogonal joints;
- Thermal expansion/ contraction of the members during the welding process the deformation of whole assembly because of post weld internal stresses (these problems can be reduced/ eliminated by intercalated welds, welding technology, preheating of welding area;
- Mechanical proprieties changing after welding; [N01]recommends a 10% reduction of allowable stresses for joint calculation;
- Welding flaws, affecting the strength of joint; this leads to high qualified workmanship;
- Expensive non-destructive control methods

As advantages over other assembly techniques [C02] specify:

- More rational shapes for subassemblies;

156

- More effective use of raw material and manufacturing time savings;
- The possibility for automation and manufacturing time reduction;
- Cheap and easy to maintain manufacturing machines;
  - The absence of fasteners
  - A superior strength for long welds

Recent studies are dedicated to replace riveting and bonding of stringers on the skin by welding. In [Z02] is concluded hot spot stress level is lower the in riveted structures, and the smallest crack growth rate related to machined and riveted structures. The problems of crack propagation through all welded structure remained.

4. CHS WELDED JOINTS WITHOUT GUSSETS

CHS (Circular Hollow Structures) are together with RHS (Rectangular Hollow Structures) very used in civil engineering (including offshore platforms and cranes)too, the joints being static, dynamic and fatigue analyzed in[K01], [P01], [W03], [W01] and [Z01].

The literature presents many kinds of planar and multiplanar joints between same type of different kind of members (hollow or open profiles). In [*02], [K01] and [W01] there are presented connections between CHS and "I" or "U" shape profiles.[K01] shows that the most usual configuration for combined loads is between same kind of members (CHS/ CHS or RHS/ RHS), other combinations being very rare. For aerospace the "T" and "K" joints present interest (Figure 3a). There is analyzed the axial load (AXL), in plane bending (IPB) and out of plane bending (OPB).

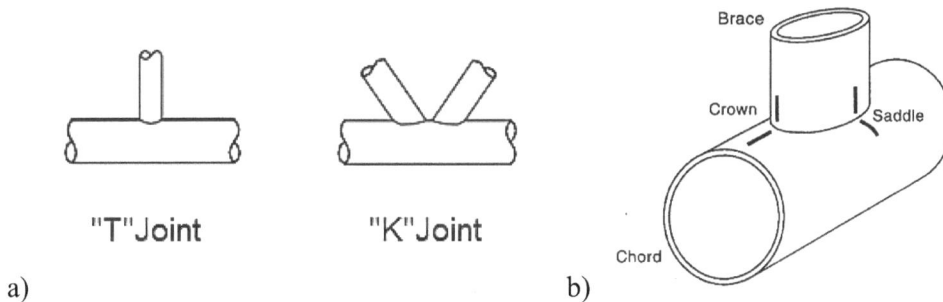

a)   b)

Figure 3: a) "T" and "K" joints[Z01]; b) Terminology of "T" joint [P01]

Figure 3 b) presents the terminology in the points of interest of "T" joint. Starting from sixties there were made a lot of studies regarding strength design and fatigue behavior of "T" connection, especially from civil engineering area. Most recent accepted results for unstiffen "T" joint are given in [Z01]; there are given values for SCF (Stress Concentration Factor) for:
- AXL – Chord (Saddle & Crown) and Brace Saddle & Crown)
- IPB – Chord (Crown) and Brace (Crown)
- OPB – Chord (Saddle) and Brace (Saddle) (Value for IPB Saddle and OPB Crown are considered negligible)

Last researches in CHS area are oriented to:
- "T" joint HSS under combined AXL and IPB loads [H01]
- Fatigue behavior of hollow structures [C05]
- Elliptical hollow structures [N01]

5. CHS GUSSETED WELDED JOINTS DESIGN

In heavy loaded civil structures, gussets are added to allow bracing connection. [C01] and [K01] present design recommendations related to CHS gussets stress gradient and limit deflection. In civil engineering gussets are also used to connect horizontal beams to columns ("seat brackets"), or connection of columns to basement plates [B01], [*02] and [M01].

In [N04], [N05], [P01] are presented studies related to the strengthen "T" joint using base plate ("chord doublers") or external collar [B01].[B01], also are presented how to use gussets for stress level lowering in "K" joint; [N04] analyzed economical repair of damaged structures using chord doublers.[N03] made a study related to reducing the stress level in "K" joint by adding gussets (Nazari and Durack) showing the SCF (Stress Concentrator Factor) decrease with 45% for axial loads, 33% for in plane bending and 18% for out of plane bending.

In literature there are given many examples of gussets; the shape, proportion and dimensions vary by author (Figure 4 and 5). Gussets are radial or tangent placed, inserted in members or not. In nodes with bracing they are used also for bracing connection.

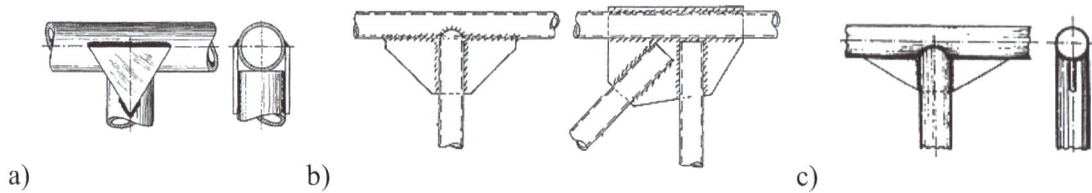

a)   b)   c)

Figure 4: Gusset examples a) triangular [P02]; b) Double [N01], [B04], c) Double [B04]

Between the wars were used connections with tubular gussets used also for wire bracing connection (Fig. 5 a). A particular application is presented in Figure 5b), welds in vertical member being made in holes made in gusset. "U" Shaped section gusset (Figure 5c) is recommended for roll cage of racing cars. The length of gusset is recommended to be three times bigger than tube diameter [*01].

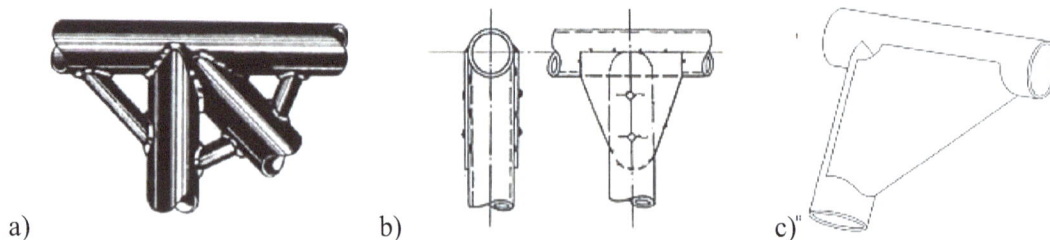

a)   b)   c)"

Figure 5: Gusset examples a) Tubular [avia-it.com]; b) Hole welded [G01]; c) "U" Shaped [*01]

According to [N07] tapered gussets need to be used in all important joints to provide gradual changes in stress level in joint members; gussets are further recommended for stress level reduction for lowering fatigue effects. [F01] recommends gussets for joints reinforcing, adding strength and rigidity, being also a safety solution. Employing gussets increase also the torsional rigidity of the whole structure.

[B04] recommends gussets especially for vibrations and/ or OPB subjected areas. In [B04] there are presented few types of gussets but there are not given any pre-dimensioning or design recommendation.

Related to 'K" joints diversity of gussets is also large. The gussets can be with external doubler (collar) (Figure 6 a), inserted in symmetry plane of joint (Figure 6 b).

Ref. [D03] recommends gussets for:

- Additional weld length provided (Fig. 6 c);

- The possibility to shorten the braces (load being carried by gusset)

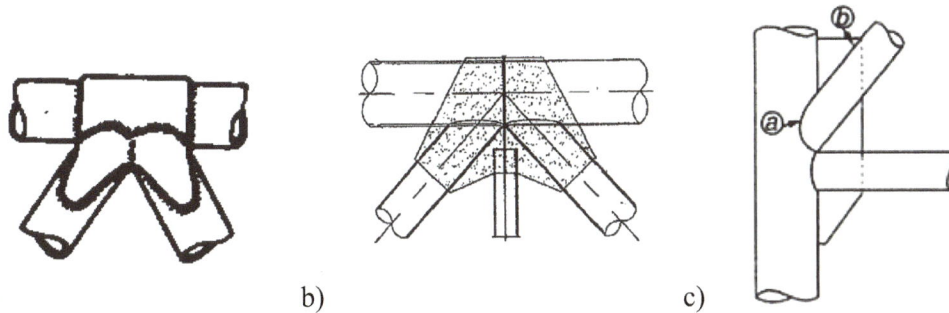

a)   b)   c)

Figure 6: "K" Joints gussets: a) With collar [I01]; b) Inserted [N03]; c) Weld (a) secured by weld (b) [D03]

Gusset free margin is recommended to be curved to decrease the stress level in braces (Fig. 7 a) [D03]. Ref. [G01] presents a curved free edge gusset with welded stiffened strip to prevent buckling (Fig. 7 b).

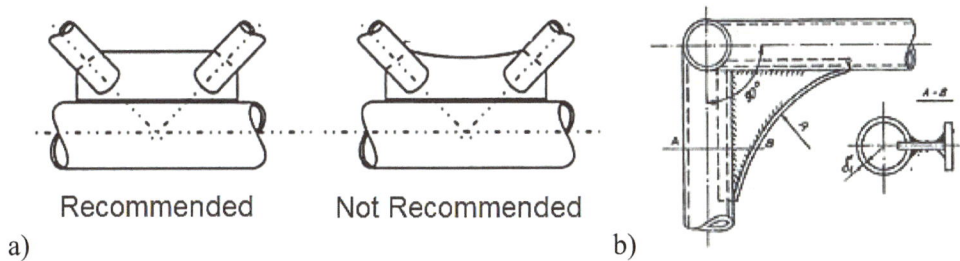

Recommended   Not Recommended

a)   b)

Figure 7: a) Curved free margin gusset [D03]; b) Flanged free margin gusset [G01];

Ref. [P03] recommends ending the members in gusset by spherical forming (Fig. 8 a). This solution is very used in aerospace applications (Fig. 8 b).

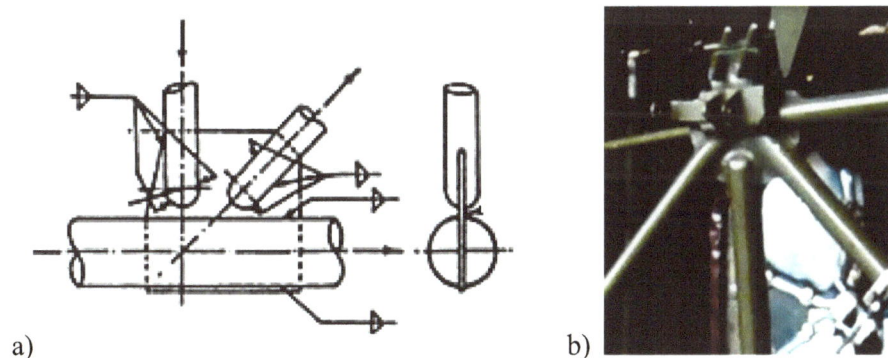

a)   b)

Figure 8: a) End formed braces [P03]; b) Gusseted node - SA315 helicopter [Deutsche Museum, Oberschleissheim];

Related to radial vs. tangent placement of gusset, ref. [B02] mention problems raised by radial assembly as cracks in brace at gusset margins and chord deformation (Figure 9 a). After [B02], correct placement of gusset is tangent to members of joint, also recommending using carefully gussets because they transfer loads but also add rigidity to joint leading to the member's failure. Therefore, before adding gussets an analysis of consequences should be made.

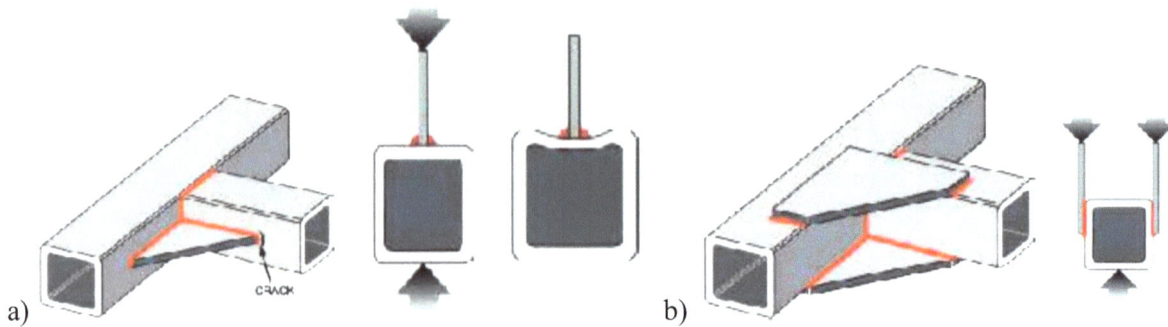

a) b)

Figure 9: a) Radial placement and b) Tangent placement of gusset [B02];

In practical design, gussets are dimensioned relative to tubes diameter, followed bya verification with a FEM analysis being performed. In lightweight structures, weight savings is related also to the shape of gussets. A correct designed gusset will decrease the stress level in the most stressed area, leading to the need of a lighter tube, thus a lighter structure. It is important to mention the existence of many welded structures free of gussets. This is due to the fact design approach in aerospace in not yet homogenous, depending a lot of the design team of manufacturer.

6. CHS GUSSETED WELDED JOINTS DESIGN

The Ref. [M04] uses for pre-dimensioning the calculated length of welds. Also based on the welds length, in [C02] are presented pre-dimensioning of gusseted joined with braces not in contact with chord and in [N06] and [S01] are presented cutted formed end tubes attached to gusset.

[T01] analyzed the stress distribution for double gusseted "T" joint, noting that stress distribution depends a lot function of joint type, its peaks (HSS) having a great influence over fatigue life.

First studies for understanding the behavior of aerospace CHS welded structures was performed by NASA in thirties [B03], being analyzed simple and gusseted tubular joints from different materials, thermal treatment of welding process. The study was focused on "T" and "K" joints, AXL and IPB loaded, all results being associated with weight savings. It was found that inserted gusset joints had best results, performances being improved by gusset cutout in member joining area.

The Ref. [H02] give an example of stress value and location and critical buckling stress calculation for a thick rib (Fig 10).

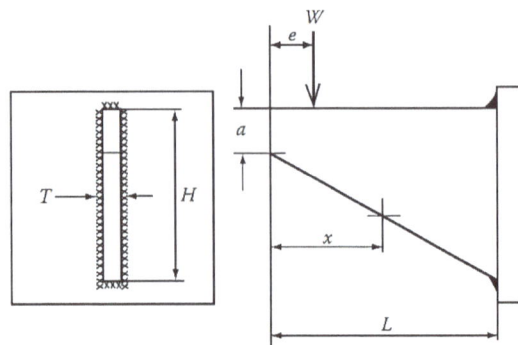

Figure 10: Trapezoidal plate plane loaded parameters [H02];

Recent studies related to gussets are focused on:
- Gusset to strap connection under axial load [J01];
- Fatigue behaviour of OPB loaded gussets [K02];

160

- Failure of tube gusset connection under AXL loading [C03], [M02], [M03], [L03], [K01], [M05] and compression [L01];
- Buckling of gussted to open profiles bracing connection [C04], [C06], [R01], [N02]
- Plane end tube to gusset connection [O02]
- Parabolic free edge gusset stress distribution [S02]

7. CONCLUSIONS

The CHS structures are present in many applications on aircraft structures, but dedicated studies for joints are more close to the steel construction area. It can be concluded that:
- Despite of the big variety of gussets for joints (number of gussets, form, section, free edge shape, dimensions, proportions, radial or tangent placement, inserted or not in joint members) there are no design recommendations for gussets form and placement function of load type or application;
- There are no design recommendations where to use or not gussets for joints reinforcement;
- There are no studies to asses gusseted joint vs. simple braced joint;
- There are not recommendations regarding pre-dimensioning or stress calculation of gussets (plane dimension and wall thickness);
- Existing studies for non stiffened joints are for civil engineering range (general over 100mm diameter), for aerospace thin wall tube structures (diameters in range 16 – 40mm) information being not available;

REFERENCES

[B01] .Blodgett O. W., Design of Steel Structures, The James F. Lincoln Arc Welding Foundation, 1976
[B02] Blodgett O. W., Using gussets and other stiffeners correctly, www.weldingdesign.com, 2005
[B03] Brueggeman W., Strength of Aircraft Joints, NACA Report No. 584, 1936
[B04] Bruhn E. F., Analysis and design of flight vehicle structures, Tri-State Offset Company, 1973
[C01] Cao J. J., et al., Design Guidelines for Longitudinal Plate to HSS Connections, Journal of Structural Engineering, 1998
[C02] Constantin E. T., Proiectarea masinilor, utilajelor şi construcţiilor sudate, Suport de curs, Universitatea din Galaţi, 1981
[C03] Cheng J. J., Kulak G. L., Gusset Plate Connection to Round HSS Tension Member, Engineering Journal, Fourth Quarter, 2000
[C04   ] Chou C. C., Chen P. J., Compressive behaviour of central gusset plate connections for a buckling-restrained braced frame, Journal of Constructional Steel Research, No. 65, 2009
[C05] Conti F., Verney L., Bignonnet A., Fatigue assessment of tubular welded connections with the structural stress approach, Fatigue Design Proceedings, Senslis France, 2009
[C06] Chou C. C., Liou G. S., Yu J. C., Compressive behaviour of dual gusset-plate connections for buckling-restrained braced frames, Journal of Constructional Steel Research, No. 76, 2011
[D03] Duggal S. K., Design of Steel Structures, Tata McGraw Hill, New Delhi, 2009
[F01] Fournier R., Fournier S., Metal Fabricator's Handbook, Penguin, 1990
[F02] Farkas J., Jarmai K., Analysis and Optimal Design of Metal Structures, Balkema, Rotterdam, 1997
[G01] Grosu I., Calculul şi construcţia avionului, Editura didactică şi pedagogică, Bucureşti, 1965
[H01] Haghpanahi M., Pirali H., HSS Determination for a Tubular T- Joint under Combined Axial and Bending Loading, Internaţional Journal of Engineering Science, Vol 17, No 3-4, 2006
[H02] Huston R., Josephs H., Practical Stress Analisys in Engineering Design, CRC Press, 2009
[I01] Iliescu P., Mitu P., Stoian G., Manualul tinichigiului structurist de aviaţie, Ed Didactică şi Pedagogică, Bucureşti, 1974

[J01]   Jensen A. P., Limit analysis of gusset plates în steel single-member welded connections, Journal of Constructional Steel Research, Nr. 62, 2006

[K01]   Kiymaz G., Seckin E., Investigation of the behaviour of gusset plate welded slotted stainless stell tubular members under axial tension, 14th Internațional Symposium on Tubular Structures Proceedings, London, 2012

[K02]   Kishiki] S., Yamada S., Wada A., Experimental evaluation of structural behaviour of gusset plate connection în BRB frame system, The 14th World Conference on Earthquake Engineering, China, 2008

[K03]   Kurobane Y., et al., Design guide for structural hollow section column connections – CIDECT/ TUV Verlag, 2004

[L01]   Lee H. D., et al, Investigation of the Tube-gusset Connection în 600MPA Circular Hollow Section, Procedia Engineering, Nr. 14, 2011

[L02]   Loftin L. K., Quest for performance. Evolution of modern aircraft, NASA Scientific and Technical Infromation Branch, Washington, D.C., 1998

[L03]   Ling T.W., et al, Investigation of block shear tear-out failure în gusset-plate welded connections în structural steel hollow sections and very high strength tubes, Engineering Structures, Nr. 29, 2007

[M01   Martin L. H., Purkiss J. A., Structural Design of Steelwork, Butterworth-Heinemann, 2008

[M02]   Martinez-Saucedo G., Packer J. A., Slotted end connections to hollow sections, CIDECT Report 8G-10, 2006

[M03]   Martinez-Saucedo G., Packer J. A., Willibad S., Parametric finite element study of stotted end connections to circular hollow sections, Engineering Structures, Nr. 28, 2006

[M04]   Mateescu D, Caraba I., Construcțiimetalice. Calcululșiproiectareaelementelor din oțel, Edituratehnică, București, 1980

[M05]   Moreau, R., Finite element evaluation of the "modified-hidden gap" HSS slotted tube-to-plate connection, Connections VII, 7th Internațional Workshop on Connections în Steel Structures, Timișoara, 2012

[N01]   Narayna K. S., et al, Static strength analysis of elliptical chord tubular T-joints using FEA, Internațional Journal of Multidisciplinary Educațional Research, Vol1, Issue 4, 2012

[N02]   Nascimbene R., Rassati G. A., Wijesundara, K., Numerical simulation of gusset plate connections with rectangular hollow section shape brace under quasi-static cyclic loading, Journal of Constructional Steel Research, 2011

[N03]   Nazari A, Durack J., Application of The HSS Method to The Fatigue Assessment of HSS Shiploader Boom Connection, 5th Australasian Congress on Applied Mechanics, Brisbane, 2007

[N04]   Nazari A., et al., Analytical Methods for Better Design and Repair of Mechanical Welded Structures, CRC Mining Technology Conference, Fremantle, 2003

[N05]   Nazari A., et al., HSS Design with parameters equations for fatigue assessment of tubular welded structure, Australian Mining Technology Conference, 2006

[N06]   Novac Gh., Proiectarea masinulor, utilajelor și construcțiilor sudate, Suport de curs, Universitatea Transilvania Brașov, 1991

[N07]   Niccoli, R. History of Flight, White Star, Italy, 2002

[O01]   Ocel J. M., Dexter R. J., Fatigue-Resistant Design for Overhead Signs, Mast Arms Signal poles and Lighting Standard, Minnesota Department of Transportation, USA, 2006

[O02]   Oliveira C., Christopoulos C., Packer J. A., High Strength Brace Connectors for use în SCBF and OCBF, SEAOC Convention Proceedings, 2011

[P01]   Packer J. A., Henderson J. E., Hollow Structural Section – Connections and trusses - Canadian Institute of Steel Corporation, 1997

[P02]   Parmley R., Illustrated Sourcebook of Mechanical Components, McGraw Hill, 2000

[P03]   Punmia B. C., Ashok K. J., Arun K. J., Comprehensive Design of Steel Structures, Laxmi Publications Pvt Ltd, 1998

[R01]   Roeder C. W., Lompkin E. J., Lehman D. E., A balanced design procedure for special concentrically braced frame connections, Journal of Constructional Steel Researchm Nr. 67, 2011

[S01]   Șerb A., Proiectarea și încercarea structurilor sudate, EdituraTehnică – Info, Chișinău, 2001

[S02]   Syed Z. I., Stress analysis of welded gusseted frames, Iowa State University, 2011

[T01]   Teodorescu C. C., Mocanu D. R., Buga M., Imnibarisudate, Edituratehnică, Bucureşti, 1972

[W01]   Wardenier J., et al., Hollow Sections în Structural Applications, CIDECT, 2010

[Z01]   Zhao X. J., et al., Design guide for circular and rectangular hollow section welded joints under fatigue loading, CIDECT/ TUV Verlag, 2001

[Z02]   Zhang X., Li Y., Damage Tolerance and Fail Safety of Welded Aircraft Wing Panels, AIAA Journal, Vol. 43, No. 7, 2005

[*01]   * * *, Roll Cage Construction, Appendix B, nasa Rally Sport, 2011, nasarallysport.com

[*02]   * * *, Eurocode 3, Part 1.8. Design of Joints, CEN, 2002

NOTES ON EVOLUTION OF AIRCRAFT STRUCTURES LATTICED BEAM JOINTS

G. Dima[1], I. Balcu[2]
[1]Nuarb Aerospace SRL, Brasov, ROMANIA, g.dima@nuarb.ro
[2]Transylvania University, Brasov, ROMANIA, balcu@unitbv.ro

Abstract: *Latticed beams are met in aerospace structures from the beginnings of aircraft, being used a big variety of materials, configurations and joints. This paper presents representative solutions used on primary structure from the first aircrafts to the moment semimonocoque structures became a standard. It is shown the evolution of latticed beam joints was not a linear and smooth process, an analysis of factors influencing and possible explanations being stated.*
Keywords: *aircraft structure, latticed beams, tubular joints, gussets, welded structure*

1. GENERAL

The latticed beam was the first structure to offer aviation manufacturers an acceptable compromise between strength, rigidity and internal storage volume for a minimum weight. Latticed beam is the most enduring structure for aircraft, being still used on light aircrafts.

One of the characteristics of aircraft beams is the complexity of structural joints. Due to the swept aerodynamic form, the internal structure needs to offer support to a complex shape. Thus, the rectangular or prismatic frames used in the beginnings of aircraft era, were replaced by trapezoidal or irregular polygon (especially for helicopters).

The replacement of wire bracing (stay wires) with diagonal members leads to triangular stiffened beams in all planes but also to very complex members connections. In main fuselage attachment points (hard points for wing, empennage, landing gear) structural joints were stiffened to withstand the concentrated loads.

A large variety in joint design can be seen depending on manufacturer, even on aircraft; it was a lack of design homogeneity or standardisation.

2. THE BEGINNINGS

Since the beginnings of aircraft history, lightweight structure strength was one of biggest challenge for pioneers together with the aerodynamic of lifting surfaces, an appropriate thrust source and accurate controls.

The first controlled successful flight was not only an fortunately attempt, being the result of a scientific approach of Wright brothers, based on a carefully study of the research of their forerunners like Sir George Cayley, Otto Lilienthal or Octave Chanute [02], [08]. Being the author of an well-known monography, Octave Chanute, an successful civil engineer (bridges and railroads) acted as consultant for Wright brothers. It was Chanute's idea to combine the biplane (intensively tested by Lilienthal) with the latticed beam, resulting in a high inertia momentum beam [04].

The Figure 1 (a) presents Lilienthal bird like wing biplane with upper plane mounted on a mast; having wooden ribs, the margins of the wing were wire attached by mast. In Figure 1 b) Wright's Flyer had the two wings of the biplane acting as two flanges of a Pratt beam; having three bays each side, those were reinforced with diagonal wire bracing, this being a lightweight innovation in order to replace the diagonal bracing [14].

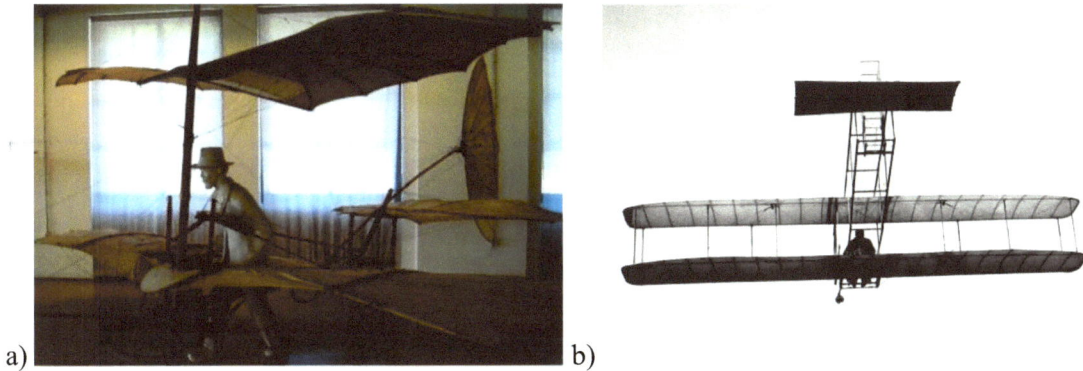

a)   b)

Figure 1: a) Lilienthal biplane [Deutsche Museum, Oberschelissheim]; b) Wright's brothers Flyer [12]

The difference between Figure 1 a) and b) is obvious, only after seven years Flyer having a minimum weight and also stiff enough to support the aerodynamic loads. The Wright brothers concept was a great leap forward, in a period when the design was "chaotic" [11] and up to 1912 to the biggest majority of prototypes ("strange machines") the fact they will take off was like gambling [11], or "cut & try" method [09]. Being a compromise between many conflicting requirements, therefore hard to be obtained, even nowadays, the design of a successful aircraft is not an exact science [07].

3. LATTICED BEAM OF EARLY AIRCRAFTS

The latticed beams of early aircrafts were composed by longerons, columns and diagonal members (wire bracings). The most used section was square but there were also exceptions (Figure 2). The longerons were initially made from ash or hickory, spruce being used later for weight saving [03].

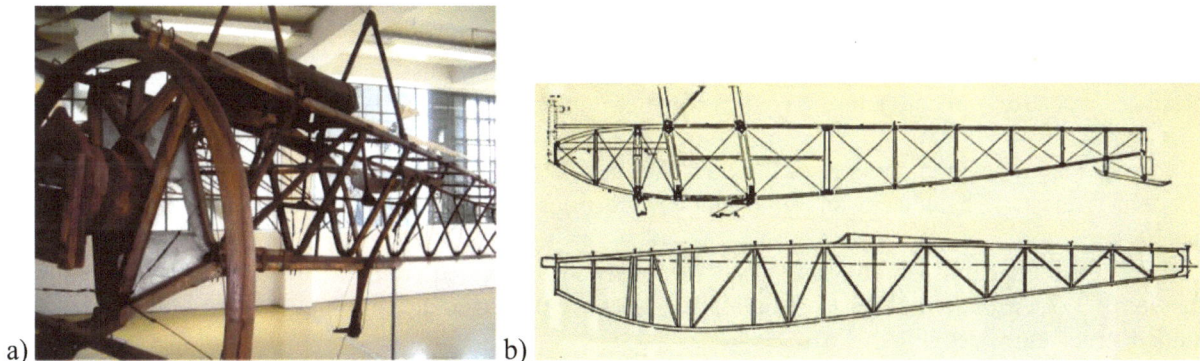

a)   b)

Figure 2: a) Vollmoeller plane with triangular fuselage section [Deutsche Museum, Oberschleissheim]; b) Side view of wire braced (Spad 13) and diagonal members fuselages (Hansa Brandemburg) [10]

Wire bracing even having low weight, needed a skilled mechanic to adjust and maintain them, having the risk of structure failure if one wire fails. For theese reasons, manufactures tend to replace wires with diagonal members, leading also to simplified structure joints (Figure 2 b).

Camm in 1919 noted that the design and type of fittings employed for connecting the latticed beam members varies greatly, being one of the distinctive constructional details of a plane, being mainly the result of desire for originality of each individual designer and had to disappear with the progress of the industry [03]. Even this aspect leads to manufacturing and productivity problems, it was propagated until nowadays.

In the early days of aviation, the fuselage fittings were made of aluminium alloy, but after in 1915 the standard fittings were from stamped steel [10].

Figure 3 shows different types of systems: "U"shaped bolts - Bleriot (a), aluminium sockets – Deperdussin (b), stamped and bent sheets with welded sockets – German Aviatik (c). The fittings were provided with eyelets for wire attachments.

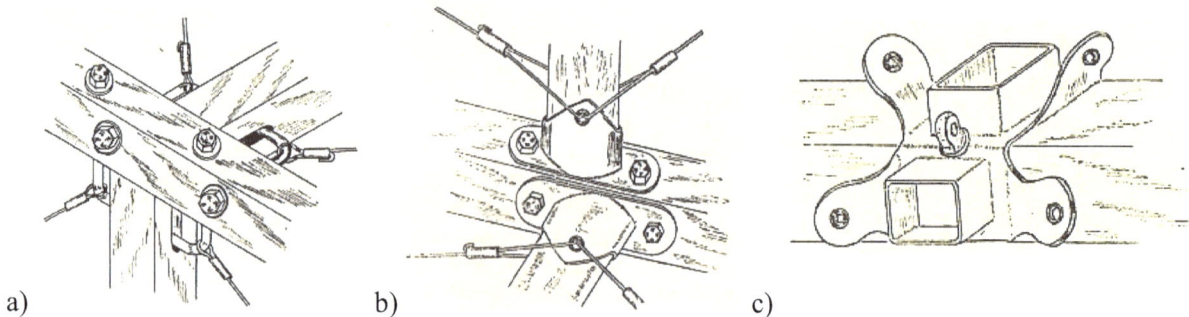

a)   b)   c)

Figure 3: Latticed beam fittings: a) Bleriot; b) Aluminium Socket; c) Stamped and welded [03]

Because of problems related to wooden/ fabric construction, manufacturers searched for solutions to eliminate the lack of durability, crashworthy and damage tolerance, flammability, anisotropy, wire bracing and so on. Between latticed beams and semimonocoque, there were a lot of concepts as:
- Beam without diagonals, with external stiffening skin;
- Beam with longerons, frames column members replaced by frames with stiffening skin;
- Beam with a system of stringers replacing the longerons, reinforced by circular soft frames and swept skin;
- Beams with frames, diagonal wire bracing and skin

As materials there were employed:
- Wood, aluminium or steel tubes for longerons, columns, frames and diagonal members
- Plywood, veneer, steel or aluminium sheet metal (plain or corrugated) for skin and frames
- Aluminium or steel for connections and fittings

It can be concluded a big mixture of members type and materials was used; in a logical approach these concepts leads to the semimonocoque, but many of this solutions were used many years after successful semimonocoque plane. Even semimonocoque appeared in 1912 (Deperdussin) showing improved features and characteristics, because of manufacturing costs it was not implemented in the design of the new aircrafts of the era.

One of the biggest requirement of the First World War was the short development time, as a response to battlefield request. There were aircrafts which were released only after 3 – 4 months after first hand sketches [07].

The transition from bamboo and wood to metal led to the requirement of new joining techniques. Starting from 1907, Anthony Fokker used on his airplanes welded structure, joining up to eight members also with hinged struts and wires (Figure4 a).

Since 1919, Camm stated that metal tubing is the most practical form in which steel can be used on aircrafts [03]. In Figure 4 b) is presented a joint with gusset like curved tublets for wire connections. There were used also spruce filled tubes to prevent local buckling [03].

The welded tubes structures were known and appreciated for accuracy and productivity since First World War, but welding techniques needed more progress in order to compete with wooden structures. Flight magazine noted in 1918, that Fokker structure welded nodes are the result of an "excellent workmanship". Early problems with welding led to the idea the welding depends a lot of the welder skills, this being propagated until present time.

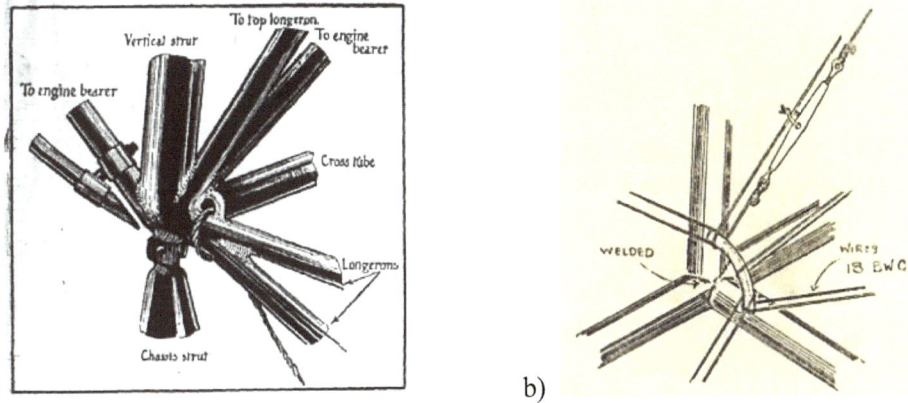

Figure 4: Welded joints on early planes: a) Fokker joint [Flight, Oct 1918]; b) Camm joint [03]

The manufacturing drawbacks due to the lack of knowledge added to welds fatigue failure determined producers to find alternate solutions as bracket or riveted joints..

4. NON-WELDED LATTICED BEAM

Riveted sheet metal structures got in to aviation very early, being employed by engineers working in Zeppelin team. Having the experience of huge airships structures, they come with the experience of lightweight latticed beams, with lightweight details. Aplying this first to flying boats (the biggest ariplanes of that time), the structural members were hollow structures or latticed beams, leading to higly complex structural nodes (Fig. 5).

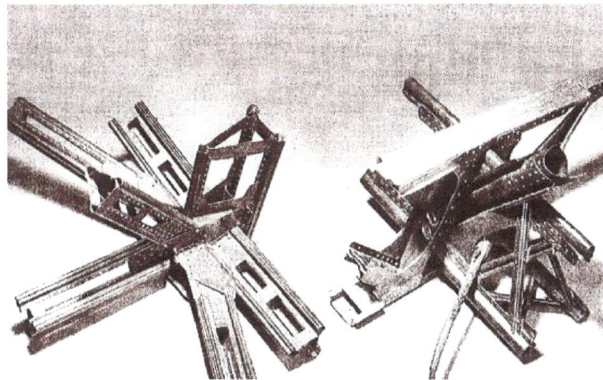

Figure 5: Hollow sections members riveted nodes of Dornier Rs.I flying boat (1915) [13]

After J1 unsuccessful attempt to use metallic welded stressed skins, Junkers concentrated himself on multiplanar latticed structures. These structures needed dedicated nodes, Junkers using as members closed and also open profiles (rarely used in aerospace) with formed ends. The joint was secured by formed steel brackets riveted together with members ends (Figure 6 a).

Other designs used machined tube end fittings welded or riveted on members. Welded was replaced because circular welds were not stiff enough (welds have to be used in order to allow shear and avoid tensile loading). Figure 6 b) shows an example with tubes connected to lugs with threaded shaft. Other example is a planar bracket joining the riveted end fork members (Figure 6 c).

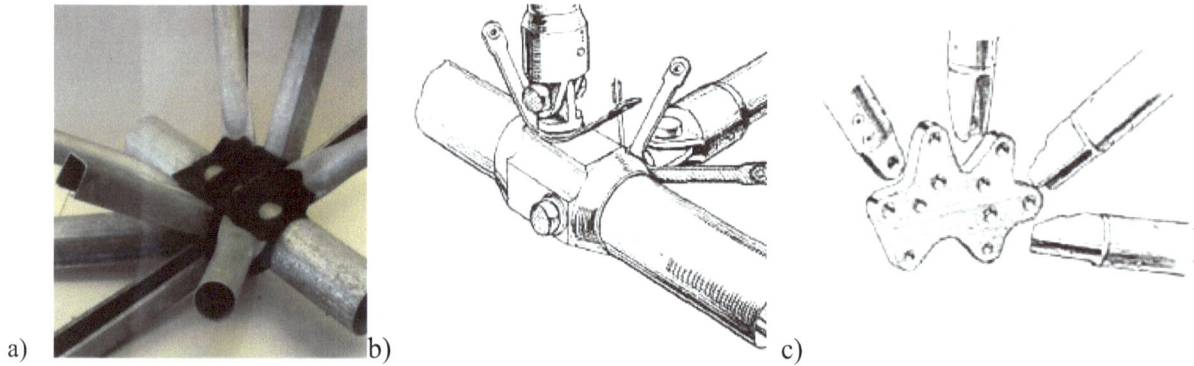

Figure 6: a) Formed joint brackets (Junkers, 1924) [Deutsche Museum, Oberschleissheim];
b) Machined ends tubes joint (Sidestrand, 1926) [Flight, Jul 1929];
c) End forks riveted tube joint [Flight, May 1930]

In 1928 Blackburn Lincock used a hibrid solution by riveting sheet metal brackets to the end of beam members. Brackets were attached by tubular rivets which in time had not satisfactory results; in present this kind of rivets are not used in structural applications (Figure 7 a). The planar columns and diagonals subassemblies were riveted, while they were screw mounted on longerons.

Short find a different solution to provide a smooth tension flow from structure attaching points to the beam members by inserting gussets in splitted end of members. The connection between members and gussets was made by the meaning of twin doublers riveted to members with blind rivets (Figure 7 b). Even if this concept is very robust and insures a long service life, it employs a big number of workmanship to every joint.

In present times blind rivets are not allowed for structural applications.

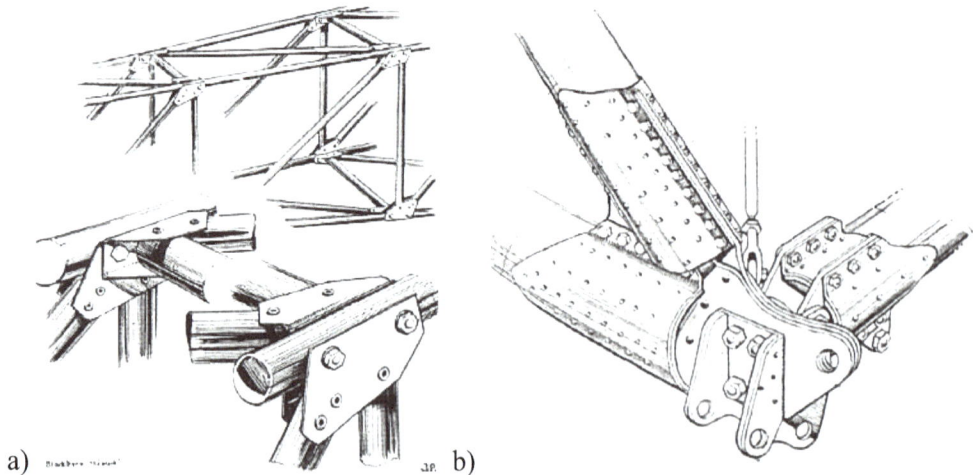

Figure 7: a) End brackets riveted to members (Blackburn, 1928) [Flight, Jul 1928];
b) Members with riveted end doubler and gussets (Shorts Valletta, 1930) [Flight, Jul 1930]

The lightweight structures employ thin walled structures, thus buckling being one of the biggest problem. To prevent buckling, Bristol used in 1929 corrugated sheet members. In Flight magasine (1929) is mentioned that corrugated sheet columns had a better buckling behaviour than circular section tubes (Figure 8 a).

Making an assesment study, the conclusions are:
- For the same wall thickness and critical general buckling force, the corrugated sheet column weight is 40% bigger than circular section tube. The local buckling critical force is 90% of the circular section tube.

- For the same mass and wall thickness, the general buckling critical force of the corrugated sheet column is 35% lower than the circular section tube. The general local buckling force is 90% form the circular section tube.
- The corrugated columns do not save weight and do not improve buckling behaviour by geometry; the single improvement can be acomplish by local hardening obtained by small bend radius of corrugations.

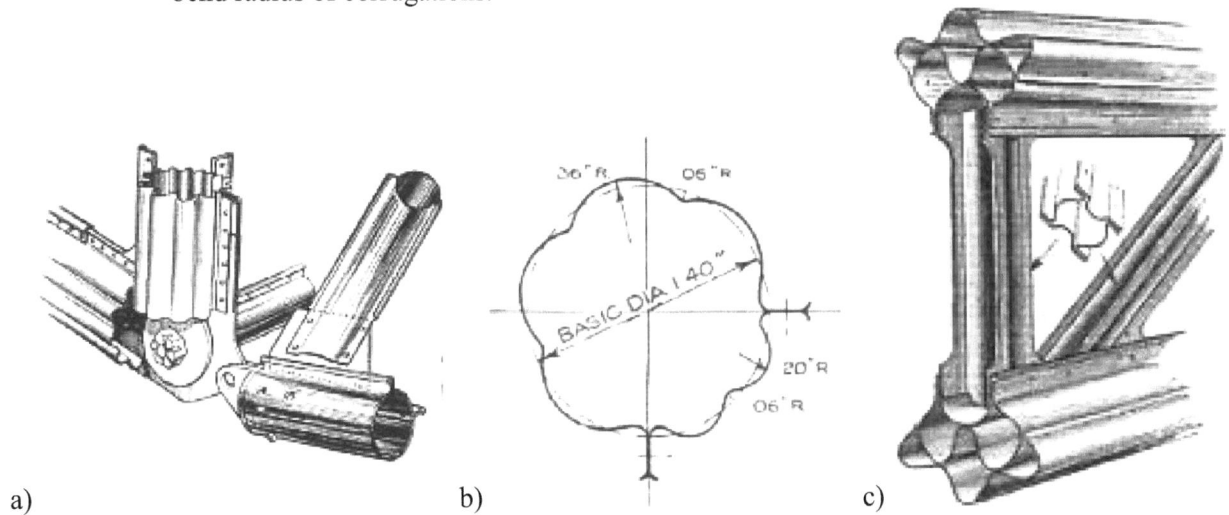

a)   b)   c)

Figure 8: Corrugated sheet members with gussets (Bristol 110A, 1929) [Flight, Jul 1929];
b) Corrugated member section [Flight, Feb 1928]; Corrugated sheet longeron (Stal 2) [Flight, Nov 1934]

In the thirties spot welding start to be used in aerospace; in Figure 8 c) is given an example from Stal 2 plane – a very complex and expensive solution. Currently the spot welding has very limited applications on aircraft primary structures.

In 1934 Shorts used riveted profiles columns to improve buckling behaviour. Gussets were used not only to join the members, being bigger than joint dimensions. The solutions were similar with metallic bridges (Figure 9). Riveted profiled members were used for many aircrafts, as usual for longerons or high loaded stringers. For closed section members (hollow structures) currently are used omega and "U" profiles.

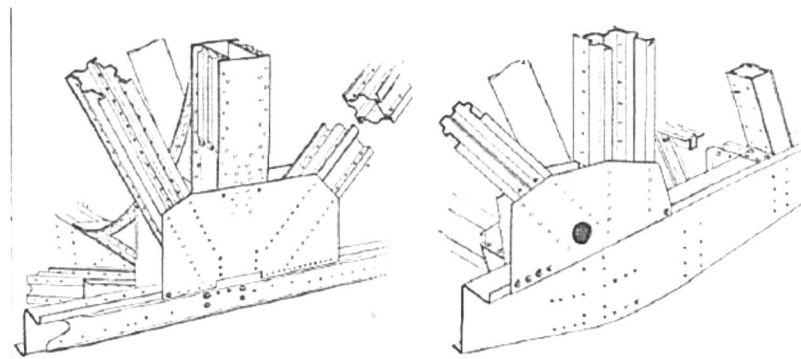

Figure 9: Riveted profiles columns with gussets (Scylla, 1934) [Flight, Apr 1934]

In 1928 Flight presents a simplified structure with gussets junctions to every node (Fig. 10). Gussets were inserted in the symmetry plane of members manufactured from two or three omega profiles riveted.

Even presents a big number of rivets and the weight is increased by the flanges of the members, the structure is simpler than the structures used before, having also a better shock behaviour (in hard landing).

The gussets help also to decrease the buckling length; a 20% reduction of buckling length leads to 50% increase of buckling force.

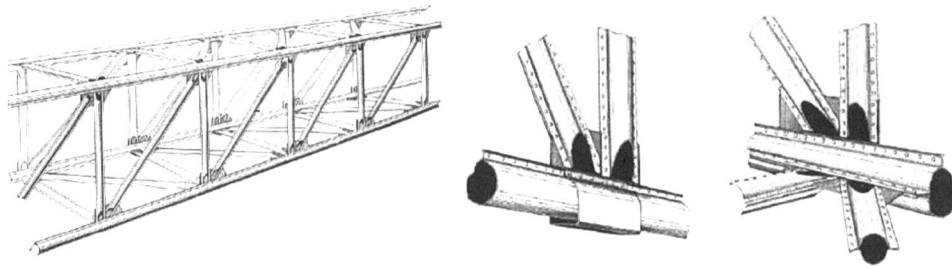

Figure 10: Latticed beam and joints of riveted omega profiles members [Flight/ Aircraft Engineer, Feb 1928]

A late application of brackets mounted latticed beam is met on Hawker Hurricane (1935). Using steel longerons and aluminium columns and bracing, it employed steel brackets both bolted and riveted to beam members. Figure 11 shows the fuselage and centre structure beam and few details of nodes with different geometry or concept leading to a big development effort.

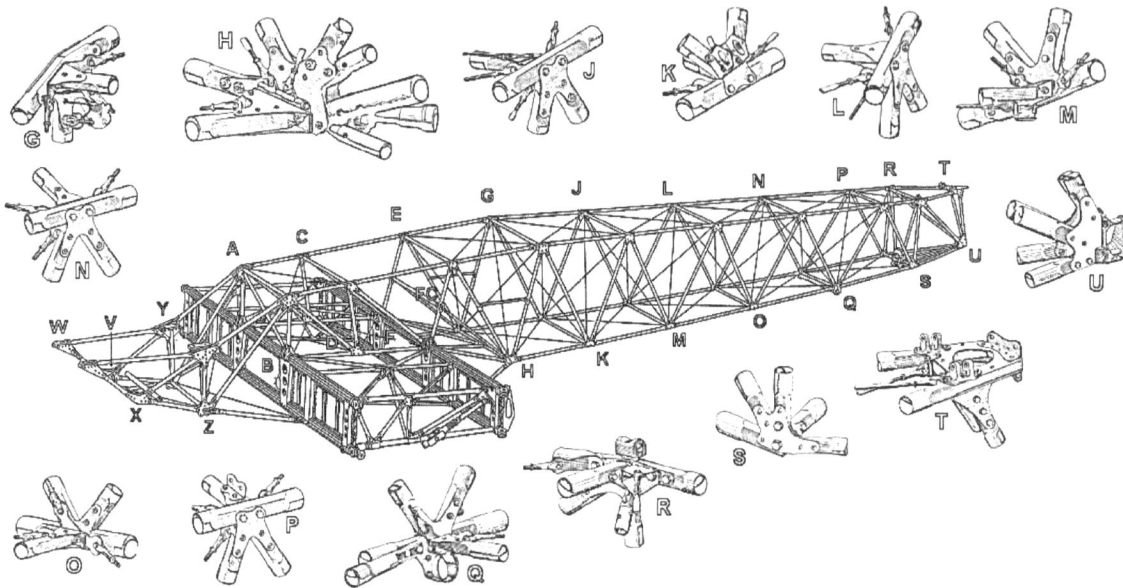

Figure 11: Central structure and fuselage (Hawker Hurricane, 1935) [militaryphotos.net]

5. WELDED LATTICED BEAM

Parallel with developing riveted or bolted concepts, progresses were made with welded joints. Intermediate concepts are shown in Fig. 12 a) and b) with doublers and gussets inserted in members end. Doubler and local stiffening plates are used also in civil engineering; in aircraft remained only members direct welded to each others, employing or not gussets.

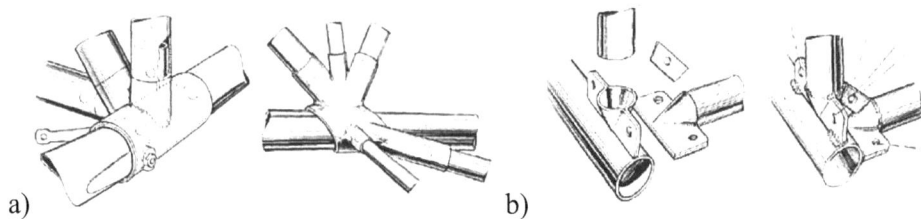

a)   b)

Figure 12: a) Doubler joints; b) Members inserted gussets joints [Flight, May 1929]

In 1930 Avro Trainer presented more new solutions which being kept and developed in time as follows:

- Bushing inserted in tubes for external members bolted attachments (Fig. 13 a);
- Formed end tubes for joining smaller diameters tuber (Fig. 13 b);
- Tangent placed gussets used also for external fittings attachement (Fig. 13 c, d)

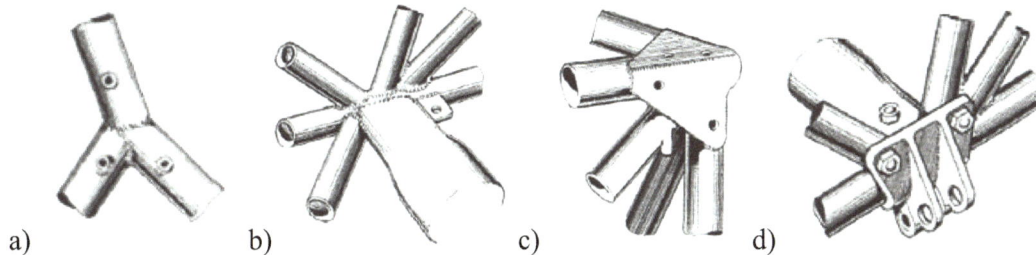

a)   b)   c)   d)

Figure 13: a) Tubes inserted bushing; b) Joined formed end tube;
c) Gusset as basement for fitting; d) Fitting attached to welded gusset [Flight, 1930]

Even in time there were a lot of alternative to welded joints for latticed beams, currently is used only welding for tubular structures.

Welded fuselage is not employed on airplanes from 40's, being extensive used for helicopters center and tail boon structures up to 70's. For primary structures, latticed beam is used nowadays only for small aircraft fuselage, the wings being made of riveted aluminium structure.

6. CONCLUSIONS

The first reliable structure for aircraft was the latticed beam, solution inspired from civil engineering. It was used for fuselage and wing, and it was a standard up to thirties, until semimonocoque (originated from ships engineering) [05], [01] finally demonstrated its superiority. Biplane wing was used until monoplane demonstrates its aerodynamic superiority and its internal structure riched a level where the beam made by two planes was no needed anymore.

The latticed beams were made by wooden members with wire bracing until steel or aluminum tubes replaced them, needing a simpler joining. The transition from latticed beam fuselage to semimonocoque was somehow superposed over the transition form wood to metal, but they were almost independent processes even they interfered a lot [06].

Welding is used from the early aircraft but in that time the technology was not mature. For this reason, for more than 30 years manufactures searched for alternative to welding, joining type leading even to members different construction.

Latticed beams are used only in limited application for fuselage, having welded nodes. Weld is used in many other application on aircrafts but not in latticed beam structures (landing gears, empennage, seats, etc).

REFERENCES

[01]   Anderson J. D., Introduction to flight, McGraw Hill, 2004
[02]   Balotescu N., s.a., Istoria aviaţiei române, Editura ştiinţifică şi enciclopedică, Bucureşti, 1984
[03]   Camm S., Aeroplane construction, Crosby Lockwood, London, 1919
[04]   Corona E., Notes on Aerospace Structures, AME 30 341, University of Notre Dame, 2006
[05]   Groh R., A Brief History of Aircraft Structures, aerospaceengineeringblog.com/aircraft-structures/, 2012

[06]   Jakab P. L., Wood to metal: The structural origins of the modern airplane, Journal of Aircraft, Vol 36, 1999

[07]   Loftin L. K., Quest for performance. Evolution of modern aircraft, NASA Scientific and Technical Infromation Branch, Washington, D.C., 1998

[08]   Niccoli, R. History of Flight, White Star, Italy, 2002

[09]   Pomilio O., Airplane design and construction, McGraw Hill, New York, 1919

[10]   Rathburn J. B., Aeroplane construction and operation, Stanton and Van Vljiet, Chicago, 1918

[11]   Vivian E. Ch., A History of Aeronautics, Harcourt, New York, 1921

[12]   Wells M., A history of engineering and structural design, Routledge, Oxfordshire, 2010

[13]   * * *, Dornier Post Sonderaurgabe, Dornier, Friedrichshafen,1984

[14]   * * * , History of Aircraft Structures and Structural Design Considerations for Contemporary Aircraft, engr.sjsu.edu

OPTIMIZATION OF SPECIFIC FACTORS TO PRODUCE SPECIAL ALLOYS

I. Milosan

Transilvania University of Brasov, ROMANIA, milosan@unitbv.ro

Abstract: *Finding the best solution from all the industrial solution is an optimization problem. The paper presents an experimantal study regarding the casting of 5 special alloys parts using. For this process it was calculate the optimization of the specific hourly productivity and the cost of each line in hand, aiming to achieve maximum benefit, using the Phases I problem of the Simplex algorithm. The number of unknowns components of calculation using the classical method is large, difficult, requiring a large volume of work and are insufficiently precise, all work was done whit the optimization by the linear programming, using a personal C-Soft.*

Keywords: *optimization, linear programming, objective function, C-Soft, special alloys*

1. INTRODUCTION

In optimization problems, we have to find solutions which are optimal or near-optimal with respect to some goals. Usually, we are not able to solve problems in one step, but we follow some process which guides us through problem solving. Often, the solution process is separated into different steps which are executed one after the other. Commonly used steps are recognizing and defining problems, constructing and solving models, and evaluating and implementing solutions [1].

In an optimization problem, the types of mathematical relationships between the objective and constraints and the decision variables determine how hard it is to solve, the solution methods or algorithms that can be used for optimization, and the confidence you can have that the solution is truly optimal.

The optimization of production processes outside materials must be made in relation to an economic criterion, so the function should be an objective indicator of economic efficiency of the process analyzed [1].

The main optimization criteria are economic, technical and economic nature.

Planning processes to solve planning or optimization problems have been of major interest in operations research [1-5]. Planning is viewed as a systematic, rational, and theory-guided process to analyze and solve planning and optimization problems.

The planning process consists of several steps:
1. Recognizing the problem,
2. Defining the problem,
3. Constructing a model for the problem,
4. Solving the model,
5. Validating the obtained solutions, and
6. Implementing one solution.

2. STANDARD FORM

A first stage of optimization is to determine the mathematical model and the second step is finding the optimum coordinates in the multifactorial space. This means determining the extreme values (maximum or minimum) optimized parameters and factors, which receives the optimized parameter values, this step is even calling optimization.

Optimizing a process in terms of technological restriction require the use of a mathematical model which contains the optimized and restrictions of type equality and inequality [1].

If the optimized function and the restrictions are linear, then linear programming is used and when the optimized function and constraints are used nonlinear programming [3, 4].

The most used method to optimize the restriction is the Simplex algorithm method [4, 5].

Simplex algorithm applies when the number of equations (m) and number of variables (n) is large and full description of the method is cumbersome.

To optimize an industrial process using linear programming, the mathematical model consists of three parts [1]: objective function, limitations and non negative restrictions problem:

a) The objective function:

$$F = \sum_{j=1}^{n} c_j x_j \; ; j = 1, 2, ..., m, m+1, .., n \tag{1}$$

where: $x_j (x_1, x_2, ...x_n)$ = system variables-process's parameters;
c_j - connection coefficients<

b) The inequality constrants problem (functional conditions of the process):

$$\sum_{\substack{j=1 \\ i=1}}^{n,m} a_{ij} x_j \geq b_i \tag{2}$$

where: $j = n$; $i = m$. (m < n)

$$\sum_{\substack{j=1 \\ i=1}}^{n,m} a_{ij} x_j \leq b_i \tag{3}$$

$$\sum_{\substack{j=1 \\ i=1}}^{n,m} a_{ij} x_j = b_i \tag{4}$$

where: a_{ij} are called coefficients technology and can be each positive, negative or null.

c) the nonnegative terms:

$$x_j \geq 0 \tag{5}$$

The relations (1) - (5) form a Canonical Linear program (CLP)

The optimization of such problems it made through the following steps [1]:
1. Establishment of Canonical Linear Program (CLP);
2. Perform linear Canonical Linear Program (CLP) in the Standard Linear Program (SLP) by adding or subtracting (depending on the shape of each inequality: $\geq$, $\leq$, =) of variable spacing (x_{ie}) or artificial variables (x_{ia}) variables are added in order to easily get value system (for finding the solution to start).
3. Next step in the resolving this optimization problem is to finding initial starting solution, by equating to 0 the next equation:
$$(n-m)_n = 0 \tag{6}$$
where: n = number of the unknown of the problem and m = number of the equations
From the relation presented, it was determine the base variables (BV) and the values of base variables (VBV).
4. Simplex table is built, starting by iteration 0 (iteration= changing of base) presented in table 1

Table 1: Simplex Tableau, iteration 0

c_j / c_i	VB	VVB	c_1 x_1	c_2 x_2	...	c_r x_r	...	c_m x_m	c_{m+1} x_{m+1}	...	c_k x_k	...	c_n x_n
c_1	x_1	x_1	1	0	...	0	...	0	$a_{1,m+1}$	...	a_{1k}	...	a_{1n}
c_2	x_2	x_2	0	1	...	0	...	0	$a_{2,m+1}$	...	a_{2k}	...	a_{2n}
.	.	.	.	.	...	.	...	.	.	...	.	...	.
.	.	.	.	.	...	.	...	.	.	...	.	...	.
c_r	x_r	x_r	0	0	...	1	...	0	$a_{r,m+1}$	...	a_{rk}	...	a_{rn}
.	.	.	.	.	...	.	...	.	.	...	.	...	
.	.	.	.	.	...	.	...	.	.	...	.	...	
c_m	x_m	x_m	0	0	...	0	...	1	$a_{m,m+1}$	...	a_{mk}	...	a_{mn}
		z_j z_0	z_1	z_2		z_r	...	z_m	z_{m+1}	...	z_k	...	z_n
		$z_j - c_j$	z_1-c_1	z_2-c_2	...	z_r-c_r	...	z_m-c_m	$z_{m+1}-c_{m+1}$	...	z_k-c_k	...	z_n-c_n

where:

BV - the base variables

VBV - the values of base variables

x_j and x_i = system variables-process's parameters;

$x_j (x_1, x_2, x_m, \ldots x_n)$ and $x_i (x_1, x_2, x_m)$; $j = n$; $i = m$. ($m < n$)

After the iteration, considering differences $z_j - c_j$, analysis is accomplished according to the shape of the program (maximum or minimum) in resolving the case [1].

This results on this study optimization by linear programming, it was verified using software in C-SOFT for rapid optimization of operating plants with limited representative sample utilization .To solve, the one-phase approach is applied [1].

3. EXPERIMENTAL RESEARCHES

A first stage of optimization is to determine the mathematical model and the second step is finding the optimum coordinates in the multifactorial space. This means determining the extreme values (maximum or minimum) optimized parameters and factors, which receives the optimized parameter values, this step is even calling optimization.

In this experiment it was intended to study the obtaining of 5 cast iron landmarks (R1-R5) using molybdenum, nichel and copper as a alloying elements of cast iron.

For this process were used in the following specifications: specific consumption for each milestone achieved daily, quantity available of alloying cast iron elements, aiming to achieve maximum benefit, using the Simplex algorithm - the Phases I problem [1, 2].

The optimization of the obtaining of 5 cast iron parts is made in relation to an economic criterion, so the function is an objective indicator of economic efficiency of the process analyzed [6, 7].

The presentation of the input data is presented in table 2.

Proceedings of COMEC 2013

Table 2: The presentation of the input data

Alloys	Specific consumption for each milestone achieved daily, [kg]					Quantity available, [kg]
	R1	R2	R3	R4	R5	
Mo Cast iron	10	20	40	-	20	20000
Ni Cast iron	20	-	20	10	20	5000
Cu Cast iron	20	-	20	20	10	10000
Benefit [Euro]	10	20	20	40	10	-

Analyzing Table 2 it mentions the following:

- specific daily consumption to achieve a specific type pieces R1, consuming 10 kg of Mo Cast iron , 20 kg of Ni Cast iron and 20 kg of Cu Cast iron, brings a benefit of 10 Euro;
- specific daily consumption to achieve a specific type pieces R2, consuming 20 kg of Mo Cast iron, brings a benefit of 20 Euro;
- specific daily consumption to achieve a specific type pieces R3, consuming 40 kg of Mo Cast iron, 20 kg of Ni Cast iron and 20 kg of Cu Cast iron, brings a benefit of 20 Euro;
- specific daily consumption to achieve a specific type pieces R4, consuming 10 kg of Ni Cast iron and 20 kg of Cu Cast iron, brings a benefit of 40 Euro;
- specific daily consumption to achieve a specific type pieces R5, consuming 20 kg of Mo Cast iron, 20 kg of Ni Cast iron and 10 kg of Cu Cast iron, brings a benefit of 10 Euro;

1) Establishment of linear canonical program (CLP)

a) The objective function (function to be optimized is the beneficial):

$$F = 10x_1 + 20x_2 + 20x_3 + 40x_4 + 10x_5 = \max \tag{7}$$

b) The restrictions problem is:

$$10x_1 + 20x_2 + 40x_3 + 20x_5 \leq 20000 \tag{8}$$
$$20x_1 + 20x_3 + 10x_4 + 20x_5 \leq 5000 \tag{9}$$
$$20x_1 + 20x_3 + 20x_4 + 10x_5 \leq 10000 \tag{10}$$

c) The non negativity conditions:

$$x_1 \geq 0 ; x_2 \geq 0; x_3 \geq 0; x_4 \geq 0; x_5 \geq 0 \tag{11}$$

2) Perform linear canonical transformation program (CLP) in the standard linear program (SLP) by adding or subtracting (depending on the shape of each inequality: $\geq$, $\leq$, =) of variable spacing (x_{ie}) variables are added in this care in order to easily get value system (for finding the solution to start).

a) The objective function

$$F = F = 10x_1 + 20x_2 + 20x_3 + 40x_4 + 10x_5 = \max \tag{12}$$

b) The restrictions problem is:

$$10x_1 + 20x_2 + 40x_3 + 20x_5 + x_{1e} = 20000 \tag{13}$$
$$20x_1 + 20x_3 + 10x_4 + 20x_5 + x_{2e} = 5000 \tag{14}$$
$$20x_1 + 20x_3 + 20x_4 + 10x_5 + x_{3e} = 10000 \tag{15}$$

c) The non negativity conditions:

$$x_1 \geq 0 ; x_2 \geq 0; x_3 \geq 0; x_4 \geq 0; x_5 \geq 0; x_{1e} \geq 0 ; x_{2e} \geq 0; x_{3e} \geq 0; \tag{16}$$

3) From the relation presented, it was determine the base variables (BV) and the values of base variables (VBV), presented in table 3.

Table 3. Base variables (BV) and values of base variables (VBV)

BV	VBV
x_{1e}	20000
x_{2e}	5000
x_{3e}	10000

The values from VBV are considered to be an admissible basic solution [1];

4) Simplex table is built, starting by iteration 0 (iteration= changing of base), presented in table 4.

Table 4: Simplex Tableau, Phase I, iteration 0

c_i \ c_J	BV	VBV	0	0	0	10	20	20	40	10
			x_{1e}	x_{2e}	x_{3e}	x_1	x_2	x_3	x_4	x_5
0	x_{1e}	20000	1	0	0	10	20	40	0	20
0	x_{2e}	5000	0	1	0	20	0	20	10	20
0	x_{3e}	10000	0	0	1	10	20	0	**20**	10
z_j		0	0	0	0	0	0	0	0	0
	$z_j - c_j$	0	0	0	- 50	- 100	- 100	- 200	- 50

Because is an maximum program optimization, it was analyzed all differents $z_j - c_j$

- establish a procedure that allows moving from one base to another;
- basic changes are made by decreasing values (problem solved is maximum) optimization function;
- stops the iteration process (moving from one base to another), when it is not possible to increase the value of optimization function.
- enter the base, the x_4 is the entering variable in the base and x_{2e} is the leaving variable of the base;
- value of **20** from the column of x_4 are called the pivot operation. The simplex algorithm proceeds by performing successive pivot operations which each give an improved basic feasible solution; the choice of pivot element at each step is largely determined by the requirement that this pivot improve the solution.
- Stops the iteration process (moving from one base to another) when it is not possible to decrease the value of optimization function, so to reach the optimal solution, F=max, with solution $x_{i\ optimum}$ and all differents $z_j - c_j \geq 0$, so it reached the optimal solution, results presented in table 5.

Table 5. :Simplex Tableau, iteration 1

c_i \ c_J	BV	VBV	0	0	0	10	20	20	40	10
			x_{1e}	x_{2e}	x_{3e}	x_1	x_2	x_3	x_4	x_5
0	x_{1e}	20000	1	0	-	10	**20**	40	0	0
0	x_{2e}	4500	0	1	-	10	0	10	0	10
40	x_4	50	0	0	-	1	0	1	1	1/2
z_j		2000	0	0	-	40	0	40	40	20
	$z_j - c_j$	0	0	-	30	- 20	20	0	10

- enter the base, the x_2 is the entering variable in the base and x_{1e} is the leaving variable of the base;
- value of **20** from the column of x_2 are called the pivot operation.

In table 6 are presented Simplex table, iteration 2.

Table 6: Simplex Tableau, iteration 2

c_J / c_i	BV	VBV	0	0	0	10	20	20	40	10
			x_{1e}	x_{2e}	x_{3e}	x_1	x_2	x_3	x_4	x_5
20	x_2	1000	-	0	-	1/2	1	2	0	0
0	x_{2e}	4500	-	1	-	10	0	10	0	10
40	x_4	50	-	0	-	1	0	1	1	1/2
z_j		22000	-	-	0	50	20	80	40	20
	$z_j - c_j$		-	-	0	40	0	60	0	1-

Because all differents $z_j - c_j \geq 0$, so it reached the optimal (maximum) solution
The mathematical results of the optimal solutins are:

$z_0 = F_{max} = 22000$ (benefit in Euro), with solutions:

$x_{2\ optimum} = 1000$ (the daily specific consumption of landmark R_2, in kg);

$x_{4\ optimum} = 50$ (the daily specific consumption of landmark R_4, in kg);

5. CONCLUSIONS

Analyzing all data taken into account, there can say the following:
- For this process it was calculate the optimization of the specific hourly productivity and the cost of each line in hand, aiming to achieve maximum benefit, using the Simplex algorithm the Phases I Method.
- In this case, because the number of components is large, the above methods are cumbersome, requiring a large volume of work, using the classical calculation.
- By using this software in place, reduce the computing time to several hours using traditional method to 2-3 minutes, getting an accurate result, respecting both the economic and technical component, without affecting the smooth running of the metallurgical process.
- The mathematical results of the optimal solutins of this application are: the benefit is 22000 Euro, producing landmak 2 and 4, consuming the quantities of 1000 and 50 kg respectively.

REFERENCES

[1] Taloi, D.: Optimization of metallurgical processes. Application in metalurgy, Didactic and Pedagogical Publishing House, Bucharest, (1987).

[2] Babu, B., V., Angira, R.: Optimization of Industrial Proceses Using Improved and Modified Differential Ev olution, Springer-Verlag, Berlin, (2008).

[3] Anderson, C., G.: Applied metallurgical process testing and plant optimization with design of experimentation software. Minerals, Metals & Materials Society, part I, 1-26, (2006).

[4] Klemes, J., Friedler, F., Bulatov, I., Varbanov, P.: Sustainability in the Process Industry: Integration and Optimization, Green Manufacturing & Sistem Engineering, Manchester, (2011).

[5] Liptak, B., G.: Optimization of Industrial Unit Processes-Second Edition, CRC Press, Boca Raton, (1998).

[6] Gui, W-I., Yang, C.-H., Chen, X.-F.,Wang. Y.-L.: Modeling and optimization problems and challenges arising in Nonferrous Metallurgical Processes. Acta Automatica Sinica, , 39(3), 197–207, (2013).

MINIMUM COST DESIGN OF A RING-STIFFENED CYLINDRICAL SHELL LOADED BY EXTERNAL PRESSURE

József Farkas[1], Károly Jármai[2]

[1] Professor emeritus, Dr.sci.techn. University of Miskolc, Hungary, altfar@uni-miskolc.hu

[2] Professor, Dr.sci.techn. University of Miskolc, Hungary, altjar@uni-miskolc.hu

Abstract: The aim of this paper is to find the minimum cost of a ring-stiffened circular cylindrical shell loaded by external pressure. The minimum cost is given by the optimum dimensions, which can be calculated by an optimization technique. The calculation shows that the cost reduction has an effect reducing the shell diameter. The decrease in diameter restricted by a production constraint, that the inner diameter should be minimum of 2 m, to allow the welding and painting within the shell. This paper describes the optimization of this kind of structure considering cost calculation, which includes not only the material, but welding and painting costs as well.

Keywords: *stiffened shell, minimum cost design, ring stiffeners*

INTRODUCTION

The cylindrical shell used in various structures, such as pipelines, offshore structures, columns and towers, bridges, silos, etc. shell is stiffened against buckling of the ring - stiffeners or stringers or perpendicular. The efficiency depends on the type of stiffening load. In many cases, the loads and brace studied in comparison with the cost of realistic numerical models and concluded by the structural design aspects of the optimized versions [1,2,3,4,5].

Since in Eurocodes [6] design method for stiffened shell buckling is not given, the design rules of Det Norske Veritas [7] are used. In this new investigation newer DNV shell buckling formulae are applied.
Optimum design of ring-stiffened cylindrical shells has been treated in [8, 9]. The results of model experiments for cylindrical shells used in offshore oil platforms have been published by Harding [10]. Cho and Frieze [11] have compared the proposed strength formulation with DNV rules, British Standard BS 5500 and experimental results.

The tripping of open section ring-stiffeners is treated by Huang and Wierzbicki [12]. Buckling solutions for shells with various end conditions, stiffener geometry and under various pressure distributions have been presented by Wang et al. [13] and by Tian et al. [14].

In Akl et al. [15] the adopted approach aims at simultaneously minimizing the shell vibration associated sound radiation, weight of the stiffening rings as well as the cost of the stiffened shell. The production costs as well as the life cycle and maintenance costs are computed using the Parametric Review of Information for Costing and Evaluation (PRICE) model (PRICE System, Mt. Laurel, N.J. 1999) without any detailed cost data.

In the optimization process the optimum values of shell diameter and thickness as well as the number and dimensions of ring-stiffeners are sought to minimize the structural volume or cost. In order to avoid tripping welded square box section stiffeners are used, their side length and thickness of plate elements should be optimized.

Besides the constraints on shell and stiffener buckling the fabrication constraints can be active. To make it possible the welding of stiffeners inside the shell the minimum shell diameter should be fixed (2000 mm). vThe calculations show that the volume and cost decreases when the shell diameter is decreased. Thus, the shell diameter can be the fixed minimum value. Another fabrication constraint is the limitation of shell and plate thickness (4 mm).

The remaining unknown variables can be calculated using the two buckling constraints and the condition of volume or cost minimization. The relation between the side length and plate thickness of ring-stiffeners is determined be the local buckling constraint. To obtain the optimum values of variables a relative simple systematic search method is used.

The cost function contains the cost of material, assembly, welding and painting and is formulated according to the fabrication sequence.

1 CHARACTERISTICS OF THE OPTIMIZATION PROBLEM

Given data: external pressure intensity $p = 0.5$ N/mm^2, safety factor $\gamma = 1.5$, shell length $L = 6000$ mm, steel yield stress $f_y = 355$ MPa, elastic modulus $E = 2.1 \times 10^5$ MPa, Poisson ratio $v = 0.3$, density $\rho = 7.85 \times 10^{-6}$ N/mm^3, the cost constants are given separately.

Unknown variables: shell radius R, shell thickness t, number of spacing between ring-stiffeners n, thus, the spacing between stiffeners is $L_r = L/n$, the side length of the square box section stiffener h_r, the thickness of stiffener plate parts t_r.

2 CONSTRAINT ON SHELL BUCKLING

According to the DNV rules [7]

$$\sigma = \frac{\gamma p R}{t} \leq \frac{f_y}{\sqrt{1 + \lambda^4}}, \lambda = \sqrt{\frac{f_y}{\sigma_E}} \tag{1}$$

$$\sigma_E = \frac{C \pi^2 E}{12(1 - v^2)} \left(\frac{t}{L_r}\right)^2 \tag{2}$$

$$C = \psi \sqrt{1 + \left(\frac{\rho_1 \xi}{\psi}\right)^2}, \psi = 4, \rho_1 = 0.6 \tag{3}$$

$$\xi = 1.04\sqrt{Z}, Z = \frac{L_r^2}{Rt}\sqrt{1 - v^2} \tag{4}$$

3 CONSTRAINT ON RING-STIFFENER BUCKLING

The moment of inertia of the effective stiffener cross-section should be larger than the required one

$$I_x \geq I_{req} \tag{5}$$

The effective shell length between ring-stiffeners is the smaller of

$$L_e = \frac{1.56\sqrt{Rt}}{1 + 12\dfrac{t}{R}} \text{ or } L_r \tag{6}$$

The distance of the gravity centre of the effective ring-stiffener cross-section (Fig. 1)

$$y_E = \frac{L_e t \left(h_r + \frac{t + t_r}{2} \right) + h_r t_r \left(h_r + t_r \right)}{3 t_r h_r + L_e t} \tag{7}$$

The moment of inertia of the effective stiffener cross-section

$$I_x = \frac{t_r h_r^3}{6} + 2 t_r h_r \left(\frac{h_r + t_r}{2} - y_E \right)^2 + h_r t_r y_E^2 + \frac{L_e t^3}{12} + L_e t \left(h_r + \frac{t + t_r}{2} - y_E \right)^2 \tag{8}$$

The relation between h_r and t_r is determined by the local buckling constraint

$$t_r \geq \delta h_r, \delta = \frac{1}{42 \varepsilon}, \varepsilon = \sqrt{\frac{235}{f_y}} \tag{9}$$

For $f_y = 355$ $\delta = 1/34$, the required t_r is rounded to the larger integer, but $t_{rmin} = 4$ mm.
The required moment of inertia

$$I_{req} = \frac{\gamma p R R_0^2 L_r}{3E} \left[1.5 + \frac{3 E y_E 0.005 R}{R_0^2 \left(\frac{f_y}{2} - \sigma \right)} \right] \tag{10}$$

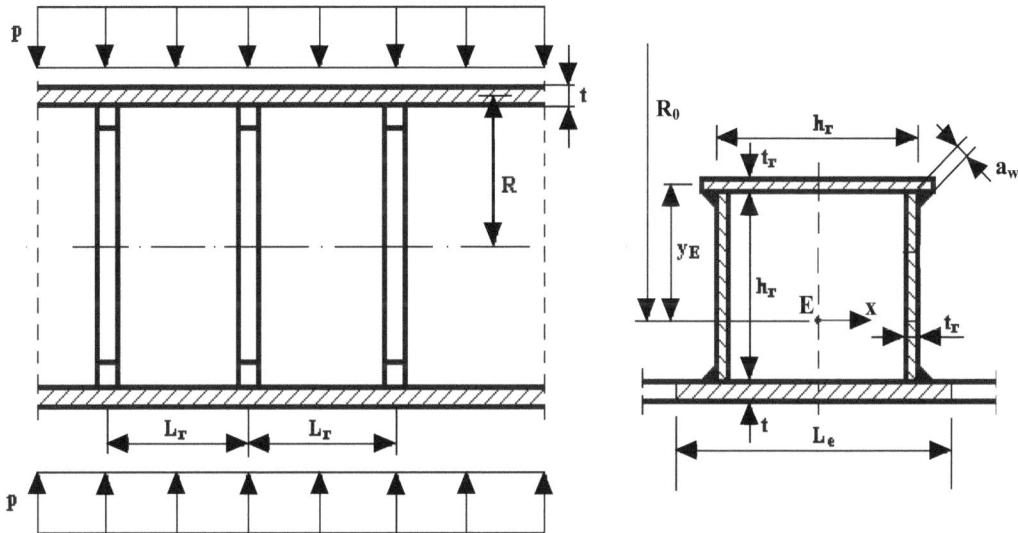

Fig. 1 Ring-stiffened cylindrical shell loaded by external pressure

4 THE COST FUNCTION

The cost function contents the cost of material, assembly, welding and painting and is formulated according to the fabrication sequence.

The cost of assembly and welding is calculated using the following formula [1, 2, 3, 5]

$$K_w = k_w \left(C_1 \Theta \sqrt{\kappa \rho V} + 1.3 \sum_i C_{wi} a_{wi}^n C_{pi} L_{wi} \right) \qquad (11)$$

where k_w [$/min] is the welding cost factor, C_1 is the factor for the assembly usually taken as $C_1 = 1$ min/kg$^{0.5}$, Θ is the factor expressing the complexity of assembly, the first member calculates the time of the assembly, κ is the number of structural parts to be assembled, ρV is the mass of the assembled structure

The second member estimates the time of welding, C_w and n are the constants given for the specified welding technology and weld type, C_p is the factor of welding position (for downhand 1, for vertical 2, for overhead 3), L_w is the weld length, the multiplier 1.3 takes into account the additional welding times (deslagging, chipping, changing the electrode).

The fabrication sequence is as follows:

(a) Welding the unstiffened shell from curved plate parts of dimensions 6000x1500 mm and of number

$$n_p = \frac{2R\pi}{1500},$$

which should be rounded to the larger integer. Use butt welds of length

$$L_{w1} = n_p L, \quad \Theta = 3, \kappa_1 = n_p, V_1 = 2R\pi L t, k_W = 1, \qquad (12)$$

the welding technology SAW (submerged arc welding)

for $t = 4$-15 mm $\quad C_{W1} = 0.1346 \times 10^{-3}$ and $n_1 = 2,$ \hfill (13a)
for $t > 15$ mm $\quad C_{W1} = 0.1033 \times 10^{-3}$ and $n_1 = 1.9,$ \hfill (13b)

$$K_{W1} = k_W \left(\Theta \sqrt{\kappa_1 \rho V_1} + 1.3 C_{W1} t^{n_1} L_{W1} \right). \qquad (14)$$

(b) Welding the ring-stiffeners separately from 3 plate parts with 2 fillet welds (GMAW-C –gas metal arc welding with CO_2):

$$K_{W2} = k_W \left(\Theta \sqrt{3\rho V_2} + 1.3 x 0.3394 x 10^{-3} a_W^2 L_{W2} \right) \qquad (15)$$

where

$$V_2 = 4\pi h_r t_r \left(R - \frac{h_r}{2} \right) + 2\pi h_r t_r \left(R - h_r \right) \qquad (16)$$

$$L_{W2} = 4\pi \left(R - h_r \right), a_W = 0.7 t_r \qquad (17)$$

(c) Welding the $(n+1)$ ring-stiffeners into the shell with 2 circumferential fillet welds (GMAW-C)

$$K_{W3} = k_W \left(\Theta \sqrt{(n+2)\rho V_3} + 1.3 x 0.3394 x 10^{-3} a_W^2 L_{W3} \right) \qquad (18)$$

where

$$V_3 = V_1 + (n+1)V_2, L_{W3} = 4R\pi (n+1) \qquad (19)$$

Material cost

$$K_M = k_M \rho V_3, k_M = 1 \text{ \$/kg} \qquad (20)$$

Painting cost

$$K_P = k_P S_P, k_P = 28.8x10^{-6} \text{ \$/mm}^2, \tag{21}$$

$$S_P = 2R\pi L + 2R\pi \left[L - (n+1)h_r \right] + 2\pi (R - h_r) h_r (n+1) + 4\pi \left(R - \frac{h_r}{2} \right) h_r (n+1) \tag{22}$$

The total cost

$$K = K_M + K_{W1} + (n+1) K_{W2} + K_{W3} + K_P \tag{23}$$

5 RESULTS OF THE OPTIMIZATION

In the following the minimum cost design is obtained by a systematic search using a MathCAD algorithm. For a shell thickness t the number of stiffeners n is determined by the shell buckling constraint (Eq. 1) and the stiffener dimensions (h_r and t_r) are determined by the stiffener buckling constraint (Eq. 5).

The search results for $R = 1851$ and 1500 (Tables 1 and 2) show that the volume and cost decreases when the radius is decreased. Thus, the realistic optimum can be obtained by taking the radius as small as possible. This minimum radius is determined by the requirement that the internal stiffeners should easily be welded inside of shell, i.e. $R_{min} = 1000$ mm. Therefore the more detailed search is performed for this radius (Table 3).

Table 1: Systematic search for $R = 1850$ mm. Dimensions are in mm. The minimum cost is marked by bold letters

t	n	$\sigma < \sigma_{adm}$ MPa	h_r	t_r	$I_x > I_{req}$ x10^{-4} mm^4	Vx10^{-5} mm^3	K \$
11	7	126<152	180	6	3352>3341	10490	18770
12	6	115<143	180	6	3530>3502	10830	18640
13	5	106<124	190	6	4245>4014	11290	18650
14	4	99<109	200	6	5050>4888	11710	**18620**
15	4	92<121	200	6	5252>4718	12400	19390

Table 2: Systematic search for $R = 1500$ mm. Dimensions are in mm. The minimum cost is marked by bold letters

t	n	$\sigma < \sigma_{adm}$ MPa	h_r	t_r	$I_x > I_{req}$ x10^{-4} mm^4	Vx10^{-5} mm^3	K \$
8	10	140<157	160	5	1745>1616	6830	13890
9	8	125<140	160	5	1590>1550	6870	13250
10	6	112<115	160	5	1995>1885	7130	**12900**
11	5	102<106	150	5	2109>2102	7480	12950
12	5	93<120	160	5	2217>2003	8050	13570

It can be seen from Table 3 that the optima for minimum volume and minimum cost are different. It is caused by the larger value of fabrication (welding and painting) cost. The details of the cost for $K = 7221$ \$ are given in Table 4.

Table 3: Systematic search for $R = 1000$ mm. Dimensions are in mm. Optima are marked by bold letters

t	n	$\sigma<\sigma_{adm}$ MPa	h_r	t_r	$I_x>I_{req}$ x10^{-4} mm^4	V x10^{-5} mm^3	K \$
5	16	150<156	110	4	402>364	3192	8338
6	12	125<141	100	4	353>296	**3177**	7631
7	9	107<123	100	4	387>336	3343	7321
8	7	94<111	100	4	419>400	3579	7244
9	5	83<90	110	4	572>557	3854	**7221**
10	4	75<82	120	4	759>703	4186	7419
11	3	68<69	130	4	982>953	4505	7598

Table 4: Details of the minimum cost in \$. (The sum of the welding and painting costs is \$4196)

K_M	K_{W1}	$(n+1)K_{W2}$	K_{W3}	K_P	K
3025	673	474	665	2384	7221

6 CONCLUSIONS

The structural volume and the cost decrease when the shell radius is decreased. Thus, the shell radius should be taken as small as possible. The minimum radius is determined by the limitation that the internal ring-stiffeners should welded into the shell ($R_{min} = 1000$ mm).

The shell thickness and the number of ring-stiffeners can be calculated using the constraint on shell buckling. In order to avoid ring-stiffener tripping, welded square box section rings are used. The dimensions of the rings can be determined from the constraint on ring-stiffener buckling. The constraints on buckling are formulated according to the newer DNV design rules.

In the cost function the costs of material, assembly, welding and painting are formulated. The welding cost parts are calculated according to the fabrication sequence. The optima for minimum volume and minimum cost are different, since the fabrication cost parts are relative high as compared to the whole cost.

The ring-stiffening is very effective, since in the case of $n = 1$ (only 2 end stiffeners) the required shell thickness is $t = 18$ mm, the volume is $V = 7144$x10^{-3} mm^3 and the cost is $K = \$10450$, i.e. the cost savings achieved by ring-stiffeners is $(10450-7221)/10450$x$100 = 31\%$.

ACKNOWLEDGEMENT

The research was supported by the TÁMOP 4.2.4.A/2-11-1-2012-0001 priority project entitled 'National Excellence Program - Development and operation of domestic personnel support system for students and researchers, implemented within the framework of a convergence program, supported by the European Union, co-financed by the European Social Fund. The research was supported also by the Hungarian Scientific Research Fund OTKA T 75678 and T 109860 projects and was partially carried out in the framework of the Center of Excellence of Innovative Engineering Design and Technologies at the University of Miskolc.

REFERENCES

[1]   Farkas J, Jármai K (1997) Analysis and optimum design of metal structures, Rotterdam, Brookfield, Balkema
[2]   Farkas J, Jármai K (2003) Economic design of metal structures, Rotterdam, Millpress
[3]   Farkas J, Jármai K (2008a) Design and optimization of metal structures, Chichester, UK, Horwood Publishing

[4]   Farkas J, Jármai K(2008b) Minimum cost design of a conical shell – External pressure, non-equidistant stiffening. In: Proceedings of the Eurosteel 2008 5th European Conference on Steel and Composite Structures Graz Austria. Eds Ofner R. et al. Brussels, ECCS European Convention for Constructional Steelwork Vol.B. 1539-1544.
[5]   Farkas J, Jármai K (2013) Optimum design of metal structures, Heidlberg, Springer Verlag
[6]   Eurocode 3 (2009) Design of steel structures. Part 1-1: General structural rules. Brussels, CEN
[7]   Det Norske Veritas (2002) Buckling strength of shells. Recommended Practice DNV-RP-C202. Høvik, Norway
[8]   Pappas M, Allentuch A (1974) Extended capability for automate design of frame-stiffened submersible cylindrical shells. Comput. Struct. 4: (5) 1025-1059.
[9]   Pappas M, Morandi J (1980) Optimal design of ring-stiffened cylindrical shells using multiple stiffener sizes. AIAA J 18: (8) 1020-1022.
[10]   Harding JE (1981) Ring-stiffened cylinders under axial and external pressure loading. Proc. Inst Civ. Engrs Part 2. 71:(Sept.) 863-878.
[11]   Cho SR, Frieze PA (1988) Strength formulation for ring-stiffened cylinders under combined axial loading and radial pressure. J. Constr. Steel Res. 9: 3-34.
[12]   Huang J, Wierzbicki T (1993) Plastic tripping of ring stiffeners. J. Struct. Eng Proc Am Soc Civ Eng 119: (5) 1622-1642.
[13]   Wang CM, Swaddiwudhipong S, Tian J (1997) Buckling of cylindrical shells with general ring-stiffeners and lateral pressure distributions. In: Proceedings of the Seventh Internat. Conf. Computing in Civil and Building Engng. Vol.1. Eds. Choi ChK et al. Seoul, Korea 237-242.
[14]   Tian J, Wang CM, Swaddiwudhipong S (1999) Elastic buckling analysis of ring-stiffened cylindrical shells under general pressure loading via the Ritz method. Thin-Walled Struct. 35: 1-24.
[15]   Akl W, Ruzzen M, Baz A (2002) Optimal design of underwater stiffened shells. Struct. Multidisc. Optim. 23: 297-310.

EXPERIMENTAL INVESTIGATION ON ENERGY DENSITY OF BIO-FUELS

Liviu Costiuc

Transilvania University, Braşov, ROMANIA, e-mail lcostiuc@unitbv.ro

Abstract: *This paper presents the results obtained in investigation on energy density of bio-fuels. There are analyzed five kinds of bio-fuel candidates like sunflower oil, waste used oil and canola oil. Measuring of energy density (heat of combustion) of bio-fuels was done using the Parr oxygen bomb calorimeter. There are presented the sample preparation, the testing procedure, the comparative results on the mixed and separated fuels and also the conclusions of the experiment. Even the energy density of bio-fuels studied decreases after mixing with diesel fuel, the fuels could be effectively used for energy generation by car engines.*
Keywords: *energy density; Parr oxygen calorimeter; biodiesel, bio-fuel*

1. INTRODUCTION

The use of so called bio-fuels is not new. Since 1990 engineers and researchers have been experimenting with using vegetable oils as fuel for a diesel engine. The rise of the price for regular fuels determined recently that it is necessary to investigate the bio-fuel properties and engine parameters for reliable operation.

The main form of used vegetable oil (UVO), waste vegetable oil (WVO) used in the UK is rapeseed oil (also known as canola oil, primarily in the United States and Canada) which has a freezing point of -10°C. However the use of sunflower oil, which gels at around -12°C,[10], is currently being investigated as a means of improving cold weather starting. Unfortunately oils with lower gelling points tend to be less saturated (leading to a higher iodine number) and polymerize more easily in the presence of atmospheric oxygen [8]. Recycled vegetable oil, also termed UVO, WVO, used cooking oil, or yellow grease is recovered from businesses and industry that use the oil for cooking.

As of 2000 [8], the United States was producing in excess of 11 billion liters (2.9 billion U.S. gallons) of recycled vegetable oil annually, mainly from industrial deep fryers in potato processing plants, snack food factories and fast food restaurants. If all those 11 billion liters could be recycled and used to replace the energy equivalent amount of petroleum, almost 1% of US oil consumption could be offset.[8]

Taxation on SVO/PPO as a road fuel varies from country to country, and it is possible the revenue departments in many countries are even unaware of its use, or feel it too insignificant to legislate. Germany used to have 0% taxation, resulting in it being a leader in most developments of the fuel use. However SVO/PPO as a road fuel began to be taxed at 0,09 €/liter from 1 January 2008 in Germany, with incremental rises up to 0,45 €/liter by 2012. However, in Australia it has become illegal to produce any fuel if it is to be sold unless a license to do so is granted by the federal government.[8]

All most diesel car engines are suitable for the use of straight vegetable oil, also commonly called pure plant oil (PPO), with suitable modifications. Even, Rudolf Diesel the fathers of the engine, the first attempts were to design an engine to run on coal dust, but later designed his engine to run on vegetable oil. Principally, the viscosity and surface tension of the SVO/PPO must be reduced by preheating it, typically by using waste heat from the engine or electricity, otherwise poor atomization, incomplete combustion and carbonization may result.

The relatively high kinematic viscosity of vegetable oils must be reduced to make them compatible with conventional compression-ignition engines and fuel systems. Cosolvent blending is a low-cost and easy-to-adapt technology that reduces viscosity by diluting the vegetable oil with a low-molecular-weight solvent.[11] This blending, has been done with diesel fuel, kerosene, and gasoline, amongst others; however, opinions vary as to the efficacy of this. Noted problems include higher rates of wear and failure in fuel pumps and piston rings when using blends.

2. MATERIALS AND METHODS

2.1. Materials

The tested bio-fuels come from main three sources: regular diesel fuel for cars, sunflower oil and canola oil. A supplemental source was used to investigate energy density because there are large quantities of waste oil, including commercial and industrial activities. Mainly, bio-fuels are prepared in samples at the Transilvania University automotive facility from Braşov County, Romania.

The Biodiesel samples were filtered and separated in two kinds of samples.

First kind is pure vegetable oil, like Biodiesel
1. Sunflower oil and Biodiesel
2. Used vegetable oil.

The second kind is a mixture of diesel fuel and vegetable oil, namely canola oil, also known as rapeseed oil.

There are three volume fractions used in the sample mixture with diesel fuel, the 5%, 10% and 20% rapeseed oil.

For each kind of sample were prepared three probes. After filtering, the samples were weighed using a precision 0.1 mg. Each sample weight was approximately 1.4 gram.

2.2. Experimental System& Calibration

The equipments used for the energy density tests were: XRY-1C Oxygen bomb calorimeter, Figure 1, XRY-1C software and Kern & Sohn ABJ 220-4M analytical balance.

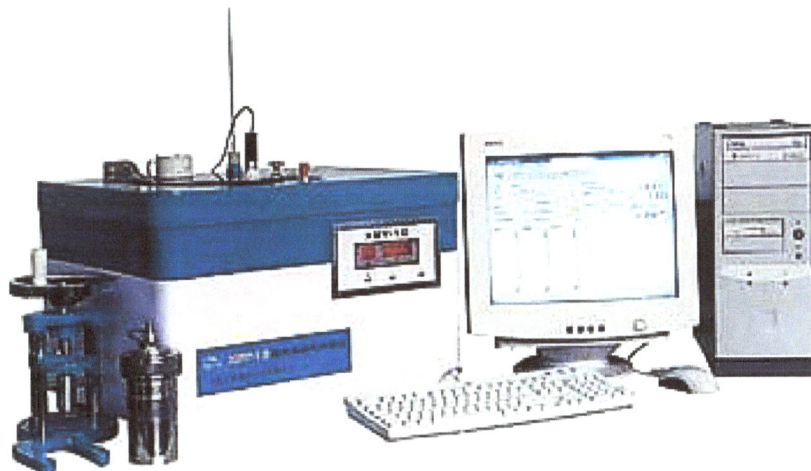

Figure 1: XRY-1C experimental system

The technical specifications for Oxygen bomb calorimeter are:
- Calorimeter capacity: 11000~15000 [J/K];
- Temperature measuring domain: 10-35[°C];
- Temperature reading resolution: 0.001 [°C];
- BIAS ≤0.2% [°C];
- Oxygen maximum pressure: 20 [MPa];
- Normal functioning temperature: 15-28 [°C]; temperature variation during experiments < 1 [°C];
- Relative humidity: < 85 [%];

Before determinations of the calorific value of samples, it is necessary to do the calibration of oxygen bomb calorimeter. This consists in a reverse procedure. Having the heat of combustion of benzoic acid as $H_e =$ 26454 [J/g], it is determined by the same kind of test the thermal capacitance of the calorimeter, W[J/K], burning in crucible the benzoic acid and knowing its mass, m_e [g] using equation (1).

$$W = \frac{H_e \cdot m_e - (Q_{wc})}{(t_f - t_i)} \qquad (1)$$

where, for calibration:

m_e - benzoic acid and knowing its mass =1.0018 [g]

Q_{wc} - the heat correction for the wire's burning and for the cotton's burning, Q_{wc} =131.36 [J]

t_f - the final temperature from the main stage, t_f = 20.538 [°C]

t_i - the initial temperature from the main stage, t_i = 18.405 [°C].

The burning process of benzoic acid in the calorimeter during calibration can be seen in Figure 2. After calibration the calculated thermal capacitance of the calorimeter is W = 12328 [J/K].

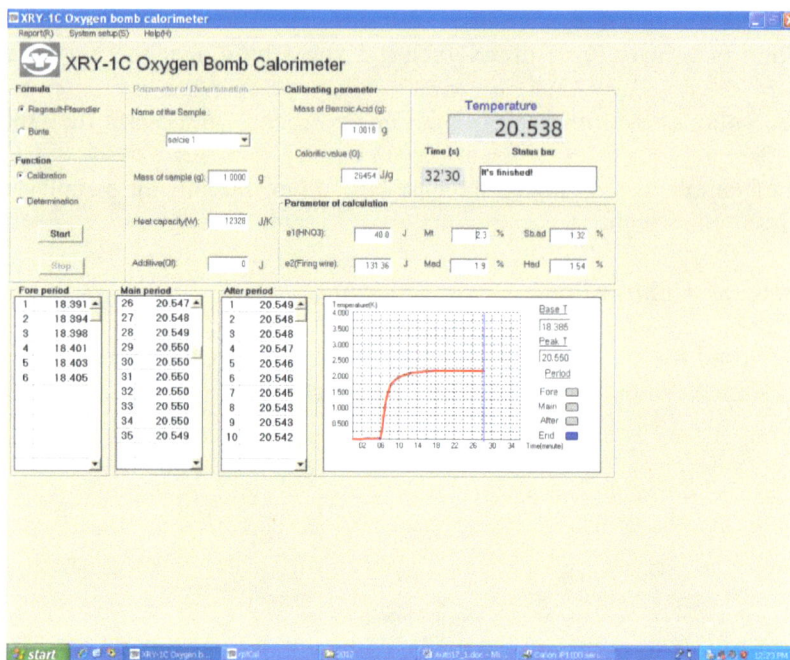

Figure 2: Calibration screen for XRY-1C.Mass of sample = 1.0018 g, cotton wire mass = 0.0061 g, burning wire mass = 0.0040 g, cotton burning heat =108.49 J, wire burning heat =22.87 J

2.3. Heat of Combustion

Because the tested material is a fuel, the energy density measurement means the determination of the heat of combustion. The heat of combustion is the energy released as heat when a compound undergoes complete combustion with oxygen in an enclosure of constant volume.

The gross calorific value at constant volume (named also as higher heating value or gross energy or upper heating value or higher calorific value) is the absolute value of the specific energy of combustion, in Joule, for the unit mass of a solid recovered fuel burned in oxygen in a calorimetric bomb under the conditions specified. The products of combustion are assumed to consist of gaseous oxygen and nitrogen (coming from the gaseous atmosphere of burning), of carbon dioxide and sulphur dioxide, of liquid water (in equilibrium with its vapor) saturated with carbon dioxide under the conditions of the bomb reaction, all at the reference temperature [4]. The equation used to calculate gross calorific value in adiabatic conditions, Qgr,ad, is presented in equation (2).

$$Q_{gr,ad} = \frac{W \cdot (t_f - t_i - t_\alpha) - (Q_1 + Q_2 + Q_3 + Q_4)}{m_c} \qquad (2)$$

where: W - calorimetric factor [J/K], determined by calibration of calorimeter

- m_c - the fuel sample's mass [g];
- t_f - the final temperature from the main stage [°C]
- t_i - the initial temperature from the main stage [°C]
- ta - the temperature correction caused by exterior heat losses [°C];
- Q_1 - the heat correction for the wire's burning and for the cotton's burning [J]
- Q_2 - the heat correction when the nitric acid is formed Q_2= 40 [J].
- Q_3 - the heat correction when the sulfuric acid is formed Q_3= 1298 m_1 [J].
- Q_4 - the heat correction for the benzoic acid's combustion, if the fuel has been mixed with this acid for plasticization.

The net calorific value at constant volume (lower calorific value) is the absolute value of the specific energy of combustion, in Joules, for the unit mass of a solid recovered fuel burned in oxygen under conditions of constant volume and such that all the water of the reaction products remains as water vapor (in a hypothetical state at 0.1 MPa), the other products being, as for the gross calorific value, all at the reference temperature [4]. With the bomb calorimeter is measured the gross calorific value. The testing procedure is presented in [1, 2, 3].

3. RESULTS AND DISCUSSION OF CALORIMETRIC ANALYSIS

For each sample, in the XRY-1C software there were introduced the input data: the mass of ignition wire in grams; the mass of cotton fuse in grams; the calorific value of wire [J/g]; the calorific value of cotton [J/g] and the mass of test sample in grams. These input data values are weighted for each sample with a precision Kern & Sohn ABJ 220-4M analytical balance by a 0.01mg resolution.

After burning the software plotted the graph temperature-time as shown in Figure 3, and calculates the gross calorific value using the Regnault-Pfaudler[2], [3] method as shown in Table 1.

Table 1: Calorimeter Temperature vs. Time for Biodiesel 1, sample 2

Determination		Time:	9/20/2012 4:43:27 PM
SampleMass[g]	**1.3819**	Mad[%] 1.9	Mt[%] 2.3
Capacity[J/K]	**12328.00**	Sb,ad[%] 1.32	Had[%] 1.54
SampleName	moto7_2	**Formula R-P**	
	Temperature [°C]		
AutoID	**Fore period**	**Main period**	**After period**
1	23.932	24.560	27.602
2	23.929	25.835	27.601
3	23.926	26.562	27.599
4	23.923	26.905	27.596
5	23.921	27.093	27.594
6	23.919	27.213	27.591
7		27.296	27.588
8		27.353	27.584
9		27.402	27.581
10		27.437	27.577
11		27.469	
12		27.494	
13		27.515	
14		27.534	
15		27.549	
16		27.561	
17		27.572	
18		27.581	
19		27.587	
20		27.592	
21		27.597	
22		27.600	
23		27.602	
24		27.603	
25		27.604	
26		27.604	
27		27.603	
Qb,ad[J/g]	**Qgr,ad[J/g]**	**Qnet,ad[J/g]**	
33402	**33224**	**32720**	

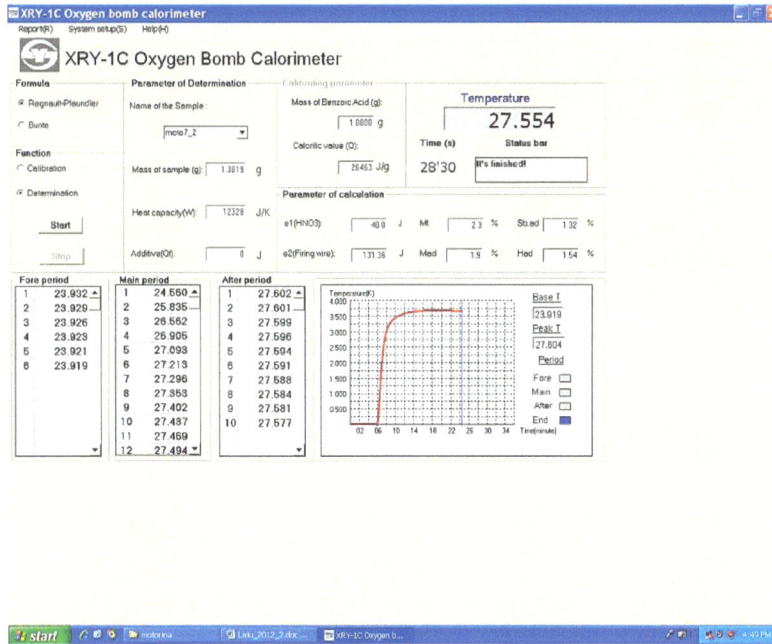

Figure 3: Measurements for Biodiesel 1-Sunflower Oil, sample 2, Mass of sample = 1.3819 g, cotton wire mass = 0.0062 g, burning wire mass = 0.0039 g, cotton burning heat =108.5 J, wire burning heat =22.86 J, thermal capacitance of the calorimeter, W = 12328 J/K

The results obtained for pure and mixtures of bio-diesel fuels are presented in Table 2. Qb,ad is burning calorific value, in adiabatic conditions, Qgr,ad is gross calorific value and Qnet,ad is net calorific value, in adiabatic conditions. These values are compared with diesel fuel which was also tested. For each kind of sample were tested three probes, the value for heat of combustion is a mean value with tolerance.

Table 2: Results for Heat of Combustion of bio-fuels

Column 1	Qgr,ad [MJ/kg]	Qnet,ad [MJ/kg]
Biodiesel 1-Sunflower Oil	33.224 ± 0.13	32.720 ± 0.13
Biodiesel 2-Waste Oil	19.805± 0.54	18.867± 0.54
Diesel Fuel	45.961± 0.14	45.404± 0.14
Diesel Fuel +5% Canola Oil	43.006± 0.38	42.461± 0.38
Diesel Fuel +10% Canola Oil	43.341± 0.34	42.795± 0.54
Diesel Fuel +20% Canola Oil	42.802± 0.22	42.259± 0.22

From Table 2 it could be seen that the heat of combustion of the Bio-fuels compared with the heat of combustion of the Diesel fuel are smaller with 27.71% for Biodiesel 1, with 56.90% for Biodiesel 2, with 6.43% for Biodiesel 3, with 5.70% for Biodiesel 4 and with 6.97% for Biodiesel 5.

It can be observed a small variation of heat of combustion value, even the Biodiesel 3, diesel fuel with 5% canola oil, the Biodiesel 4, diesel fuel with 10% canola oil, and the Biodiesel 5, diesel fuel with 20% canola oil, around 43.00 MJ/kg, which is about 6.4% smaller value than usual Diesel fuel.

Table 3: Heat of Combustion for fuels [4],[7]

Fuel	Qgr,ad [MJ/kg]	Qnet,ad [MJ/kg]
Crude oil	45.60	42.60
Gasoline	46.60	43.50
Diesel fuel	45.70	42.80
Jet Fuel, JP-4	46.60	
Methanol	22.90	20.10
Ethanol	29.80	26.90
Liquefied petroleum gas	50.10	46.60
Liquefied natural gas	55.20	48.60
Liquid hydrogen	141.00	120.00
Residual oil	42.20	39.50

An interesting result is the increasing of the heat of combustion value for the Biodiesel 4, diesel fuel with 10% canola oil comparing with the bio-fuels with 5% and 20% canola oil.

The results are in agreement with the data from the literature, as it could be seen in Table 3. Because the heat of combustion of bio-fuels, such as Biodiesel 3, Biodiesel 4 and Biodiesel 5 is close to Diesel fuel heat of combustion, the use of these fuels is justified. But, if it is considered the price for Diesel fuel and the price for vegetable oil than the use of bio-fuels is tempting, even a decrease with 30% on heat of combustion for Biodiesel 1.

3. CONCLUSION

The calorimetric analysis reveals a comparable heat of combustion of the mixed vegetable oil with diesel fuel to be used as fuel. The calorific power of those materials has a reasonable level, compared to those of petroleum derivates. The Biodiesel 3, Biodiesel 4 and Biodiesel 5 have a comparable calorific power compared with common diesel fuel, because the polymers in these kinds of products are present especially as resins for composite materials, rather than single material.

Even for pure vegetable oil, the calorific power is still greater than those of different sorts of coals, the smaller amount of sulphur compounds makes them as environmentally friendly and the price of acquisition can be a way to use these bio-fuels to obtain a reasonable amount of energy.

REFERENCES

[1] European Committee for Standardization, Solid recovered fuels - Methods for the   determination of calorific value, DD CEN/TS 15400:2006, 2006.
[2] *** - SR ISO 1928/95 – Determinarea puterii calorifice superioare şi inferioare a   combustibililor solizi.
[3] *** - DIN 51900-1/2000 Determining the gross calorific value of solid and liquid fuels using the bomb calorimeter and calculation of net calorific value - Part 1: General information.
[4] V.B. Ungureanu, Gh.Băcanu, D. Şova, V.Sandu, L.Costiuc - Termodinamica. Aplicatii   practice/ Thermodynamics. Practical works. Editura Universitatii Transilvania, Brasov, 2010, 404 pag. ISBN 978-973-598-832-6
[5] Costiuc, L., Popa, V., Serban, A., Lunguleasa, A., Tierean, M.H., Investigation on heat of   combustion of waste materials, Proceedings of the International Conference on Urban   Sustainability, Cultural Sustainability, Green Development Green Structures and Clean Cars,   Malta, September 15-17, 2010, Published by WSEAS Press, ISSN: 1792-4781, ISBN: 978-   960-474-227-1, pag. 165-168.

[6] Costiuc, L., Lunguleasa, A., Improving measurement accuracy of biomass heat of combustion using an oxigen bomb calorimeter, Bulletin of the Transilvania University of Brasov, vol.2(51)-2009, ISSN 2065-2119(print), ISBN- 978-973-598-521-9, pag. 467-474.

[7] Walters, R.N., Hackett, S.M., Lyon, R.E., Heats of combustion of high temperature polymers, http://large.stanford.edu/publications/coal/references/docs/hoc.pdf.

[8] http://en.wikipedia.org/wiki/Vegetable_oil_fuel, accesed 20/09/2013

[9] Lunguleasa, A., Costiuc, L., ş.a. Combustia ecologică a biomasei lemnoase. Editura Universităţii TRANSILVANIA din Braşov, 2007. print+CD, ISBN 978-973-598-194-5

[10]   Altin, R., S. Cetinkaya, and H. S. Yucesu. 2001. The potential of using vegetable oil fuels as fuel for diesel engines. Energy Conversion and Management 42, no. 5 (March): 529-538. doi:10.1016/S0196-8904(00)00080-7

[11]   Knothe, Gerhard (2001). "Historical Perspectives on Vegetable Oil-Based Diesel Fuels" (PDF). Inform 12 (11): 1103–1107. Retrieved 2009-06-24

TOPOLOGY OPTIMIZATION BY A QUASI-STATIC FLUID-BASED EVOLUTIONARY METHOD

László Daróczy[1], Károly Jármai[3]

[1] PhD. student, University of Magdeburg "Otto von Guericke", Germany,

[2] Professor, Dr.sci.techn. University of Miskolc, Hungary, altjar@uni-miskolc.hu

Abstract: *The current article proposes a new algorithm for topology optimization based on a fluid dynamics analogy. The new algorithm possesses characteristics similar to the most well-known methods, as the ESO/BESO method working with discrete values and the SIMP method (using OC or MMA) working with intermediate values, as it is able to work both with discrete and intermediate densities, but always yields to a solution with discrete densities. It can be proven mathematically that the new method is actually a generalization of the BESO method, and when using appropriate parameters it will operate exactly as the BESO method. The new method is less sensitive to rounding errors than the BESO method when using iterative solvers and is able to give alternative topologies to well-known problems. The article presents the basic idea, the optimization algorithm, and compares the results of three cantilever optimizations to the results of the SIMP and BESO method.*

Keywords: *topology optimization, evolutionary method, truss structures*

1. INTRODUCTION

The classical problem of compliance minimization is used to present the idea of the new method. According to the well-known formulation:

$$\min_{\mathbf{x}} \quad C = \frac{1}{2}\mathbf{f}^T\mathbf{u}$$
$$\text{s.t.} \quad \mathbf{Ku} = \mathbf{f} \quad , $$
$$0 < x_{\min} \leq x_i \leq 1$$
$$\sum x_i V_i - V_0 \cdot f = 0$$

(1)

where $\mathbf{K}$ is the global stiffness matrix, V_0 is the volume of the full domain, $\mathbf{f}$ is the external force acting on the structure, $\mathbf{x}$ is the design variable (density of the elements), $\mathbf{u}$ is the displacement, f is the volume ratio to be satisfied, x_i is the density of cell i, $x_{\min}$ is the minimal density of the cells, V_i is the volume of the finite element cell i.

To calculate the elemental stiffness matrix the SIMP (Solid Isotropic Material with Penalization) method is used, where the Young's modulus is calculated according to Zhou & Rozvany [1], Bendsoe & Sigmund [2] as

$$E(x_i) = (x_i)^p E_0,$$

(2)

where E_0 is the Young's modulus of the solid material, and p is the penalty factor.

The sensitivity number is calculated using the formulation of BESO (Bidirectional Evolutionary Structural Optimization) method to be consistent with the definitions [3]. We can note that there is no fundamental difference compared to the definition of the sensitivity number used by SIMP, as the equation is changed only by a constant ($-1/p$):

$$\alpha_i = -\frac{1}{p}\frac{\partial C}{\partial x_i} = -\frac{x_i^{p-1}}{2}\mathbf{u}_i^T\mathbf{K}_i^0\mathbf{u}_i$$

(3)

In this formulation (3) is proportional to the increase of the mean compliance resulting from the removal of element i. If we want to minimize the compliance (C) we have to maximize the sensitivity number of the valid elements (so we have to delete elements from the solid region, which would result in low increase of the compliance, if deleted). From here the optimized variable will be denoted by α, as with the use of sensitivities, we are actually maximizing it instead of the minimization of the compliance.

In the followings the BESO method will be reviewed shortly to be able to present the important similarities and differences between the BESO and the new algorithm. As all topology optimization algorithms, the BESO algorithm starts with the definition (see Figure 2) of the problem, followed by the discretization of the domain and with the definition of boundary conditions. Afterwards, the iterative algorithm performs either the defined number of cycles, or until convergence criteria are reached. In each cycle we perform a Finite Element Analysis followed by the calculation of sensitivity numbers. At this point it is very important to ensure a mesh-independent solution, for which reason mesh-independence filter and historical stabilization filter are applied to the sensitivities [4]. As BESO uses a different approach compared to SIMP method, a different method has to be used for handling the volume constraint. Opposed to the SIMP method, the volume constraint is not immediately applied, but rather step by step and in each cycle elements are removed or added to ensure the volume constraint of the actual step ($\sum x_i V_i - V_0 \cdot f_j = 0$, where j is the iteration number). After this step the current cycle is finished, and a new one begins.

2. QUASI-STATIC QUASI-FLUID APPROACH

Our new approach is based on a resemblance taken from the nature (as e.g. simulated annealing algorithm). Fluids usually tend to move away from high-pressure regions to lower pressure regions or from higher values of a potential field to lower levels (e.g. waterfall) to create equilibrium. This behaviour can be used for an optimization process. If we want to minimize a scalar-field, we simply need to define the pressure of the fluid to be higher in regions with higher scalar values, so it will move away from it. In the case of maximization however, the rule is the inverse.

After solving the equation $\mathbf{Ku}=\mathbf{f}$ for an intermediate solution of the topology optimization process, a quasi-static quasi-fluid simulation step will be performed (further on called as QSQF). The following concepts need to be defined:

Density of the fluid continuum (the design variable itself): At the beginning of the QSQF step $\rho_f = x$, therefore $\rho_f \in [0,1]$ must be satisfied. However, following the QSQF step $x \neq \rho_f$, instead we will introduce a *historical density-damping* scheme.

$$p_f^{new} = x = H_D p_f^{old} + \left(1 - H_D\right) p_f^{calc},\qquad(4)$$

where H_D is a *historical density- damping* coefficient for stabilizing the solution (which must be within the range of [0,1]).

The idea behind this formulation is that the fluid continuum can move extremely quickly in the presence of huge pressure differences. However due to the $\rho_f = x$ definition the optimized solid structure should be updated in a coupled manner with the fluid continuum. As this would require vast computational resources, we apply instead a quasi-static approach, where the optimized solid structure and fluid continuum is updated in a segregated way. To do this however we need to make sure, that no sudden change can happen inside the continuum. This is actually similar to the averaging scheme applied to sensitivities by BESO method.

Potential field: $U(\alpha)$. The potential field acting on the fluid continuum is a function of the sensitivities. This function defines whether we are maximizing or minimizing. However without the loss of generality from here on we will only consider minimization.

Equation of state for the fluid continuum: $p_f(\rho_f)$. This function defines the connection between the pressure and density of the fluid (compressible fluid). To prevent fully void regions, the density must be between the defined values (x_{min} and 1). Additionally, the pressure must be positive.

Equilibrium equation for the fluid:

$$p_f(\rho_f) + U(\alpha) = const \tag{5}$$

This equation means that the sum of the energy stored by the potential field and the energy resulting from the pressure of the fluid is constant in every point (see Figure 1). The term quasi-fluid comes from the fact that this is not an equation for a real fluid, but rather for a continuum behaving *similarly* to fluids. It is worth noting that the hydrostatic equation is very similar to the previous form:

$$`p_f(\rho_f) + U_{gravity} = p_f + \rho gz = const \tag{6}$$

In the followings we will denote *const.* in Eqn. 5 by *EquilibriumLevel*, as it represents an important parameter of the method.

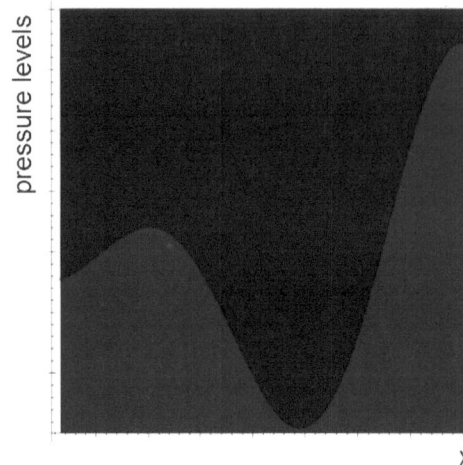

Figure 1: Quasi-static equilibrium state, Grey=Potential energy, Black=Pressure of the fluid continuum

3. WORKFLOW OF ALGORITHM AND APPLIED FILTERS

Although the main idea is easy to understand now, presenting the workflow of the optimization process is still necessary. The comparison of the BESO and QSQF method's workflow is summarized on Figure 2.

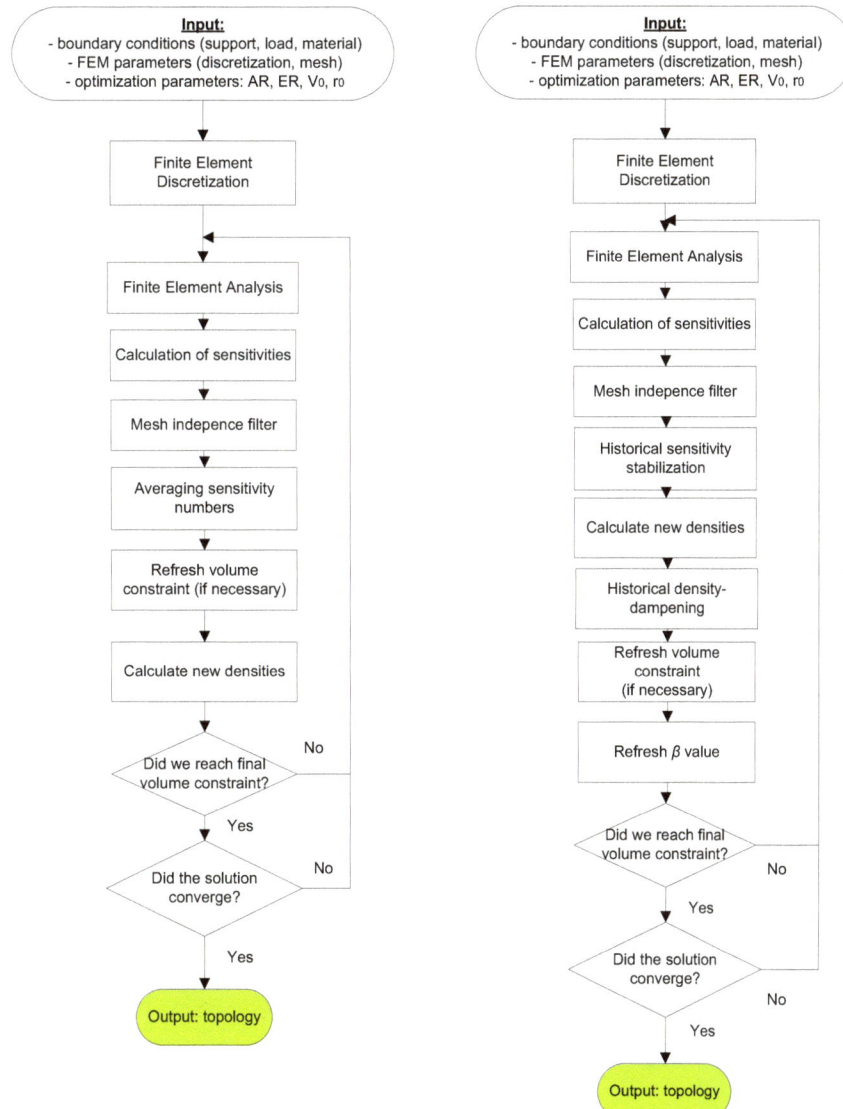

Figure 2: Proposed workflow of QSQF (quasi-static quasi-fluid) optimization (right side) compared to BESO method (left side)

One can see that although we used a completely different approach and analogy from nature, the two algorithms have very similar workflow, with the only difference being that the new algorithm is capable of working with intermediate densities. However at $\beta=\infty$ it yields to the BESO algorithm, thus it is a generalization of BESO method.

4. TEST EXAMPLE

In the followings the results of the new method are presented and compared to the results of the previous methods, SIMP&OC and BESO, using some classical problems. The FEA was solved using preconditioned conjugate gradient method with the final residual error always in the range of 10^{-6}-10^{-10} N (3D problems) and 10^{-8}-10^{-10} N (2D problems), depending on the problem. The FEA code was successfully validated against an example of a bent cantilever using ADINA R&D Inc. ADINA®.

Although the chosen problem is well-known and well-researched basic example of the optimization and thus we cannot expect to achieve huge improvement, we found it important to validate the algorithm against these tests. For the presented cases the new method has achieved the well-known solutions without error and for two cases it could even provide, although only slightly, but better solutions with different topologies compared to the literature. We consider this a major achievement, as these problems have been examined for decades.

All tests were run for 200 steps to ensure that no sudden change happens in the later iterations and C_{200} is presented along with $j_{1\%}$ and/or $j_{2\%}$, where C_j is the compliance in step j and

$$j_{k\%} = j, \text{for which } \left| \frac{C_j - C_{200}}{C_{200}} \right| \leq k/100 \qquad (7)$$

The convergence criterion is reached when the compliance is within ±1 or ±2% of the compliance at step 200. This criterion was proposed in order to ensure that the final solution was really reached and not only slow convergence occurred, Except for the benchmarking of the software normal convergence criterion should

be used (e.g.) $\left| \frac{C_j - C_{j+1}}{C_{j+1}} \right| \leq k/100$).

4.1 A bent cantilever

As on the field of linear elasticity, where the example was tested, the resulting structure does not depend on the magnitude of the load or Young's modulus, publications and books use many times small but easily comparable load and Young's modulus values. Here in *Example 1* we will use E=1 MPa, v=0.3, 160 x 400 mm domain with 160x40 discretization. The load is F= −1 N, the volume constraint V_f =0.5, while x_{min}=0.001, r_{min}=3.0 mm (Figure 3). The calculated values corresponded to the values given by Huang & Xie [5] for both SIMP&OC and BESO method with the current in house code (Figures 4-7). The $C_{200}, j_{1\%}$ are given for all cases, and $j_{2\%}$ values for the examples with convergence history.

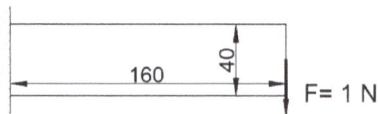

Figure 3: Problem 1

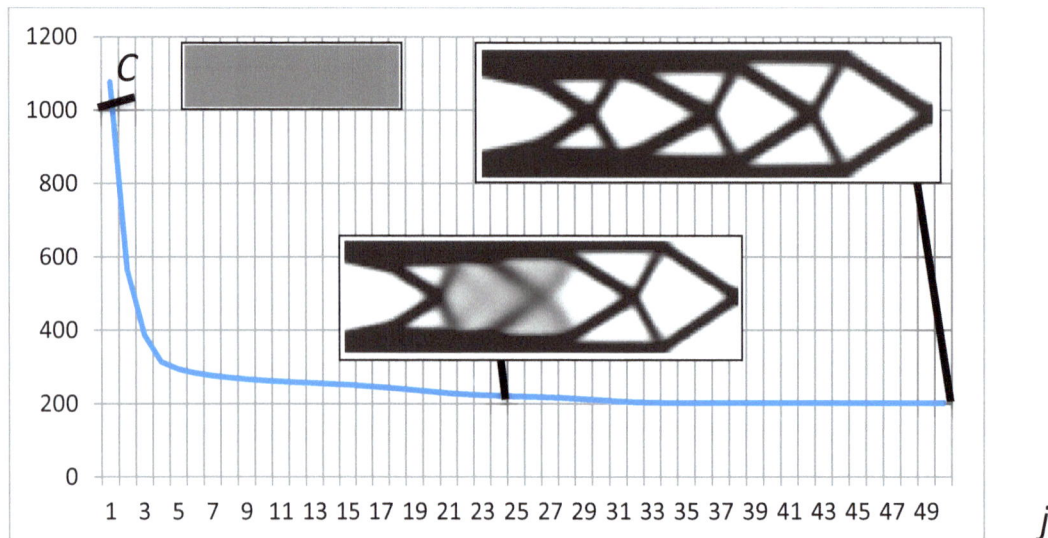

Figure 4: SIMP&OC (C_{200}=201.2 Nmm; $j_{1\%}$=33; $j_{2\%}$=33)

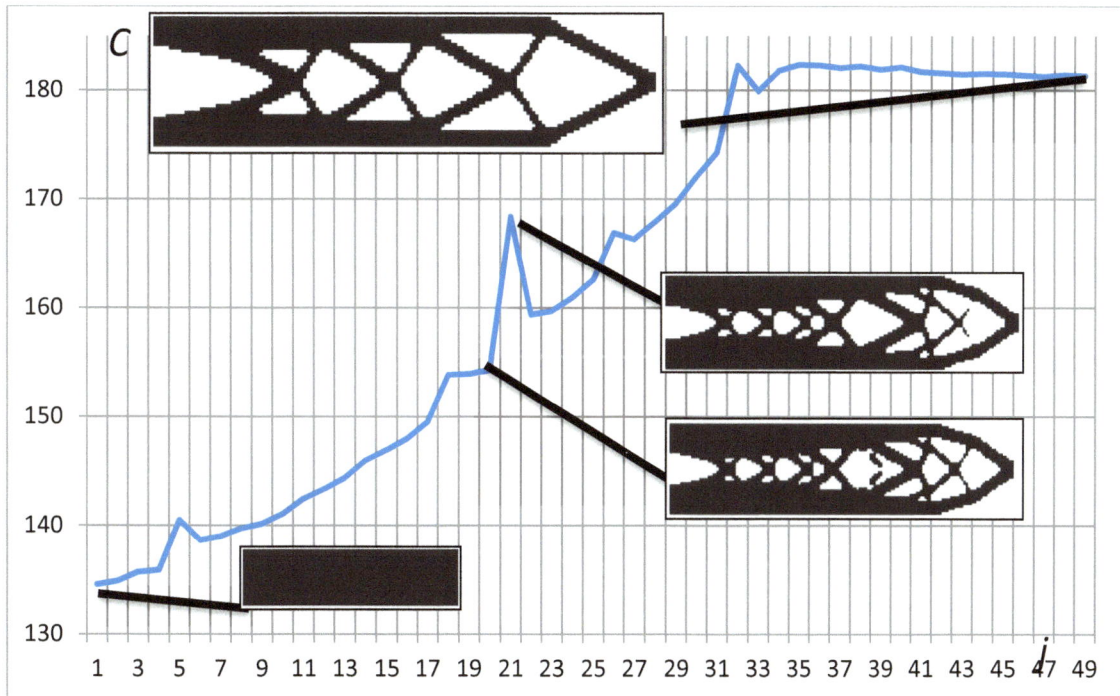

Figure 5: BESO (C_{200}=181.4 Nmm; $j_{1\%}$=32; $j_{2\%}$=32) ER=2.0%; AR_{max}=50.0%

It is important to point out that all algorithms reach the $j_{1\%}$ state at almost the same speed, but they need significantly more time to reach the C_{200} value: the SIMP method reaches the presented value only at j=188, the BESO at j=46, and the QSQF at j=93. However it is important to note, that the result given by QSQF is 0.22% lower than the result of BESO method, but with a different topology! The convergence history is given for Figures 4-6, while Figure 7 only presents that using a different β value the QSQF method is also able to give the previously known topology. Depending on the supports and loadings, there can be asymmetric solutions as well Cheng & Liu (2011).

Note: At SIMP the values are higher due to the presence of intermediate densities.

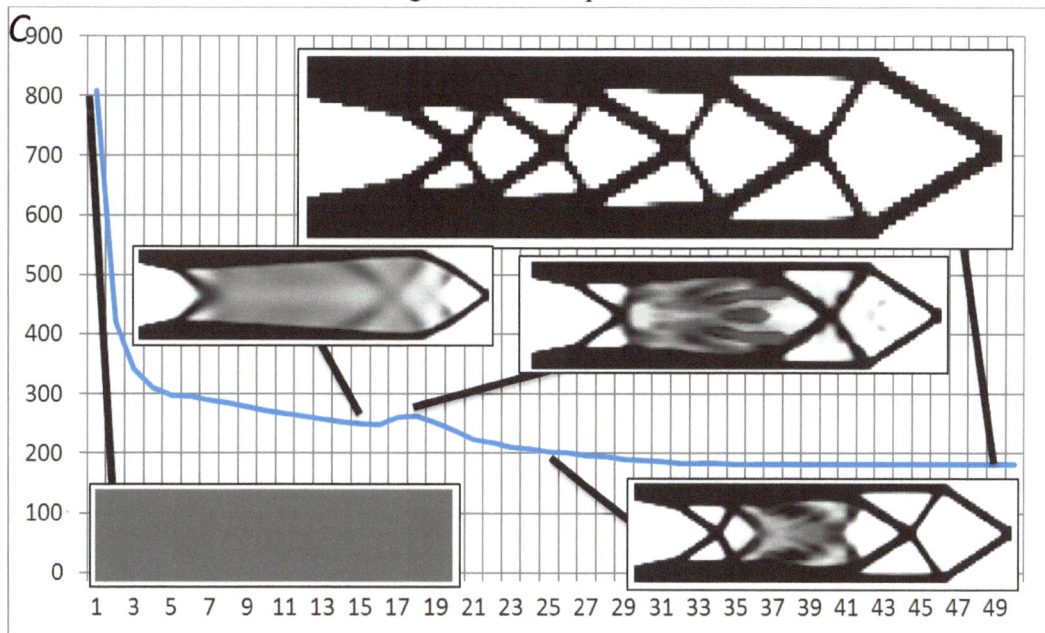

Figure 6: QSQF (C_{200}=181.0 Nmm; $j_{1\%}$=35; $j_{2\%}$=32, pcw.) V_0=0.55; ER=1.5%; H_s=0.5; H_d=0,5

(it.<30); β=4,6,8…

Figure 7: QSQF (C_{200}=184 Nmm; $j_{1\%}$=52, inv.pow) V_0=0.7; ER=1.5%;H_s=0.5; H_d=0,4 (it.<40); β=4,5,6…

5. RESULTS AND CONCLUSIONS

In this article we have proven, that the BESO method can be derived using an analogy from nature. Moreover we proposed a new algorithm, which is generalization or extension of the BESO method, as it is able to work both with intermediate densities, but in special cases ($\beta=\infty$, H_D=0) it behaves like the BESO method. Moreover due to the introduced new parameters it provides flexibility and more options than BESO or SIMP. Although at first is seems that the introduction of new parameters makes the decision maker's task more difficult, but we want to emphasize, that most parameters are only present to help advanced users in their research. Usually, H_D=0.5; H_s=0.5; β_0=4 and *piecewise linear fuzzyfication functions* are the recommended, so we only have to choose ER_{max}, AR_{max}, V_f and β_{inc}.

The main advantage of QSQF is not that it is much faster compared to SIMP or BESO method, but the different path to the solution. As the algorithm converges through designs containing intermediate densities, it is less sensitive to rounding errors (i.e. with the use of iterative solvers for the FE model) and instead of immediately deleting bars, it slowly makes them disappear, which also removes the local peaks in the convergence history (see Figure 5-6). Moreover, as topology optimization at the moment only serves as a starting tool for the design, in real world applications engineers usually use them as an intuition, and therefore they may need more alternatives for the same problem. Just note, that the QSQF method was able to find a slightly better but different topology on Problem 1 (see Figure 6 compared to Figure 4).

We would like to point out again that the main advantages of the algorithm are the generalization of the BESO method, and the possibility to provide many different, but equally good solution to the same problem, thus giving alternatives to the engineers, or the possibility to choose more aesthetic solutions, which is a more and more urging need in the field of civil engineering and mechanical engineering, where the attractivity of a product is defined more and more by its design and aesthetics, rather than structural simplicity and simple functionality.

Although extensive testing is still needed, the method can already be applied to many cases successfully. In our future work we would like to extend the analysis of the QSQF method to not only compliance minimization problems, and perform an extensive comparison to BESO and SIMP method. We would like to find other analogies as well taken from nature (e.g. the use of gravitational fields). An ambitious plan would be to find a generalization, which includes ESO, BESO SIMP and QSQF algorithm too, but the existence of such an algorithm is an open question

ACKNOWLEDGEMENTS

The research was supported by the TÁMOP 4.2.4.A/2-11-1-2012-0001 priority project entitled 'National Excellence Program - Development and operation of domestic personnel support system for students and researchers, implemented within the framework of a convergence program, supported by the European Union, co-financed by the European Social Fund. The research was supported also by the Hungarian Scientific Research Fund OTKA T 75678 and T 109860 projects and was partially carried out in the framework of the Center of Excellence of Innovative Engineering Design and Technologies at the University of Miskolc.

REFERENCES

[1]   Zhou, M., Rozvany, G. I., N. (2001): On the validity of ESO type methods in topology optimization. *Structural and Multidisciplinary Optimization*, Vol. 21, No. 1, pp. 80-83.

[2]   Bendsoe, M. P., Sigmund, O. (1995): *Topology Optimization – Theory, Methods and Applications.* Springer.

[3]   Xie, Y.M., Steven. G.P. (1993): A Simple Evolutionary Procedure for Structural Optimisation, *Computers and Structures*, Vol 49, No 5, pp 885-896, 1993.

[4]   Huang, X., Xie, Y. M. (2010): *Evolutionary Topology Optimization of continuum Structures – Methods and Applications.* Wiley.

[5]   Huang, X., Xie, Y. M. (2010): A further review of ESO type methods for topology optimization, *Structural and Multidisciplinary Optimization*, Vol. 41, No. 5. pp. 671–683, DOI 10.1007/s00158-010-0487-9

CFD INVESTIGATION OF THE FLOWS IN ONE-STAGE BLOWER AGGREGATE

László KALMÁR [1], Béla FODOR [2]

[1] Associate Professor, University of Miskolc, Miskolc, HUNGARY, aramka@uni-miskolc.hu
[2] Assistant Professor, University of Miskolc, Miskolc, HUNGARY, aramfb@uni-miskolc.hu

Abstract: *The paper deals with the numerical investigation of the flows in radial-flow one-stage blower-aggregate. The main aim of this numerical study is to predict the relevant operating behaviour of the blower and to determine detailed information about the flow characteristics inside that. The calculated distributions of flow characteristics in the blower determined by ANSYS-FLUENT are available to decide whether the main functional parts of the blower are working properly, or not.*

The CFD model developed in this case for carrying out the numerical simulation is taken account of the heating effect of the electro-motor by applying specific heat sources inside the calculation domain. The calculated average values of flow characteristics are usable to determine the characteristics of the calculated operating points of the blower. The discrete points of the predicted performance curves can be compared with measured data obtained by experimental tests of the blower-aggregate for their validation.

Keywords: Blower-aggregate, CFD methods, main flow characteristics, performance curves.

1. INTRODUCTION

The investigated blower aggregate noted by BA_0 can be seen in Figure 1 and the same aggregate is shown in Figure 2 in a disassembled state to introduce the main parts of the blower. The first step of our numerical investigation was to create the complete computational domain of the blower-aggregate. It has been produced in the commercial pre-processing tool ANSYS-GAMBIT [1]. The entire three-dimensional computational domain of the blower aggregate is covered by the casing of the aggregate as shown in Figures 1 and 3.

Figure 1: Photo of the investigated blower aggregate BA_0

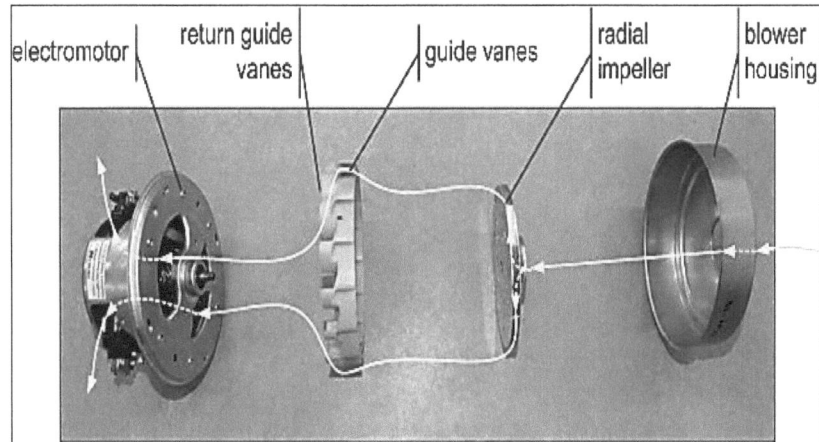

Figure 2: Main components of the blower aggregate BA_0

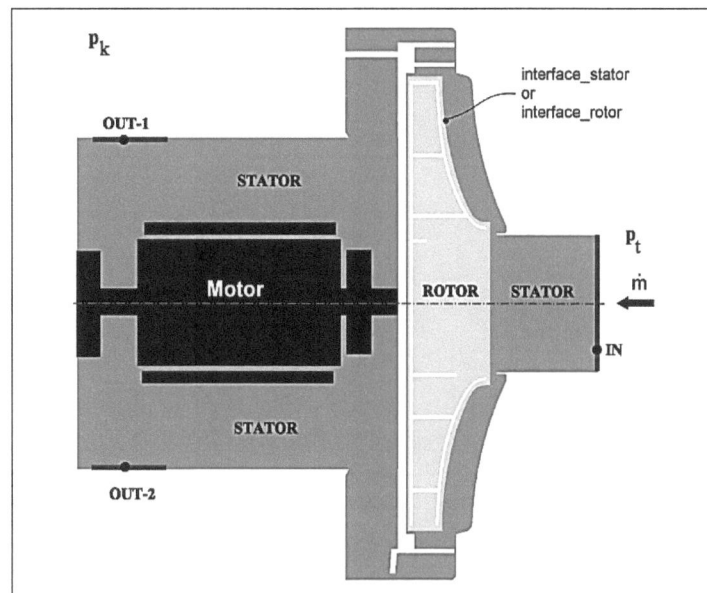

Figure 3: The main cross-section of the entire 3D computational domain of the blower aggregate

At the inlet cross section of the blower – to produce relatively homogenous velocity distribution along the inlet cross section during carrying out the numerical simulation – a cylindrical short pipe section with circular cross section was connected to the inlet section of the blower aggregate BA_0. At the outlet sections of the blower BA_0 can be seen on the cylindrical surface of electrical motor casing (see Figures 1-3). The total 3D computational domain (see Figure 3) contains the rotating impeller (see Figure 4) and the stationary guide vanes and the stationary return guide vanes (see Figure 5-6) of the blower, which can be found inside the blower casing.

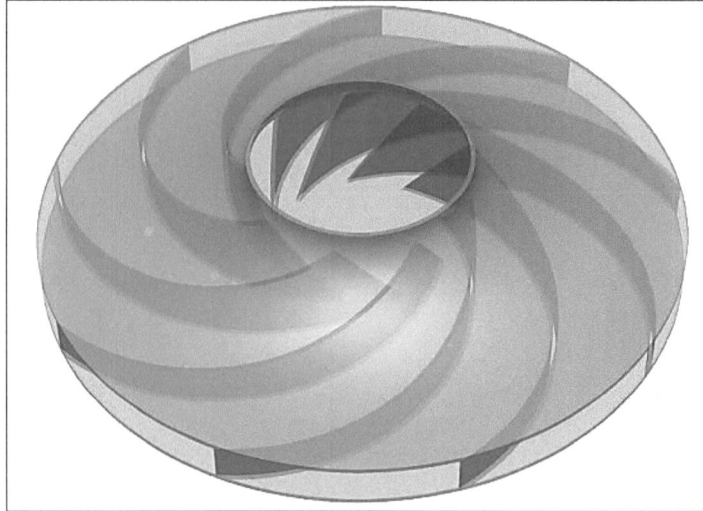

Figure 4: The 3D computational model of the impeller of the blower aggregate BA_0

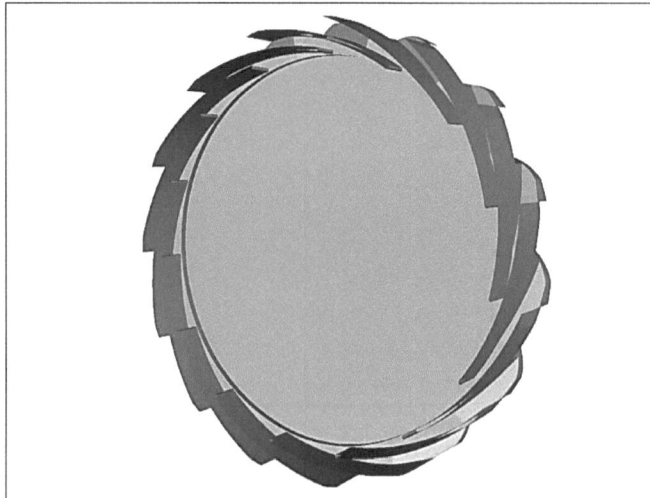

Figure 5: The 3D computational model of the stationary guide vane of the blower aggregate BA_0

Figure 6: The 3D computational model of the stationary return guide vane of the blower aggregate BA_0

The air flows into the blower BA_0 throughout the inlet section and arrives into the impeller, then flows across it which increases the total energy of the flowing air. After that first, the air flows in the guide vanes at impeller side then flows in the return guide vanes on the back side. Finally the air flows through the pressured side of the blower across the electrical motor and leaves the blower throughout the outlet sections.

To carry out the numerical simulation of flow in blower BA_0, the total computational domain has to be divided into sub domains, as shown in Figure 3. Two types of important sub-domains have to be detached because of their operations: the rotational sub-domain (named ROTOR) is the sub-domain of the blower impeller and the stationary sub-domains (named STATOR) which are bounded by the walls of the blower, the guide vanes and return guide vanes. The main aim of this numerical investigation is to determine the relevant operating characteristics of the blower by applying CFD numerical methods.

2. COMPUTATIONAL PROCEDURE

The finite volume method is applied to determine the characteristics of the flow problems by FLUENT. That is why before starting to run the code all the sub-domains have to be divided into finite volumes. In other words, we have to mesh the total computational domain. When carrying out this procedure we have to pay extra attention to the geometrical characteristics of each finite element. The mesh generation was carried out with the commercial code ANSYS-GAMBIT. To get information about the quality of mesh elements it is the most convenient way to display the actual values of the cells skewness. In our case 19,03 million cells were developed and the measured maximum value was equal to 0.8947, which means that the quality of the meshing is acceptable to perform the CFD computation procedure for the flow in blower aggregate.

3. COMPUTATIONAL RESULTS

For all the cells of the total computational domain the "density based implicit Gauss-Seidel" numerical solver was used during the numerical solution process assuming unsteady flow. Because during the operations of blower relatively high velocities and pressures increase take place, it was also supposed that the fluid was compressible and viscous. In this way the standard k-ω SST turbulent models and perfect-gas law were applied in our simulations. The heat sources inside the computational domain were also applies to simulate the heating effects of the electro-motor. The computational results obtained by ANSYS-FLUENT are illustrated below.

The initial working parameters of our calculation were the mass-flow rate at inlet of blower aggregate and the rotation speed of the blower impeller, by applying these initial parameters the actual pressure difference between the outlet and inlet sections of the blower was calculated.

By using the determined values of the flow parameters, *one possibility is* to show local distributions of flow characteristics (e.g. velocity, or pressure distributions, relative streamlines inside the impeller, streamlines inside stationary guide vane and return guide vane, etc.) which is very important to the designers to get detailed information about the flow structures locally along planes in flow domain [3-4].

The *other possibility is* to determine average values in cross sections by using the calculated results by FLUENT. In this case, because the page-limit of the paper only the variations of average flow characteristics are introduced below.

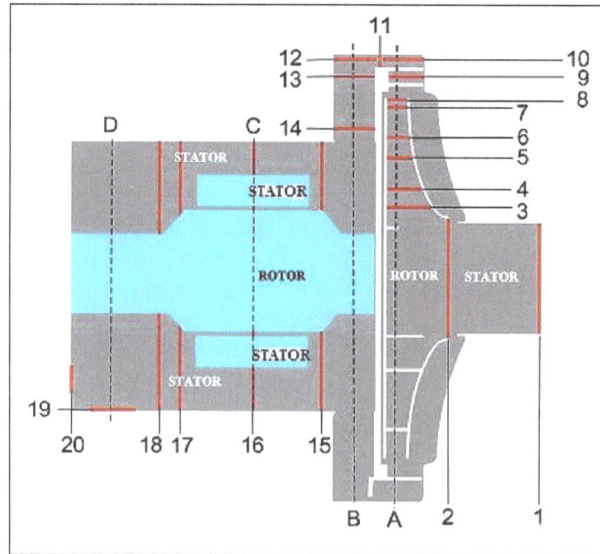

Figure 7: Positions of cross sections 1-19 of the flow and four planes of **A**, **B**, **C** and **D**

Figures 8-12 show the variations of averaged flow parameters (they are in our case: average mass-flow, absolute and dynamic pressures, density, and temperature) determined by surface integration of physical fields concerning to the denoted 19 different sections of the flow in direction of the main flow inside the blower aggregate BA_0 (see Figure 7). Knowing the variations of these averaged parameters is very important, if we want to get useful information about the operational characteristics of the blower. All the variations of these averaged flow parameters were determined for operation state at mass flow rate 0.033 kg/s for the blower aggregates BA_0.

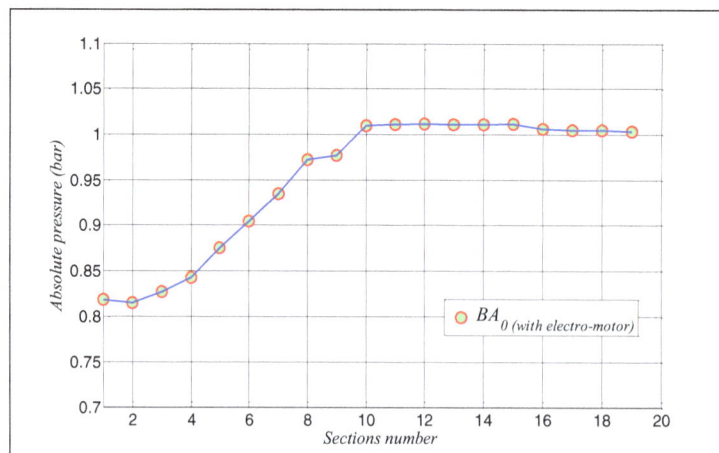

Figure 8: Variation of absolute pressure in the blower aggregates BA_0 at mass flow rate 0.033 kg/s

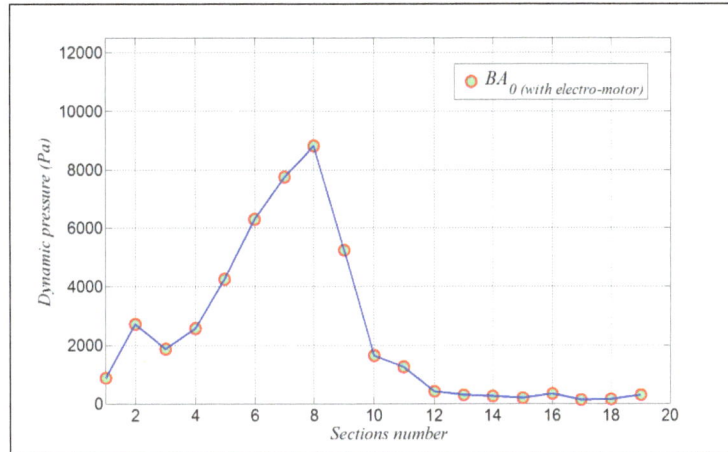

Figure 9: Variation of dynamic pressure in the blower aggregates BA_0 at mass flow rate 0.033 kg/s

The cross sectional average values of absolute and dynamic pressures, density and temperature were determined by surface integration for the denoted 19 cross sections and the variations of these flow characteristics are shown in Figures 8-11. The values of mass flow rate concerning these cross sections are also calculated and shown in Figure 12.

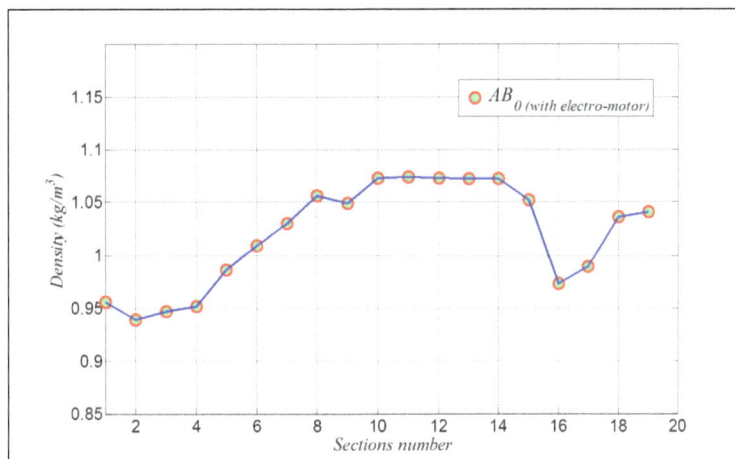

Figure 10: Variation of air density in the blower aggregates BA_0 at mass flow rate 0.033 kg/s

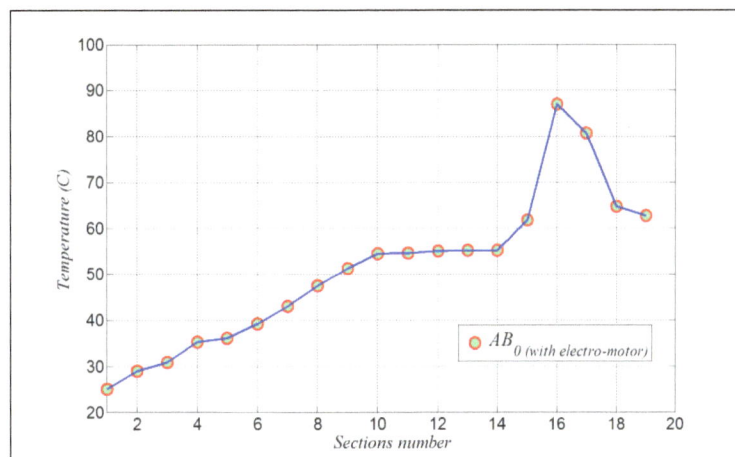

Figure 11: Variation of air temperature for the blower aggregates BA_0 at mass flow rate 0.033 kg/s

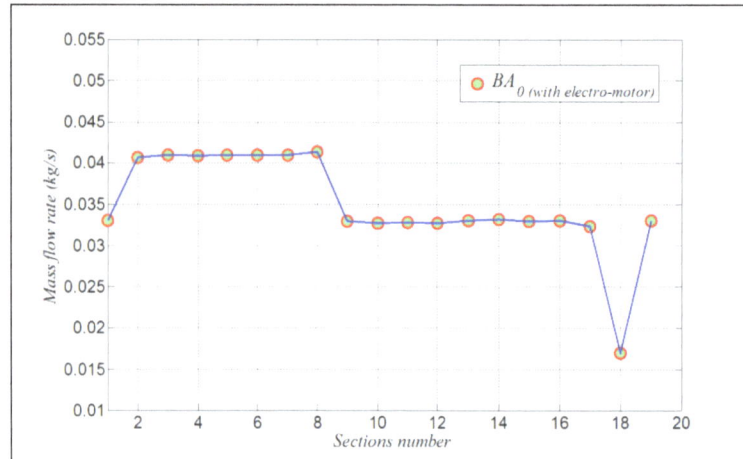

Figure 12: Variation of mass flow rate in the blower aggregates BA_0 at mass flow rate 0.033 kg/s

In this way in Figures 8-10 inside the impeller (between sections 2 and 8) the variations of absolute and dynamic pressure show the energy increase through the impeller. Inside stationary guide vane (between sections 8 and 10) the increase in cross sectional area causes a sudden drop in the dynamic pressure and a small increase in the absolute pressure. Inside the electro-motor (between sections 15 and 16) the heating effect of the motor causes temperature increasing. In Figure 12 between sections 1 and 8 the relatively large increase (or between sections 8 and 9 the relatively large decrease) caused by leakage.

For validation of our numerical calculation we compared the characteristics of the calculated operating point determined by actual value 0.033 kg/s of the mass flow rate with the measured characteristic curve of the blower aggregate BA_0 given by laboratory tests [2], which are shown in Figure 13. A very good agreement can be observed in this figure.

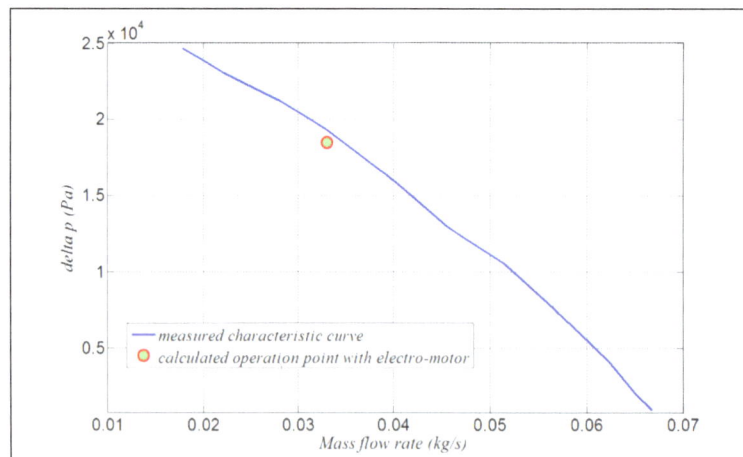

Figure 13: Contrasting of calculated working point and the measured characteristic curve of the blower aggregate BA_0

4. CONCLUSION

This study has demonstrated that numerical flow simulations are able to provide useful information about the global characteristics of the flow inside the blower aggregate. By knowing the calculated average operation parameters of the blower aggregate in phase of design state of the blower it is possible to find the best blower among the designed variations. By studying the local flow structures [3-4] determined by using the simulation results we can get very important information about local flow problems (e.g. the sizes and local positions stagnations or vorical zones) also supporting the design process.

ACKNOWLEDGEMENT

The work was carried out as part of the TÁMOP-4.2.1.B-10/2/KONV-2010-0001 project in the framework of the New Hungarian Development Plan. The realization of this project is supported by the European Union, co-financed by the European Social Fund.

REFERENCES

[1] Ansys Inc., ANSYS FLUENT 12.0 User's Guide, Canonsburg, PA, USA, 2009.

[2] Lakatos, K., Szaszák, N., Mátrai Zs., Soltész, L., Szabó, Sz.: Experimental Development of Guide Vanes and Return Guide Vanes of a Mini Blower, Proceeding of MicroCAD International Computer Science Conference, Miskolc, pp. 65-72, 2009.

[3] Kalmár, L., Janiga, G., Soltész, L.: Characterisation of Different One-Stage Blower Configurations Using 3D Unsteady Numerical Flow Simulations, REZANIE I INSTRUMENT v Technologicseszkij Szisztemax, Minisztersztvo Obrazovanija i Nauki, Molodezsi i Szporta Ukraini, 81' 2012, pp. 112-119., (ISSN 2078-7405), Harkov, 2012.

[4] Kalmár, L., Janiga, G., Fodor, B., Soltész, L., Varga, Z.: CFD Simulation of the Flow in One-Stage Radial-Flow Blower Aggregate, 11th Conference on Power System Engineering, Thermodynamics & Fluid Flow, ES-2012, pp.1-8., (ISBN 978-80-261-0004-1), Pilzen, Czech Republic, 2012.

FRACTAL DIMENSION OF CHROMATIN REGIONS IN HISTOLOGICAL PICTURES REVEALS THE PRESENCE OF EPITHELIAL TUMOURS

Liviu GAIȚĂ[1], Manuella MILITARU[1], Gabriela POPESCU[1]

University of Agronomic Sciences and Veterinary Medicine of Bucharest, Romania, clinica@ortovet.ro

Abstract: *This study examines the hypothesis that the fractal dimension of chromatin regions in histological pictures has a systematic and measurable variation when tumors emerge in a tissue. We used 194 histological pictures of healthy tissue and of tissue of the same nature with tumoral changes from 24 cases of dogs and cats treated for malignant and benign epithelial tumors: carcinoma, seminoma, adenoma, trichoblastoma, epitelioma.*

Fractal analysis was performed on pictures reduced to selected chromatin areas. The results indicate that for chromatin pictures at x40 magnification the fractal dimension is significantly increased when tumoral changes are present on more than 20% of the picture area. The largest effect on fractal dimension was identified for mammary gland carcinoma.

Keywords: *pathology, fractal analysis, histology, oncology*

1. INTRODUCTION

In spite of huge progress made lately in laboratory and information technology, there are parts of the pathology practical work that are still entirely dependent on the heavy involvement of the human expert. The histopathologic diagnostic is an example: the microscopic images of tissue from the suspected lesion are examined thoroughly, many of the cell and tissue architecture features are assessed against classification criteria, and a conclusion is drawn on the status of a normal or pathologic state, on the type, and on the severity of the lesion – when that is confirmed.

This work is done by a highly qualified expert, hence the inherent subjectivity and vulnerability to errors is unfortunately combined with the responsibility derived from the formulated diagnostic and with the pressure of time, which rarely allows for second or third opinions, as therapy needs to start or be adjusted as soon as possible.

One of the tools identified as a possible contribution to alleviating this constraint is fractal analysis applied on the histological pictures captured with a digital camera. The basis for this approach is the fractal aspect of biological objects, which was revealed by Mandelbrot himself [1], and stands out in an obvious manner in most various circumstances and over several scale sizes. It has been confirmed [2] in macroscopic and microscopic morphology of organisms (as highlighted by various imagistic techniques), in the dynamics of physiological parameters, in DNA sequences, in population dynamics.

The potential usefulness of fractal analysis for diagnostic purposes was advocated as early as 1997 by Cross [3] as the fractal dimension seems to be systematically impacted by the pathologic changes of morphology at cellular and tissular level. Einstein [4] brought extensive evidence of the capacity of fractal analysis to identify the changes that occur in the nuclear chromatin when pathologic processes occur; that paper outlined also some of the most effective ways in which the fractal dimension, as a synthetic numeric measure associated with characteristics of a lesion, can be integrated in heuristic or statistical models to facilitate diagnostic and prognostic.

A number of positive results were published, confirming the potential usefulness of the fractal analysis in diagnostic and prognostic, in various fields of human medicine [16]. A particular domain where fractal analysis is considered useful is oncology [6], [7]. The aim of this study was to check if the fractal dimension is useful for the identification of cancerous lesions in histology samples from dogs and cats. Only scarce relevant results were published for fractal analysis applied in the cito- and histo-pathology of these species. [8], [9], [10], [11].

2. MATERIALS AND METHODS

In studying the effectiveness of the fractal analysis as a pathology tool, an important constraint is the wide variability of the cases on which it is intended to be tested and applied. To simplify the problem and in hope of a clarification we selected cases of organs affected by epithelial tumours and organised the histology images in two groups:

- *Group 1*, images in which morphology features indicative of cancer were present on more than 20% of the total area of the image;
- *Group 2*, images in which morphology features indicative of cancer were absent, i.e. present on less than 20% of the total area of the image.

The morphology features indicative of cancer were the histology features revealed by H&E staining which are traditionally used to diagnose the tumours in dogs and cats according to WHO classification. Images in both groups were selected from the same patients, thus limiting the impact of factors other than the presence/absence of tumoral changes on the fractal dimension of the pictures. Briefly, the control group was made of pictures of healthy tissue of the same nature from the same patients.

We analysed samples from 24 patients, 19 dogs and 5 cats, with benign and malign epithelial tumours: mammary gland carcinoma, other carcinoma, mammary adenoma, epitelioma, seminoma, mammary fibroadenomatosis, mammary adenoma, trichoblastoma, trichoepitelioma, hemangiopericitoma. They received treatment at the ORTOVET clinic in Bucharest and the histology diagnostic was made in all cases in the Pathology Laboratory of the FMV-USAMV in Bucharest, between May 2011 and March 2012 [12]. The fractal analysis was performed on 194 images, 142 images in *Group 1* and 52 images in *Group 2*.

The digital images were captured on Olympus BX41 microscope with its built-in camera and with Olympus Cell^B software. Here we present results based on the set of pictures made with the x40 magnification objective.

The choice of method and parameters for image processing and for fractal analysis made use of outputs from previous research on optimising the procedure for a good sensitivity of the resulting fractal dimension to subtle changes in the examined tissue [13].

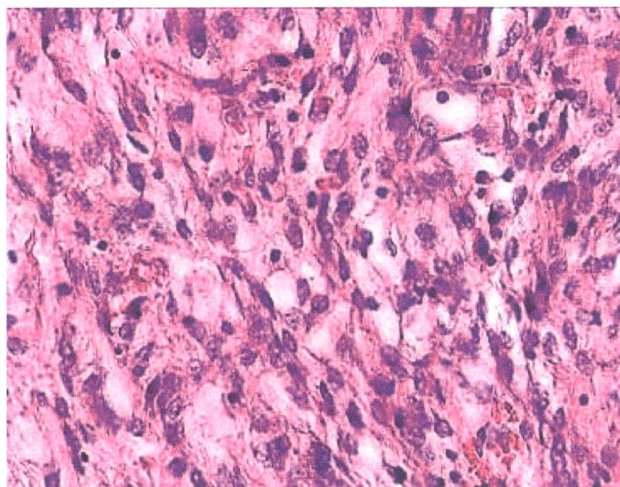

a

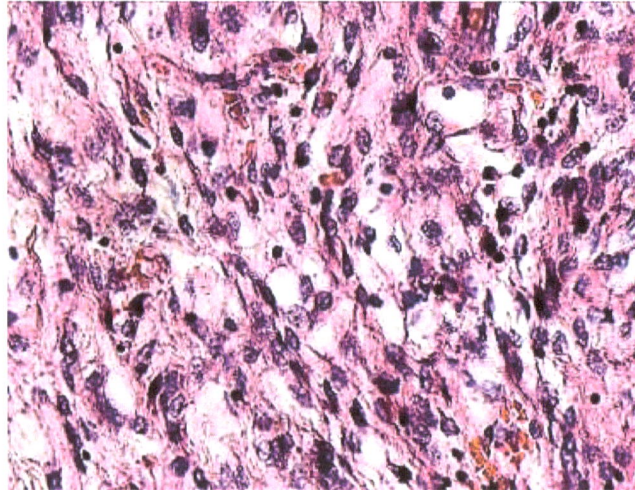

b

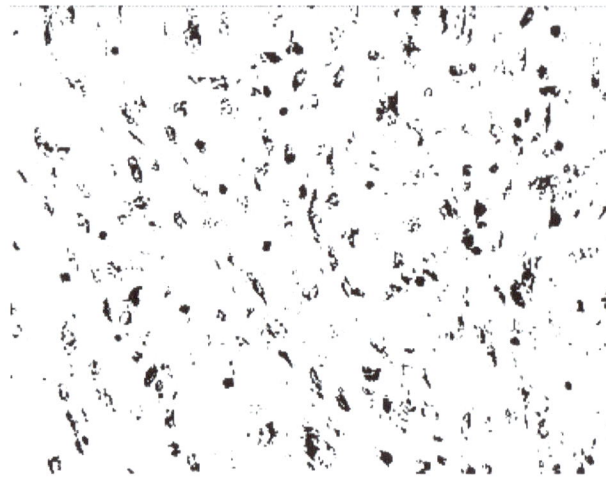

c

Figure 1: Picture processing: (**a**) original picture, (**b**) result of balance on hue, contrast, brightness, and saturation, followed by directional sharpen, (**c**) result of segmentation by colour mask to extract chromatin regions, followed by conversion to black and white

There were three major steps carried out in preparing the picture for the fractal analysis.

(i)   *The balance* of the picture, regarding hue, contrast, brightness and saturation. Directional sharpen was also applied.

(ii)   *The segmentation* of the picture. It was made by a colour mask that identified and selected the chromatin regions in the picture, the rest of the picture being deleted.

(iii)   Finally, *the conversion* of the picture – limited now to chromatin areas - to a grey palette. Once the parameters of the balance procedure and of the colour mask were chosen, the images were batch processed with Corel© PhotoPaint.

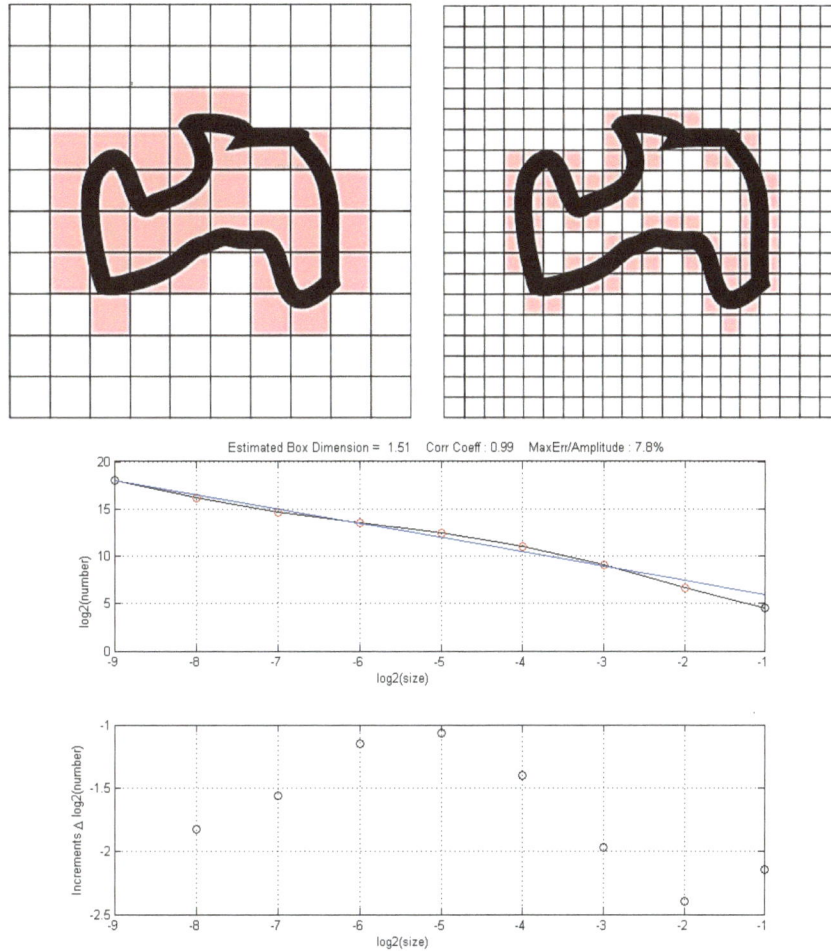

Estimated Box Dimension = 1.51   Corr Coeff : 0.99   MaxErr/Amplitude : 7.8%

Figure 2: Computing the fractal dimension of an image by the box method, using FracLab

The fractal analysis of the gray-scale images was performed using FracLab 2.05, developed by Research Center INRIA Saclay - Île-de-France. The method chosen to compute the fractal dimension was the box method, with the regression curve drawn by the least squares method (Figure 1). The box method provides a very good approximation of the Hausdorff dimension [...]:

$$D_H = \lim_{\varepsilon \to 0} \frac{\log N(\varepsilon)}{\log \frac{1}{\varepsilon}} \tag{1}$$

The statistical analysis of the results was made using StatsDirect 2.7.9.

3. RESULTS AND DISCUSSION

The results in Table 1 suggest that the fractal dimension of the chromatin regions is greater in the images where tumoral changes are present over more than 20% of the total area.

Table 1: Fractal dimension FD of chromatin regions
Group 1: tumour lesions on less than 20% of the total image area
Group 2: tumour lesions on more than 20% of the total image area

Group	Number of images in the group	Average FD	Standard deviation	CI 95% for average FD	Maximum FD	Minimum FD
Group 1	52	1.641981	0.026268	1.649294 ...1.634668	1.7083	1.592
Group 2	142	1.67831	0.040498	1.685029 ...1.671591	1.7685	1.5945

The Fisher test showed a significant difference between the variances in the normal and tumoral groups and the appropriate variant of the Student test was used. The resulting equivalent number of degrees of freedom for the considered population was 140.04762 and the value for t=7.292236. The probability that the difference between the averages of the groups is a random event, p is less than 0.0001. Mann-Whitney test confirmed the results. The statistical power for a 5% significance is over 99.99%. The confidence interval ±95% for the difference -0.036329 between the averages of the two groups is [-0.046155, -0.026503]. The 95% confidence intervals of the averages of the two groups are not overlapping (Figure 3).

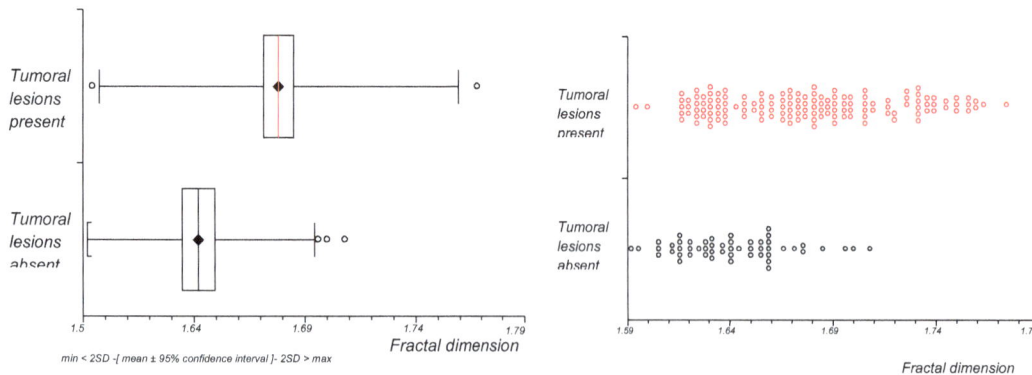

Figure 3: Spread and box-and-whisker plots for the fractal dimension: the ±95% confidence intervals when tumours are absent and, respectively present, do not overlap

On the subset of cases of mammary gland carcinoma, the gap increased between the 95% confidence intervals of the average fractal dimension in the presence, respectively absence of tumoral lesions. The statistical significance of the difference between the means was also greater, as the variances were similar, hence the standard Student test was acceptable.

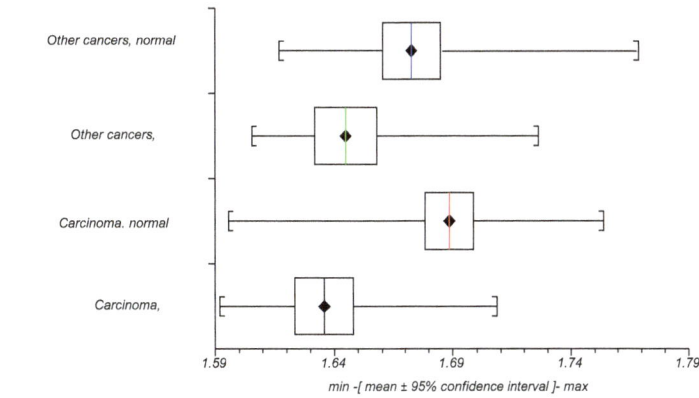

Figure 4: Fractal dimension of chromatin regions in normal/tumoral tissue for patients with mammary gland carcinoma and for patients with other types of cancer

The fact that the 95% confidence intervals of the averages of the two groups are not overlapping sustains the possibility to define distinct domains for the values of the fractal dimension that can be associated with the presence and, respectively, the absence of tumoral lesions in the histology image that is analysed.

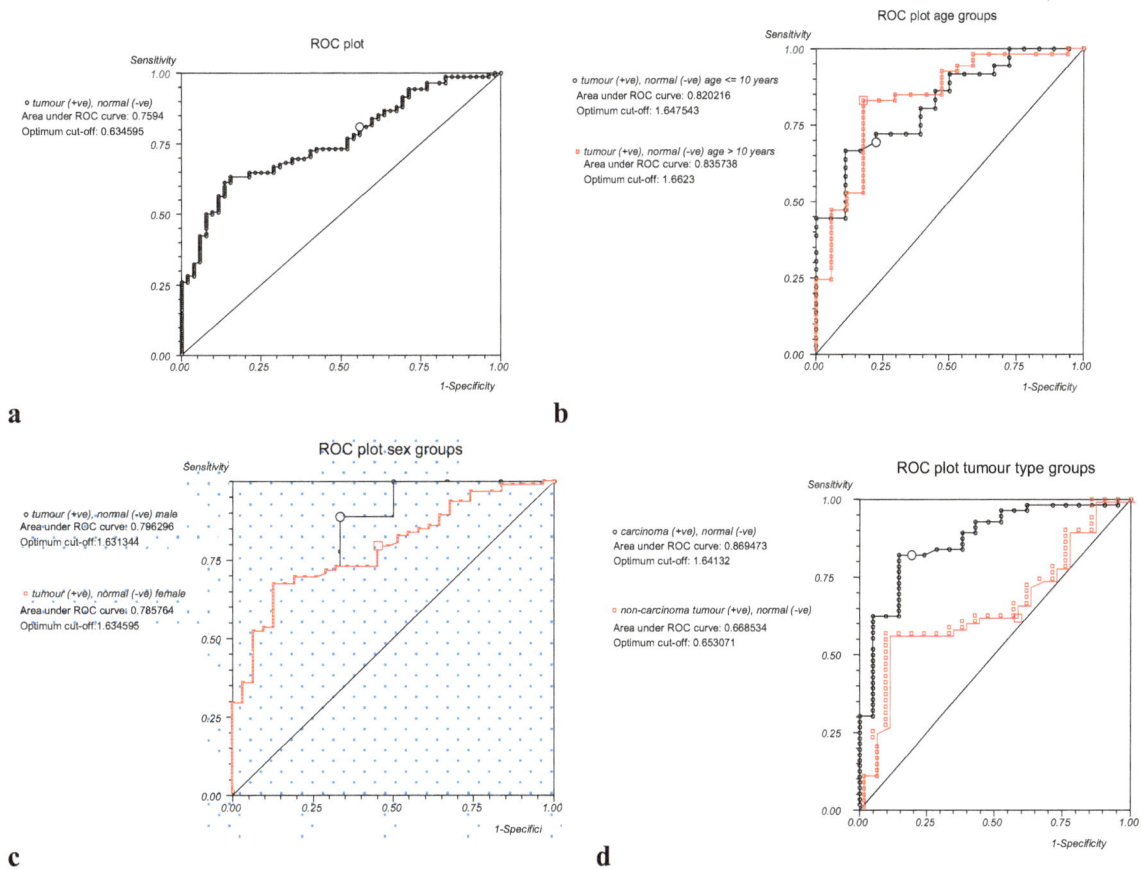

Figure 5: ROC plots for (**a**) the entire group studied, (**b**) separate age groups, (**c**) separate sex groups, and (**d**) separate tumour type groups

A ROC curve analysis was performed to illustrate and assess this possibility (Figure 5). T he same analysis was repeated also for divisions into sex groups and age groups. Looking at the area under the ROC curve [17], it is remarkable that those divisions improved, especially the age division, the relevance of the fractal dimension as an indicator for the presence of tumours, in spite of the reduced size of data set for each group. It suggests that future research that could aim at defining normal and pathological domains for the fractal dimension should consider defining such domains separately for groups segregated by relevant criteria (age and sex among them). The results presented here show that the correlation between the fractal dimension and the presence of lesions is strong enough so that the required sample size for a relevant statistical conclusion is attainable with reasonable effort.

Another possible approach to make use of the added value of the fractal dimension as a diagnostic tool is to directly corroborate it with other clinical, paraclinical, and therapeutic data in integrated models like artificial neural networks applied for diagnostic and prognostic.

4. CONCLUSIONS

The fractal dimension of chromatin regions in histological pictures varies systematically when epithelial tumour lesions occur, hence it can be used as a diagnostic tool. For some types of cancer, like the mammary gland carcinoma, there are indications that normal and pathological ranges could be defined for the fractal dimension of chromatin areas computed in standardised conditions. Using segregating relevant criteria like age and sex can improve the sensitivity and selectivity of this potential diagnostic instrument.

REFERENCES

[1]   Mandelbrot B., The fractal geometry of the nature, Freeman, San Francisco, 1993
[2]   Losa G. A., The fractal geometry of life, Rivista di biologia, Vol 102 (1), pp. 29-59, 2009
[3]   Cross S. S., Fractals in Pathology, Journal of Pathology, Vol. 182, pp.1-8, 1997
[4]   Einstein A. J., Wu H.-S., Sanchez M., Gil J., Fractal characterization of chromatin appearance for diagnosis in breast cytology, The Journal of Pathology, Vol. 185:4, pp. 366-381, 1998
[5]   Losa G. A., Do complex cell structures share a fractal-like organization?, 6eme Congres Europeen de Science des Systemes, Paris, 2005
[6]   Baish J. W., Jain R. K., Fractals and cancer, Cancer Research, Vol. 60, pp. 3683-3688, 2000
[7]   Amuda P.F. F., Gatti M., Junior F. N. F., Amuda J. G. F., Moreira R. D., Junior L. O., Arruda L. F., Godoy M. F., Quantification of fractal dimension and Shannon's entropy in histological diagnosis of prostate cancer, BMC Clinical Pathology, 13:6, 2013
[8]   De Vico G., Cataldi M., Maiolino M., Beltraminelli S., Losa G. A., Fractal analysis of canine trichoblastoma, Fractals in Biology and Medicine, Birkhauser Basel, 2005
[9]   De Vico, Peretti V., Losa G. A., Fractal organization of feline oocyte cytoplasm, European Journal of Histocemistry Vol. 49 (2), pp. 151, 2005
[10]   Simeonov R., Computer-assisted fractal analysis of spontaneous canine mammary gland tumours on cytologic smears, Trakia Journal of Sciences, Vol. 5(1), pp.65-88, 2007
[11]   Simeonov R., Simeonov G., Nuclear cytomorphometry in feline mammary gland epithelial tumours, Veterinary Journal, 179 (2): 296-300, February 2009
[12]   Popescu G., Computer assisted morphometry and fractal analysis – complementary methods for the diagnostic and prognostic of epithelial tumours, DVM dissertation guided by Prof. Militaru M., UASVM-FVM, Bucharest, 2012
[13]   Gaiţă L., Militaru M., Performance optimization of fractal analysis used for the microscopic pathology diagnostic, Scientific Research Contributions to Veterinary Medicine Advance, Symposium, UASVM-FVM, Bucharest, 2011

[14] Gaiţă L., Militaru M., Fractal Analysis Used for Identifying the Impact of Drugs on Hepatic Tissue, Journal of Comparative Pathology, Vol. 148, Issue 1, January 2013

[15] Gaiţă L., Militaru M., Current situation in using fractal analysis for tumour diagnostic and prognostic, Rev. Rom. Med. Vet. 3/2010

[16] Klonowski W., Stepien R., Stepien P., Simple fractal method of assessment of histological images for application in medical diagnostics, Nonlinear Biomedical Physics, 2010, 4:7

[17] Zweig M. H., Campbell G., Receiver-operating characteristic (ROC) plots: a fundamental evaluation tool in clinical medicine, Clinical Chemistry, Vol. 39:4, pp. 561-77, 1993

TOWARDS PERSONALIZED VEHICLE SAFETY

Luděk Hynčík[1], Luděk Kovář[2]
[1] University of West Bohemia, Plzeň, CZECH REPUBLIC, hyncik@ntc.zcu.cz
[2] MECAS ESI s. r. o., Plzeň, CZECH REPUBLIC, ludek.kovar@mecasesi.cz

Abstract: *Human body models start to play an important role in safety system design. Comparing to dummies, virtual human models can be scaled to represent wide spectra of population. The paper describes the performance of virtual biomechanical human models as a support for design and optimization of not only passive and active safety systems used in various modes of transport. The accent is stressed on scaling where human subject of different mass, age and gender that can be created automatically. The difference in response for various subjects in impact scenario is shown.*
Keywords: *impact, human, virtual model, scaling, safety*

1. INTRODUCTION

The traffic accidents cause one of the highest numbers of severe injuries. The numbers of deaths or fatally injured citizens prove that the traffic accidents and their consequences are still a serious problem to be solved. The statistics show the decreasing number of accidents in the past years [9], but the decrease is still necessary to be speeded up regarding also the socioeconomic aspects of the problem [3].

A lot of effort is devoted to both passive and active safety systems development. The transportation standards usually define safety requirements by regulations (e.g. ECE-R94, 96/79/EC and ECE-R95, 96/27/EC in Europe) with specific dummies (e.g. Hybrid III 50% and Eurosid II).

The dummies include sensors for monitoring accelerations, loads and other signals and each dummy is developed for a specific scenario but there are limitations of these dummies like only specific body size (5%, 50% and 95%) or calibration just for a specific test. Taking into account that the consequence of the traffic accident is highly influenced by the stature of the body, the virtual human body models start to play significant role because they can be scaled or even personalized towards a particular population or even particular person.

2. METHOD

Contemporary mathematical methods enable extensive analyses of technical problems by numerical approach. Many industry fields apply virtual prototyping towards new product development including virtual testing, where the whole process is modelled by computer software. The safety approach considers specific and biofidelic human body models to be developed and validated [1]. There are several approaches to be used in virtual human body modelling [5]. Whilst multi-body approach (MBS) performs in fast calculation times but with very limited application fields, detailed finite element models might predict very detailed level of behaviour including real fractures or ruptures occurring in the human body after external impact. That is why hybrid models benefiting of both approaches based on deformable segments linked together as by MBS is further used with the advantage of full articulation and very short calculation time.

2.1. Hybrid model

For the purposes of scaling, the authors are developing VIRTHUMAN [8], a special hybrid model, that benefits from both approaches combining them in an appropriate manner. The model is useful for dynamical simulations including interactions with safety systems. The basic model used benefits from the hybrid approach combining so called "*superelements*" linked to the MBS structure by nonlinear springs. Between the superelements, there are straps of elements without any response in order to keep continuous surface during articulation. The structure of the model is displayed in Figure 1.

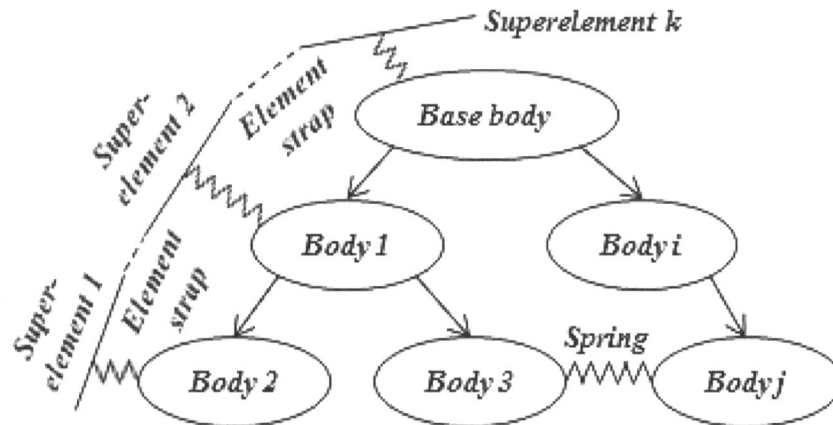

Figure 1: The hybrid approach

The reference geometry of the surfaces is chosen from the European database CAESAR in order to be close enough to Hybrid III 50 % and Eurosid II dummies as far as body dimensions are concerned. The MBS is defined as an open-tree structure that is required for scaling.

2.2. Scaling

The scaling algorithm is developed on the basic parameters of height, mass and age [6]. For the given age and gender, the height interval is proposed. For chosen height, the mass interval is proposed. Then, the model is scaled based a set of over 13.000 Czech sportsmen (males and females) measured by the Charles University in Prague (Faculty of Sciences, Department of Anthropology and Human Genetics) during the Czechoslovak Spartakiada in 1985 [3]. Flexibility scaling affects all joints in the human body by a multiplier based on the so-called "*flexindex*" [1].

2.3. Impact tests

To assess the scaled developed models, 2 impact (sled) tests were chosen [7]. The first one concerned 13 years old boy cadaver further referred as H7613DOT. The total height was 156 cm and the total mass was 50 kg that were the basic parameters for scaling. Other anthropometrical dimension after automatic scaling were also been checked in order to confirm the scaling process, see
Figure 2 and Table , where acceptable difference is approved.

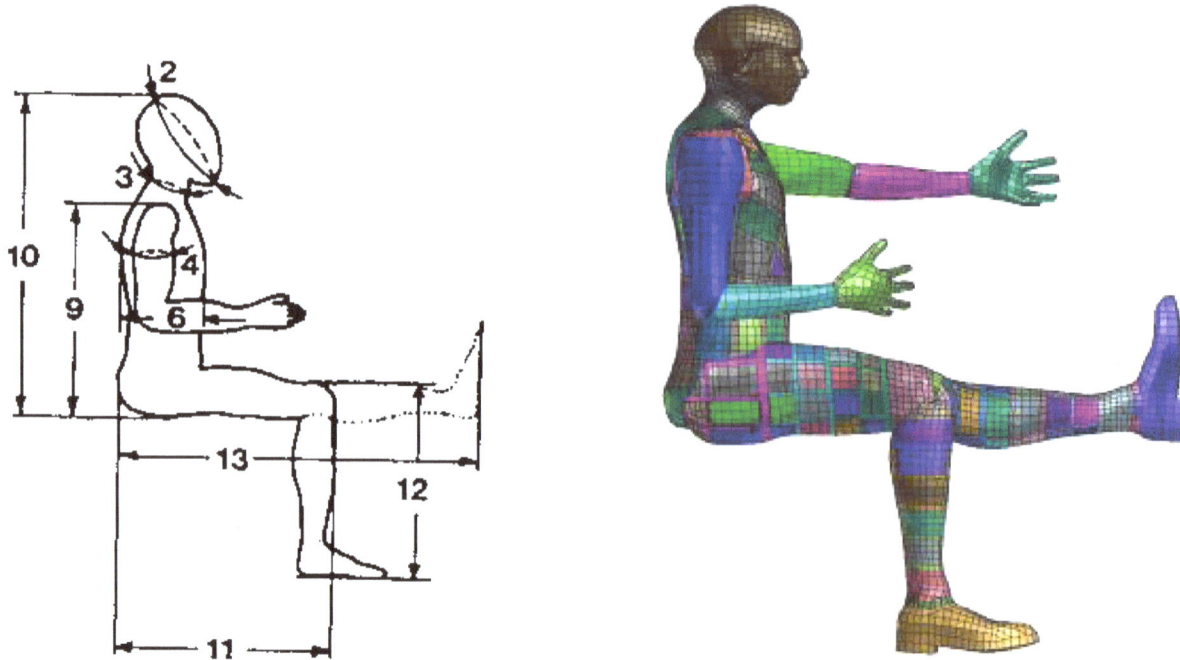

Figure 2: Human anthropometry (left) and scaled model to 13 years old boy (right)

Table 1: The anthropometric data of 13 year old boy (* denotes dimensions according to Figure 2)

Dimension	Cadaver	Scaled model	Difference [%]
Buttocks-shoulder (9*) [cm]	60	56	-7
Sitting height (10*) [cm]	81	78	-4
Pelvis-heel (13*) [cm]	82	94	15

The second test concerned 47 years old female cadaver further referred as H9311DOT. The total height was 169 cm and the total mass was 76 kg that were the basic parameters for scaling. Other anthropometrical dimension after automatic scaling were also been checked in order to confirm the scaling process, see
Figure 2 (left) and Table 2, where acceptable difference is approved.

Table 2: The anthropometric data of 47 year old female (* denotes dimensions according to Figure 2)

Dimension	Cadaver	Scaled model	Difference [%]
Buttocks-shoulder (9*) [cm]	145	141	-3
Sitting height (10*) [cm]	92	91	-1
Pelvis-heel (13*) [cm]	99	101	2

Both models were positioned into the sled test environment (see Figure 3) according to [7] and the same deceleration pulse as during the experiment was applied. The female test concerned also driver airbag.

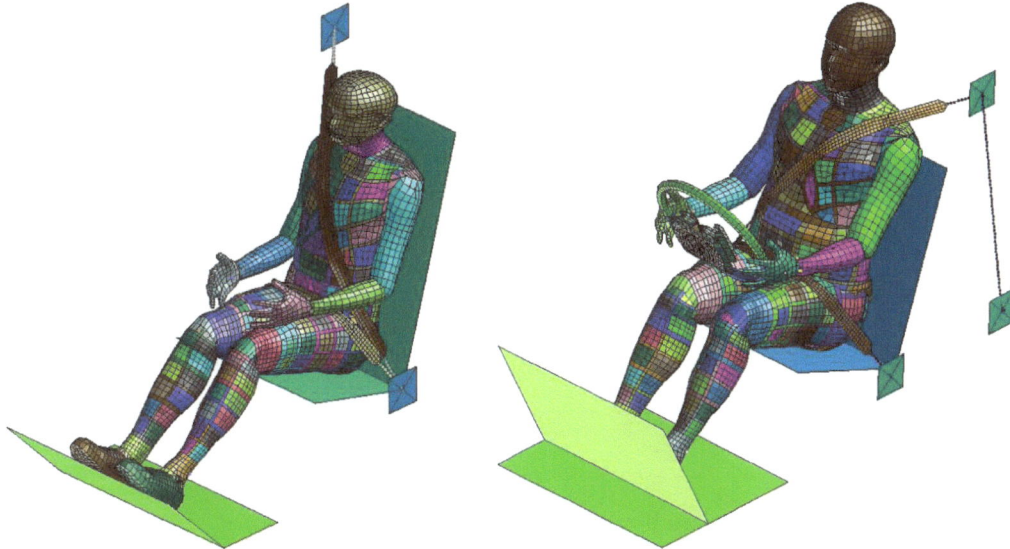

Figure 3: 13 years old boy (left) and 47 years old female (right) in sled test environment

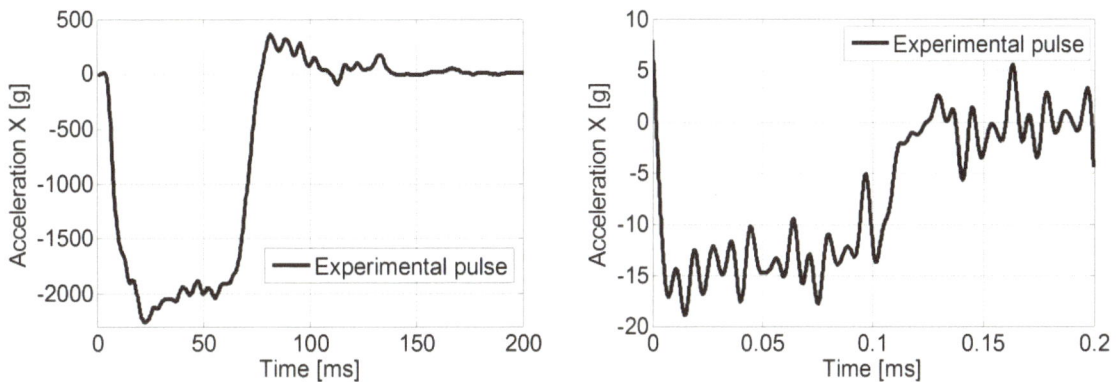

Figure 4: Deceleration pulses (left: 13 years old boy, right: 47 years old female)

3. RESULTS

Table 1 and Table 2 show that even based on few parameters, the scaling algorithm is able to develop subjects corresponding to the real anthropometry. Since the height and the mass are the scaling parameters, they are precisely the same for the scaled models. The 13 years old boy model fits to the 90th percentile, whilst the 47 years old female fits to the 92nd percentile within the anthropometrical data set [3].

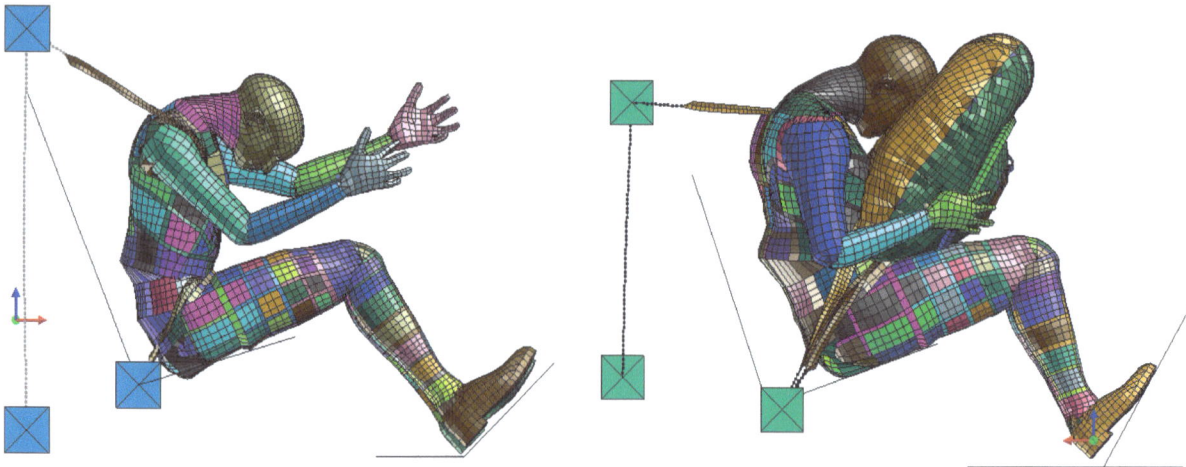

Figure 5: 13 years old boy, right: 47 years old female, both at 100 ms

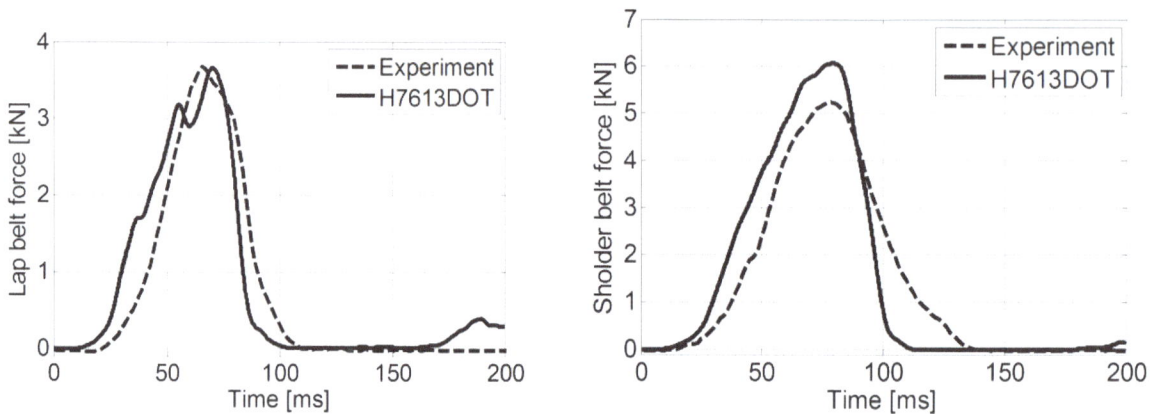

Figure 6: Lap belt (left) and shoulder belt (right) forces compared to experiment

The correct biofidelic performance of the models mentioned above was proven comparing the results to the available experimental data. Both the 13 years old boy model and cadaver were loaded by a deceleration pulse (see Figure 4 left) from the initial velocity equal to 41 km/h. The crucial entity influencing the whole response is the belt force. The time dependent belt forces are shown in Figure 6. Figure 7 shows the head acceleration in the local coordinate frame fixed to the head.

For the 13 years old boy model (H7613DOT), both the lap and shoulder belt forces correspond well to the experimental signal. The peak of the lap belt is reached around 70 ms and the peak force is around 3.5 kN (Figure 6 left). The peak of the shoulder belt force comes at around 75 ms and it is over 5 kN for the experiment and slightly over 6 kN for the simulation (Figure 6 right). The slight difference is acceptable.

The acceleration measured on head is also well corresponding including the head injury criterion that is very similar between the simulation and the experiment. The absolute value of peak of the frontal acceleration (Figure 7 left) is slightly over comparing to the experiment that is acceptable. The acceleration magnitude (Figure 7 right) corresponds well to the experiment except the experimental peak at 80 ms that is probably caused when the head hits the sternum. However, the head injury criterion (Figure 7 right) shows good correlation.

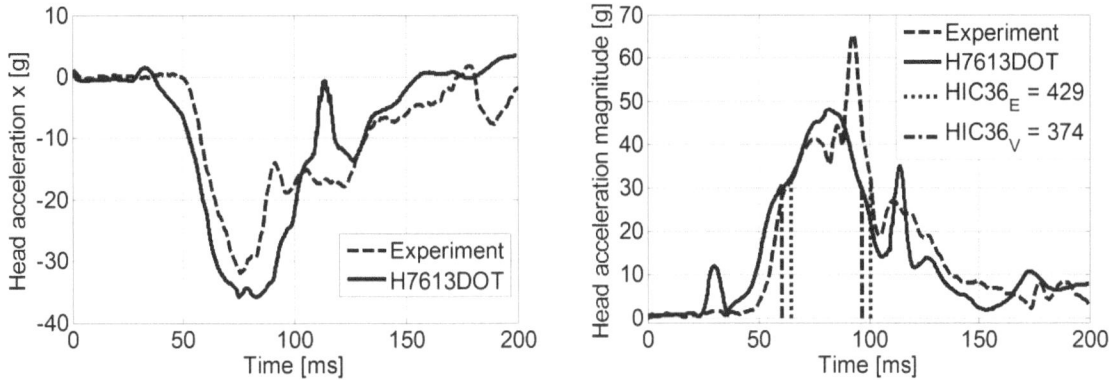

Figure 7: Head acceleration in the frontal direction (left) and head acceleration magnitude (right)

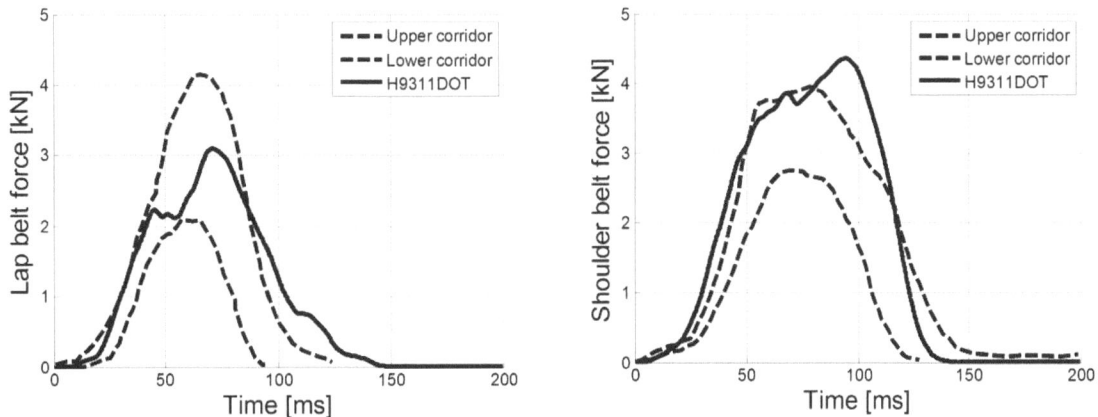

Figure 8: Lap belt (left) and shoulder belt (right) forces

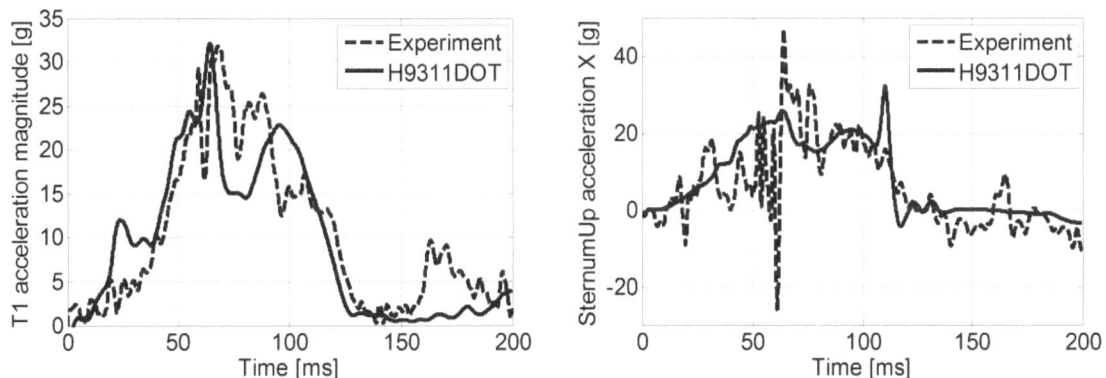

Figure 9: T1 acceleration magnitude (left) and upper sternum acceleration (right)

Both the 47 years old female model and cadaver were loaded by a deceleration pulse (see Figure 4 right) from the initial velocity equal to 49 km/h. Since the belt forces signal was not available, the initial response of the model was tuned in order the belt forces being in corridors from similar test. The tuning process means setting the parameters like friction coefficient and contact parameters.

For the 47 years old female model (H9311DOT), the lap belt peak force is 3 kN whilst the shoulder belt peak force is over 4 kN. As was mentioned above, the corridors do not correspond to this test; they are just use to tune the kinematics of the model.

The frontal direction acceleration measured on the vertebra T1 (Figure 9 left) fits quite perfectly to the experimental signal as well as the upper sternum acceleration in the frontal direction (Figure 9 left).

4. CONCLUSION

The paper summarized up-to-date knowledge in automatic developing virtual human models of various sizes. The present method developed by authors enabling anthropometrical scaling of multi-body based human models is used. Hybrid approach based model performing in fast calculation with reasonable injury description was used and scaled to young boy and older women.

The correct response for the scaled models was shown to validate the scaling process. For 2 specific cadavers of different age, size and gender (13 years old boy and 47 years old female), the scaled models were developed and the sled test performance was tested and compared to experiment..

The paper proved that it is necessary to take into account particular body sizes in order to obtain a proper behavior of the body including injury assessment during the impact. The scaled models are useful and powerful tool for virtual assessment of human body behavior under loading regarding the wide spectra of population.

ACKNOWLEDGMENT

The work is co-financed by the Technology Agency of the Czech Republic within the project TA01031628 "Scalable human models for increasing traffic safety".

REFERENCES

[1] Araújo, C.G., Flexibility assessment: normative values for flexitest from 5 to 91 years of age, Arq. Bras. Cardiol. 90(4): 257-263, 2008, doi: 10.1590/S0066-782X2008000400008.

[2] Ayache N., Computational Models for the Human Body, Elsevier, 2004.

[3] Bláha, P., Anthropometry of Czech and Slovak Population from 6 till 55 years, 1985.

[4] Daňková A., Ekonomická stránka dopravních nehod, *Dopravní inženýrství*, vol. 2, 2007. http://www.dopravniinzenyrstvi.cz/clanky/ekonomicka-stranka-dopravnich-nehod

[5] Hynčík, L., Kovář, L., Dziewoński, T., Baudrit, P., Virtual human models for industry, Proceedings of COMAT 2012.

[6] Hynčík, L., Čechová, H., Kovář, L., Bláha, On Scaling Virtual Human Models, SAE Technical Papers 1, 2013, doi: 10.4271/2013-01-0074.

[7] Kallieris, D., Four cadaver tests by using 3-pt belt and 3-pt belt – airbag combination, NHTSA report, contract No. DTNH 22-89-D-07012, September 1993.

[8] Maňas, J, Kovář, L., Petřík, J., Čechová, H., Špirk, S., Validation of human body model VIRTHUMAN and its implementation in crash scenarios, Proceedings of TMM 2012, doi: 10.1007/978-94-007-5125-5_46.

[9] Skácal L., Hloubková analýza mezinárodního srovnání dopravní nehodovosti v ČR, 2007. http://www.czrso.cz/index.php?id=402

THEORETICAL ENGINE DESIGN SOLUTION TO MINIMIZE CONSUMPTION AND POLLUTION

Assist. Eng. Călin Itu [1]

[1] *Transilvania* University of Brasov, Brasov, ROMANIA, calinitu@yahoo.com or calinitu@unitbv.ro

Abstract: In this paper, we attempt to describe a possible design solution of the engine with turning arm in order to bring some information with respect to engine working efficiency. The modifying of the functional parameters such as: piston stroke, displacement or ratios of compression will lead to obtain some different energy engine parameters (power, torque) as well as different ecology engine parameters (consumption, emission).

The analyses were made using specific module of the PTC software (Pro/E Mechanism.

Keywords: *engine, optimization, piston stroke and ratio of compression*

1. INTRODUCTION

The evolution of the automotive which became one of the most important vehicles, as well as the enhancement of the car park from the last time were imposed a continuous development process for different parameters or characteristics (increase power, lower consumption, increase the durability or reliability) of the sub-assembly that power propelled the automotive.

The power unit with the highest development level in the actual context that directly contributes to automotive propulsion is the engine with internal combustion. The grand manufacturers of the automotives are focused the activity on the different functional parameters or constructive parameters of the engine in order to obtained more power with lower consumption or lower emission of the exhaust gas (CH, NOx, CO, e.g.).

The efficiency of the engine with internal combustion depends on cycle characteristic processes. The assessments with regards to engine efficiency can be made based on constructive parameters, such as: engine displacement, stroke – bore ratio, ratio of compression or based on operational parameters (engine management, forming and controlling mixture air/combustible, ignition timing, e.g.).

The increasing pressure that acts on the engine manufacturers regarding to minimize combustible consumption or gas exhaust emissions, it make them to pass in production flux the news constructive solutions for actual engines that are being in the market – gasoline and diesel engines.

One of the actual constructive solutions with satisfied results of the consumption or emissions is the engine with turning arm and variable compression ratio – VCR. His concept is emerged from 2001 and the promoter is Dr. Joe Erlich. This engine has appeared on the automotives being in marketplace from this millennium (more exactly, SAAB has developed this concept) and the VCR technology could be one of the keys that would lead on to obtain some better performance both partial charges and full charges.

Beside the increase engines performance goal focused, the research-development process has to search the solutions in order to minimize the expensive for development and optimization of the product. It can be achievable by means of computers that replace classical tools and making achievable efficiency and useful methods.

In the analysis and simulation mechanical systems domain there are a lot of programs used, such as: classical programming languages (FORTRAN, PASCAL, C), design programs (CATIA, PRO/E, EUCLID, SOLID WORKS), MBS software (ADAMS, FLEXUS, DYMES) and FEA software (ANSYS, PATRAN, NASTRAN)

The analysis, modeling and simulation were made by means of Pro/Mechanism module of the PRO/E software.

2. FUNCTIONAL DESCRIPTION OF ENGINE MECHANISM

2.1. ENGINE MECHANISM WITH TURNING ARM

The kinematics schema for engine mechanism with turning arm is shown in figure 1. Also, in figure 2, it is shown the virtual model.

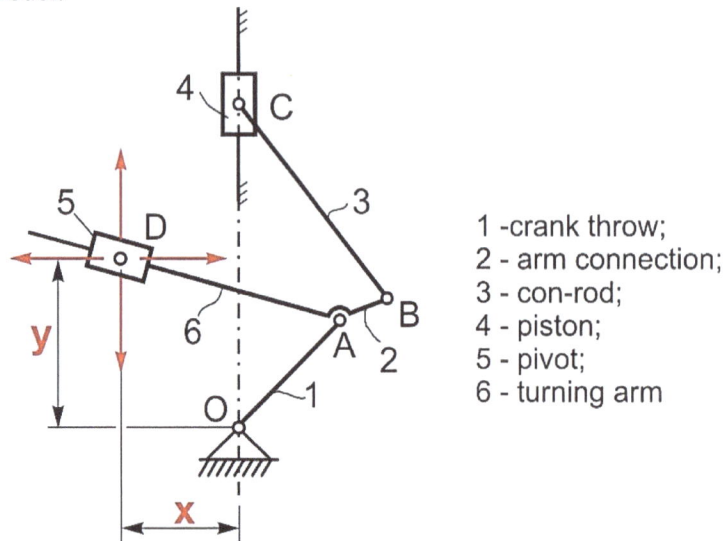

1 -crank throw;
2 - arm connection;
3 - con-rod;
4 - piston;
5 - pivot;
6 - turning arm

Figure 1 - Kinematics schema for engine with turning arm

This new concept is an operating principle of the conventional engine carrying on that converts translational motion of piston in rotating motion of crankshaft by con-rod.

The difference between a conventional engine and this engine analyzed consists in how con-rod is assembled in mechanism: for conventional engine the con-rod is directly connected to crankshaft and for engine with turning arm, the con-rod is connected to crankshaft through arm 2. This arm (2) is jointly with turning arm 6. This turning arm slides in the bearing (5), and it can made possible OB length varying and piston stroke, also. When the engine mechanism running, the point of turning (D) that is position by means of x and y variables can be moved on the vertical and vertical direction thus position of the turning arm is changed. In this way, it can be obtained different values of the compression ratio and piston stroke and also can be obtained an optimization of the engine working regimes depending on charge.

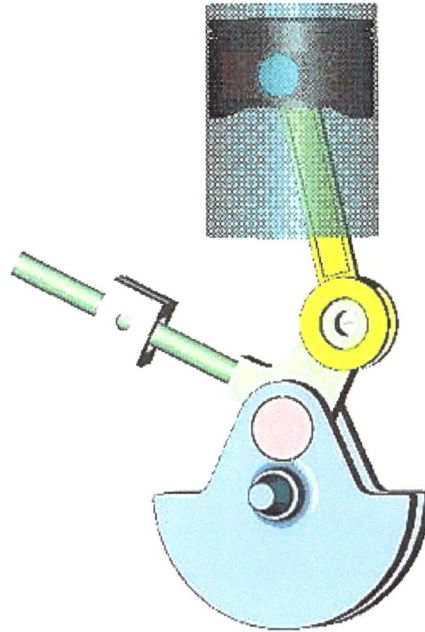

Figure 2 - Virtual model of engine with turning arm

Based on figure 3b, for the engine with turning arm, the trajectory of the con-rod big end obtained on a complete crankshaft rotation is ellipse unlike circular trajectory obtained on a conventional engine. This elliptical trajectory will determine an increased of piston motion time between top dead center and bottom dead center comparison by conventional engine.

More holding piston close to the T.D.C. (top dead center) create more friendly conditions in order to carry on the ignition process at constant volume with more efficiency.

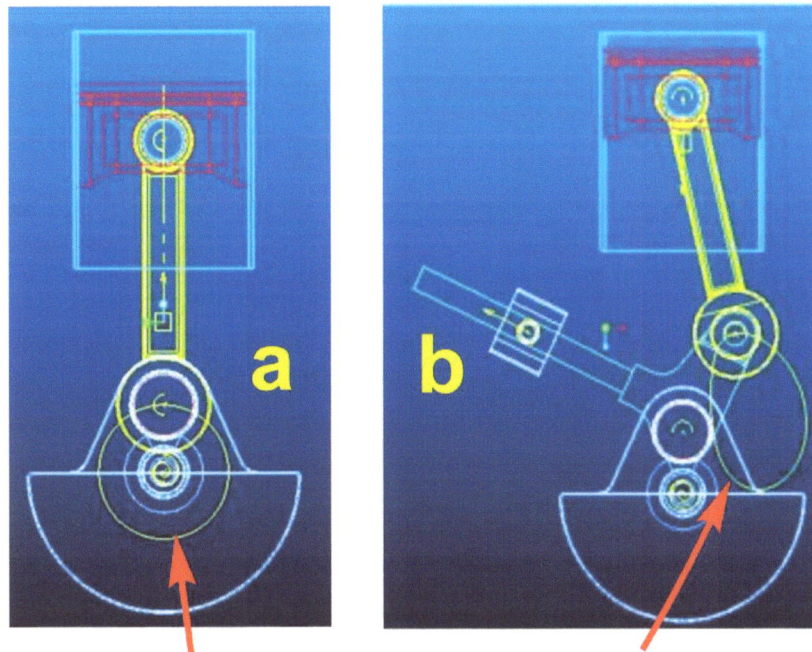

Trajectory of the con-rod big end for conventional engine is circle.

Trajectory of the con-rod big end for engine with turning arm is ellipse

Figure 3 - Comparison trajectories of the both engine analyzed

Also, on the T.D.C. the force that acts on piston generated a torque motor because the connecting rod small end has an eccentric position with respect to crankshaft axis what it means there is torque motor on the all-downward stroke time detent process [2].

Also, at the engine with turning arm, the admission stroke and exhaust stroke are longer than conventional engine, in this way the gas exchange process is favored. [2]

The elliptical trajectory of the con-rod big end at engine with turning arm will generate an increasingly of piston stroke, displacement engine and ratio of compression comparison by conventional engine. This aspect, it can be observed in figure 4. The conventional engine piston stroke obtained (curve 1) is shorter than engine with turning arm piston stroke (curve 2). A modifying of piston stroke will have implication about displacement engine and ratio of compression.

The compression ratio parameter (ε) has a directly influenced on some important parameters of engine, such as: filling grade, pressure of the end compression, temperature of the end compression and residual gas coefficient (ratio between gas quantity remaining in cylinder from the previous cycle and live gas quantity aspirate, the both expressed in kilo mol).

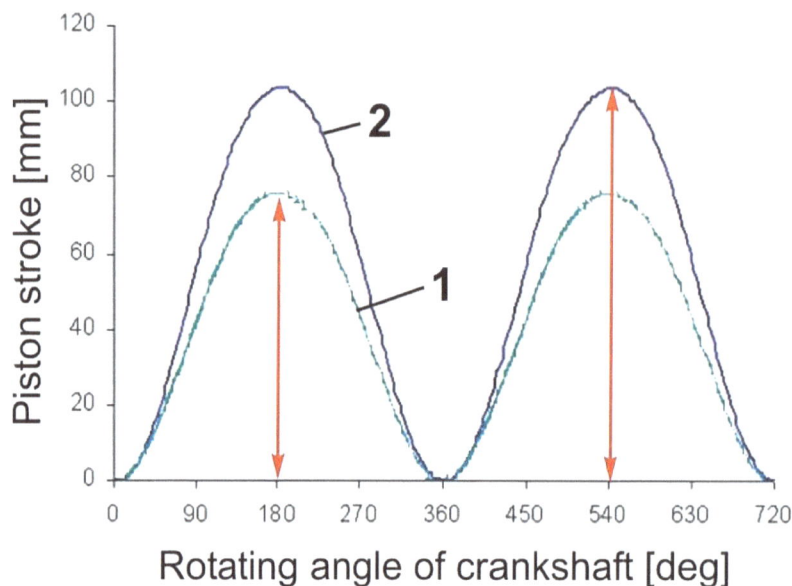

Figure 4 - Evolution of piston stroke in terms of angle of crankshaft

The advantages of this engine type consist in addition to high-density power and lower emission to silent and uniform working. The companies are interested of this engine type (General Motors, Ford, Daimler Chrysler and MG Rover) and makes plans in order to industrialize it in the next two years [2].

3. CONCLUSIONS

As it was specified, the using of the modeling and virtual simulation software offers a quickly and efficiency control of geometric and functional system parameters.

By means of simulation software used (Pro/Mechanism), it was analyzed the output kinematics parameter (piston stroke) and implicitly of the others that are depending on it (displacement engine, ratio of compression), starting from input kinematics parameter (crank throw position).

It can be concluded that from constructive and functional viewpoint, based on parameters determined from virtual simulation, the engine with turning arm presents advantages with respect to a conventional engine.

Beside mentioned, the solution of the engine mechanism with turning arm can be adopted on the any type of engine with gasoline or diesel with two or four cycles engine, for any type of engine dimensions.
VCR technology offers the largest potential improvement in part-throttle fuel efficiency and CO_2 emissions when compared to other competing technologies.

Also, VCR technology can offer torque enhancement at low rpm when boost systems are least effective.

The main obstacles to adoption of VCR are incompatibility with major components in current production and difficulties of combining VCR and non-VCR manufacturing within existing plant. As environmental pressure on the automobile increases and investment plans for new products are put in place, the justification for VCR will become more evident.

An UK-based international engineering is claiming that its new design for a combustion engine has the potential to revolutionize the global market for all internal combustion engine applications, with 40% improvement in fuel efficiency and 50% fewer emissions compared to conventional engines.

It is very interesting, what Dr. Joe Erlich (the promoter of engine mechanism with turning arm) said about conventional engine: "I have always felt that the conventional crank and con-rod was flawed. It wasted too much energy and does not optimize combustion."

Taking into account to over 160 million combustion engines are made every year, this revolutionary engine can be represented a very good solution for the actuality.

REFERENCES

[1]   B. Grundwald, Teoria, construcția și calculul motoarelor pentru autovehicule rutiere, Editura Didactică și Pedagocică, București – 1969;

[2]   Ioan Mircea Oprean, Automobilul modern, Editura Academiei Române, București – 2003;

[3]   Gheorghe Bobescu, Cornel Cofaru, Motoare pentru automobile și tractoare – vol I, Editura Tehnică, Chișinău – 1996.

[4]   Abăitancei D., Bobescu Gh., Motoare pentru automobile, Editura Didactică și Pedagogică, București – 1975.

[5]   Radu Mărdărescu, Victor Hoffmann, Dan Abăitancei, Motoare pentru automobile și tractoare

EXPERIMENTAL INVESTIGATIONS UPON CONTACT BEHAVIOR OF BALL BEARING BALLS PRESSED AGAINST FLAT SURFACES

Mănescu Tiberiu Jr. [1], Gillich Gilbert-Rainer[2], Mănescu Tiberiu Ştefan[3], Suciu Cornel[4]
[1] Eftimie Murgu University, Reşiţa, ROMANIA, tibi.jr@yahoo.com
[2] Eftimie Murgu University, Reşiţa, ROMANIA, gr.gillich@uem.ro
[3] Eftimie Murgu University, Reşiţa, ROMANIA, t.manescu@uem.ro
[4] Ştefan cel Mare University, Suceava, ROMANIA, suciu@fim.usv.ro

Abstract: *Experimental methods that offer point by point information regarding the investigated surfaces are preferred in the study of mechanical contacts. Such a method, advanced in [1-6], consists of investigating the contact model obtained by pressing a metallic punch against a flat, optically transparent, surface, by aid of laser profilometry. The variation of surface reflectivity is measured and used to accurately determine the shape and dimensions of contact areas. The present study employed the abovementioned method in order to experimentally investigate the contact behavior of various ball bearing balls, when pressed against the flat surface of a thick sapphire window with different normal loads. Experimental results were compared to theoretical predictions and good agreement was found between results.*
Keywords: *experimental investigations, mechanical contacts, contact area, laser profilometry, reflectivity*

1. INTRODUCTION

The present paper illustrates experimental investigations conducted on spherical punch-flat surface contact models, by aid of reflectivity. The experimental testing consists of mapping by aid of laser profilometer of the contact surface generated when pressing an equivalent punch against an elastic half-space. For the investigated contact model, the punch is bound by an equivalent surface that incorporates both surface geometries while the elastic half-space is modeled by a thick, optically transparent, plate. As the present investigations aimed to study steel-on-steel contacts, the transparent plate was made of sapphire. The material choice is due to the similarity between sapphire and steel elastic properties. Thus, the longitudinal elasticity modulus for sapphire is $E = 345\ GPa$, while for steel, it is $E = 210\ GPa$, and the sapphire's Poisson's ratio ($v = 0,29$) is also similar to the one for steel ($v = 0,3$). The maximum contact stresses tolerated by sapphire are also in the same size order as for steel ($\sigma_c \approx 2\ GPa$ for sapphire and $\sigma_c \approx 2,5\ GPa$ for bearing steel).

When a light beam falls onto the interface between two optically different media, it is partially reflected back to the first medium and partially transferred to the second medium, as illustrated by Figure 1.

When the presented contact model is scanned by aid of a laser profilometer, the light wave generated by the optical sensor passes through the sapphire plate and then either meets a sapphire-metal interface (corresponding to points inside the contact area) or a sapphire-air interface (corresponding to points outside contact area). As shown in [1-8], the light is reflected differently from in the two situations, which permits to generate 3D representations of the deformed punch surface and accurately asses contact area shape and dimensions.

As demonstrated in [3] *when the light is reflected by an absorbing medium, the reflected light intensity decreases with the increase of the refractive index of the primary transparent medium.* For the considered contact model, the refractive index for sapphire is higher than for air and lower than the one corresponding to metal. This translates, in terms of the above mentioned property, in a lower reflectivity measured at the sapphire-metal interface than the one determined when the light meets the separation between air and sapphire.

This property allows accurate evaluation of the contact area, if the variation of surface reflectivity is known. Most modern laser profilometers can determine this parameter along with the measurement of surface microtopography.

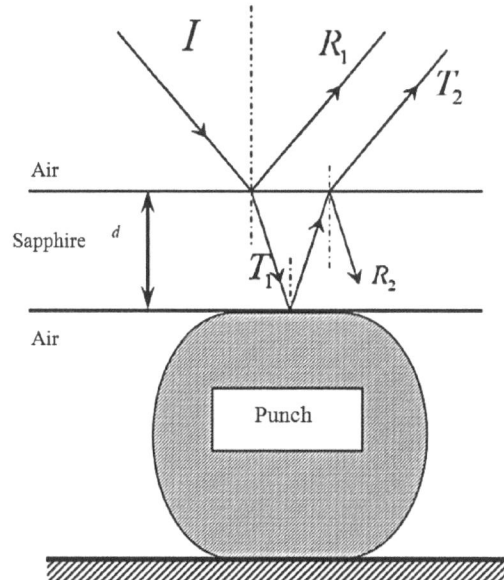

Figure 1: Reflection–refraction of light when the punch-sapphire window contact model is investigated by aid of laser profilometry

2. EXPERIMENTAL SET-UP AND EQUIPMENT

The main instruments employed to determine micro and macro characteristics of real surfaces are profilometers. Depending on the way data regarding the investigated surface is collected, these instruments can fall into two major categories: contact profilometers and non-contact profilometers. The present research was conducted by aid of a μScan® laser profilometer manufactured by NanoFocus, equipped with a CLA 10 chromatic optical sensor. This optical profilometer, belonging to the non-contact category, allows 2D and 3D measurements of real surfaces, and can be used for various applications, both for research and industrial purposes. The main component of the profilometer is its optical sensor, which can be moved along the vertical axis of the system, thus ensuring that the light is focused on the investigated surface. In order to place the investigated specimens under the optical sensor, the profilometer is equipped with a positioning unit (x-y axes) with sample stage.

The present investigations were conducted on contact models obtained by pressing various bearing balls against thick, sapphire discs. Several levels of normal load were applied to the contact, by aid of an experimental device as the one described in [7] and illustrated schematically in Figure 2.

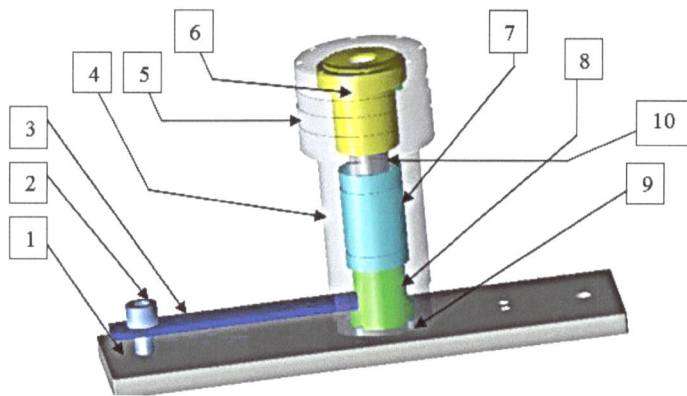

Main components:

1 – Base plate
2 – Normal load screw
3 – Normal load elastic lamella
4 – Body
5 – Top nut
6 – Sapphire holder bush
7 – Ball bush
8 – Normal load pad
9 – Fastening screws
10 – Plunger

Figure 2: Experimental device for normal load application

The experimental device employed for the present investigations was designed to allow application of purely normal forces to contacts created between a ball, or another body, and a thick sapphire disc. The normal load application system consists of an elastic lamella, (3), supported on a cylindrical roller, thus forming a lever. The force applied to one end of the elastic lamella by fastening of the screw (2), is then multiplied by the lever system and applied to the plunger (10) by means of another roller. The plunger presses against the punch-support subassembly, which in its turn presses against the sapphire window, thus loading the contact (any translation of the sapphire window is blocked by the upper nut (5)).

In order to accurately determine the applied normal force, the strain of the elastic lamella is measured by aid of two 10/120 LY 11 Hottinger strain gauges placed on opposite sides of the lamella and linked to a model P3 Strain Indicator and Recorder, manufactured by Vishay, using a half-bridge connection. Figure 3 illustrates the experimental device used for the present research is, together with a detailed view of the strain gage placement on the elastic lamella.

Figure 3: experimental device for the study of mechanical contacts under normal load by aid of laser profilometry and placement of the strain gauges used to evaluate applied forces

3. RESULTS AND DISCUSSIONS

The present study aimed to experimentally evaluate stress and strain states generated in ball bearing balls without apparent surface defects when in contact. To that end, the above-presented experimental method based on surface reflectivity assessment was employed. The experimental tests were conducted on three bearing balls, having different diameters. The surfaces of these spherical punches were first mapped by aid of laser profilometry and their dimensions were accurately determined. It was found that the used metallic spheres have respective mean curvature radii of $2,002mm$, $3,9513mm$ and $6,35mm$. For exemplification, a 3D representation of the surface microtopography corresponding to the ball having a $12,775mm$ diameter, is shown in Figure 4 a. Figures 4 b and 4 c illustrate two reciprocally perpendicular profiles of the surface, and their approximation by circular profiles, which allows for measurements of the surface curvature. The mapped area was $1 \times 1mm$, and the measurement was conducted at a resolution of $2 \times 2\mu m$.

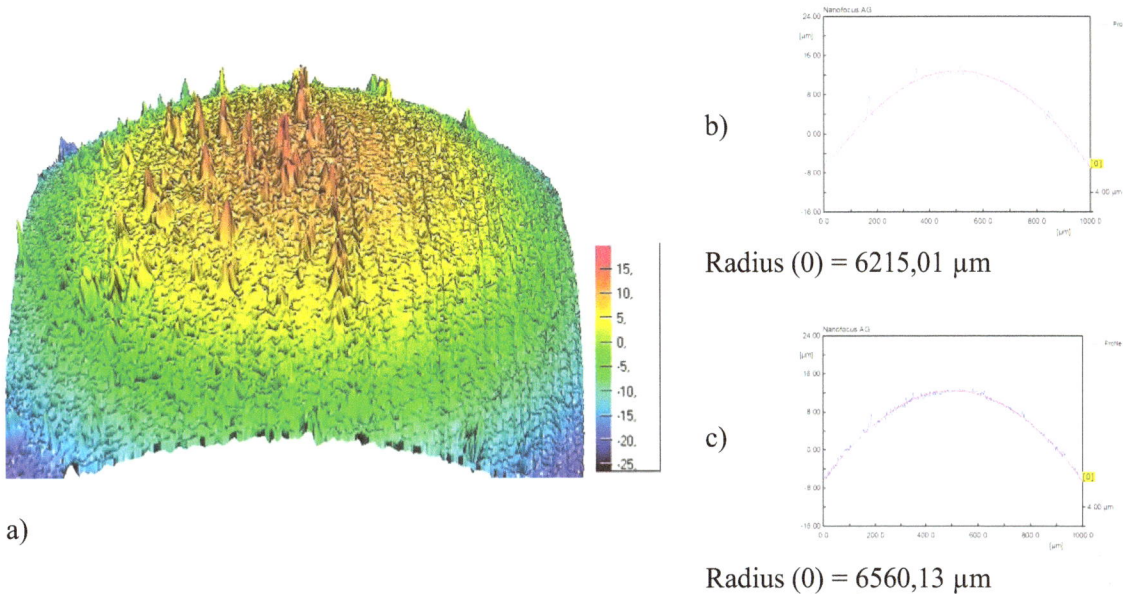

a)

b)

Radius (0) = 6215,01 µm

c)

Radius (0) = 6560,13 µm

Figure 4: Surface microtopography and reciprocally perpendicular profiles for the $12,775mm$ steel ball

Using the presented experimental equipment, the three bearing balls were placed in contact with a $3mm$ thick sapphire window. The obtained contacts were subjected to various normal loads and contact regions were mapped by aid of laser profilometry. The experimental results consist of both 3D and plane representations of the contact surface reflectivity inside and near the ball-sapphire contact area. These plots can be further interpreted by aid of the specific functionalities for data display and processing available from the profilometer software. If further analysis is necessary the measurement data can be exported to other data processing software.

Typical experimental measurements of surface reflectivity in the contact proximity were represented by aid of three-dimensional plots as illustrated in Figures 5-7. In order to highlight the contact area shapes and dimensions, plane representations of surface reflectivities were plotted, as shown in Figure 8.

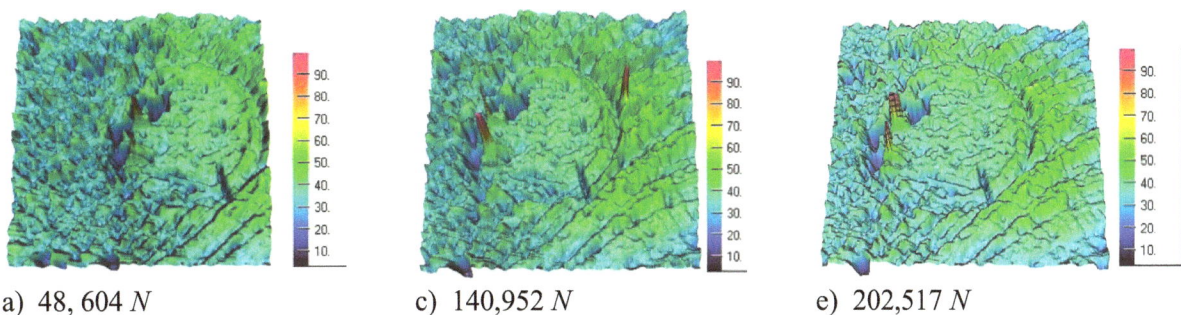

a) $48,604\ N$ c) $140,952\ N$ e) $202,517\ N$

Figure 5: 3D representation of surface reflectivity of the contact region for a $4,004mm$ steel ball and a flat sapphire window at various normal load levels

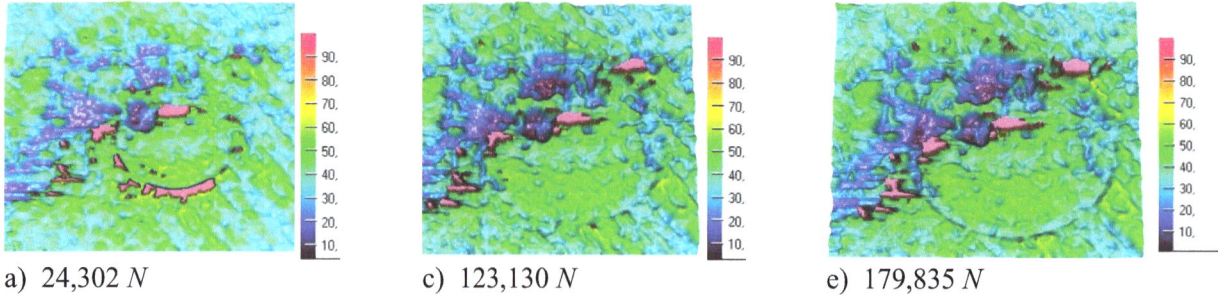

a) 24,302 N c) 123,130 N e) 179,835 N

Figure 6: 3D representation of surface reflectivity of the contact region for a 7,927mm steel ball and a flat sapphire window at various normal load levels

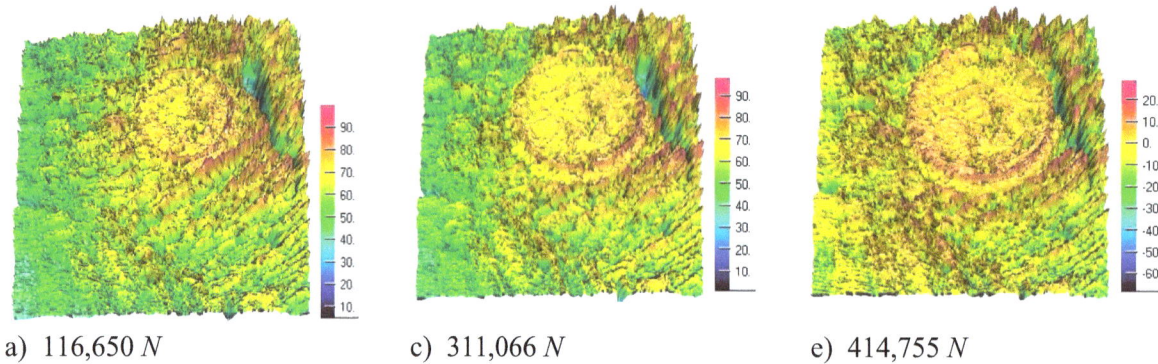

a) 116,650 N c) 311,066 N e) 414,755 N

Figure 7: 3D representation of surface reflectivity of the contact region for a 12,775mm steel ball and a flat sapphire window at various normal load levels

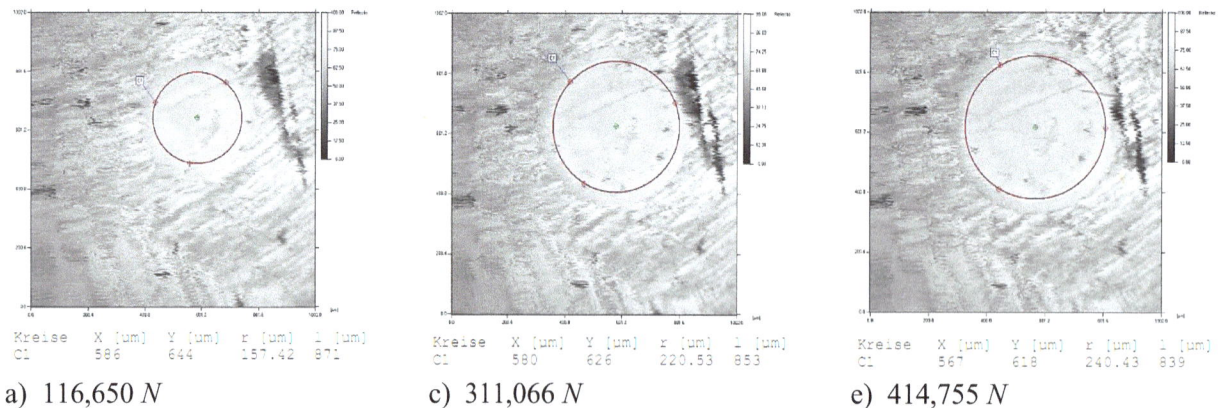

a) 116,650 N c) 311,066 N e) 414,755 N

Figure 8: Plane representation by shades of grey for the surface reflectivity of the contact region for a 12,775mm steel ball and a flat sapphire window at various normal load levels

Using the experimentally measured contact areas as input data, the corresponding maximum contact pressures were determined. For verification, the experimentally obtained results for contact area radii and maximum generated contact pressures were compared to theoretical values, as illustrated by figures 9-12. The theoretical predictions were obtained by application of the classical Hertz formulae to the experimentally measured surface curvature radii and applied normal forces.

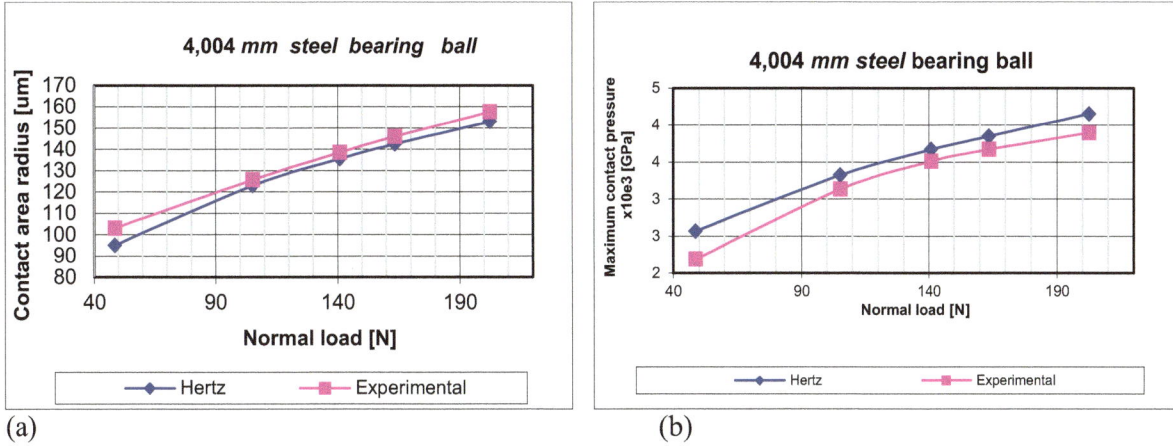

Figure 9: Contact area radius (a) and maximum contact pressure (b) evolutions with load for the 4,004 *mm* steel ball pressed against a thick sapphire window

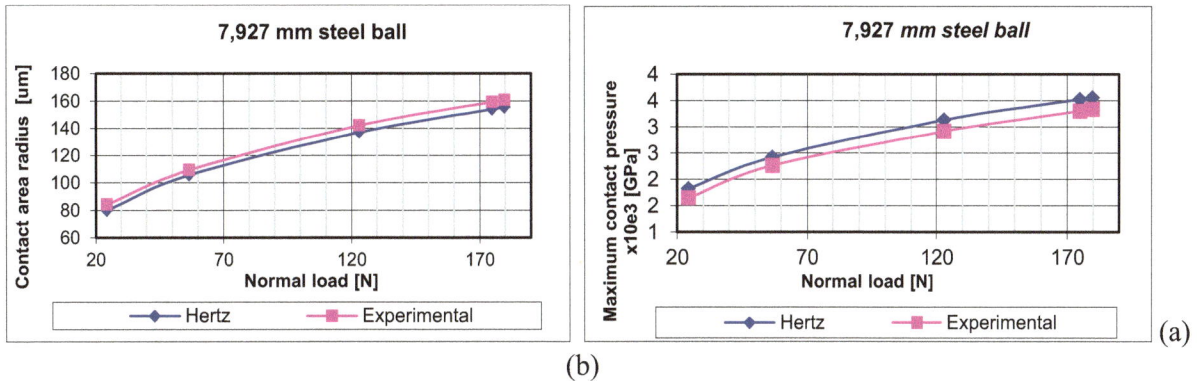

Figure 11: Contact area radius (a) and maximum contact pressure (b) evolutions with load for the 7,927 *mm* steel ball pressed against a thick sapphire window

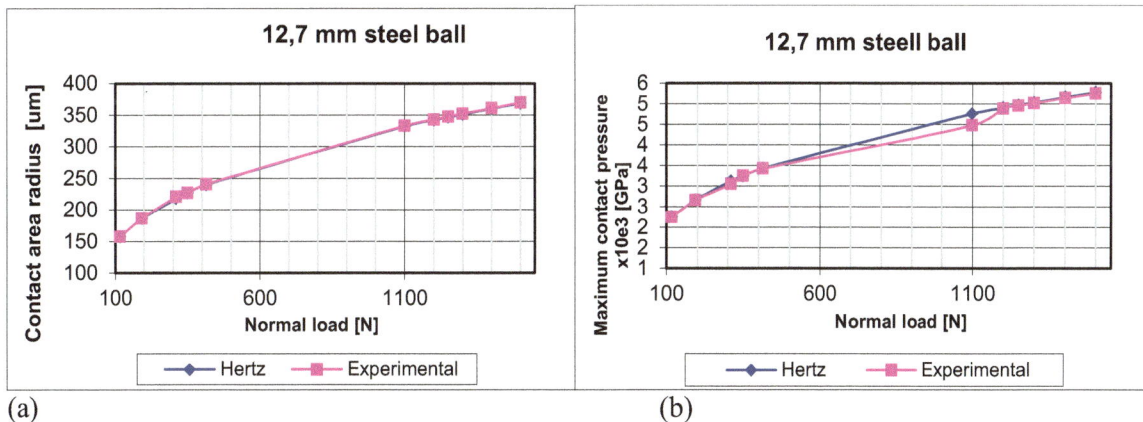

Figure 12: Contact area radius (a) and maximum contact pressure (b) evolutions with load for the 12,775 *mm* steel ball pressed against a thick sapphire window

The experimental measurements were compared to theoretical predictions and the mean deviation between the two data sets was plotted against load variation, as illustrated in Figures 13 and 14. It can be noticed that the deviation is more important in the case of small diameter punches subjected to low level loads and it takes lower and almost constant values as dimensions and forces increase. It can be concluded that the employed method is more accurate in the case of larger diameter punches subjected to more significant loads.

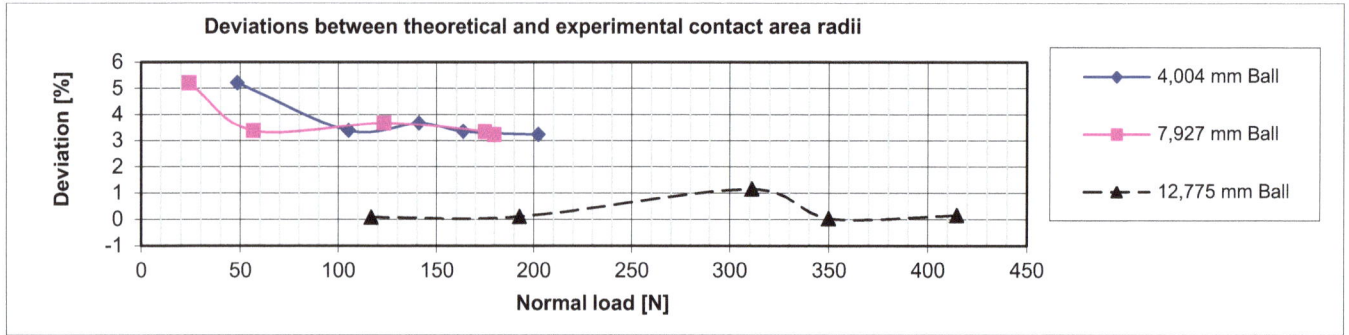

Figure 13: Variation of the relative deviation between theoretical predictions and experimental measurements of contact area radius, plotted against load variation

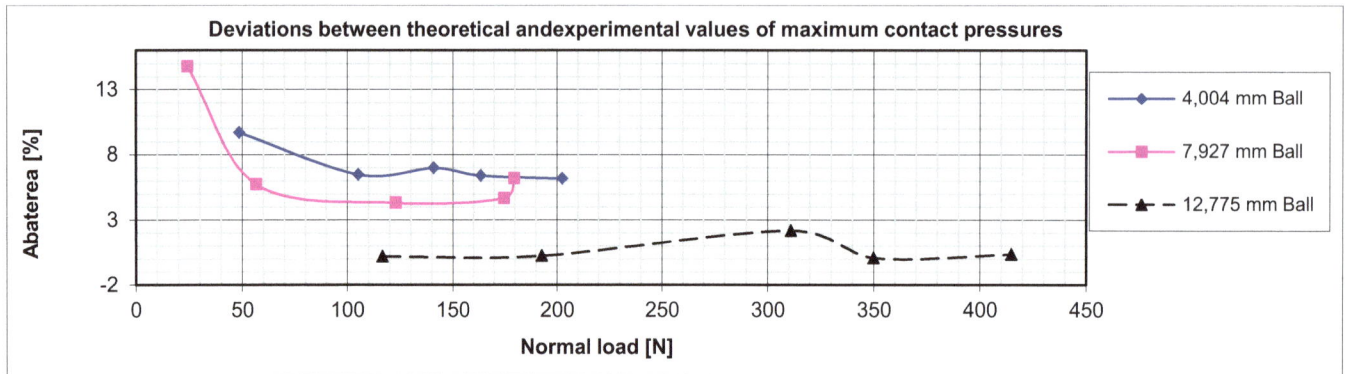

Figure 14: Variation of the relative deviation between theoretical predictions and experimental measurements of maximum contact pressure, plotted against load variation

4. CONCLUSIONS

The experimental investigations presented herein can be summarized by some conclusions, as follows:
- ✓ The present work consisted of investigating spherical punch-flat surface contacts by aid of reflectivity. It was experimentally verified that, as advanced in literature, in the case of a metal-on-sapphire contact, surface reflectivity is lower inside contact area than outside it, which allows for accurate evaluation of contact area shape and dimensions;
- ✓ Several tests were conducted for various punch diameters, and different normal forces. The experimental results can be visualized by aid of either three-dimensional representations or topographic representations in shades of grey of the reflectivity variations corresponding to the contact surface. These plots can be further processed and analyzed using the facilities offered by the profilometer software:
- ✓ After proper systematization of the obtained results, a comparative analysis was conducted between experimental measurements and theoretical predictions upon contact parameters and very good agreement was found.

REFERENCES

[1]. Diaconescu, E.N., A New Tool for Experimental Investigation of Mechanical contacts. Part I: Principles of Investigation Method, VAREHD9, Suceava, 1998, 243-248.
[2]. Diaconescu, E.N., Glovnea, M.L., A New Tool for Experimental Investigation of Mechanical contacts. Part II: Experimental Set-up and Preliminary Results, VAREHD9, Suceava, 1998, 249-254.

[3]. Glovnea, ML, Efectul discontinuităților geometrice de suprafață asupra contactului elastic, Teza de doctorat, Universitatea Suceava, 1999.

[4]. Diaconescu, E.N.and Glovnea, M.L., 2000, "Validation of reflectivity as an experimental tool in contact mechanics", VAREHD 10, Suceava, pp. 471 – 476

[5]. Diaconescu, E.N., and Glovnea, M.L., 2002, „Evaluation of Contact Area by Reflectivity", Proc. 3rd AIMETA International Tribology Conference, Italy, on CD.

[6]. Diaconescu, E. N., Glovnea, M. L., 2006, "Visualization and Measurement of Contact Area by Reflectivity", ASME J. Tribol., 128, pp. 915–917

[7]. Suciu, C., Comoritan, V., 2006, "Preliminary Experimental Investigations Upon Junction Growth Under Tangential Loading"–13th VAREHD INTERNATIONAL CONFERENCE, Suceava, Oct. 6-7,

[8]. Suciu, C., Român, I., " Experimental Investigations Upon Circular Contacts Between Ethylene Vinyl Acetate Bodies By The Aid Of Reflectivity", THE ANNALS OF UNIVERSITY "DUNĂREA DE JOS" OF GALAȚI, FASCICLE VIII, 2011 (XVII), ISSN 1221-4590, Issue 2 - TRIBOLOGY, pp.102-105

MECHANICAL BEHAVIOR OF HEMP-BASED COMPOSITE SUBJECTED TO IMPACT TEST

Maria Luminita SCUTARU[1], Marius Baba[2]

[1] *Transilvania University of Brasov, Department of Automotives and Mechanical Engineering, 29 Eroilor Blvd, 500036, Brasov, Romania, lscutaru@unitbv.ro*

[2] *Transilvania University of Brasov, Department of Automotives and Mechanical Engineering, 29 Eroilor Blvd, 500036, Brasov, Romania, mariusbaba@unitbv.ro*

Abstract: *This paper presents a study regarding the impact testing of some composite laminate panels based on polyester resin reinforced with hemp fabric. The effects of different impact speeds on the mechanical behavior of these panels have been analyzed. The paper lays stress on the characterization of this composite laminate regarding the impact behavior of these panels by dropping a weight with low velocity.*
Keywords: *Composite material, Low-velocity impact, Hemp fibers, Impact testing*

1. INTRODUCTION

Composite materials are used in a wide range of applications; however, they are used with prudence in applications where transverse loadings appear, for instance, loadings given by transverse impact with low velocity. The damage resistance is connected to the material's capability to minimize the failures' effects given by impact, while damage tolerance is given by the material's capability to maintain its properties even after failures' appearance in material. Usually, these properties are called residual properties.

One of the difficulties regarding the properties and evaluation of composites is, ironically, an advantage, namely, the capability to allow users to tailor their properties to suit the design needs[1],[2]. In applications, the use of composites based on natural fibers is yet limited at the so-called non-structural components such as inner components of cars. One of the main reasons for this limitation consists in the sensitivity of these composites at impact and the difficulty in critical evaluation regarding damages caused at impact.

2. METHOD

The research has been carried out only for composite panels reinforced with hemp fabric. All composite panels presenting a rectangular shape and being underpinned on all edges and have the same material's structure: *five layers of thermosetting resin reinforced with hemp fabric.*

The plies sequence has been carried out in the hand lay-up process using a roll for resin impregnation of hemp fibers. Finally, the structure's thickness has been 4 mm. The laminate panel has been maintained at room temperature for two weeks from which five specimens of rectangular shape (150 x 100 mm) have been cut. The specimens have been subjected to impact by dropping a weight according to the standard ASTM-D5420-98a .

The impact testing by dropping a weight is used to characterize the dynamic behavior of a material. The experimental setup consists in a two column frame and a weight which can be lifted and released in free fall with minimum friction by sliding along columns under own weight (Fig. 1). The indentor presents a hemispherical head with a 16 mm diameter and its mass is equal to 1.9 kg. This indentor hits the middle of the rectangular specimen. The accelerometer is fixed on the upper part of the indentor and the signal (acceleration) is taken over in computer by help of an acquisition device type NI USB 6521 BNC.

The Hemp specimens were supported on all rigid plate (a test machine) and fixed price margins four screws through the rubber top pieces easily tightened by hand. Behind the projectile was set screw an accelerometer to measure the acceleration of the projectile and that the contact force during impactului.
Rezultatele were recorded using a purchasing card.

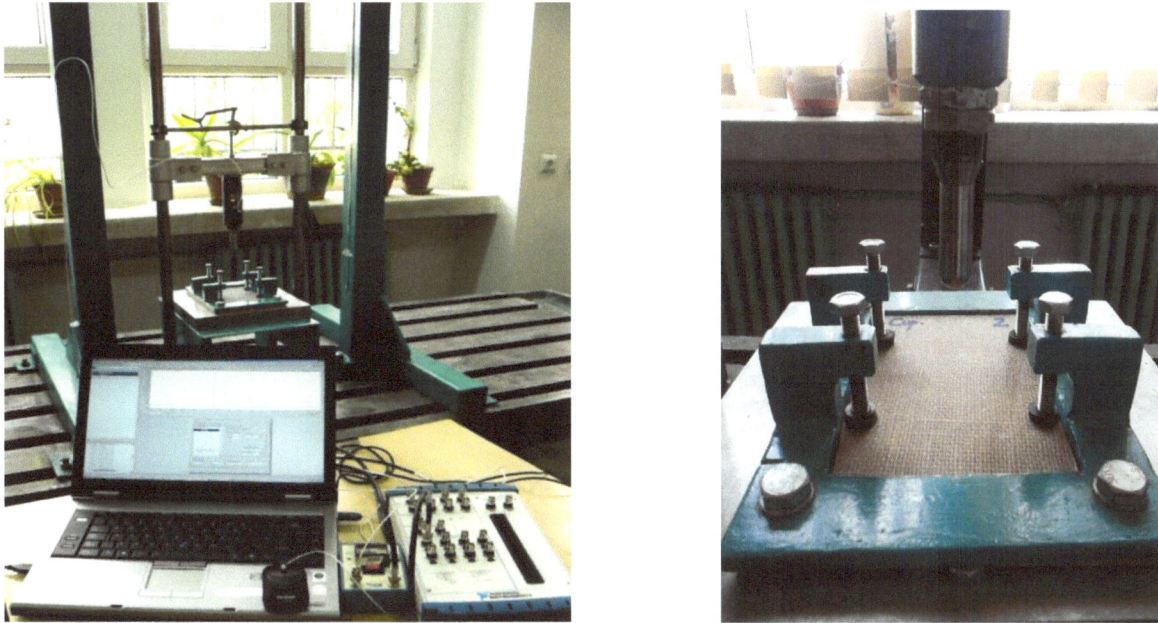

Figure 1:. Impact device and specimen trapped in device

Using this kind of testing, some data regarding the mechanical properties of a material can be obtained, namely:
- The energy, U, absorbed during impact;
- The variation of impact force, F, at the impact moment;
- The variation of indentor's displacement, d, versus time, etc.

In the case of impact testing by weight falling, the only measured feature versus time is the contact force, F(t), exerted by the weight which falls on specimen while the specimen's deflection is determined as a function of time by numerical integration of the indentor's motion equation. The acquisition of experimental data (acceleration) as well as computing the response parameters described above have been carried out using a block diagram conceived by the LabView program.

3. EXPERIMENTAL RESULTS

Using the LabView program, the variation of force "F" has been computed at the impact's moment after the signal has been recorded with the accelerometer through the acquisition device. In the same way, following distributions have been represented:• The variation of impact force F versus the displacement δ at the impact's moment;
- The variation of indentor's displacement δ at the impact's moment;
- The variation of energy U at the impact's moment.

The samples were tested at a speed of 1 m/s, 2 m/s 3 m/s or 4 m/s. (figure 2). The results obtained after the impact testing of hybrid carbon-hemp composite laminate are presented in table 1 as well.

Table:1. Results of impact testing

Specimen	Results			Observations
	Specimen thickness [mm]	Falling height [m]	Impact speed [m/s]	
1	4	0.815	4	Total break
2				
3				
4				
5		0.458	3	Break at first ply
6				
7		0.203	2	Trace on specimen is slightly marked
8				
9		0.05	1	Trace on specimen is not visible
10				

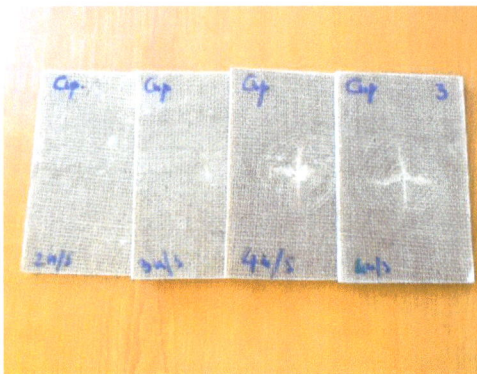

Figure 2: Hemp fabric specimens after the impact loading

They also force-displacement curves were obtained at the point of contact and was determined BVID corresponding energy level (maximum kinetic energy projectile that damages are not visible to the naked eye). Specimen hemp (Cnp) (v=1m/s)

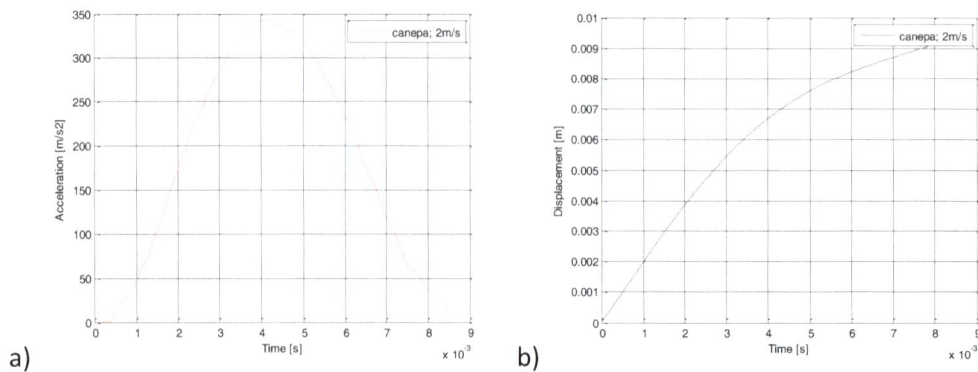

Figure 3: The variation of acceleration) of the projectile and the driving b) of the projectile on impact when the specimen

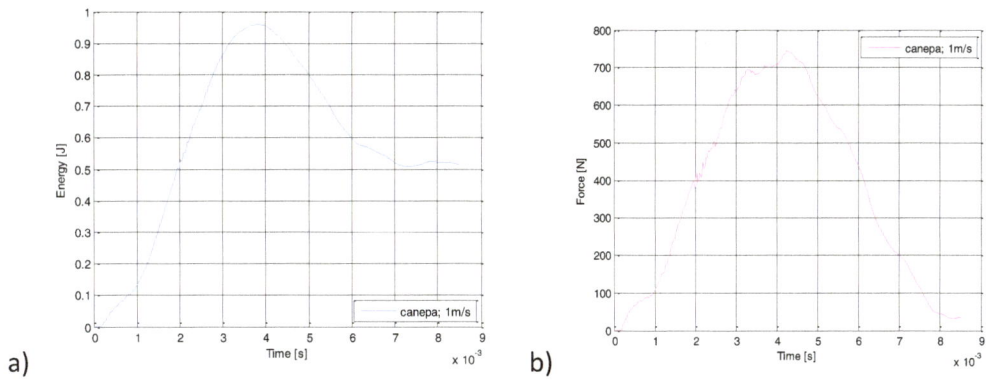

a)   b)

Figure 4: Variation of energy absorbed by the specimen) and force-time variation in the impact Specimen hemp (Cnp) (v=2m/s)

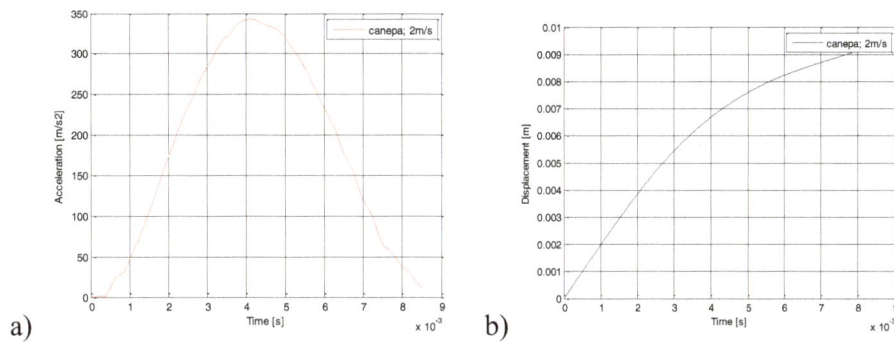

a)   b)

Figure 5: The variation of acceleration) of the projectile and the driving b) of the projectile on impact when the specimen

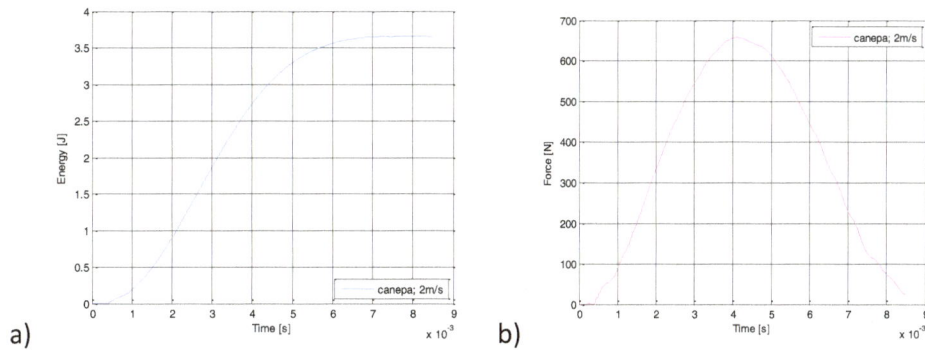

a)   b)

Figure 6: Variation of energy absorbed by the specimen) and force-time variation in the impact Specimen hemp (Cnp) (v=3m/s)

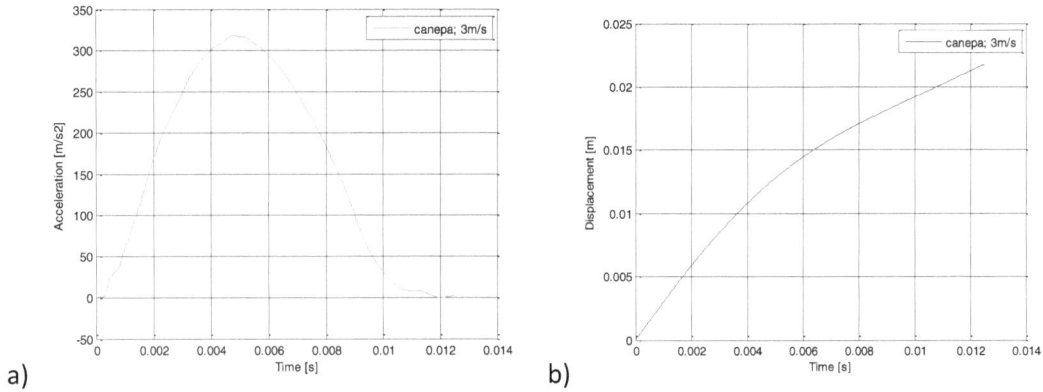

a) b)

Figure 7: The variation of acceleration) of the projectile and the driving b) of the projectile on impact when the specimen

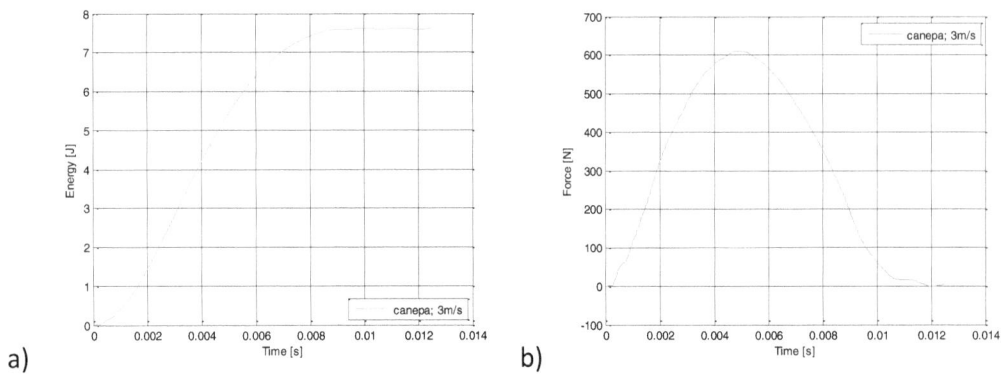

a) b)

Figure 8: Variation of energy absorbed by the specimen) and force-time variation in the impact Specimen hemp (Cnp) (v=4m/s)

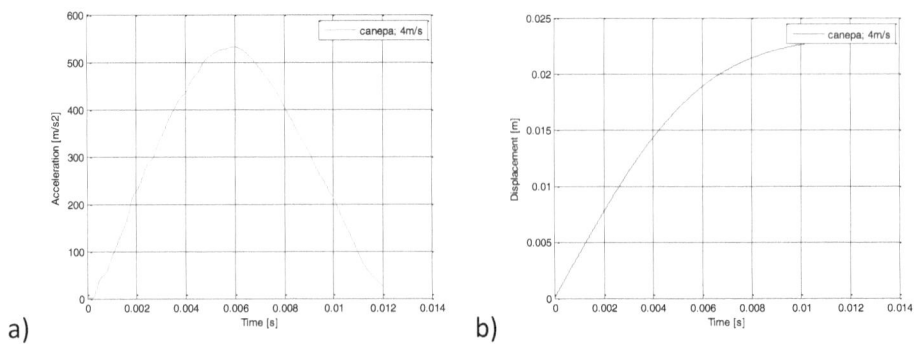

a) b)

Figure 9: The variation of acceleration) of the projectile and the driving b) of the projectile on impact when the specimen

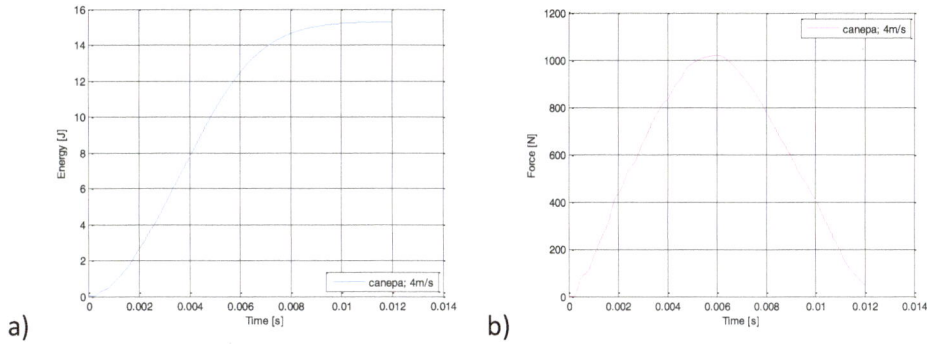

Figure 10: Variation of energy absorbed by the specimen) and force-time variation in the impact

Given the many possibilities of breaking the problem impact behavior of composite materials is complicated enough, an alternative predictive methods to study the impact. Based on the results of dynamic analysis could highlight using finite element stress distribution σ_x, σ_y and of the τ_{xy} type occurring in the four layers of composite fabric made of hemp.

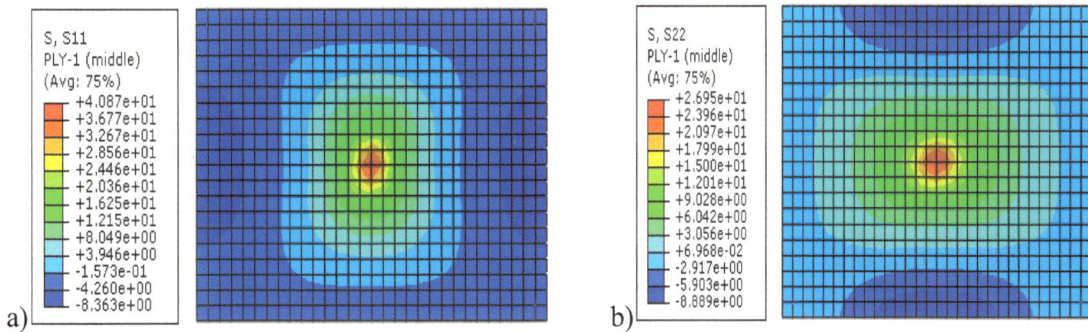

Figure 11: The distribution of σ_x stress a) and σ_y b) appears in the first layer of the composite fabric made of hemp

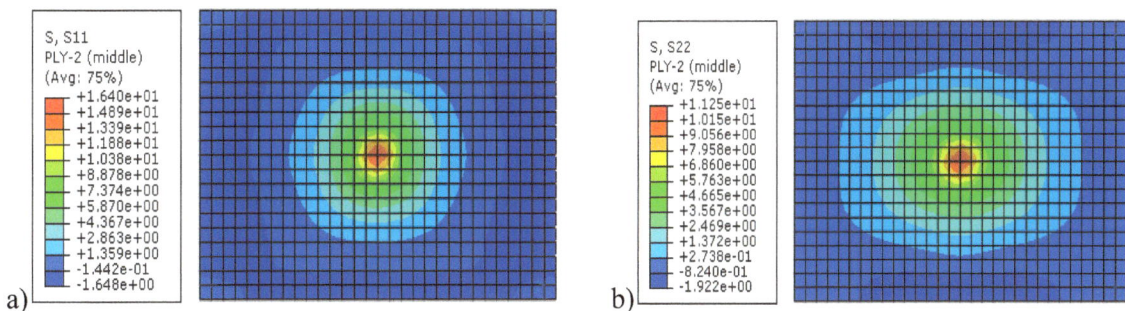

Figure 12: The distribution of σ_x stress a) and σ_y b) appears in the second layer of the composite fabric made of hemp

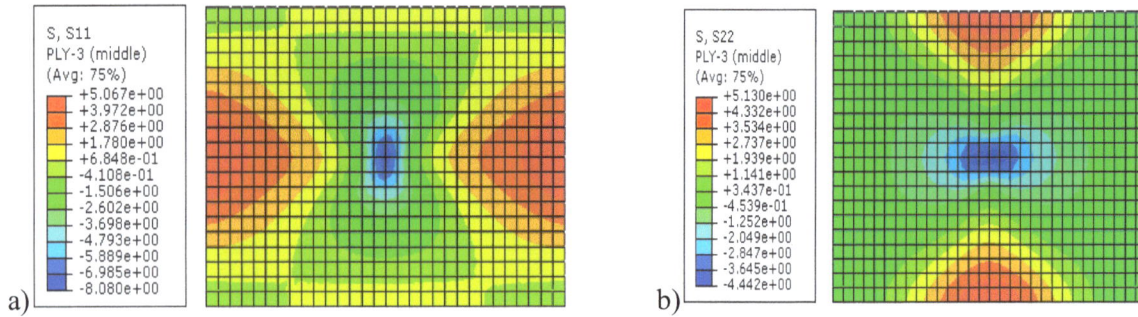

Figure 13: The distribution of σ_x stress a) and σ_y b) appears in the third layer of the composite fabric made of hemp

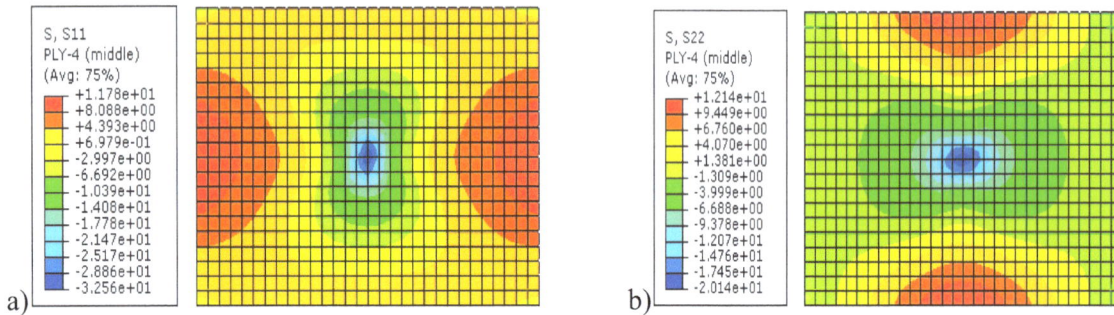

Figure 14: The distribution of σ_x stress a) and σ_y b) appears in the last layer of the composite fabric made of hemp

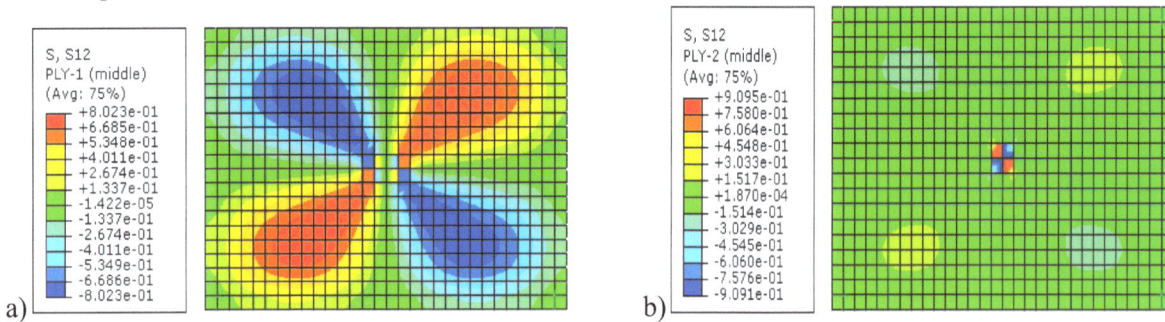

Figure 15: The distribution of τ_{xy} stresses occurring in the first layer a) and the second layer b) of the composite fabric made of hemp

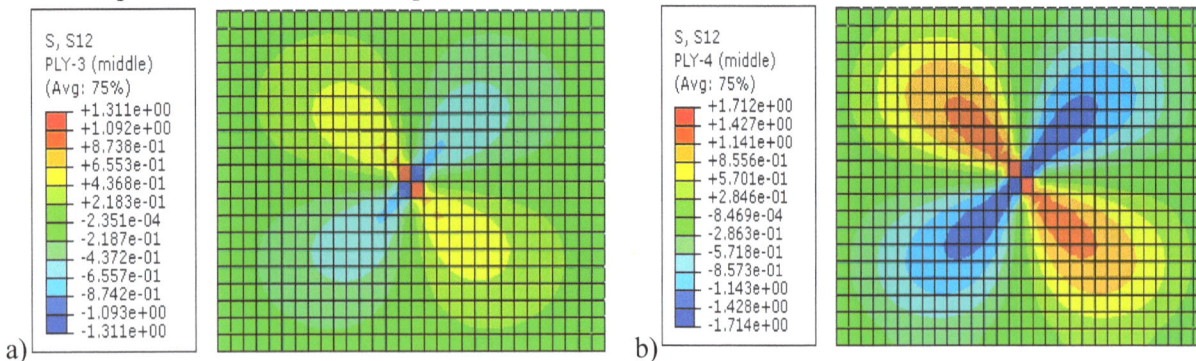

Figure 15: The distribution of τ_{xy} stresses occurring in third layer a) and the last layer b) of the composite fabric made of hemp

4. CONCLUSION

Analysing this curves we can say that these curves present leaps and inflexions due to the presence of delaminations. Integrating the area under the loading curve until the maximum value of the force (according to the first failure) the energy required to initiate the failure can be obtained. At the impact's moment, the energy accumulates in time and is direct proportional with the force and increases until reach a constant landing. After reaching a maximum value of the force, this decreases in time, the energy U being absorbed in material and the force's decrease took place after reaching the landing of U.

In general, at the composite laminates the energy is frequently absorbed by creating some delamination surfaces called delamination breaks that lead to the strength and stiffness decrease.

Analysing the specimens after impact testing can be noticed that the failure areas localized on the specimens' surface are smaller than those localized on the backfront. This leads to the conclusion that the cracks have been propagated from the place where the intedor hits to the backfront of the panel.

REFERENCES

[1] N.D. Cristescu, E.M. Craciun, E. Soos, Mechanics of elastic composites, Chapman & Hall/CRC, (2003);

[2] H. Teodorescu-Draghicescu, S. Vlase, Computational Materials Science, 50(4), 1310 (2011);

[3] H. Teodorescu-Draghicescu, S. Vlase, Upper and Lower Limits in the Elastic Properties of Low-Shrink Sheet Molding Compounds, 23rd International Congress of Theoretical and Applied Mechanics (ICTAM2012), Beijing, August 19-24, China Science Literature Publishing House, pp. 136 (2012);

[4] Vlase, S.Teodorescu-Draghicescu, H. Calin, M. R.Serbina, L. Simulation of the elastic properties of some fibre-reinforced composite laminates under off-axis loading system. OPTOELECTRONICS AND ADVANCED MATERIALS-RAPID COMMUNICATIONS,Vol.5, Issue 3-4, pp.424-429, (2011).

FLEXURAL RIGIDITY EVALUATION OF COMPOSITE SANDWICH PANEL OF CARBON-HEMP

Maria Luminita SCUTARU[1], Marius Baba[2], Janos TIMAR[3]

[1] *Transilvania University of Brasov, Department of Automotives and Mechanical Engineering, 29 Eroilor Blvd, 500036, Braşov, Romania,* lscutaru@unitbv.ro

[2] *Transilvania University of Braşov, Department of Automotives and Mechanical Engineering, 29 Eroilor Blvd, 500036, Braşov, Romania,* mariusbaba@unitbv.ro

[3] *Transilvania University of Braşov, Department of Automotives and Mechanical Engineering, 29 Eroilor Blvd, 500036, Braşov, Romania,* jamcsika_timar@unitbv.ro

Abstract: *The paper presents the most important mechanical properties determined in a simple tensile test on a 0.4 mm thickness compozit of carbon-hemp impregnated with epoxy resin, used as skins for an advanced ultralight sandwich composite structure with expanded polystyrene as core. The sandwich panel is subjected to flexural load-unload tests. This kind of fabric presents very good mechanical properties and is suitable to reinforce a quite large range of epoxy resins. The aim of using this fabric is to obtain thin structures with complex shapes and high stiffness for the automotive industry. The flexural load-unload tests show an outstanding stiffness of the whole sandwich panel. Various specimens' thickness has been used.*

Keywords: *Composite material, Hemp fibers, Hybrid composite, epoxy-resins, flexural-rigidity, sandwich*

1. INTRODUCTION

The carbon fibre-reinforced epoxy resins are used extensively to build composite structures with an outstanding specific weight/strength ratio. Such structures, usually called laminates, present a relative poor tensile stiffness and the flexural stiffness remains at a low level due to low sensitivity at flexural loads of the carbon fibres, especially of the unidirectional reinforced ones [1]-[3]. In general, composite laminates are manufactured from thin layers called laminae. These laminates present a quite low stiffness and flexural rigidity. A solution could be the increase of the layers but this leads to the disadvantage of increasing the overall weight as well as the resin and reinforcement consumption. For pre-impregnated composites, to predict their elastic properties, homogenizations and averaging methods can be used [4]-[5]. A better solution to increase the overall stiffness of a composite laminate is to use a biocomposite in the structure. This kind of composite can be a hemp-composite and presents the advantage to absorb the excessive resin. A composite laminate with this kind of embedded core material presents following main advantages: weight saving, stiffness increase, quick build of the structure's thickness, saving of resin and reinforcement as well as an increased possibility to obtain a better surface finish.

2. METHOD

The research has been carried out on eight composite panels presenting a rectangular shape and being underpinned on two edges. All specimen have the same material's structure:
- a layer of thermosetting resin reinforced with carbon fabric;
- a layer of thermosetting resin reinforced with hemp fabric;
- a layer of thermosetting resin reinforced with carbon fabric;
- a layer of thermosetting resin reinforced with hemp fabric.

The plies sequence has been carried out in the hand lay-up process using a roll for resin impregnation of carbon and hemp fibers. Finally, the structure's thickness has been 4 mm. The laminate panel has been maintained at room temperature for two weeks from which eight specimens have been cut. Composite specimen were made using the Romanian standard SR EN ISO 14125 since 1998, materials-plastic composites reinforced with fibers Determination of bending. This part of ISO 14125 is based on ISO 178 and handles fiber reinforced plastics. Keep the test conditions relevant to the system of carbon fiber reinforced and extended in the ISO 178 test conditions includes both the three-point test method (method A) and four-point test (Method B), and conditions for composites based on carbon fibers.

3. EXPERIMENTAL RESULTS

The equipment used is a testing machine at constant speed. The testing machine three-point bending test is produced by Lloyd's Instruments, UK, being a car guy LR5K Plus, which provides a maximum force Fmax = 5 kN. The testing machine presents the following characteristics:

- Force range: 5 kN;
- Speed accuracy: <0.2%;
- Load resolution: <0.01% from the load cell used;
- Analysis software: NEXYGEN Plus.

The samples were subjected to bending with a constant speed of 5 mm / min until fracture or until the tension (load) and deformation (elongation) has reached a predetermined value (fig.1)

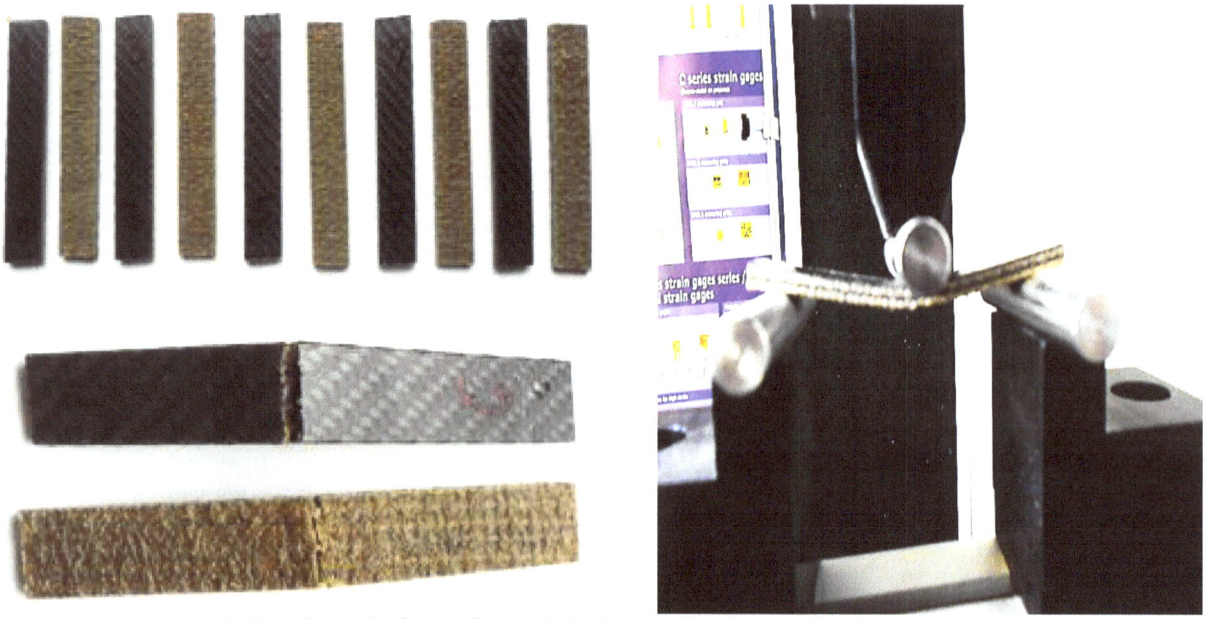

Figure 10: Specimen before, after and during application to the three-point flexural

During the test, the load measured by the specimen and its elongation. Also we have accurately measured the size of each specimen: specimen cross section and width. These dimensions were introduced as input data in the computer connected to the test machine having NEXYGEN software that retrieves data from experimental testing machine and process statistics.

The bending test results for carbon-hemp hybrid composite C-Cnp figure2-4 are presented based on centralized processing tab.1

Tab.1 The mechanical properties of the composite hybrid C-Cnp following applications flexural

Samples No.	Load at Max. Load	Max. Bending Stress at Max. Load	Flexural Rigidity	Young's Modulus of Bending
	[kN]	[MPa]	[Nm^2]	[MPa]
Specimen No. 1	0,403687	161,474802	0,29297599	3662,19989
Specimen No. 2	0,43507311	174,029243	0,35647281	4455,91017
Specimen No. 3	0,34740662	138,96265	0,35821571	4477,69641
Specimen No. 4	0,35514589	142,058355	0,33281249	4160,15614
Specimen No. 5	0,3507151	140,286038	0,35060906	4382,61321
Specimen No. 6	0,74215661	296,862645	0,4029541	5036,92621
Specimen No. 7	0,54645625	218,5825	0,22014731	2751,84138
Specimen No. 8	0,61859262	247,437048	0,26177819	3272,22742

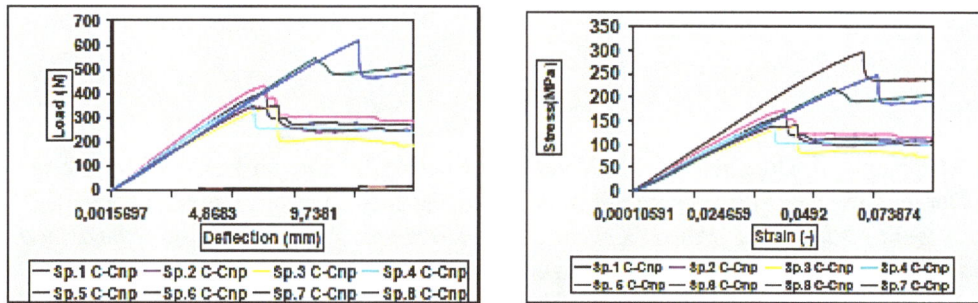

Figure 2. Load-deflection distribution and Stress-strain distribution of the composite hybrid C-Cnp

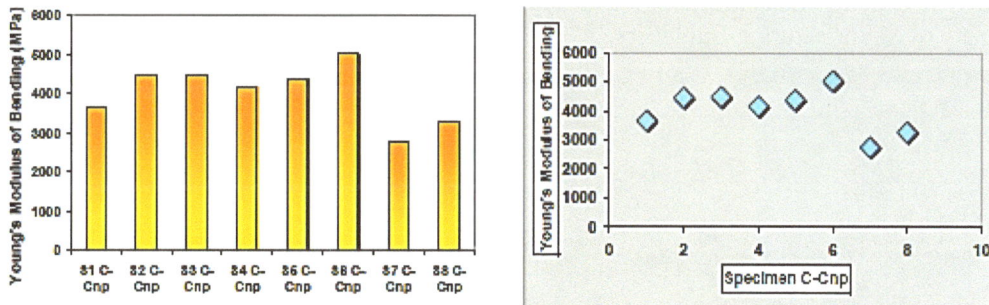

Figure 3. Young's modulus of bending distributions of the composite hybrid C-Cnp

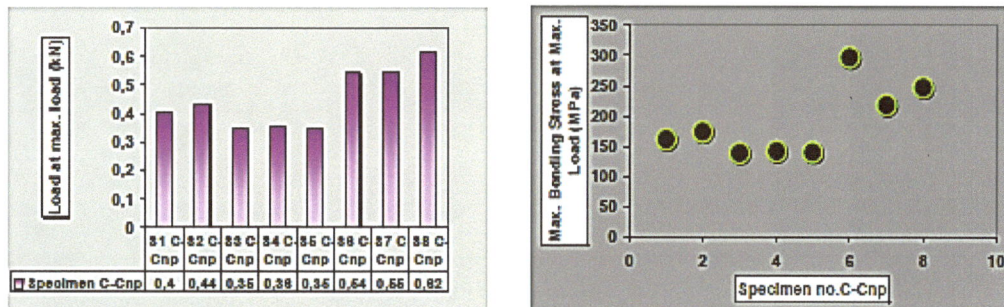

Figure 4. Maximum load distribution and maximum stress of the composite hybrid C-Cnp

4. CONCLUSION

The carbon fibres are suitable to fit special structures and devices for the future car due to their excellent thermal and electric conductivity. Also composites based on carbon and hemp fit special structures and devices for auto industry due to their excellent thermal and electric conductivity as well for their good force at break distribution as a function of Young's modulus.

The comparison between the flexural rigidity of the structure obtained experimentally and that obtained through the theoretical approach shows a good agreement between the experimental data and the theoretical approach. The following conclusions can be drawn:

In the first analysis of composite hybrid graphics when C-Cnp can easily be seen that if this modulus dispersion is situated in a range of values in the range 2500-5500 MPa corresponding to a range between 0.2 - 0.5 nm2 stiffness.

If the distribution recorded force max is between 0.4-0.6 kN the stress-strain distributions in all types of compozit hemp-carbon present a non-linear tendency.

Almost all of these prepregs subjected to three-point bend tests have presented a fall of their stiffness at certain strain values For instance, in case of, specimen no. 1-6 this decrease is in the range 4,8683-9,7381

REFERENCES

[1]   D. B. Miracle, R. L. Donaldson, *ASM Handbook Volume 21: Composites.* ASM International, 2001;

[2]   I. M. Daniel, O. Ishai, *Engineering of Composite Materials.* Oxford University Press, 2nd ed., 2005;

[3]   J. R. Vinson, *Plate and Panel Structures of Isotropic, Composite and Piezoelectric Materials, Including Sandwich Construction.* Springer, 1st ed., 2005;

[4]   H. Teodorescu-Draghicescu, S. Vlase, "Homogenization and Averaging Methods to Predict Elastic Properties of Pre-Impregnated Composite Materials," *Computational Materials Science,* vol. 50, issue 4, pp. 1310-1314, Febr. 2011;

[5]   A. B. Strong, *Fundamentals of Composites Manufacturing: Materials, Methods and Applications.* Society of Manufacturing Engineers, 2nd ed., 2007.

COMPARATIVE ANALYSIS OF THE CRACKING RATE FOR A STAINLESS STEEL LOADED AT 213K TEMPERATURE

Roşca, V.[1], Miriţoiu, C.[1], Geonea, I.[1], Romanescu, Alina[1]

[1] Mechanics Faculty, University of Craiova, Romania, rosca_valcu@yahoo.com

Abstract: *The existance of a variable (cyclic) loading over a part or an subassembly may lead to the crack appearance in its body. The crack will spread unit it will reach a critical length leading to the instant specimen fracture. An important parameter that can control the fatigue fracture is the crack propagation rate marked as da/dN. This is the advancing length of the crack in a loading cycle. There were proposed, by various scientists from the researching field, many empiric relations, that have resulted from the experiments, that follow the fatigue fracture phenomena.*

In this paper a comparative analysis of the cracking speed will be made by using three mathematical models: Paris formula, Walker relation and Donahue relation. The experiments were made on CT specimens, with side notch, from a stainless steel 10TiNiCr175 type. The loading temperature was 213K (meaning - 60°C), and the loading was made for three types of asymmetry factors: R=0.1, R=0.3 and R=0.5. During the loadings, some primal quantities were taken into account: the variation of the cracking length a_i and the corresponding cycles number N_i. With these values there was calculated the variation of an important parameter in the Fracture Mechanics, ΔK– the stress intensity factor , and respectively the crack growth rate da/dN by polynomial method and the three presented models. With the obtained models some graphics were drawn representing the da/dN parameter variation and there were made comparisons between the four used formulas.

Keywords: *crack, fracture, stress intensity factor, crack growth rate, asymmetry coefficient*

1. INTRODUCTION

Some products or parts that are included in the composition of some equipments or aggregates from the chemical, food or extractive industry work in low temperature or cryogenic environments. The used materials must have a good behavior at these temperatures, must not modify their physical-mechanical properties during their activity. That is why it is necessary to permanently follow their structural integrity.

A material used to design these parts is the 10TiNiCr175 stainless steel, V2A class. Beside the working environment temperature, a considerable influence over the material strength is the loading type: static, dynamic and variable character. In this last case of loading, the loading degree is very important, namely the loading asymmetry factor R, ie R= σ_{min}/σ_{max}, where σ_{min} and σ_{max} represent the minimum and maximum stresses, where the stress state varies.

During the working time, because of some material, constructive, environment or loading factors, micro-cracks may appear which can increase up to a critical value producing the final fracture of the product.
In addition to the "Strength of Materials" classical calculus, there can be made a complex analysis of the materials breaking state using the notions from "Fracture Mechanics".

The main parts that control the material facture process of a fatigue loaded product are the cracking rate marked with da/dN or da/dt and the stress intensity factor variation marked with ΔK [1], [6]. We mark with: a the crack length, N is the loading cycle number and by t is marked the reference time, at which the crack variation da is reported. A general calculus relation for the stress intensity factor, abbreviated with SIF, it is founded in many references and has the form of relation (1) [6], 3.6 formula/pp. 32:

$$\Delta K = C \cdot \Delta \sigma \cdot \sqrt{a}. \tag{1}$$

We can see that this relation simultaneously included the loading stress σ and the crack length a, and C is a parameter that can be determined by using many relations, being dependent on the crack domain and geometry.

An important graphic of the cracking rate variation da/dN in relation with the stress intensity factor variation ΔK is presented in figure 1, [3], [4], [6]. In this graphic we can distinguish three domains:

    I.   The crack initiation domain;

    II.   The area of the crack stable propagation, which can be controlled and followed;

    III.   The brutal and fragile cracking domain, which cannot be controlled.

There are highlighted two very important values for SIF: threshold stress intensity factor ΔK_{th}, respectively the critical stress intensity factor or the fracture tenacity K_c, figure 1.

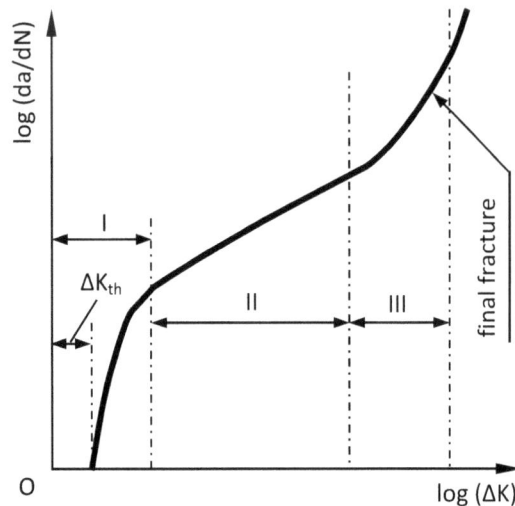

Figure 1: Cracking Rate Versus Stress Intensity Factor

(bilogarithmically coordinates)

In order to determine the propagation cracking speed, da/dN, some researchers have presented several empirical formulas that have resulted from experimental data. Mostly target the second domain of the sigmoid curve, figure 1. For our paper we have limited at the analysis of four studying models, marked in this way:

    1.   The sequential polynomial method according to the ASTM E647 standard [7], [4], [6];

    2.   Paris P.C.formula, [3]- pp. 204, [6]-pp.42;

    3.   Walker K. formula [3]- pp.209;

    4.   Donahue J.R. formula [3]- pp. 209.

For the models presented above, the mathematical calculus relations for the propagation cracking rate are:

- Paris formula:
$$\frac{da}{dN} = V_2 = C_2 \cdot \left(\Delta K \right)^{m_2} ; \tag{2}$$

- Walker formula:
$$\frac{da}{dN} = V_3 = C_3 \cdot \frac{\left(\Delta K \right)^{m_3}}{\left(1 - R \right)^{r_3}} ; \tag{3}$$

- Donahue formula:
$$\frac{da}{dN} = V_4 = C_4 \cdot \left(\Delta K - \Delta K_{th} \right)^{m_4} \tag{4}$$

The factors C_2, m_2, C_3, m_3, C_4, m_4 and r_3 are material constants and are obtained at the experimental data processing inserting the condition that a part from the second domain, figure 1, to be successively approximated with the (2), (3) and (4) relations.

2. EXPERIMENTS AND RESULTS PROCESSING

From a bar strip, made by 10TiNiCr175 stainless steel, several CT type C-R model specimens were processed, with side notch, figure 5.3, pp. 86/ [4]. The same loading force is applied on circumferential direction and the crack propagation takes place on the radius R direction, figure 5.7/ pp. 89/ [4]. The specimens were tested on a hydraulic pulsing device, with a freezing chamber, figure 5.8/pp. 92/ [4], [5]. The testing temperature was 213K (-60°C) and there were used three asymmetry factors: R=0.1, R=0.3 and R=0.5, meaning eccentrical tensile fatigue loadings, positive oscillating. During the loading, after a first-crack in which the threshold stress intensity factor ΔK_{th} was achieved, that corresponds to the crack length $\mathbf{a_0}$, it was passed in the stable propagation domain (II) and it was marked successively the crack length variation $\mathbf{a_i}$, respectively the number of the loading cycles N_i. For the computation of the crack's length an extensometer with elastic lamellae was used, (figure 5.12/ pp.97/ [4], [5]) with a conversion result using the elastic compliance method [2], pp. 862-873, [4], pp. 96.

The values sets (a_i, N_i) experimentally obtained are processed using the sequential polynomial method according to the ASTM E-647 standard: [4]- pp.82, [6]- pp.146, [7], determining the growth cracking rate with the formula 4.10/pp.83/[4]:

$$\frac{da}{dN} = V_1 = \frac{A_1}{C_2} + 2 \cdot A_2 \cdot \frac{N_i - C_1}{C_2^2},$$ (5)

(see relations (2), (3), (4)/[5])

Then, there will be determined the stress intensity factor variation ΔK for any value of the crack length a_i: [3]/pp.66, [5]/ pp.83, [6]/pp.146,

$$\Delta K = \frac{\Delta F}{B\sqrt{W}} \cdot \frac{2 + \frac{a}{W}}{\left(1 - \frac{a}{W}\right)^{\frac{3}{2}}} \cdot \left[-5,6 \cdot \left(\frac{a}{W}\right)^4 + 14,72 \left(\frac{a}{W}\right)^3 - 13,32 \left(\frac{a}{W}\right)^2 + 4,64 \frac{a}{W} + 0,886 \right],$$ (6)

where:

ΔF- is the loading force variation, in N;

B- is the specimen thickness, in mm;

W- is the specimen active width, in mm.

It is continuously followed an elastic loading using the condition a/W≥ 0,2 , and ΔK will result in [N·mm$^{-3/2}$]. Complying with the methodology presented in the introduction, there will be determined the speeds V_2, V_3 and V_4. With the obtained values, there will be drawn the next curves:

- The cracking rates V_1, V_2, V_3 and V_4 in relation with the crack length variation $\mathbf{a}$, for the asymmetry factor R=0.1, in the same graphic, figure 2;
- The same thing for R=0.3, figure 3;
- the same thing for R=0.5, figure 4;
- the propagation cracking rates V_1, V_2, V_3 and V_4 in relation with the stress intensity factor ΔK, for the asymmetry factor R=0.1, on the same graphic, figure 5;
- the same thing for R=0.3, figure 6;
- the same thing for R=0.5, figure 7;
- $\mathbf{V_1}$ speed, according to the ASTM E-647 standard, in relation with the crack length variation $\mathbf{a}$, for the three asymmetry factors (R=0.1, R=0.3 and R=0.5), on the same graphic, figure 8;
- V_1 speed versus SIF ΔK variation, for the three asymmetry factors, simultaneously, figure 9.

3. COMMENTS. OBSERVATIONS

After a general analysis of the cracking variation rates curves according to the four variants versus the crack length variation **a**, and versus SIF variation ΔK, it can be said that on the area of the crack stable propagation (second domain – figure 1), the empirical models used in this paper approximate very well the fracture phenomenon.

For the asymmetry factor R=0.1, the crack evolution **a** is analyzed from 10,75 mm up to 15,5 mm, where the propagation is in the linear-elastic limits and the cracking rates vary from $118,9 \cdot 10^{-6}$ m/cycle , for V_1, up to $470,03 \cdot 10^{-6}$ m/cycle for Walker model ,V_3, figure 2. Reported to SIF, for the same cracking rates domain, the KΔ factor variation is in the limits of: 776 Nmm$^{-3/2}$ up to 1170 Nmm$^{-3/2}$, when it is obtained the breaking tenacity (K_c= 1170 Nmm$^{-3/2}$), figure 5.

By analyzing the loading factor R=0.3, the crack length varies between 11,25 mm and 15,75 mm, and the speeds domain ranges between $22,4 \cdot 10^{-6}$ m/cycle and $152,7 \cdot 10^{-6}$ m/cycle, figure 3 and figure 6. According to these limits, the ΔK SIF variation is produced between 632 Nmm$^{-3/2}$ and 934 Nmm$^{-3/2}$, figure 6. There can be noticed that, at the beginning of the second domain, the propagation according to the Donahue model is more slowly, and in the second part, according to the Walker model, the propagation is more quickly, figure 3 and figure 6.

The asymmetry factor R=0.5 is placed for crack lengths between 10,75 mm and 13,5 mm. The crack rates increasing domain (V_1, V_2, V_3 and V_4) vary between $26,9 \cdot 10^{-6}$ m/cycle by polynomial method (V_1) and $55,2 \cdot 10^{-6}$ m/cycle for the Paris model (V_2). Accordingly, the stress intensity factor ΔK increase from 428 Nmm$^{-3/2}$ up to 532 Nmm$^{-3/2}$. In this case, slope variation curves of da/dN crack growth rate are lower than the previous cases, figure 4 and figure 7.

About the figure 8 and figure 9, there was presented only the V_1 rate variation versus the crack length **a**, figure 8, respectively versus ΔK SIF, figure 9, for the three asymmetry factors R=0.1, R=0.3 and R=0.5, simultaneously. It is observed that an increase of the R factor leads to a decrease of the V_1 cracking rate, figure 8, and in the same time a decrease of the ΔK SIF variation, figure 9. Variation limits for the a lengths, ΔK factors and V_1 length were remembered above.

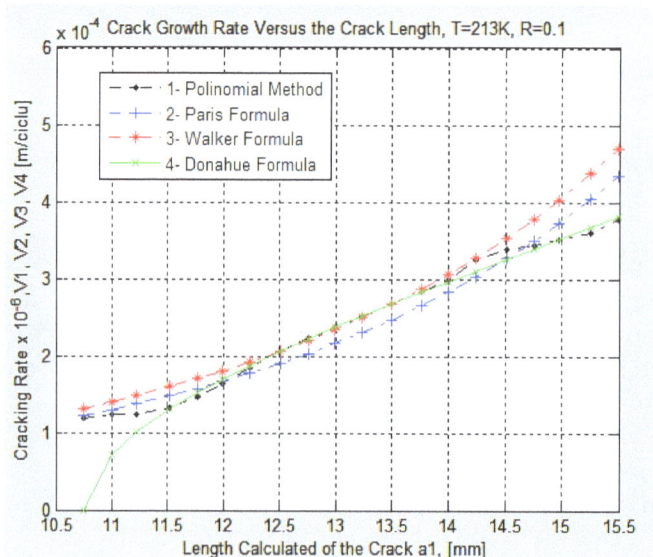

Figure 2: Crack Growth Rates Versus the Crack Length for R=0,1

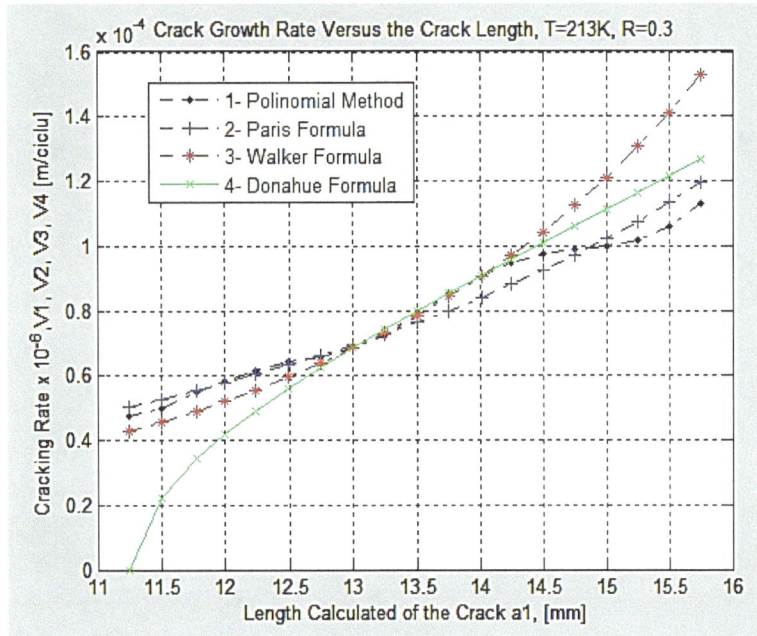

Figure 3: Crack Growth Rates Versus the Crack Length for R=0,3

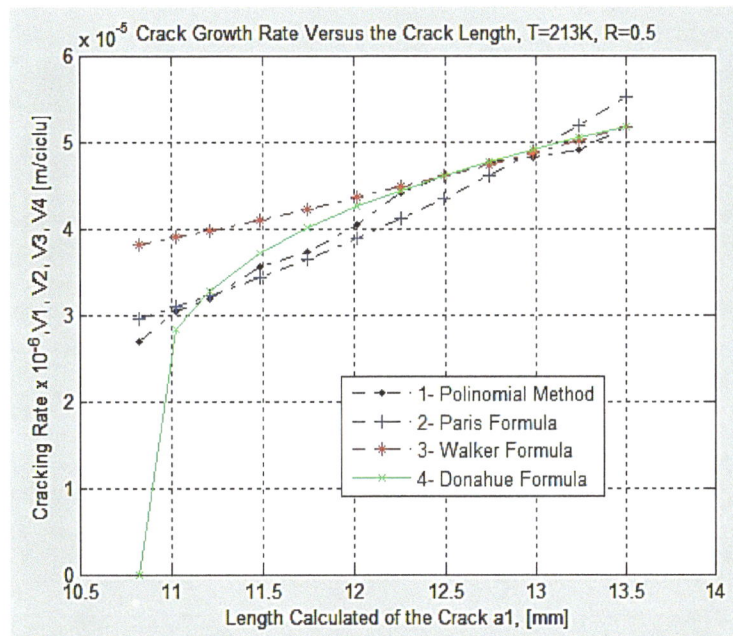

Figure 4: Crack Growth Rates Versus the Crack Length for R=0,5

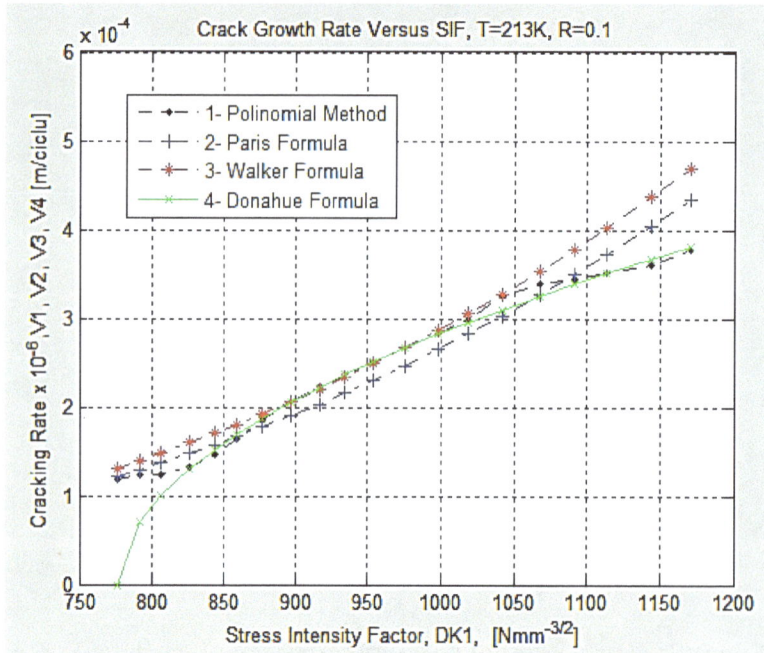

Figure5: Crack Growth Rates Versus SIF for R=0,1

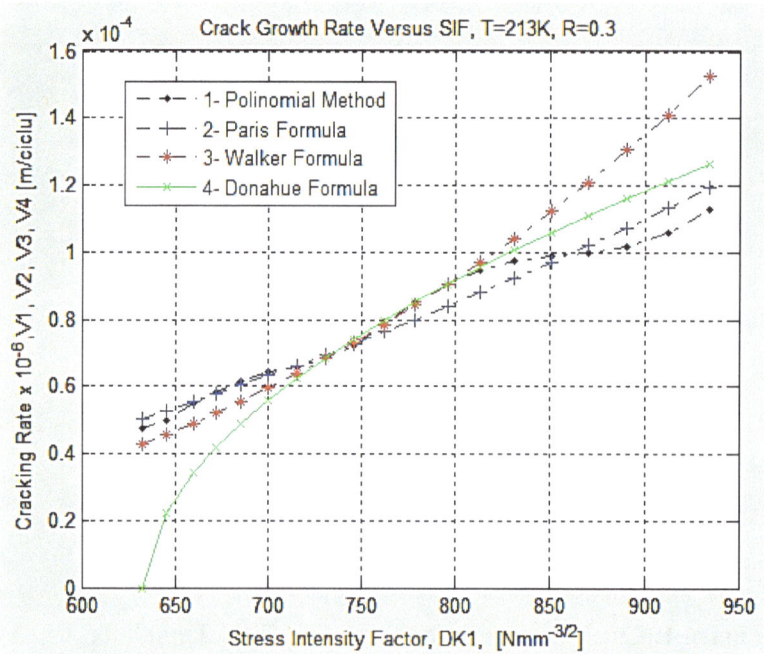

Figure 6: Crack Growth Rates Versus SIF for R=0,3

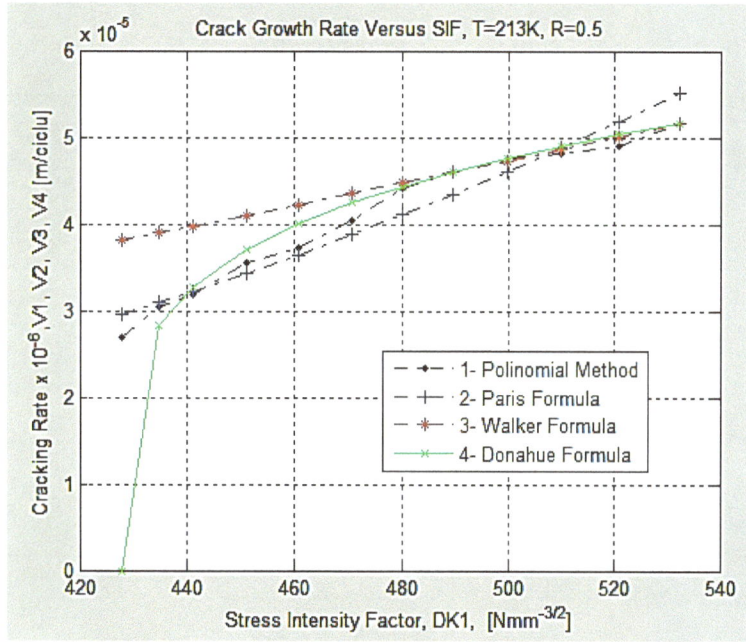

Figure 7: Crack Growth Rates Versus SIF for R=0,5

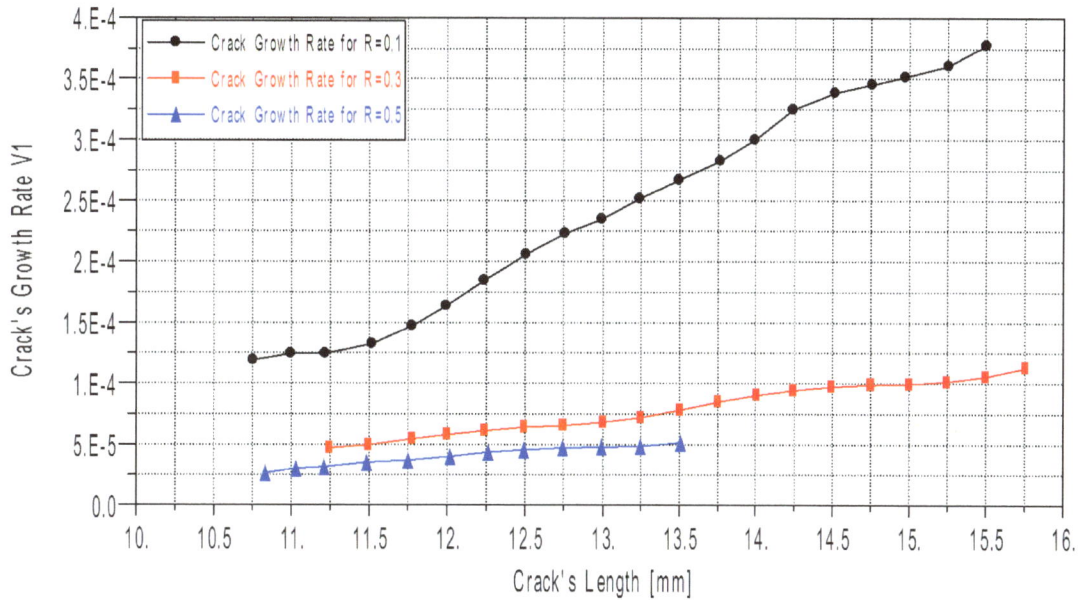

Figure 8: Crack Growth Rate **V1** (ASTM) Versus Crack's Length **a**

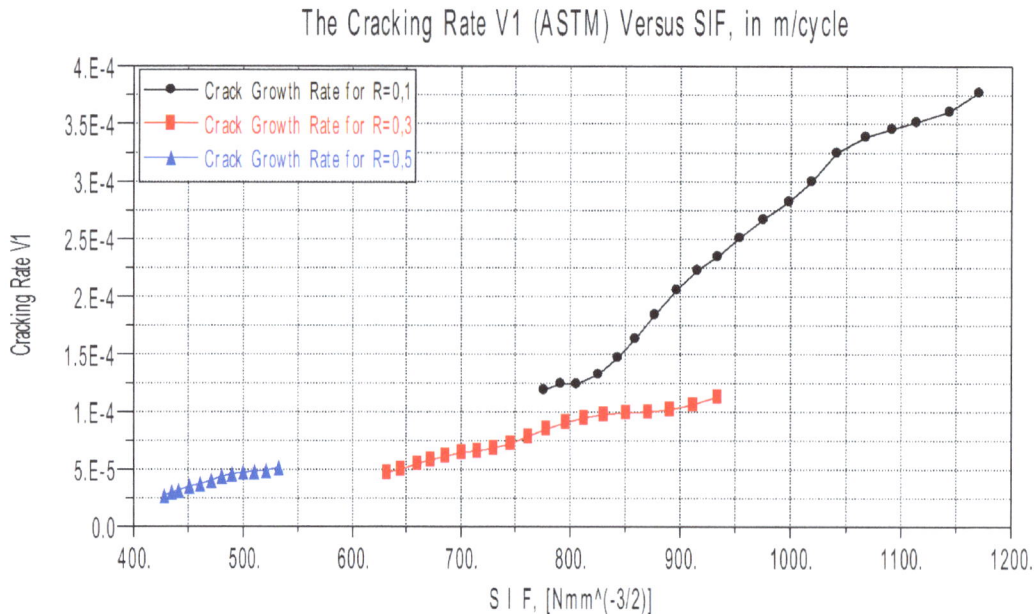

Figure 9: Crack Growth Rate **V1** (ASTM) Versus SIF Δ**K**

REFERENCES

[1]   Dumitru, I., Bazele Calculului la Oboseală, Editura Eurostampa, Timişoara, 2009;

[2]   Mc Henry, H.I., A Compliance Method for Crack Growth Studies at Elevated Temperatures, Journal of Materials, Tom 6, Nr.4, December, 1971, p.862-873;

[3]   Pană, T., Pastramă, St., D., Integritatea Structurilor Mecanice, Editura Fair Partners, Bucureşti, 2000;

[4]   Roşca, V., Contribuţii la Studiul Oboselii Monoaxiale la Temperaturi Scăzute, Teză de Doctorat, Universitatea Politehnica Bucureşti, 1997;

[5]   Roşca, V., Miriţoiu, C., Geonea, I., Romanescu, Alina, Empiric Models for Study the Crack Increase Speed in Steel R520, International Conference of Mechanical Engineering, ICOME 2013, Craiova, Romania, Tom.I, Universitaria Publishing House, p.273-280;

[6]   Rusu, O., Teodorescu, M., Lascu-Simion, N., Oboseala Metalelor, vol. 1- Baze de calcul, vol. 2 – Aplicaţii ingineresti, Editura Tehnică, Bucureşti, 1992;

[7]   ASTM E-647-95 Standard Test Method for Measurement of Fatigue Crack Growth Rates, American National Standard.

AN EXPERIMENTAL DENSIFICATION METHOD BY COMPRESSING THIN VENEERS (0,3-1,2 MM)

Sava Rodica[1], Lihteţchi Ioan[2]
[1] Transilvania University, Braşov, ROMANIA, sava@unitbv.ro
[2] Transilvania University, Braşov, ROMANIA, lihtetchi@unitbv.ro

Abstract: *Thin plywood with special destinations is one of the wood based composites used in any application that needs high quality wooden sheet material. They are composed from thin veneers (0.3 mm-1, 20mm), which must meet certain requirements to compose these laminated wood assortments, functionally suited. This paper presents an original experimental method of densification by compression of veneers made of aboriginal wooden species.*
Finally, this method improves the quality of plywood.
Keywords: *compression, veneers, thickness, density, plywood.*

1. INTRODUCTION

The thin plywood is defined as a layered product, $1 \div 3$ mm thick, made of technical veneer with thickness of $0.3 \div 1.2$ mm in aboriginal species of wood with homogeneous structure, glued with phenolic resin films.

Since our country does not manufacture these types of plywood, it was necessary their experimental production, in order to determin their physico-mechanical characteristics to use them later.

The technology for these types of plywood is similar to the production technology for outer plywood, however, presents a number of specific parameters, among them, very important being the choise of the correct veneer thickness whithin the plywood plate structure.

In order to determine the correct veneer thickness, thickness of which depends in the end plate thickness of plywood, this paper presents an original experimental method of densification by compression of veneers during the pressing process; the method contributes to improve the quality of the plywood boards.

2. THE METHOD USED FOR EXPERIMENTAL TESTS

The phenomenon of wood densification under the action of heat and pressure is due to plastic properties of wood. Moisture in wood and water intake in the layers of veneer coming from adhesive, in liquid or vapor phase, contributes to wood lamination. Another reason for maintaining the compressed state is the adhesive, especially in the interface wood - glue. The adhesive penetrates into the wood structure, especially in the pores and intercellular spaces after polymerization and keeps wood in deformed state. This phenomenon is particularly noticeable as the wood density is lower and the pressure is higher.

The plywood density is greater than or about the same as wood species from which it comes, when the humidity is the same. The reason is represented by the share of the adhesive in the plywood weigth as well as the wood densification effect under the action of temperature and pressure required to achieve the layered structure.

The treating of veneers by compression (mechanical pressure), was experimentally made on an universal machine type WE-10, for strength determining, available in "Transilvania" University of Braşov, the Faculty of Wood Industry.

The specimens used were cut in square shape with sides of 60 mm and veneers of beech, birch, alder and lime were taken, for which indicatives were assigned.

Five samples of veneer were used for each analyzed wood species.

Figure1. Universal testing machine WE-10A, equipped with two flat plates to compress veneers

How:
- Each specimen was 0.01g accurately weighed;
- Initial thickness of each specimen was measured in five points, as shown in Figure 1.
- Initial density was calculated for each specimen;
- Each specimen was set between flat plates of the testing machine, as in Figure 1.
- Each specimen was compressed, with appropriate forces following the compression values: 0.4, 0.6, 0.8, 1.0, 1.2 N/mm^2. The duration of maintaining each pressure value was 3 minutes;
- Specimen thickness values were measured after each cycle of compression in the initial established points;
- The final density of each specimen was calculated, after the last cycle of compression, ie 1.2 N/mm^2.

The calculation
- start and end density of each was calculated with the relationship:

$$\rho = \frac{m}{L \cdot b \cdot h} \cdot 10^6 \quad [\text{ kg/m}^3] \tag{1}$$

- the compression force F were calculated by the relation:

$$F = PS \cdot A \ [daN] \tag{2}$$

where:

PS - compression pressure = $(0.4 \div 1.2)$ N/mm^2
A – compression area = 36 cm^2

- The arithmetic mean of initial and final density of specimens of the same type was calculated.

Veneers densification was calculated with the formula:

$$c = \frac{h_i - h_f}{h_i} \cdot 100 \qquad [\%] \tag{3}$$

where:

h_i - initial thickness of the veneer, in mm

h_f - final thickness, after the last cycle of compression, $12 daN/cm^2$ in mm.

3. EXPERIMENTAL RESULTS

Summary results on densification veneers, are shown in Figure 2 and Figure 3.

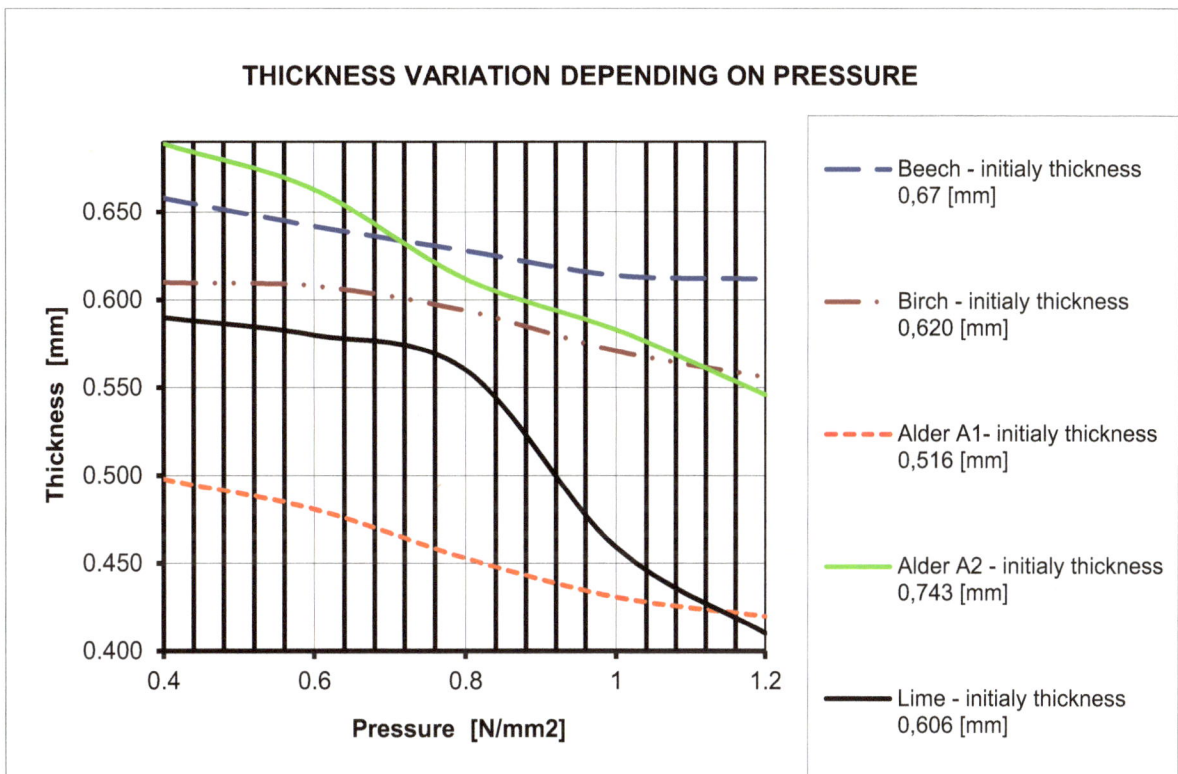

Figure 2. Thickness variation

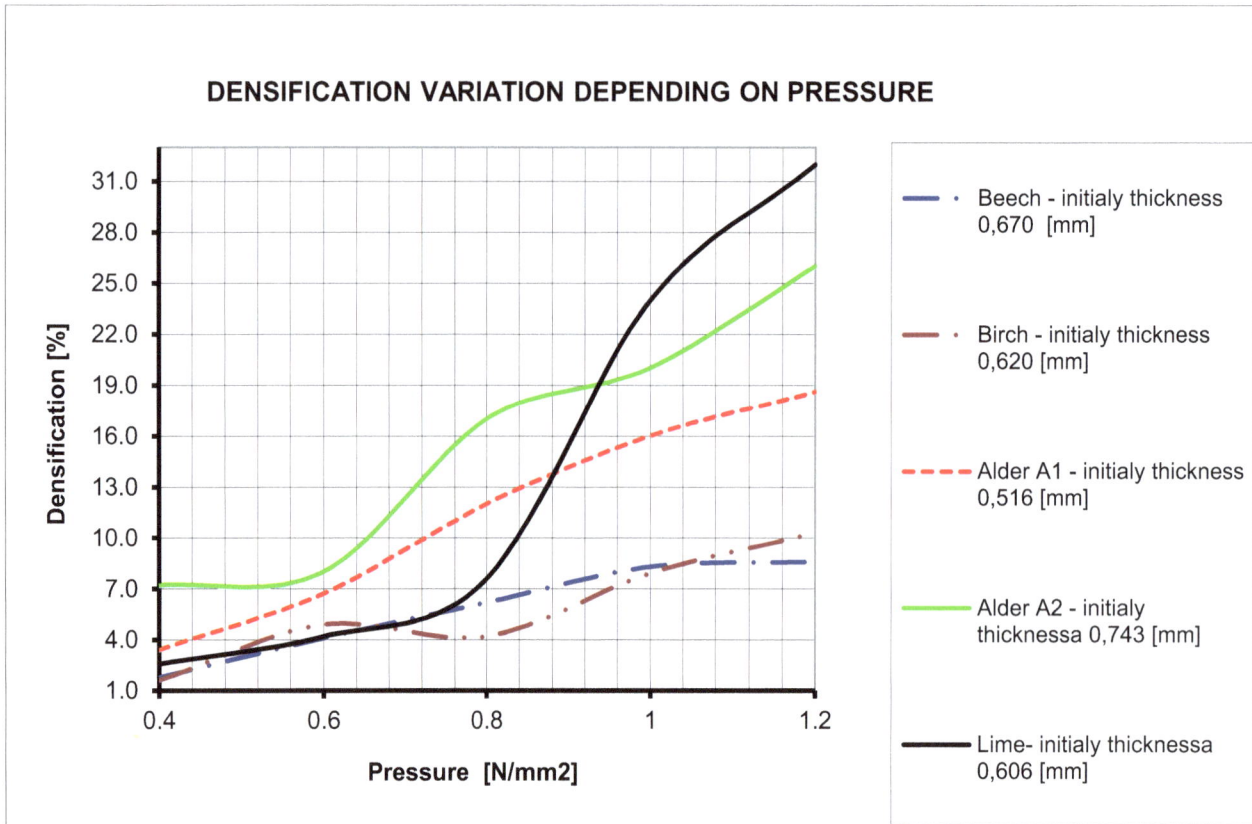

Figure 3: Densification variation

The measurements on densification by compression of the veneers, also served to the establishing of the influence that densification has on the volumic mass (apparent density) of plywood. The results regarding the density growing after densification are summarized in Table 1.

Table 1: Density growing after densification

No.	Species	Volumic mass [kg/m³]		Density growing [%]
		initially	finaly	
1	Alder	322	456	1,41
		337	479	1,42
2	Birch	435	480	1,103
3	Beech	604	666	1,09
4	Lime	328	492	1,44

Veneers compression results in a density increase with values that depend on the density of each species of which are cut.

3. CONCLUSION

Analyzing data from the summary Figure 2 and Figure 3:
- The wood species that compresses the least is beech (8.6% to 1.2N/mm2), closely followed by birch, which compresses only 2% stronger than beech, this result is due to close values of their mass displacement, it fits in the category Hardwood;
- The highest percentage of compression is for veneers lime, due to the low density value (540-610kg/m3) at a pressure of $1.2N/mm^2$, degree of compression is 32%, thus exceeding the limit value. According to the literature, the range of variation of the veneer thickness by compression should not exceed 25% of its thickness, losses of more than 25%, being incompatible with the economicity conditions required by practice;
- Veneer thickness variation under the effect of compression was 39% lower at the beech and birch veneers than alder and lime.
- Analyzing summary data in Table 1:
- Lime and alder veneers suffer a sharp increase in density, between 1.41 ÷ 1.44%, due to a low initial density, making them part of the soft species. In time, the density of beech and birch veneers increases much less that is between 1.09 ÷ 1.10%, which makes them part of the hardwood category.
- In practice, plywood made from compressed veneers have a density of about 3% higher than standard plywood, the degree of densification of compressed veneer plywood is 0.92% comparing to 0.67% as the usually manufactured plywood have;
- Preliminary compression of veneers affects one of the most important property that decides the quality of plywood, that is the shear strength of gluing. Thus, the resistance of plywood gluing obtained from individually compressed veneer is about 37% higher than that of the playwood in the current production. The fact is explained by the improved quality of gluing surfaces, resulting in a continuous adhesion of the film of glue.
- Another advantage of the preliminary compression of the veneer is the influence on the roughness of the plywood faces, reduced with about 19%. The compression of the face veneers may replace the plywood grinding operation for certain uses.

REFERENCES

[1]   Istrate V., Mitişor A., Gligor A., Paraschiv N., The manufacturing of beech plywood with compressed veneer, „Industria Lemnului" Magazine, no. 1, 1976.
[2]   Neamţu M., Research to establish the optimal structure and thickness for the formwork plywood, glued with phenoplac, „Industria Lemnului" Magazine, no.12, 1968.
[3]   Mitişor A., Istrate V., Technology of Veneer, plywood and wood fibre boards, Editura Tenhnică, 1985, Bucureşti.
[4]   Sava, R., Study regarding the physical, mechanical and technological features of the thin plywood (1-3 mm) in aboriginal wooden species, Doctorate thesis, 2005, Brasov.

VARIABLE CONDUCTANCE HEAT PIPE MODEL FOR TEMPERATURE CONTROL OF PROCESSES

Ungureanu I. Virgil-Barbu[1]
[1] University Transilvania, Brasov, ROMANIA, virbung@unitbv.ro

Abstract: *The paper presents a model for an accurate and precise method of automatic control engineering for temperature control of processes. The method requires variable conductance heat pipe with non-condensable gas added. It is solved the system of equations who describes the heat transfer processes obtaining the transfer function of the regulating system. The transient behavior is analyzed for the system with conventional variable conductance heat pipe operating in open loop and feedback controlled variable heat pipe.*
Keywords: *heat pipe, variable conductance heat pipe, heat transfer equations, transfer function of automatic systems)*

1. INTRODUCTION

The basic heat pipe is a closed container which has evacuated all non-condensable gases and contains a capillary structure (wick) and a small amount of a vaporizable fluid [1]. The heat pipe employs a boiling-condensing cycle and the capillary pumps in order to return the liquid. Figure 1 shows schematically the basic heat pipe operating system. The essential components of a heat pipe are the sealed container, the wick and the suitable working fluid: liquid in equilibrium with its own vapour. Thus, the heat pipe is composed by three zones: evaporator, adiabatic and condenser. When heat is applied along the evaporator zone, the local temperature is raised slightly and part of the working fluid evaporates at the wick surface adjacent to this zone. Simultaneously, the condenser zone is slightly cold. Because of the saturation conditions this temperature difference causes a vapour pressure difference and a vapour flowing from evaporator zone to the condenser zone, through the adiabatic zone. The heat absorbed is commensurate by the heat flow rate of vaporization. During steady state operation, the first principle of thermodynamics requires that the amount of heat absorbed by the working fluid is identical with the heat released. The wick provides a flow part for the liquid return and is responsible for the pumping.

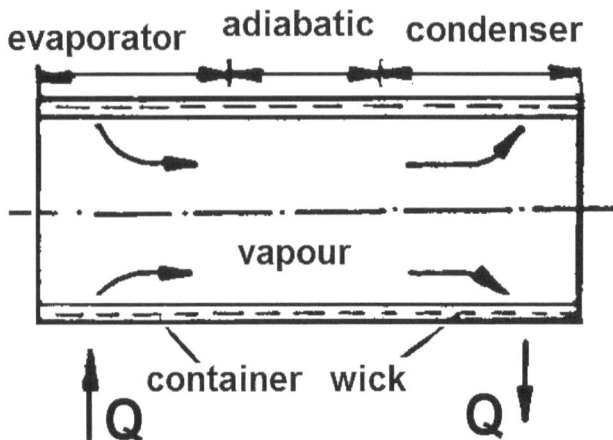

Figure 1: Heat pipe

The operating temperature ranges are cryogenic (0 to 150 K), low temperature (150 to 400 K) medium temperature (400 to 750 K) and high temperature (over 750 K). The working fluids are usually chemical elements or simple organic gases, polar molecules or halocarbons (with some restrictions), and liquid metals.

Then vapour temperature drop along the overall heat pipe is very small because the small pressure drops. Thus, the boiling-condensing cycle is essentially an isothermal process. Furthermore, the temperature losses between the heat source and the vapour on the one hand and between the vapour and the heat sink on the other hand can be very small by proper design. Therefore, the first feature of the heat pipe is that it can be designed to transport heat between two heat sources with different temperature with a very small temperature loss.

In the body forces field, the condense returns to the condenser section without the wick. But the wickless gravity assisted heat pipe contained the disadvantage that the condense returns against the vapour flow and establish an entrainment limit lower than in the case of the basic heat pipe.

The heat pipes are classified in two general types: conventional heat pipes and variable conductance heat pipes. The conventional heat pipe is a device with a very high thermal conductance with no fixed operating temperature. Its temperature varies according to variations in the heat source or heat sink. But the device can be designed in order to maintain a constant temperature of a volume. It was first realized by blocking a portion of the condenser with a non-condensable gas (figure 2). For a precise control of the temperature it is needed the addition of a reservoir downstream the condenser zone. A feedback controlled variable heat pipe can realize a quasi-absolute temperature control by heating/cooling the reservoir or gas mass control. A greater temperature control is obtained for space devices [2].

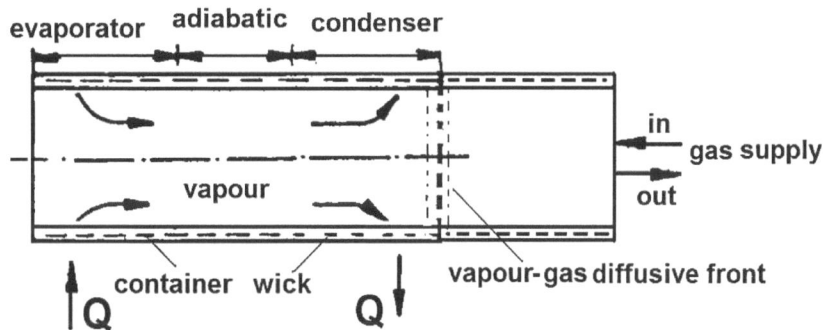

Figure 2: Variable conductance heat pipe

The most familiar variable conductance heat pipe systems include passive or active controlled system, both having the capability to control the source of heat (electronic devices, solar collectors, technologic processes, chemical reactors, etc.) at the evaporator end.

Sauciuc et al. [3] analyzed the operation of a variable conductance heat pipe for temperature control of solar collectors. The experimental results indicated that the starting point of variable conductance heat pipe is significant dependent of the amount of non-condensable gas and the superheat required for boiling.

The temperature control of the technologic processes is a slow process. The capacity of heat storage and the thermal resistances between some parts of the process are distributed along the entire heat flow.

2. MATHEMATICAL MODEL FOR THE TEMPERATURE CONTROL

A mathematical model for temperature control of the electronic components mounted in a satellite was presented in [4]. The temperature control was realized by temperature control of the reservoir, and so the non-condensable gas volume. For technologic processes, the model must to take account of the heat transfer process-heat pipe and the thermal inertia of materiel from the working space. The above physical model has not a reservoir. However, the temperature control of the technological processes in addition with a gas controlled variable conductance heat pipe was realized by two cocks: the first, by addition and the second by exhaust the gas. Thus, the gas buffer amount and subsequent the operating temperature of the heat pipe is controlled.

Supposing an exothermic process controlled by feedback variable conductance heat pipe(s). The condenser zone was in thermal contact with the cooling medium. Thus, the heat transfer is produced between the process and the cooling medium by the succession of three zones: the evaporator zone, working fluid vapour column and condenser zone of the heat pipe.

The heat flow rate transferred is:

$$\dot{Q} = G \cdot \left(T_p - T_s \right), \qquad (1)$$

(1)

were G is the global thermal conductance, T_p - the process temperature and T_s - the sink temperature.

3. DIFFERENTIAL EQUATIONS OF THE CONTROLLED SYSTEM

Thus, it can obtain four equations of thermal balance in the unsteady heat transfer regime for the process:

$$C_p \frac{dT_p}{dt} + G_{pe} T_p = Q_p + G_{pe} T_e, \qquad (2)$$

evaporator zone:

$$C_e \frac{dT_e}{dt} + \left(G_{pe} + G_{ev} + G_{ec} + G_{es} \right) \cdot T_e = G_{pe} T_p + G_{ev} T_v + G_{ec} T_c + G_{es} T_s, \qquad (3)$$

vapour column:

$$V_c \rho_v r_v \frac{dl_a}{dt} + \alpha_v \frac{dT_v}{dt} + \left(G_{ev} + G_{vc} l_a + G_{vg} \right) \cdot T_v = G_{ev} T_e + G_{vc} l_a T_c + G_{vg} T_s, \qquad (4)$$

and condenser zone:

$$C_c T_c \frac{dl_a}{dt} + C_c l_a \frac{dT_c}{dt} + \left(G_{ec} + G_{vc} l_a + G_{cs} l_a \right) \cdot T_c = G_{ec} T_e + G_{vc} l_a T_v + G_{cs} l_a T_s, \qquad (5)$$

adding an equation for mass balance of the gas inside the variable conductance heat pipe:

$$\frac{dl_a}{dt} = \beta_v \frac{dT_v}{dt} - \beta_s \frac{dT_s}{dt} - \beta_m \frac{dm_g}{dt}, \qquad (6)$$

in which are notted:

$$\alpha_v = V_c \left(l_e + l_a \right) \left(r_v \frac{d\rho_v}{dT_v} + \rho_v \frac{dr_v}{dT_v} \right); \qquad (7)$$

$$\beta_v = \frac{1 - l_a}{p_v - p_{vs}} \cdot \frac{dp_v}{dT_v} \cdot \frac{dT_v}{dt}; \qquad (8)$$

$$\beta_s = \frac{R m_g}{V_c \left(p_v - p_{vs} \right)} + \frac{1 - l_a}{p_v - p_{vs}} \cdot \frac{dp_{vs}}{dT_s}; \qquad (9)$$

$$\beta_m = \frac{R T_s}{V_c \left(p_v - p_{vs} \right)} \cdot \frac{dm_g}{dt}. \qquad (10)$$

In the transient heat transfer regime there are into account thermal capacities, C. The subsequent symbols are used too: T - temperatures, r_v - evaporation heat, ρ_v - vapour density, R - non-condensable gas constant, l_a - relative length of the active condenser region (active length and overall length of the condenser ratio), V_c - maximum volume of the condenser, p_v, p_{vs} - vapour pressure of the working fluid in the active and respectively blocked zone of the condenser, m_g - mass of gas. Subscripts are: p - process, e - evaporator zone, a - adiabatic zone, c - condenser zone, g - gas, s - sink.

The system of equations that describes this process is linear if it is considered that factors of main variables are constants having the values from the steady state regime.

Applying the Laplace transformation and considering null initial condition it is obtained a system with five linear equations having the unknown X_e, X_v, X_c, X_l, representing the Laplace's transform of the evaporator, vapour, condenser zones, and respectively the active length of the condenser. The regulated value of the temperature T_p is represented in the complex plane by X_p, the direct regulated magnitude (mass of gas) represented by X_m while disturbances are considered the heat flow of the exothermic process Q_p and the sink temperature T_s represented by X_Q and respectively X_s. The complex variable is s.

$$\left(C_p s + G_{pe}\right)\cdot X_p - X_Q = G_{pe}X_e, \tag{11}$$

$$\left(C_e s + G_{pe} + G_{ev} + G_{ec} + G_{es}\right)\cdot X_e = G_{pe}X_p + G_{ev}X_v + G_{ec}X_c + G_{es}X_s, \tag{12}$$

$$\left(V_c\rho_v r_v\right)\cdot s \cdot X_l + \left(\alpha_v \cdot s + G_{ev} + G_{vc}\cdot l_a + G_{ec} + G_{vg}\right)\cdot X_v = G_{pe}X_p + G_{ev}X_v + G_{ec}X_c + G_{es}X_s \tag{13}$$

$$\left(C_c T_c\right)\cdot s \cdot X_l + \left(C_c l_a \cdot s + G_{ec} + G_{vc}\cdot l_a + G_{cs}l_a\right)\cdot X_c = G_{ec}X_e + G_{vc}l_a X_v + G_{cs}l_a X_s, \tag{14}$$

$$s \cdot X_l = \beta_v \cdot s \cdot X_v - \beta_s \cdot s \cdot X_s - \beta_m \cdot s \cdot X_m, \tag{15}$$

The coefficients of variables X_p, X_Q, X_e, X_v, X_c, X_s, X_m and X_l are transfer functions. The above equations can be arranged in the form:

$$X_p = H_{e1}X_e + H_{q1}X_q, \tag{16}$$

$$X_e = H_{p2}X_p + H_{v2}X_v + H_{c2}X_c + H_{s2}X_s, \tag{17}$$

$$X_v = H_{e3}X_e + H_{c3}X_c + H_{s3}X_s - H_{l3}X, \tag{18}$$

$$X_c = H_{e4}X_e + H_{v4}X_p + H_{s4}X_s - H_{l4}X_l, \tag{19}$$

$$X_l = H_{v5}X_v - H_{s5}X_s - H_{m5}X_m, \tag{20}$$

Eliminating the unknown X_l, X_e, X_v and X_c it obtain the characteristic equation of the process that operating in open circuit:

$$X_p = A\cdot X_m + B\cdot X_Q + C\cdot X_s, \tag{21}$$

in which coefficients are:

$$A = \frac{\left(m_1 s + m_0\right)\cdot s}{n_4 s^4 + n_3 s^3 + n_2 s^2 + n_1 s + n_0}; \tag{22}$$

$$B = \frac{q_3 s^3 + q_2 s^2 + q_1 s + q_0}{n_4 s^4 + n_3 s^3 + n_2 s^2 + n_1 s + n_0}; \tag{23}$$

$$C = \frac{t_2 s^2 + t_1 s + t_0}{n_4 s^4 + n_3 s^3 + n_2 s^2 + n_1 s + n_0}. \tag{24}$$

The coefficients of polynomial are calculated with equation:

$$n_4 = e_1 d_1; \ n_3 = e_2 d_1 + e_1 d_2 - e_{10} d_3; \ n_2 = e_3 \cdot d_1 + e_2 d_2 - e_{10} d_4 - e_{11} d_3;$$

$$n_1 = e_4 d_1 + e_3 d_2 - e_{10} d_5 - e_{11} d_4; \ n_0 = e_4 d_2 - e_{11} d_5;$$

$$m_1 = d_{10} e_{10} + d_1 e_{12}; \ m_0 = d_{10} e_{11} + d_2 e_{12}; \tag{25}$$

$$q_3 = d_1 e_5; \ q_2 = d_1 e_6 + d_2 e_5 + d_6 e_{10}; \ q_1 = d_1 e_7 + d_2 e_6 + d_7 e_{10} + d_6 e_{11}; \ q_0 = d_2 e_7 + d_7 e_{11};$$

$$t_2 = e_8 d_1 + e_{10} d_8; \ t_1 = e_9 d_1 + e_8 d_2 + e_{10} d_9 + e_{11} d_8; \ t_0 = e_9 d_2 + e_{11} d_9.$$

The constants $d_1 ... d_{10}$ and $e_1 ... e_{12}$ are calculated with relations:

$$d_1 = b_1; \ d_2 = b_2 + \frac{b_6 a_7}{a_8}; \ d_3 = \frac{b_6 a_1}{a_8}; \ d_4 = b_3 + \frac{b_6 a_2}{a_8}; \ d_5 = b_4 + \frac{b_6 a_3}{a_8}; \ d_6 = -\frac{b_6 a_4}{a_8};$$

$$\tag{26}$$

$$d_7 = b_5 - \frac{b_6 a_5}{a_8}; \ d_8 = b_7; \ d_9 = b_8 + \frac{b_6 a_6}{a_8}; \ d_{10} = b_9;$$

$$e_1 = \frac{c_1 a_1}{a_8}; \ e_2 = \frac{c_1 a_2 + c_2 a_1}{a_8}; \ e_3 = \frac{c_1 a_3 + c_2 a_2}{a_8} - c_3; \ e_4 = \frac{c_2 a_3}{a_8} - c_4; \ e_5 = \frac{c_1 a_4}{a_8}; \ e_6 = \frac{c_1 a_5 + c_2 a_4}{a_8};$$

$$\tag{27}$$

$$e_7 = \frac{c_2 a_5}{a_8} + c_5; \ e_8 = \frac{c_1 a_6}{a_8} + c_8; \ e_9 = \frac{c_2 a_6}{a_8} + c_9; \ e_{10} = \frac{c_1 a_7}{a_8} + c_6; \ e_{11} = \frac{c_2 a_7}{a_8} + c_7; \ e_{12} = c_{10},$$

The constants $a_1 ... a_8$, $b_1 ... b_9$ and $c_1 .. c_{10}$ are calculated with relations:

$$a_1 = \frac{C_e C_p}{G_{pe}}; \ a_2 = C_e + \frac{C_p \left(G_{pe} + G_{ev} + G_{ec} + G_{es} \right)}{G_{pe}}; \ a_3 = 2 G_{pe} + G_{ev} + G_{ec} + G_{es};$$

$$\tag{28}$$

$$a_4 = \frac{C_e}{G_{pe}} - c_4; \ a_5 = \frac{G_{pe} + G_{ev} + G_{es}}{G_{pe}}; \ a_6 = G_{es}; \ a_7 = G_{ev}; \ a_8 = G_{ec};$$

$$b_1 = \alpha_v + V_c \rho_v r_v \beta_v; \ b_2 = G_{ev} + G_{vc} l_a + G_{vg}; \ b_3 = \frac{G_{ev} C_p}{G_{pe}}; \ b_4 = G_{ev}; \ b_5 = -\frac{G_{ev}}{G_{pe}};$$

$$\tag{29}$$

$$b_6 = G_{vc} l_a; \ b_7 = V_c \rho_v r_v \beta_s; \ b_8 = G_{vg}; \ b_9 = V_c \rho_v r_v \beta_m;$$

$$c_1 = C_c l_a; \ c_2 = G_{ec} + G_{vc} l_a + G_{cs} l_a; \ c_3 = \frac{G_{ec} C_p}{G_{pe}}; \ c_4 = G_{ec}; \ c_5 = -\frac{G_{ec}}{G_{pe}};$$

$$\tag{30}$$

$$c_6 = -C_c T_c \beta_v; \ c_7 = G_{vc} l_a; \ c_8 = C_c T_c \beta_s; \ c_9 = G_{cs} l_a; \ c_{10} = C_c T_c \beta_m.$$

It can be observed that the system is represented only by three transfer functions.

The coefficients from these equations are real. Thus, these magnitudes are calculated with thermal balance equations in a steady state regime obtained from equations (2)...(10) if temporal derivatives are cancelled.

4. EQUATIONS OF THE CONTROL SYSTEM

Considering the case of an ideal proportional-integral-derivative controller device it can obtain the transient response in the complex plane chart for the reference signal, exothermic flow and sink temperature variations:

$$X_p = H_p \cdot X_p^r + H_Q \cdot X_Q + H_S \cdot X_S, \tag{31}$$

in which coefficients (transfer functions) are:

$$H_p = \frac{k_p k_E m_1 \left(s + \dfrac{m_0}{m_1}\right) \cdot (s + k_{0T})(T_D s^2 + s + k_I)}{p_5 s^5 + p_4 s^4 + p_3 s^3 + p_2 s^2 + p_1 s + p_0} ; \tag{32}$$

$$H_Q = \frac{(s + k_{0T})(q_3 s^3 + q_2 s^2 + q_1 s + q_0)}{p_5 s^5 + p_4 s^4 + p_3 s^3 + p_2 s^2 + p_1 s + p_0} ; \tag{33}$$

$$H_S = \frac{(s + k_{0T}) \cdot (t_2 s^2 + t_1 s + t_0)}{p_5 s^5 + p_4 s^4 + p_3 s^3 + p_2 s^2 + p_1 s + p_0} . \tag{34}$$

in which X_p^r is the Laplace's transform of the reference input.

The coefficients of characteristic equation are:

$$p_5 = n_4; \quad p_4 = n_3 + k_{0T} n_4; \quad p_3 = n_2 + k_{0T} n_3 + k_T k_p k_E m_1 T_D; \quad p_2 = n_1 + k_{0T} n_2 + k_T k_p k_E (m_1 + m_0 T_D);$$

$$p_1 = n_0 + k_{0T} n_1 + k_T k_p k_E (m_0 + m_1 k_I); \tag{35}$$

$$p_0 = k_T k_p k_E m_0 k_I + k_{0T} n_0,$$

where k_p is the proportional control factor (factor range ratio), k_I - integrating constant, T_D - derivative time of the control device, k_E - proportional constant of the actuator, k_T - proportional constant of the temperature transducer and k_{0T} - inverse of the time constant of the transducer.

It can be observed that the tranzient response in the complex plane can be obtained by three transfer functions applyied on the reference signal X_p^r, disturbance created by the exotermic reaction heat flow X_Q, and respectively by the sink temperature variation X_S.

5. INVESTIGATION OF THE ANALYTIC CONTROL

The characteristic equation of the system is the denominator from equations (32)...(34), and poles of transfer functions are obtained by .

After some trials for real cases, the above equation can have two real roots and two complex conjugated roots, or all roots are real. The system stability is always assured because all roots of the characteristic equation are in the the left half-plane. Applying the theorem of limit in complex, the stationary deviation at an unitary step-load change is null (if it is neglected the disturbance) and is finite for each deviation.

It is proposed an example of the tranzient response of the system for an unitary step-load signal applyied to the reference signal and dispurbances, for example: k_r/s, k_q/s and k_s/s.
It is obtained:

$$X_p = H_p \cdot \frac{k_r}{s} + H_Q \cdot \frac{k_Q}{s} + H_S \cdot \frac{k_S}{s}, \tag{36}$$

This equation is decomposed in simple fractions and then, applying the inverse Laplace's transformation it can obtain the transient response of the system:

$$T_p = (c_{p0} + c_{Q0} + c_{S0}) + (c_{p1} + c_{Q1} + c_{S1}) \cdot e^{x_{p1}t} + (c_{p2} + c_{Q2} + c_{S2}) \cdot e^{x_{p2}t} + (c_{p3} + c_{Q3} + c_{S3}2) \cdot e^{x_{p3}t} + (c_{p4} + c_{Q4} + c_{S4}) \cdot e^{x_{p4}t} + (c_{p5} + c_{Q5} + c_{S5}) \cdot e^{x_{p5}t}, \tag{37}$$

in which $x_{p1} \ldots x_{p5}$ are the roots of characteristic equations (if they are real), and coefficients are obtained by decompozition in partial fractions. If two from the three roots of the characteristic equation are complex conjugated the tranzient response in not aperiodic.

For the numerical analysis of the control is realized a computer program in order to solving the transfer function.

The real roots of polynomial are obtained with Bierge-Viete algorithm.

Figure 3 presents the transient response of the system controlled with a variable conductance heat pipe in open loop. It can remark that the transient response is too slow.

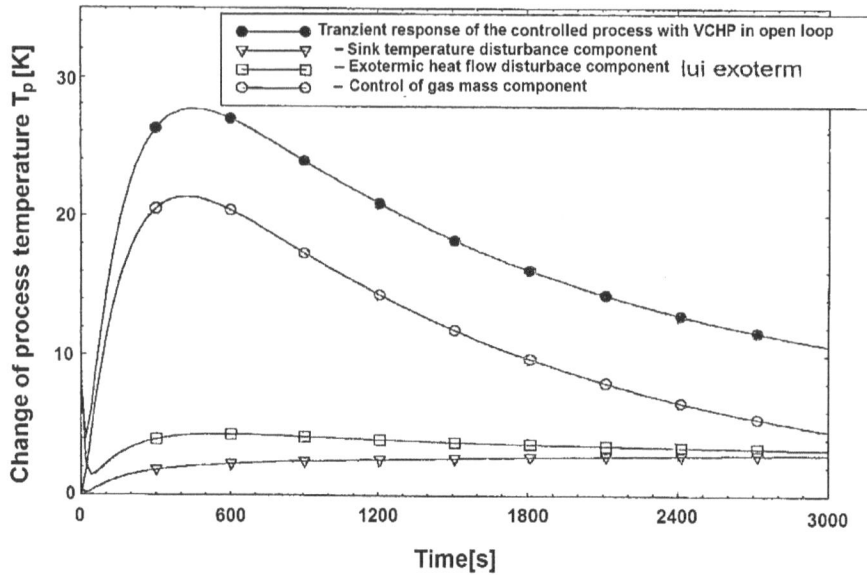

Figure 3: Transient response of the system with variable conductance heat pipe operating in open loop

Figure 4 presents a transient response of the system with a feedback controlled variable conductance heat pipe.

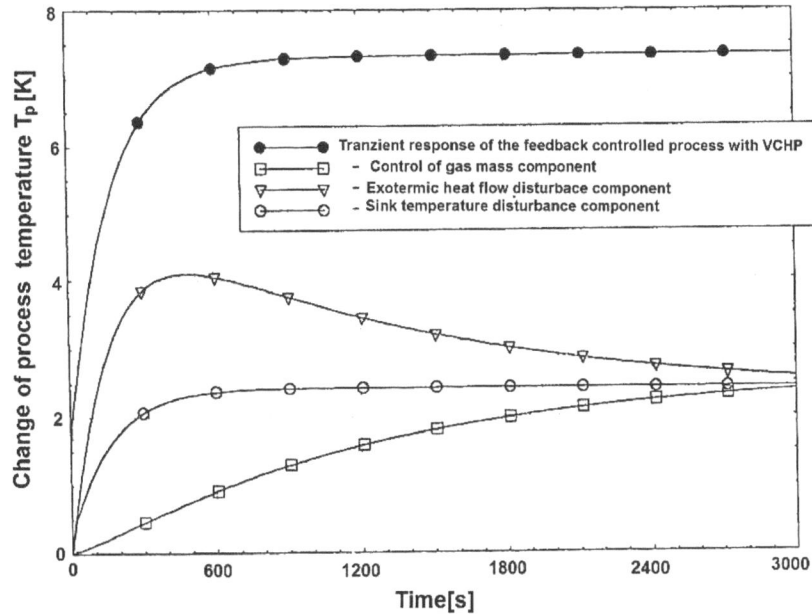

Figure 4: Transient response of the system with variable conductance heat pipe operating in open loop

If we compare the results from figures 3 and 4 it can observed improved performances of the system provided with variable conductance heat pipe. Thus, the system obtains high precision (from 30K to 8K) and reaction velocity (after 600s the regime is right established). Also, it can remarks an increase of the productivity because the control actuates on heat transfer surface and no affects the process speed progress.

3. CONCLUSION

The temperature control of the technologic processes is a slow process. The capacity of heat storage and the thermal resistances between some parts of the process are distributed along the entire heat flow. The temperature control of the technological processes in addition with a gas controlled variable conductance heat pipe was realized by two cocks: the first, by addition and the second by exhaust the gas.

It is obtained four equations of thermal balance in the unsteady heat transfer regime for the process. The system of equations that describes this process is linear if it is considered that factors of main variables are constants having the values from the steady state regime. Applying the Laplace transformation and considering null initial condition it is obtained a system with five linear equations and the system is represented only by three transfer functions. The coefficients from these equations are real. Considering the case of an ideal proportional-integral-derivative controller device it can obtain the transient response in the complex plane chart for the reference signal, exothermic flow and sink temperature variations.

It can observe that the tranzient response in the complex plane can be obtained by three transfer functions applyied on the reference signal, disturbance being created by the exotermic reaction heat flow, and respectively by the sink temperature variation.

This equation is decomposed in simple fractions and then, applying the inverse Laplace's transformation it can obtain the transient response of the system.

A greater temperature control is obtained using the feedback variable conductance heat pipes than the passive system.

REFERENCES

[1] Fetcu D., Ungureanu V., Tuburi termice, Editura Lux Libris, Brasov, 1999.

[2] Brost O., Groll, Mack H., High temperature lithium heat pipe furnace for space applications. An investigation of temperature stability and reproductibility. Proceedings of 7th International Heat Pipe Conference, Minsk, 1990.

[3] Sauciuc I, Akbarzadeh A., Johnson P., Temperature control using variable conductance closed two-phase heat pipe. International Communications in Heat and Mass Transfer, 1996.

[4] Furukawa, M., Analysis for sequential temperature control of variable conductance heat pipes, Proceedings of the 7th International Heat Pipe Conference, Minsk, 1999.

COMPARATIVE NUMERICAL ANALYSIS OF AN ARMOR PLATE UNDER EXPLOSION

V. Nastasescu[1], Gh. Barsan[2]
[1] Military Technical Academy, Bucharest, ROMANIA, nastasescuv@gmail.com
[2] Land Forces Academy, Sibiu, ROMANIA, ghbarsan@gmail.com

Abstract: *This paper presents, in a very synthetic manner, some aspects regarding developing of the blast wave calculation. The essential aspects are presented, starting with experimental formulas and finishing with the newest numerical procedures. These procedures are used for numerical analysis of the behavior of some armor plates. The armor plates are made in many constructive solutions, depending on the aim. Our work is referring to the behavior of the armor plates under explosion, especially the behavior of the armor personnel carrier plates under mine explosion. Two constructive solutions are analyzed and compared with the plane armor plate. Finally, this work presents numerical solutions, models and conclusion which can be used for the improving of the protection against mines, available for armor personnel carrier as well as for others structures having such requirements.*
Keywords: *blast wave, armor plate, explosion effects, blast wave parameters*

1. INTRODUCTION

The calculation of the blast wave effects upon structures and as well as the blast wave parameters is a problem of a large theoretical and practical interesting, especially in the last twenty years. Probably, the interest for the computation of the blast wave effects upon structures and human beings started together with the first explosions used in military targets, but in a real and consistence sense, the interesting of the researchers began after the Second World War (WW2); the terrorist threats, of the last twenty years, determined an intension of the researching, using specially experimental and numerical methods.

The vulnerability of the light armored vehicles as well as of the personnel to the blast mines is studied by all researching ways. Such studies involve an improving of the calculus, but in the same time, an improving of the constructive solutions of the armor plates. This aspect is referring to the materials, but in the last time, different shapes and constructive solutions are taken into account.

This paper presents, in a very synthetic manner, some aspects regarding the development of the blast wave calculation. A numerical study of a new constructive solution is carried out and the results are presented in a comparative way with respect to a common constructive solution of a plane plate having a constant thickness.

2. CALCULATION OF THE BLAST WAVE PARAMETERS

The blast wave calculation or explosion calculation is referring to two aspects, which can be solved separately or together: blast wave parameter calculation and blast wave effect calculation. The blast wave parameter calculation is done all times, but some times this calculation is not directly invoked (in some blast effect calculation). No matter the subject of calculation, the experimental way is a very necessar one, being a truth criterion. The experimental approach methods are not presented in this paper, being too important and a large subject.

Experimentally and analytically, determination of the blast wave profile is presented in the Figure 1, in a point at a standoff distance of the explosion initiation place [2], [3]. Friedlander first time modeled the ideal blast wave profile in 1946, but nowadays the modified Friedlander ecuation, relation (1), is mostly used.

Some notations used in relation (1) can be watched in the Figure 1; t_a is arrival time and a and b are decay parameters, being different for positive and negative pressure.

$$P(t) = P_{max}\left(1 - \frac{t}{t_+}\right)e^{\left(-a\frac{t}{t_+}\right)} + P_{min}\left(1 - \frac{t}{t_-}\right)e^{\left(-b\frac{t}{t_-}\right)} \qquad (1)$$

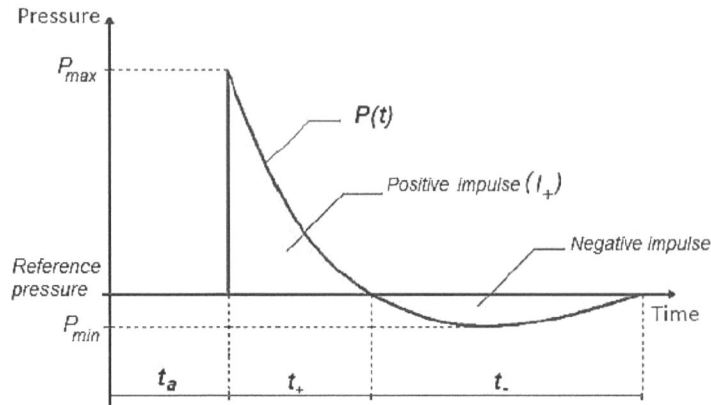

Figure 1: Blast wave profile, in time

2.1. Empirical Calculus of the Blast Wave Parameters

Almost all researchers are referring to spherical charge detonated in air, when the explosive has a spherical shape and is placed somewhere above the ground at a distance h. In the case of an explosion at ground level (surface explosion), the explosive is considered like a hemispherical charge and the parameters could multiplied with a factor grater 1 and less 2. An universal normalized description of the blast effects and parameters can be given by scaling distance Z *(Hopkinson-Cranz)* relative to the ratio:

$$Z = \left(\frac{E}{P_0}\right)^{\frac{1}{3}} \qquad (2)$$

where E is the released energy [J/kg] and P_0 is the ambient pressure [Pa]. Experimentally, and from the above relation, all air blast effects and parameters follow the same scaling law, expressed by the scaled distance Z:

$$Z = \frac{R}{W^{1/3}} \qquad (3)$$

where R is the is the distance [m] from the explosion point to a considered point (where the parameters or effects of blast waves are calculated) and W is the charge weight [kg]. Thus the scaled distance Z is independent of the type of explosion: nuclear *(Glasstone & Dolan-1977)* or non-nuclear *(Friedlander-1946, Brode-1955, Newmark & Hansen-1961, Baker-1983, Bulmash & Kingery-1984, Mills-1987, Beshara-1994, Mays & Smith-1995, Randers-Pehrson & Bannister-1997, Henrych, Held, Kinney & Grahm, Sadovskiy, Bajic and many others).*

The main parameters of an explosion are: peak positive over pressure (P_{pos}; P_{max}), positive duration (t_{pos}; t_+), negative (under) pressure (P_{neg}; P_{min}), negative duration (t_{neg}; t_-), wave decay parameter (b), the impulse (I) which can be referred to positive (I_+), negative (I_-) or total time period. All these parameters and others, can be referring to the incident (direct) pressure (P_i) or to the reflected pressure (P_r). By the mechanism of wave formation, the reflected pressure parameters are higher than incident pressure parameters, they occur practically instantaneously and they influence the damage characteristics of the blast wave. Practically, $P_r = P_{max}$. A very important parameter, specially for blast wave effect evaluation, is the impulse corresponding to the positive duration I_+, which has the expression:

$$I_+ = \int_{t_+} P(t)dt \qquad (4)$$

Therefore, the impulse, applied on the surface unit, is defined by the area under pressure-time curve (Figure 1) on the t_+ domain. Many empirical formulas for I_+ calculation exist. One of them, by Kinney, has the form of relation (5), where I_+ [Pa*s], Z [m/kg$^{1/3}$], W [kg].

$$I_+ = \frac{0.067\sqrt{1+(Z/0.23)^4}}{Z^2\sqrt[3]{1+(Z/1.55)^3}} 100 \cdot \sqrt[3]{W} \qquad (5)$$

The literature [3], [4], [5] presents some nomogrames for the calculation of all blast wave parameters, for those two main cases: explosion in free air and explosion on the surface.

Practically, for all blast wave parameters, empirical formulas exist and many authors exist too. For maximum pressure P_{max}, Brode proposed the relations (6) and (7).

$$P_{max} = \frac{0.975}{Z} + \frac{1.455}{Z^2} + \frac{5.85}{Z^3} \qquad 0.10 < P_{max} < 10 \qquad (6)$$

$$P_{max} = \frac{6.7}{Z^3} + 1 \qquad 10 < P_{max} \qquad (7)$$

All these wave blast parameters can be plotted, versus scaled distance Z or real distance R and so, a value can easily determined and some useful conclusions can be obtained.

2.2 Numerical Calculus of the Blast Wave Effects and Parameters

There are some numerical approaching ways, each of them having advantages and disadvantages regarding computationals facilities and/or spent computer time. Regardless of the way, for a right numerical modeling, adequate material models must be used. Some professional programs contain, in their material library, dedicated material models. Mat_High_Explosive_Burn with Jones-Wilkins-Lee (JWL) or Jones-Wilkins-Lee-Baker (JWLB) equations of state (EOS) for modeling of the charge, Mat_Null with Linear_Polynomial or Gruneisen equations of state for modeling of the air; these material models with their EOS are available in LS-DYNA code [6]. When the blast wave effects, upon a structure, are the aim of the numerical computation, properly material models for structure modeling have to be used. Next to these aspects, a maximum importance has the finite element formulation. Thus, for explosive and air modeling Arbitrary Lagrangian Eulerian (ALE) formulation is used. The structure modeling, no matter the aim study, Lagrangian formulation is used. As finite element calculus model is concerned, some approaching ways exist.

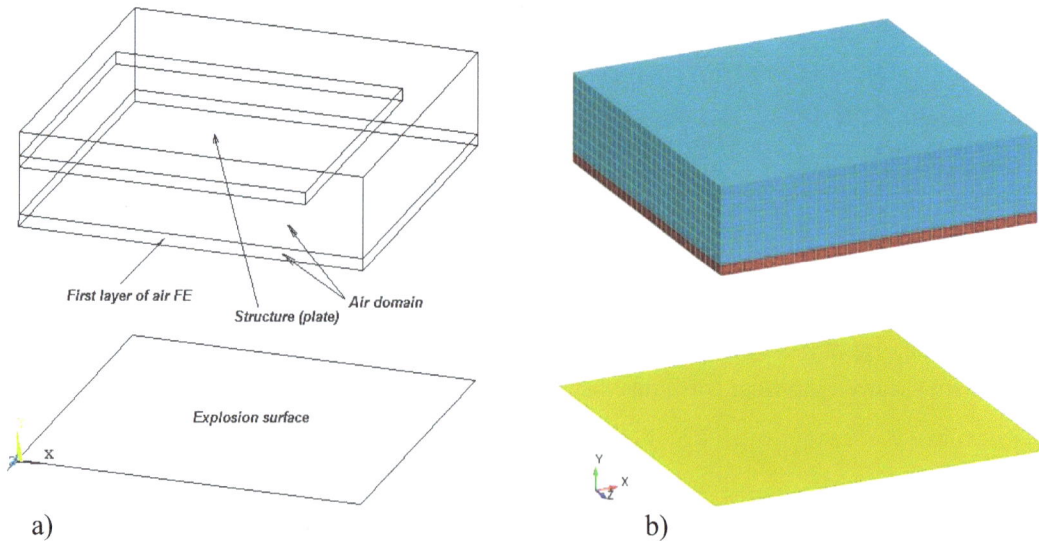

a)   b)

Figure 2: Domains and FE model (1/4 simplified model)

We used the most recent way and perhaps the most efficient way, when using the procedures Load_Blast, Load_Brode or Load_Blast_Enhanced (LBE), the modelling of the charge is avoided. We also used a new concept – Multi-Material Arbitrary Lagrange Eulerian (MM-ALE) – in which the user can model the action of more material (resulting or moved by explosion), upon a structure [1]. In Figure 2, we can see the specific domains of modelling as well as the finite element model. Taking into account the symmetry, only 1/4 of model is used (Figure 2). The air domain includes the structure, but it has smaller dimensions, being around the structure (one or more structures). The fluid-structure interaction (FSI) is modelled by a special procedure, named Constrained_Lagrange_in_Solid, available in LS-DYNA code [6].

3. COMPARATIVE NUMERICAL ANALYSIS OF AN ARMOUR PLATE

The vulnerability of light and heavy armoured vehicles to anti-vehicular blast mines is strong closed by all characteristics of the armour plate, especially of the hull or floor plate. Among all characteristics of an armour plate, an important aspect is the constructive solution of the armour plate. In this work, such a constructive solution is discussed in a comparative way.

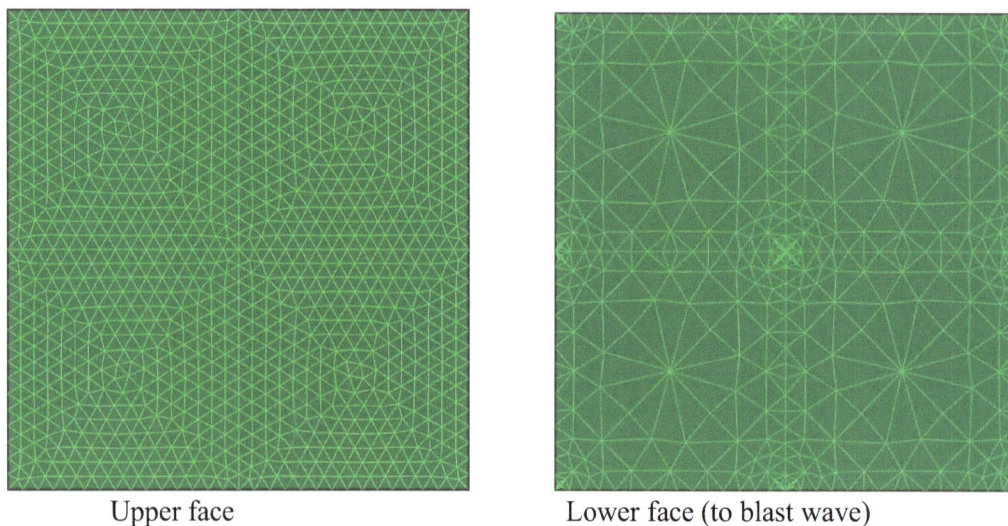

Upper face   Lower face (to blast wave)

Figure 3: Finite element model of the armour plates

Figure 4: The side profile of those two versions of armour plates

Reference constructive solution is a plane armour plate with a thickness of 2 cm; the analysed constructive solution is represented by an armoured plate having the same maximum thickness (2 cm), but the face exposed to mine explosion has a number of pyramids. The mesh (Figures 3 and 4) was the same for both constructive solutions (made by parametric describing of the geometry). In the modelling version presented in the Figures 3 and 4, the average dimension of the finite elements (4-node tetrahedron element) is 3 cm. The height of pyramids is 1 cm and their square base is 25 cm. The finite elements of air (8-node solid element) have a cube-shaped of 2 cm dimension. The results of our numerical analysis are synthetically presented in the following figures.

4. COMPARATIVE NUMERICAL RESULTS

The results presented here are referring to a square armor plate with 1m side length, placed at standoff distance of 50 cm above an explosive (Pentolite). This charge is placed just under the plate center and at the level of soil. The numerical calculus model is presented in the Figure 2. The calculus time was the same for those two constructive solutions, namely 0.001 second. In this paper, the charge mass was 27.84 kg for whole structure (only 6.96 kg for 1/4 simplified model). The explosive was not in a spherical or semispherical shape; it was in solid brick form with dimension of 0.16x0.16x0.16 m. The armor plate was clamped on its sides.

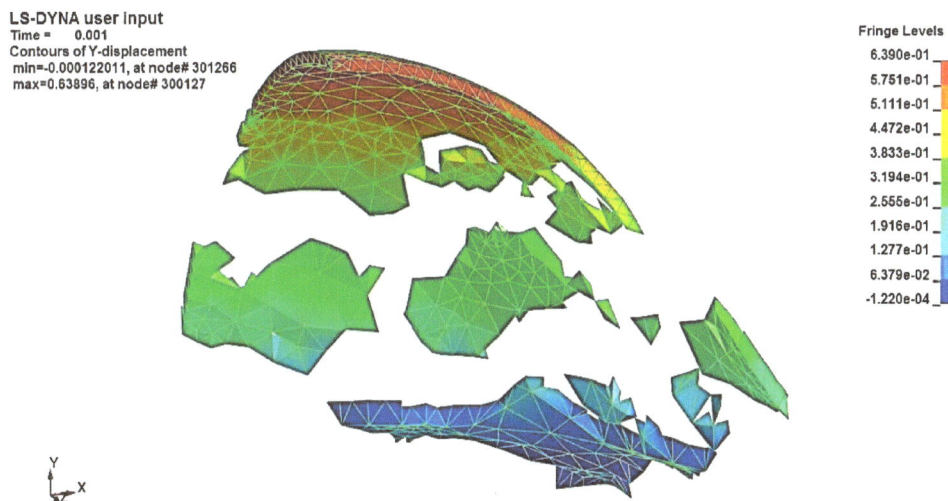

Figure 5: UY displacement for the plane armor plate

As is shown in the figures 5 to 10, the armor plate, for both constructive solutions, was broken by explosion, but some important parameters tell us which the constructive solution is better.

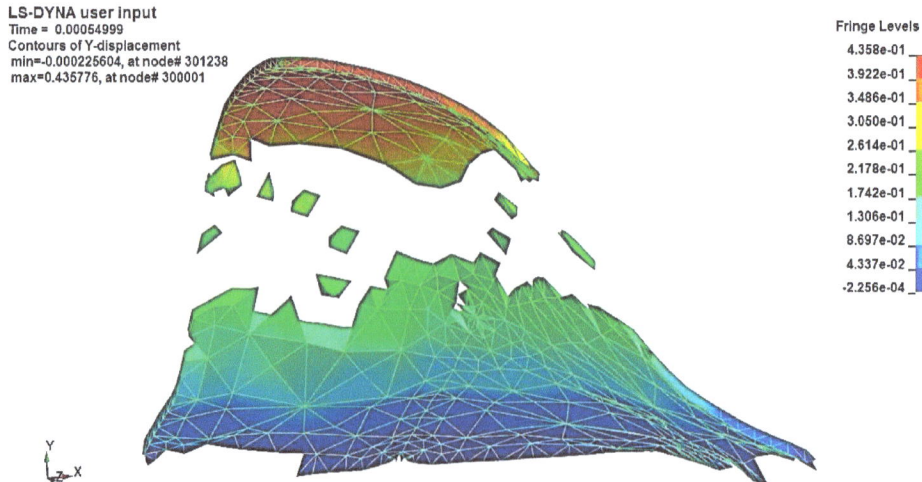

Figure 6: UY displacement for the pyramided armor plate

The studied parameters for a comparative analysis of the blast wave effect were: the maximum displacement (UY), the maximum velocity (VY), the pressure field, total energy of the plate, kinetic and internal energy etc.

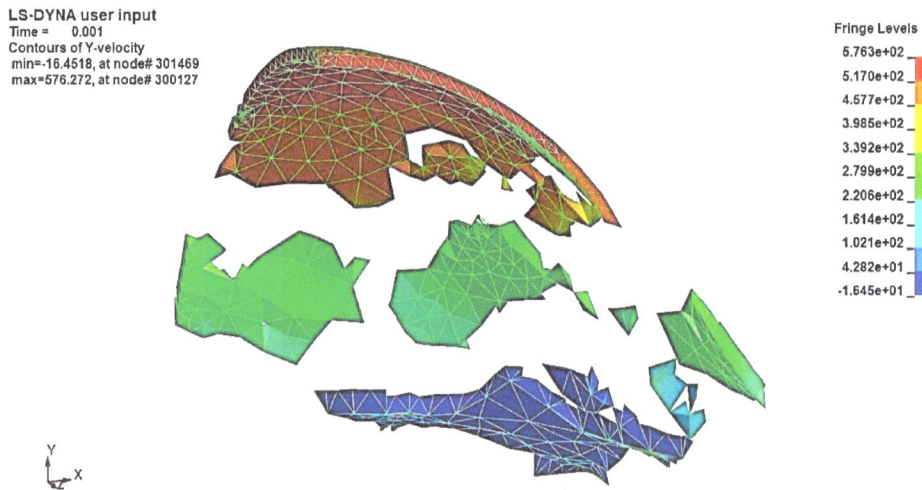

Figure 7: VY velocity for the plane armor plate

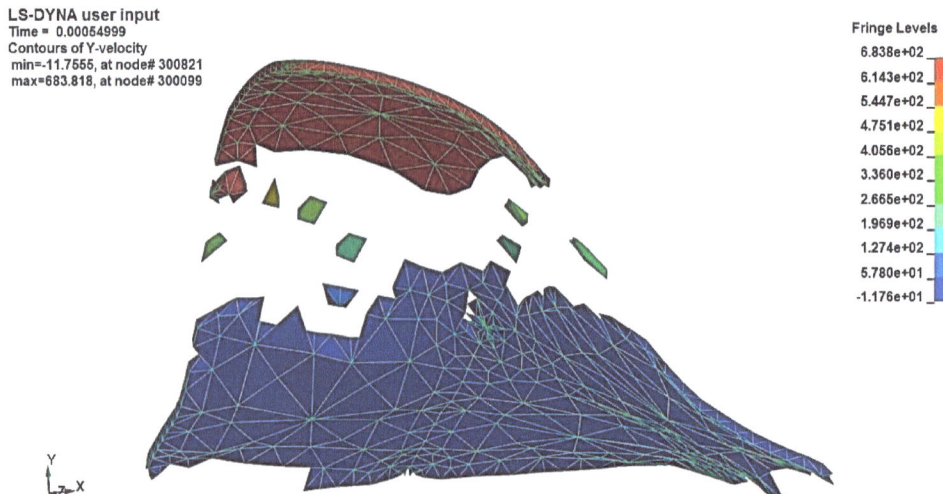

Figure 8: VY velocity for the pyramided armor plate

Our results consist in more analyzed parameters, but not of all are presented in this paper. For an easily comparing of the results, next to above presented parameters, only the pressure field is also presented in the following figures.

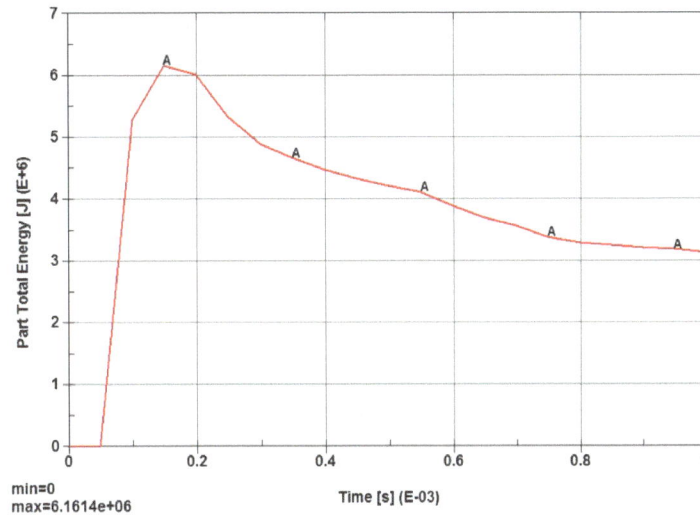

min=0
max=6.1614e+06

Figure 9: Total energy of the plane armor plate

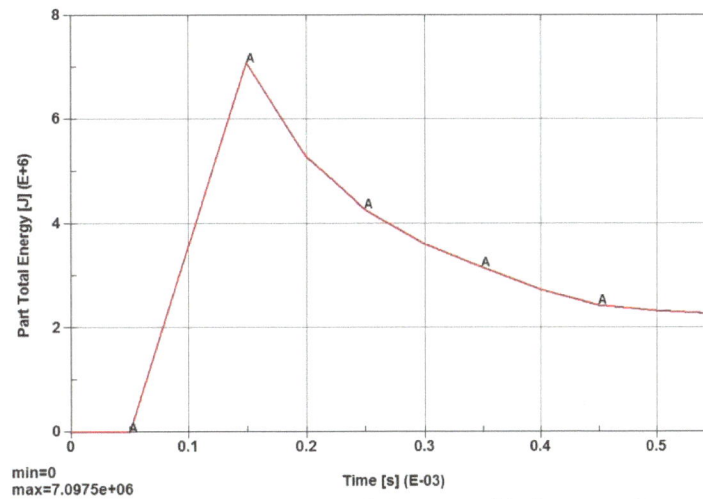

min=0
max=7.0975e+06

Figure 10: Total energy of the pyramided armor plate

5. CONCLUSION

The results presented in this paper clearly show that the pyramided armor plate can be o better constructive solution (comparatively with a plane armor plate) for the floor of the personnel armor vehicles and just for others parts of such vehicles. This conclusion is based on our numerical results and on our understanding of the phenomena. Therefore, if we calculate the armor plates mass, we will see that the pyramided armor plate is easier with 33%. The maximum UY displacement of the pyramided armor plate is less (31%) than maximum UY displacement of the plane armor plate, having a lower mass. This aspect can be explained by an increasing of the energy absorbing capacity (15%). The maximum VY velocity of the pyramided armor plate is greater (18%) than maximum VY velocity of the plane armor plate. The results seem to be very important and useful and we will continue our researching, including the optimizing the pyramid height.

REFERENCES

[1]   Benson, D., J., *A mixture theory for contact in multi-material eulerian formulations,* Comput. Meth. Appl. Mech. Eng. 140 (1997) 59–86

[2]   Kinney, G.,F., Graham, K., J., *Explosive Shocks in Air,* Springer-Verlag, New York, 1985

[3]   Martin Larcher, *Pressure-Time Functions for the Description of Air Blast Waves,* European Commission, Joint Research Centre, Institute for the Protection and Security of the Citizen, 2008

[4]   Meyer, R., Kohler, J., Holmberg, A., *Explosives,* 5th ed., Wiley-VCH, Verlag GmbH, 2002

[5]   Needham, C., E., *Blast Waves,* Springer-Verlag Berlin Heidelberg 2010

[6]   *** *LS-DYNA KEYWORD USER'S MANUAL,* Version 971, May 2007

THE MECHANICAL RESPONSE OF TEXTILE COMPOSITE MATERIALS TO DYNAMIC IMPACT TESTS

Vasile Ciofoaia[1]
[1] Transilvania University, Brasov, ROMANIA, ciofoaiav@unitbv.ro

Abstract: This paper presents an analysis of the results obtained from the dynamic impact of composite materials reinforced with woven fabrics. The specimens used were manufactured by reinforcing an epoxy resin with woven fabric EWR300 made of E-glass fibers.

The specimens were made though different manufacturing methods. The hand lay-up technology is used to prepare the specimens with different pressures (low and high pressure) in the molding step. The energy method and experimental tests are employed to determine the response of the composite material subjected to low velocity impact. The composite specimens were subjected to the flexural test (the three points method - Charpy test) and at the impact experiments of impactor in low velocity impact on the rectangular plates. The results obtained were compared taking into account the two kinds of manufacturing methods.

Keywords: fabric composite, manufacturing, dynamic test, mechanical properties.

1. INTRODUCTION

The composite materials are engineered materials made from two or more constituent materials with significantly different physical or chemical properties and which remain separate and distinct on a macroscopic level within the finished structure [9, 13]. Most of composite materials are anisotropic and heterogeneous. These two characteristics apply to the composite materials since the material properties are different in all directions and locations in the body. These properties are in contrast to those of any common isotropic material, such as for example steel, which has identical material properties in any direction and location in the body.

The common textile composites are classified according to the fiber architecture and include braided, woven, knitted stitch bonded and non-woven materials. The composite materials reinforced with woven fabric have complex structure and sophisticated micromechanical models and it is necessary to predict their elastic properties in all directions. Thus, the difficulty in analyzing the stress-strain relationship of composite materials becomes greater.

The composite materials reinforced with woven fabric have recently received considerable attention, due to their structural advantages of high specific-strength and high specific-stiffness as well improved resistance to impact, crash and fatigue [9]. When compared to unidirectional composites [13], the interlacing of fiber bundles in textile composites prevents the damage progression and hence provides an increase to impact toughness. Besides their advantageous mechanical properties, composite materials reinforced with woven fabric are easy to handle and have excellent formability and hence are widely employed in aircraft, boat and defense industry. Due to their high specific stiffness and strength, fiber reinforced polymer composite materials have long been used in the aerospace industry [1, 9, 12, 13, 14, 15].

Other notable engineering applications include pressure vessels and waste water pipes and fittings. The composites have the limitations of high cost, low damage tolerance and impact resistance. The increased application of composites has required a new method for predicting their elastic mechanical properties. For these reasons, in this paper, some structural properties of textile composites will be investigated.

Other papers [14, 15] discuss the developments in the modeling and characterization of fabric reinforced composite materials and structural components. The interlaminar mode I fracture toughness of wood laminated composite materials has been evaluated using the critical strain energy release rate associated with the onset of crack growth in double cantilever beam specimens (DCB) [2]. The structural optimization of composite materials loaded in an aggressive environment [3] and the water effects on composite made of E-glass fabrics woven fabrics was presented in [4,5,6,7]. The papers [8, 12, 14] present the computation of mechanical elastic constant of woven fabric composites. The goal of this paper is to determine the characteristics of composite materials reinforced with woven fabrics by using the experimental investigation developed in the dynamic impacts tests. The activities include prediction of the elastic properties of the specimens manufactured by reinforcing an epoxy resin with woven fabric EWR300 made of E-glass fibers and surface inspection of textile composites.

2. MECHANICAL BEHAVIOUR OF WOVEN COMPOSITE REINFORCEMENTS

The ballistic impact definition can be found in several works [1, 7, 10, 11, 12]. The term ballistic impact is used in the case of an impact resulting in complete penetration of the composite materials reinforced with woven fabric while non-penetrating impact referred to low velocity impact. Overall, other than this impact, stress wave propagation has no effect through the thickness of the laminate for the case of low velocity impact. As the projectile hits the target, the compressive and shear waves propagate outward from the impact point and reach the back surface, reflecting back afterwards. After several reflections through the thickness of the laminate, the plate motion is generated. The damage established after the plate movement is called low velocity impact [1]. However, there is also a threshold velocity which distinguishes low and high velocity impact. As implied by [1], 20 m/s is a transition velocity between two different types of impact damage and it allows a definition of high and low velocity impacts. The main focus of this paper is to study the response of composite materials reinforced with the woven fabric when impacted at low velocity by using experimental tests.

By Fracture Mechanics studies we determined the G_c, energy per unit area or the control energy release rate at crack initiation. The Charpy scheme has been adopted to measure G_c by using sharp notches and measuring energy at both low and high rates. The simple method of determining G_c is via fracture load F, and the specimen compliance C(a) via

$$G_c = \frac{F^2}{2b}\frac{dC(a)}{da} \qquad (1)$$

where b is the thickness and a is the crack length. If the elastic behavior is assumed, the load is related to the energy at the fracture U via $U=F^2C/2$ and hence G_c may be find from U,

$$G_c = \frac{U}{bC}\frac{dC}{da} = \frac{U}{bh\Phi(a/h)}, \qquad (2)$$

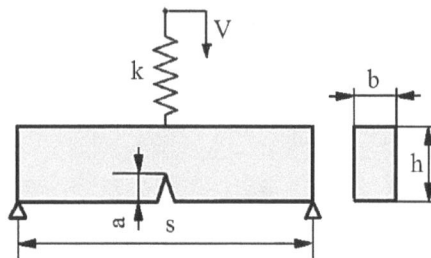

Figure 1 The Charpy three point bend test

where

$$\Phi(a/h) = \frac{C}{dC/d(a/h)}, \qquad (3)$$

is a calibration function for energies which may be deduced from C(a/h) and is evaluated for the Charpy three point bend specimen [12].

3. EXPERIMENTAL TESTS

In the impact experimental tests, the target was a square woven composite plate firmly clamped on the edges.. The specimens used were manufactured by reinforcing an epoxy resin with woven fabric EWR300 made of E-glass fibers. The hand lay-up technology is used to prepare the specimens with different pressures (low and high pressure) in the molding step. Using a digital microscope, we have captured pictures of the specimens made for trying to shock with Charpy pendulum, to analyze the structure of composite material manufactured by the two types of manufacturing technologies by pressing. To that end, it presents photos of composite structure of E Glass /polyester Colpoly 7233 manufactured by the method of the hand lay-up technology

- by pressing at low pressure (fig. 2);
- by pressing at high pressure (fig. 3).

Analyzing the figures 2 and 3, we can make a few observations:

- When using high pressure manufacturing tehnology (fig. 2), the reinforced glass layers are better consolidated, have better defined borders inthe digital photographs taken with the digital microscope.
- When using low pressure manufacturing tehnology, the cut specimens from composite plates (fig. 1) the glass layers are harder to identify even if the images are magnified with a zoom factor of 200 x (fig. 1, e și f).

Figure 2 The composite material Glass R/ polyester 7233 (low pressure manufacturing)

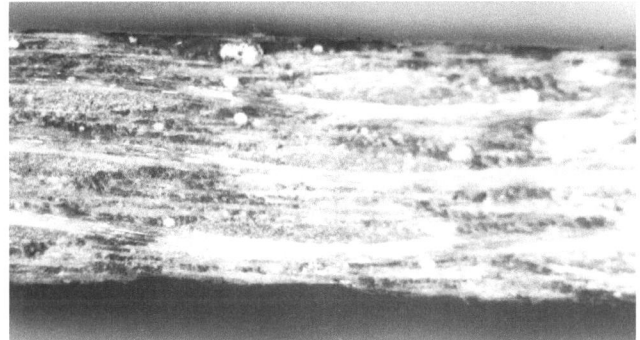

Figure 3 The composite material Glass R/ polyester 7233 (high pressure manufacturing)

The effects of the pressing manufacturing technology on the mechanical behaviour in an attempt to shock with Charpy pendulum were obtained using pieces of parallelepiped shape, sized 80 mm x 10 mm x 6 mm in accordance with European standards EN ISO 179-1 (2001) for the case of reinforced plastics.

Table 1 – The results of impact absorbed energy U for breaking the test-piece

Composite material/ The hand lay-up technology by pressing	Sample code epruveta	A [mm²]	U [J]	U/A [kJ/m²]	Composite material/ The hand lay-up technology by pressing	Sample code	A [mm²]	U [J]	U/A [kJ/m²]
E Glass/ Colpoly 7233 polyester / Low pressure (1)	P211	32.01	2.90	90.60	E Glass / Colpoly 7233 polyester / high pressure (2)	N46	34.22	4.13	120.69
	P212	35.28	3.91	110.83		N47	32.70	4.06	124.16
	P213	33.25	3.25	97.74		N48	34.50	4.20	121.74
	P214	35,36	3.42	96.72		N49	30.52	3.75	122.87
	P215	33.33	3.34	100.21		N50	32.48	3.79	116.69
	P216	34.30	3.44	100.29		N51	27.75	3.47	125.05
	P217	34,20	3.28	90.63		N52	29.00	3.63	125.17
	P218	32.10	3.12	97.20		N53	28.08	3.45	122.86
	P219	36.04	3.56	98.78		N54	30.16	3.78	125.33
	P220	27.84	2.98	107.50		N55	29.12	3.52	120.88
	Average value			99.50		Average value			122.54

The dimensions of the section were recorded for each specimen before the test of impact and we then computed the cross area. Test specimens were then subjected to the Charpy Test. The impact was generated by turning the pendulum hammer until the height h. When it was released, the hammer has described an arc, hitting the target sample, and after breaking the test-piece, reaching the rebound height h '. The difference between the initial potential energy and the potential energy of the impact is a measure of the energy required to break the test-sample. This quantity is called the energy bursting in the Charpy Test and is denoted by U. Were analyzed 10 samples made from two types of samples from different technology training by pressing (with the low pressure or high pressure) the two ways mentioned previously.

The results obtained in shock with Charpy pendulum have been summarized in table 1.

Analyzing the results, it follows that the resilience, the ratio between the breaking energy U and the cross-sectional area of the cut, is greater in the case of specimens made from composite material formed by high pressure pressing. In this case, the average value of resilience (impact resistance) was 122,54 kJ\/m2 (table 1), with 23,16% higher than the average value of 99,50 kJ\/m2 (table 1).

The second mode of testing was done on composite plates subjected to low velocity impact. The experimental evaluation of composite boards to conduct concentrated loads were studied on two types of standard specimens reinforced with glass fiber fabrics applied to concentrated loads that are applied dynamically.

The test-sample dimensions were 150 mm x 100 mm x 4 mm and the supports and the loading application were made as shown in Figure 1. The test samples were obtained by two different manufacturing processes mentioned above.

We then subjecting to resiliency testing the denoted 1-pieces numbered from 190-220 and the second groups of pieces, type 2 are numbered 30-50. For the application of the load has been used a device designed for that purpose, rendered in Figure 2.

The weight of the projectile was about 1.2 kg and the request was made at speeds ranging between 1 and 5 m/s. We have tested on at least two pieces for each of the selected drop height. We measured the height of fall and the acceleration during the impact. The integration was determined through variation of the speed and the movement during contact results and variance energy transferred to the plate. Figure 4 shows from the abutments and plates. Figure 5 presents the location of the accelerometer that able to transmit video signal for processing. The signal processing was performed with LabView. All graphics have been processed through Matlab code.

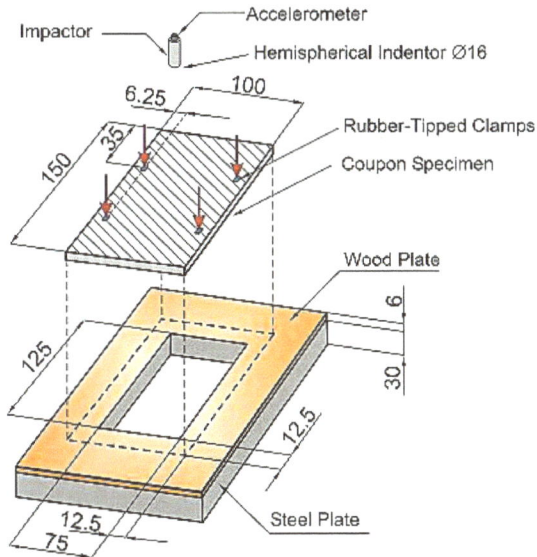

Figure 4 The supports of the test-piece and application of concentrated load

Figure 5 The positioning of the fixing screw the accelerometer

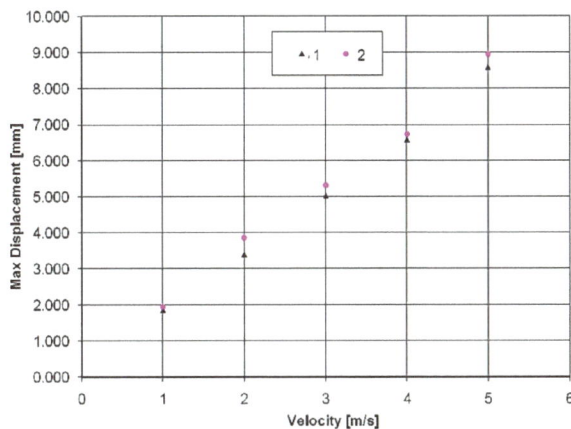

Figure 5. The average recorded displacements depending on the speed of the impactor for each type of test tube

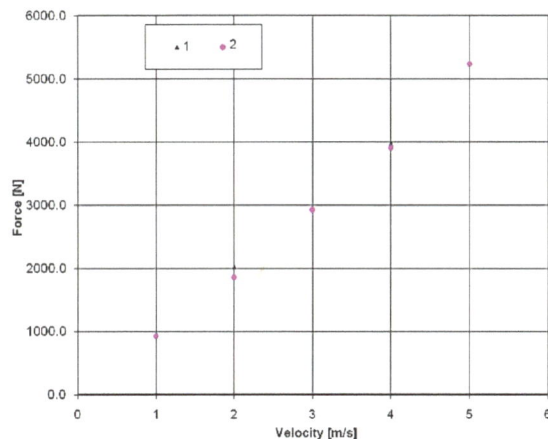

Figure 6. The average values of maximum contact forces recorded at the moment of impact depending on the speed of the projectile for each type of test specimen

Figure 7 Damage in the case of the test-piece introduced 209 (1 type) the impact of the projectile velocity of 2 m/s

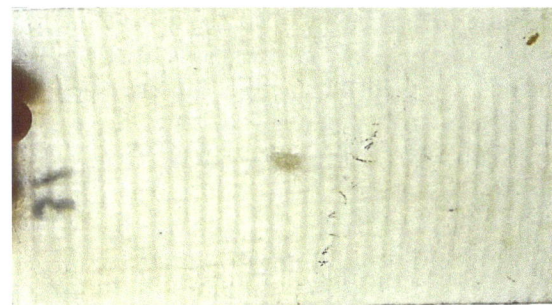

Figure 8 Damage in the case of the test-piece introduced 31 (2 type) the impact of the projectile velocity of 2 m/s

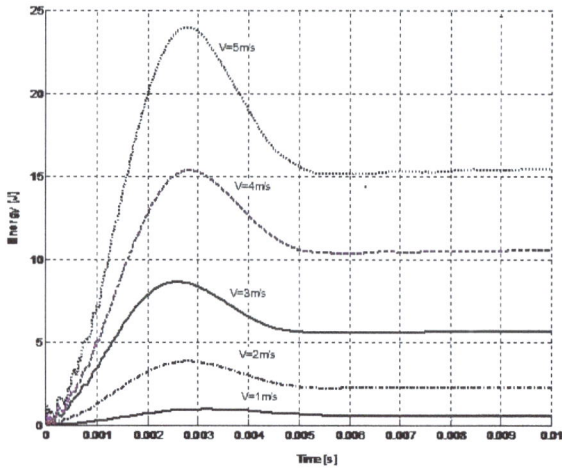

Figure 9 Variation of energy transferred to the moment of impact (pieces of type 1)

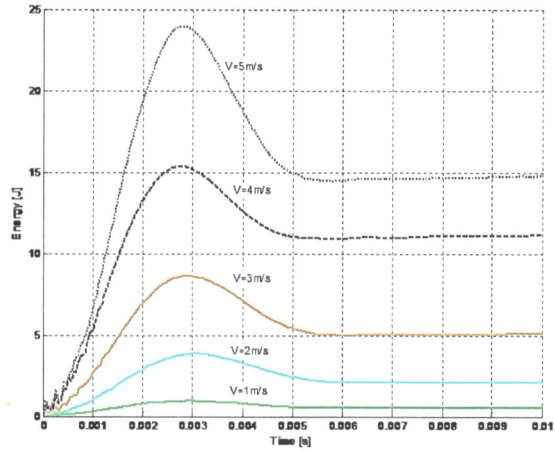

Figure 10 Variation of energy transferred to the moment of impact (pieces type 2)

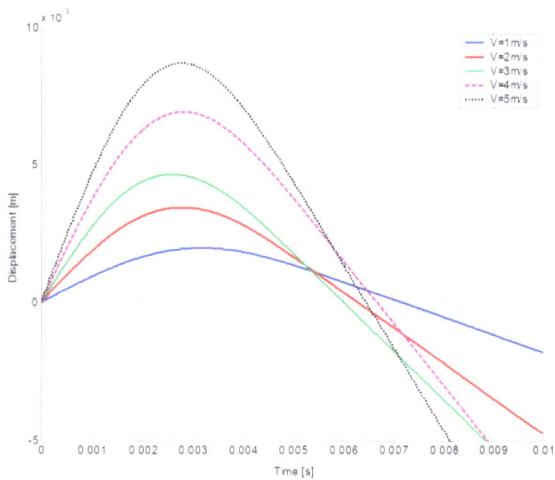

Figure 11 Travel-time curves recorded during the impact (pieces of type 1)

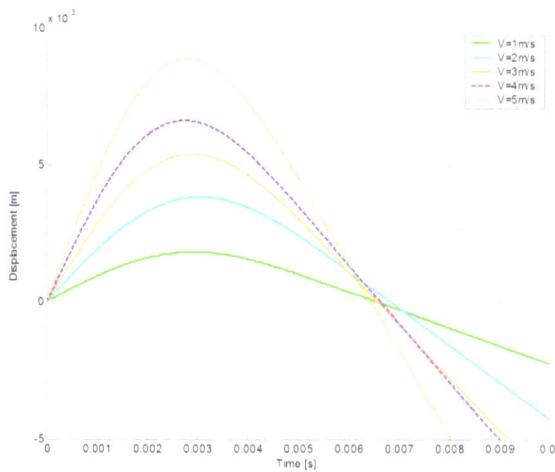

Figure 12 Travel-time curves recorded during the impact (pieces of type 2).

In figures 7, 8 are given recorded damage pieces of type 1 and 2, the initial impact with a speed of 2 m/s projectile. Between energy variations after time 0.005 s certain differences apeared. The projectile's weight remained constant and height of fall increased progressively (corresponding to initial velocity of 1m/s, 2m/s, 3 m/s and 5 m/s). At high velocities the contact force decreases and the composite plate starts to penetrate. If the speed of the project grows then it will grow and the maximum displacement of the leaf as shown in figures 11 and 12.

3. CONCLUSION

In this paper we tried to present the importance of the manufacturing of woven composite materials on the mechanical characteristics. Considering the obtained results presented in figures 5 to 12, we can draw the following conclusions:

- the resilience is greater in the case of specimens made from composite material formed by high pressure pressing. The average value of resilience (impact resistance) was 122,54 kJ\/m2 (table 1), with 23,16% higher than the average value of 99,50 kJ\/m2.
- type 2-pieces have been smaller displacements at approximately the same values of maximum contact forces during impact,
- visible damage occurred for initial projectile velocity of 2 m/s; In the case of type 1 specimens we observed occurrence of matte areas in which matrix has been damaged. Applied load determined the appearance of delaminations and radial fissures which start from the contact zone. In the case of type 2 specimens, the visible matte area corresponding to the matrix and the delaminations damage in the right point of application of the load, the radial cracks are very small.
- the damage incorporated was higher in the case of type 1 specimens; the matte area on the opposite face of the impact is greater in the case of these pieces, compared to that for the type 2 pieces.
- in order to increase the resistance of the composite materials, we could use an outer layer of sacrifice to allow the concentrated load applied to disipate, thereby reducing the generated damage.
- in case of low velocity (and, therefore, low-energy) impacts, strength reduction and eventual failures are dominated by inter-laminar debonding processes, customarily termed delamination.

REFERENCES

[1]   Abrate S., Impact on composite structure. Cambridge University Press, 2001

[2]   Baba, M. N. Dogaru, Fl. Curtu. I. Experimental determination of interlaminar fracture toughness of wood laminated composite specimens under DCB test. Materiale Plastice, 2010.

[3]   Cerbu Camelia, - Researches concerning structural optimization of some members made of composite materials loaded under aggressive environmental effects, Doctoral thesis, University "Transilvania" of Brasov, Romania, 17 decembrie 2005.

[4]   Cerbu Camelia., Ciofoaia V., Curtu I., Visan A., The effects of the Immersion Time on the Mechanical Behaviour in Case of the Composite Materials Reinforced with E-glass Woven Fabrics. Materiale Plastice, 46 nr.2 p.201-205, 2009.

[5]   Cerbu Camelia, Ciofoaia V.,Teodorescu_Draghicescu H., Rosca I.C., Water effects on the composites made of E-glass woven fabrics. Proceeding of International Conference Advanced Composite Materials. COMAT 2008, Brasov, 10-11 octombrie 2008, p.310-313.

[6]   Cerbu Camelia, Ciofoaia V., Curtu I., The Effects of the manufacturing on the mechanical characteristics of the E-glass/epoxy composites. Procceding of the 12th International Research / Expert Conference "Trends in the Development of machinery and associated tehnology" TMT 2008 Istanbul (Turkey) 26-30 august 2008p.229-232.

[7]   Ciofoaia V., Modelarea si simularea comportarii la factori mecanici si de mediu agresiv a materialelor compozite intarite cu textile. Project ID_191, no. UEFISCU 225/2007.

[8]   Ciofoaia V., Cerbu Camelia, Dogaru Fl., On the determination of the mechanical elastic constants of textile composite. COMAT 2010.

[9]   Choo V.K.S., Fundamentals of composite materials. Knowen Academic Press, Inc., Dover, Delaware USA, 1990.

[10] Dogaru Fl., Baba M.N., Analytical study of the CFRP laminated plates subjected to low velocity impact. Proceeding of International Conference Advanced Composite Materials Engineering "COMAT 2008"Brasov, 9-11 octombrie 2008, p.126-129.

[11] Dogaru Fl., Response of the CFRP rectangular plates subjected to low velocity impact. International Conference "Advanced Composite Materials Engineering" COMAT 2006 19 - 22

October 2006, Brasov, Romania.

[12] François D., Pineau A., From Charpy to Present Impact Testing. Elsevier, 2002.

[13] Gay D., Hoa S.V., Tsai S.W., Composite materials. Design and applications. CRC Press Washington D.C. 2003.

[14] Jones R.M., Mechanics of composite materials. Taylor & Francisc. Printed in USA, 1999.

[15] Morozov E.V., Mechanics and analysis of fabric composites and structures AUTEX Research Journal, vol.4, No.2, June 2004.

[16] Vasiliev V.V., and Morozov E.V., Mechanics and Analysis of Composite Materials. Elsevier Science, 2001.

MEASUREMENT OF COEFFICIENTS OF FRICTION OF AUTOMOTIVE LUBRICANTS IN PIN AND VEE BLOCK TEST MACHINE

Venetia S. Sandu

Transilvania University, Brasov, ROMANIA, e-mail venetia.sandu@unitbv.ro

Abstract : The paper deals with experimental mechanics in the field of fluid lubricants, investigating frictional characteristics of automotive engine and gear oils using experimental tests on a pin and vee block test machine (Falex). The test is standardized according to ASTM D3233 and represents a relevant and rapid measurement of coefficients of friction in a tribometer. Three engine oils and three gear oils were tested on the same loading-time procedure being analyzed the coefficients of friction according to SAE class of viscosity, degree of use (fresh or aged) and manufacturer. The calculations at different speeds and loads proved that coefficients of friction comply with the Stribeck rule. The research work succeeded to set up a procedure for operation on Falex machine, both for research and educational purposes.

Keywords: automotive lubricants, friction coefficient, tribology, pin and vee block test machine, Stribeck curve

1. INTRODUCTION

The road vehicles consume in average one-third of fuel energy to overcome friction which in case of passenger cars is divided four main contributors: engine (35%), transmissions (15%), tires (35%) and brakes (15%) [1]. The identification of friction loss sources as well as the methods to reduce them has a tremendious effect on global primary resources, especially on fuels and lubricants. Friction reduction methods are difficult to apply as operation modes, speeds, loads and working temperatures are variable in broad ranges.

The potential activities to reduce friction in engines and transmissions are related to type of lubricants and lubrication regimes, the expected benefits being reducing fuel and lubricant consumptions and emissions, wear, costs of maintenace operations and increasing energy efficiency and reliability.

Besides the improvement of surface quality (advanced coatings and special surface texturing), an important friction loss source is linked to the tribological characteristics of the lubricants. The objective of this paper is to investigate friction coefficients of engine and transmission lubricants in a special purpose tester and to compare load-carrying and friction reduction capability, the effect of viscosity and aging.

2. TESTING METHODS

The accurate measurement of friction reduction in real life condition is very complicated, time consuming and expensive, so most of the tribological measurements use special purpose tribometers which evaluate coefficients of friction within specific regimes of lubrication.

As a general rule, the lubrication should keep two surfaces in relative motion apart, for a minimum wear. A regime of lubrication depends on specific parameters of the application (load, speed, temperature, geometry of the bearing surfaces), but also on fluid lubricant viscosity and lubricity.

If it is considered as a parameter of the lubrication regime the lubrication oil film thickness, four different regimes of lubrication can be defined on Stribeck curve which impose different demands on lubricants: hydrodynamic lubrication (HL) is the one in which the relative motion of the sliding surfaces keeps a continuous fluid film separating the surfaces, elasto-hydrodynamic lubrication (EHL) in which there is a separation oil film, but elastic deformation of surface and oil viscosity are important, boundary lubrication (BL) when surfaces are in contact and chemical and physical properties of the film are dominant and finally, mixed lubrication (ML) is the lubrication characterised partly by direct contact of asperities and partly by EHL and BL.

The chosen tribometer and method should determine lubricant performance in the same lubrication regime as in real life and to measure the load-carrying properties and ability to protect against scuffing.

2.1. Tribometer

The Falex pin and vee block method is used to evaluate lubricant behavior in metal to metal applications, in standardized tests for measurement of extreme pressure [2] and wear properties of fluid lubricants [3], wear and load carrying capacity of solid film lubricants [4].

The equipment is described in figure 1. A test pin (journal) rotates at a constant speed against two vee blocks, all immersed in 60 ml of lubricant.

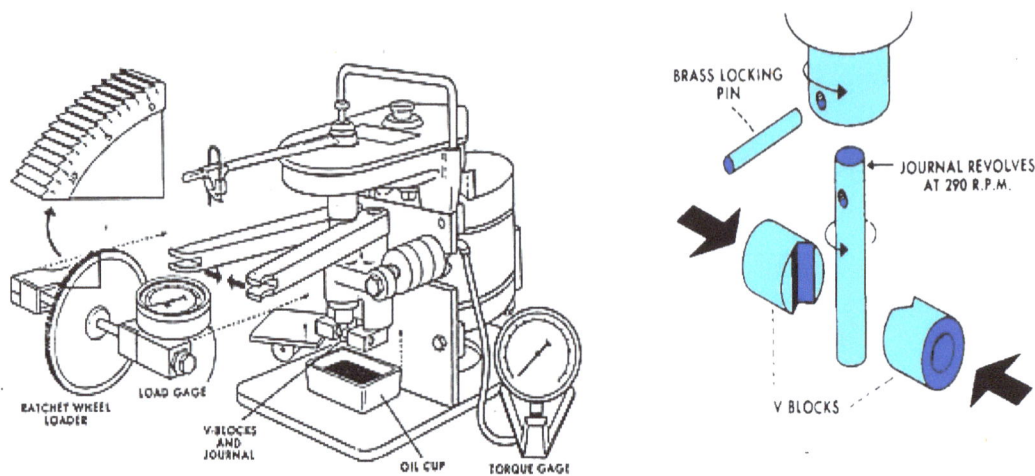

Figure 1: Falex pin and vee block test machine [2]

There are four contact lines between the pin diameter and the vee blocks when a load is applied through a mechanical gauge by means of a ratchet wheel and eccentric arm.

2.2. Lubricants

The lubricants supposed to be tested belong to commercial well-known manufacturers and are marked with M1, M2 and M3. Six lubricants were tested, three engine lubricants and three transmission lubricants, four were fresh, not used at all, and two were changed according to vehicle manufacturer recommendations.

For further discussions there were measured their densities and viscosities, which are presented in table 1, along with manufacturer code, class of viscosity and condition.

Table 1: Lubricants characteristics

Lubricant type	Class SAE viscosity	Density [kg/m^3]	Viscosity (20°C)[mm^2/s]	State
Engine	5W40M1	852	215.3	fresh
Engine	5W40M2	850	207.6	fresh
Engine	5W30M2	860	-*	aged, 10 000 km
Transmission	75W80M1	884	116.2	fresh
Transmission	75W80M1	888	-*	aged 30 000 km
Transmission	80W90M3	899	463.3	fresh

*The viscosity could not be measured with Gibson Jacobs method because the aged lubricant was opaque.

The density was measured by weighting a given volume of lubricant in a graduated cylinder, and then by dividing mass to volume. The viscosity was measured [5] on the hydrostatic test bench Cussons using the falling ball viscosimeter Gibson Jacobs, which measures the time in which a falling standardized ball travels a given distance in a graduated cylinder containing the tested lubricant.

The test was performed at the same temperature of the lubricants, at 20°C and kinematic viscosity v was calculated using equation (1).

$$v = \frac{d^2 g(\delta - \rho)F}{18v\rho} \tag{1}$$

with d - ball diameter [mm], g - gravitational acceleration [m/s^2], δ-sphere density [g/cm^3], ρ- lubricant density [g/cm^3], F- dimensionless correction factor, v- falling speed of the ball [mm/s].

2.3. Testing procedure and calculations

During the operation of the tribometer an external load is applied on the vee blocks upon the journal, all these parts being immersed in a lubricant bath. Mechanically, the direct load P_d, expressed in pounds or Newtons is decomposed in a normal force F_n and a tangential friction force F_f.

As the pin is driven at constant speed, a friction torque M_f is produced and the coefficient of friction between pin and vee blocks can be calculated based on equations of force and torque conservation, as illustrated in figure 2.

Applying the torque conservation upon the pin, it yields the following equations (2):

$$\sum M_O = 0 \, , \, M_f - 4F_f \cdot \frac{d}{2} = 0 \text{ and } F_f = \frac{M_f}{2d} \tag{2}$$

Applying the force conservation on x direction upon the vee block , it yields the following equations (3):

$$\sum F_x = 0 \, , \, -P_d + 2F_n \cdot cos\,45° = 0, \text{ and } F_n = \frac{P_d}{2\,cos\,45°} \tag{3}$$

Finally, the friction coefficient, μ, is calculated with formula (4):

$$\mu = \frac{F_f}{F_n} = \frac{\dfrac{M_f}{2d}}{\dfrac{P_d}{2\cos 45^\circ}} = \frac{M_f \cdot 2\cos 45^\circ}{2d \cdot P_d} = 2.9724\frac{M_f}{P_d}$$

(4) with M_f

expressed in inch-pounds and P_d expressed in pounds.

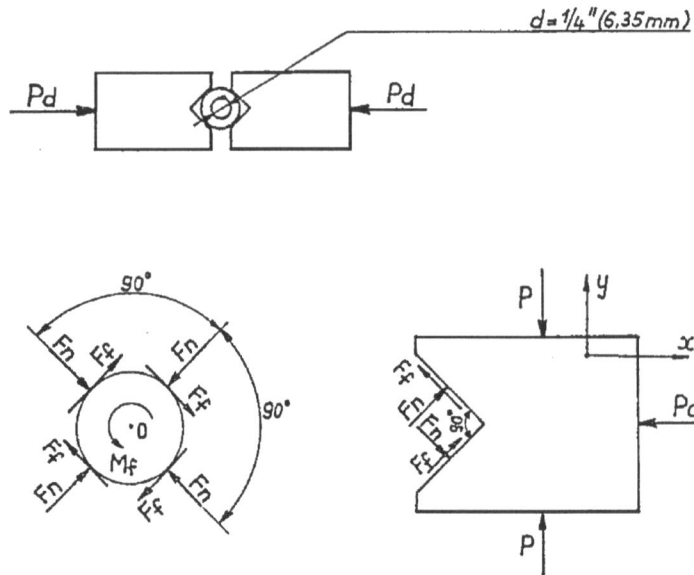

Figure 2: Scheme for the calculation of friction coefficient

Previous to load measurements, as the vee blocks are not perfecty flat, a run –in test was performed adapted from standard [2] ; the duration was of 15 minutes at the direct load applied of 300 lbs. The assessment of tribological properties is detailed in a program load–time considering similar tests performed in [6] for commercial engine oils,with the configuration of direct load P_d applied in time, according to figure 3.The test was repeated three times for each lubricant.

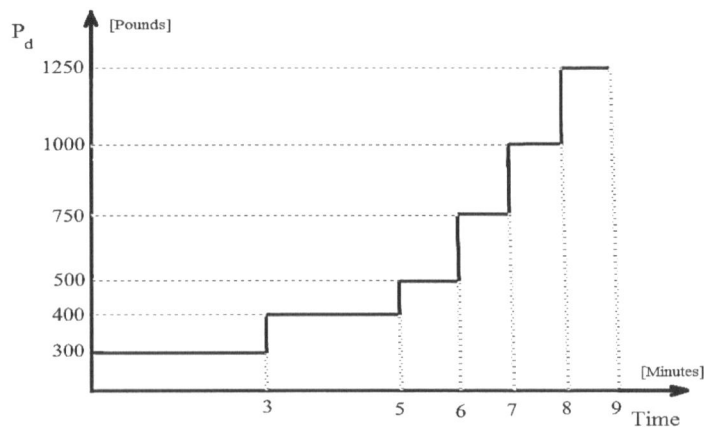

Figure 3: Load versus time - testing program

3. RESULTS AND DISCUSSIONS

3.1 Engine lubricants

The calculations of the coefficients of friction for engine lubricants were performed for the same loads and are illustrated in figure 4, having the loade expressed in pound-force and coefficient of friction being dimensionless.

The measured points were lined using third order polynomial regression. The variation of the coefficients of friction with load is similar for two fresh oils, the tribological behavior being better for manufacturer M1 than M2. Having a lower viscosity at ambient temperature, the lubricant 5W30, due to aging and contaminants has an increased coefficient of friction at higher loads.

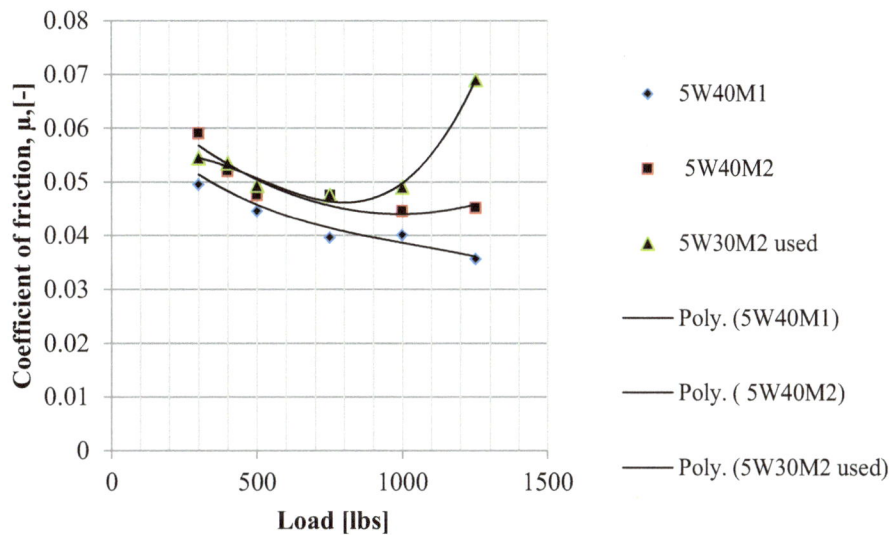

Figure 4: Coefficient of friction versus load for engine lubricants

3.2 Transmission lubricants

The coefficients of friction for transmission lubricants were calculated for the same loads as in previous case being illustrated in figure 5. The values are close to those of engine lubricants, in the same range (0.035 - 0.07). The coefficients of friction have a small increase with the class of viscosity, the increase of friction being higher for used lubricants, having the same behavior as that of engine lubricants.

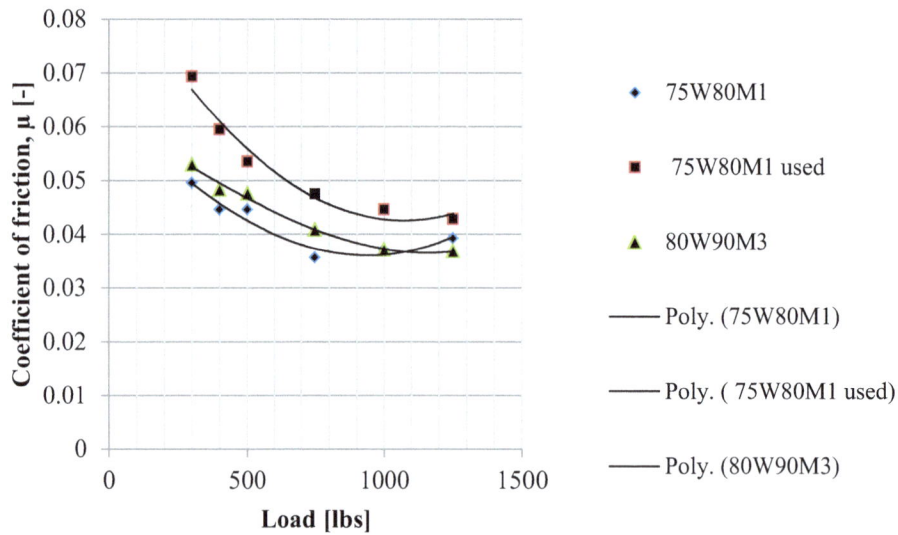

Figure 5: Coefficient of friction versus load for transmission lubricants

3.3 Failure load test

It is important to measure the failure load in the test, which is defined as the minimum load at which occurs a welding within test components. As a result the coefficient of friction and friction torque increase sharply. At that moment the pins and vee blocks are broken or damaged, as seen in figure 6. The lower value of the load at which the lubricant withstood represents the limit load carrying.

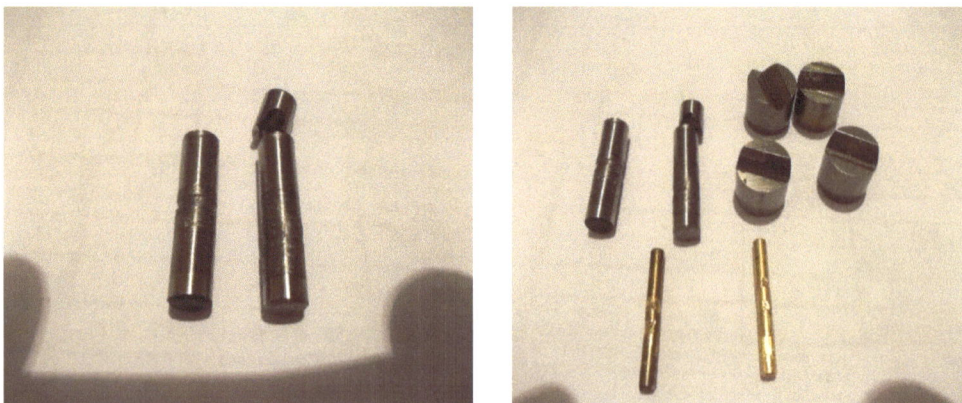

Figure 6: Scuffed pins and vee-blocks

The values of the failure loads for engine lubricants 5W40M1, 5W40M2 and 5W30M2 were 1500 lbs at a torque of 26 in.lbf.

3.4 Discussion on Stribeck curve

The tribological properties of fluids are widely represented in form of a graphical dependence known as Stribeck curve [7,8]. The variables are dimensionless, coefficient of friction μ and Stribeck parameter S, the latter being defined according to formula :

$$S = \frac{\eta \cdot v}{P_d} \tag{5}$$

with η - dynamic viscosity of the tested lubricant, v - relative speed of the pin and vee blocks, P_d - direct applied load reported to unit length of contact lines between pin and vee blocks.

For one of the tested lubricants, namely 75W80M1 transmission oil, there were calculated the position of the measurement points on Stribeck curve. The dynamic viscosity was calculated considering the lubricant density and its kinematic viscosity at temperature of the oil in testing cup, the constant rotational speed of the tribometer was turned into tangential speed, by multiplication with the radius of the pin, r. Finally, the Stribeck parameter becomes inverse proportional to direct load P_d and the correlation with coefficient of friction is illustrated in figure 7.

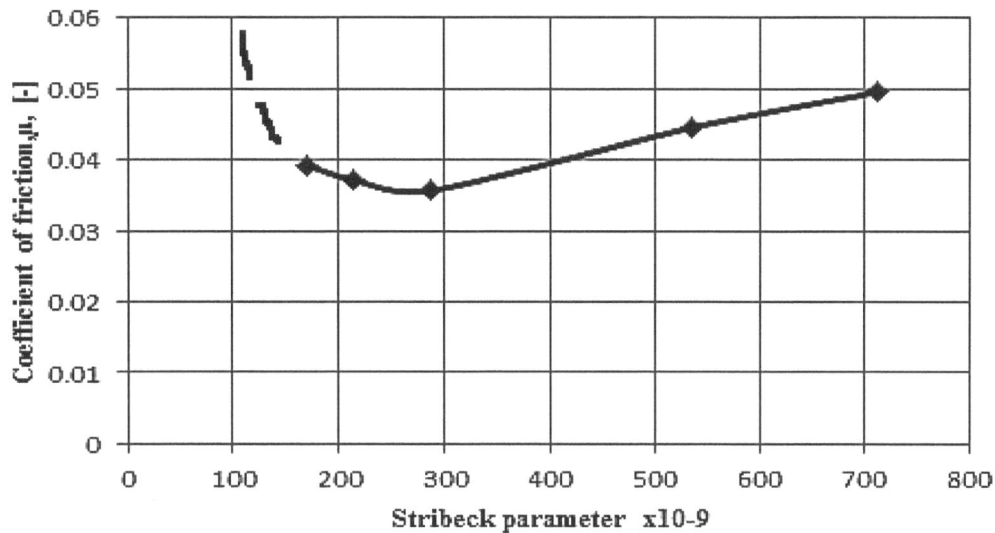

Figure 7: Stribeck curve profile for transmission lubricant

That curve expresses the friction variation within all types of lubrication areas, aforementioned described (BL, ML, EDL and HL).The left dotted line represents the area of BL and ML and the right continuous line represents the area of EDL and HL. For the transmission lubricant, the hydrodynamic lubrication begins in the point of the lowest friction coefficient when the oil film supports completely the carrying load.

CONCLUSIONS

The investigation of engine and transmission lubricants revealed that the values of coefficients of friction are close and for both types, the higher the viscosity class, the higher are the coefficients of friction.

More important than the effect of the viscosity is the effect of aging and contamination which increased considerably the coefficients of friction of both lubricants. The coefficients of friction met the profile of Stribeck curve, being identified the areas of mixed lubrication and hydrodynamic lubrication.

The work has also an educational gain, succeeding to implement a procedure for laboratory work in tribology of automotive lubricants, at Transilvania University.

ACKNOWLEDGEMENTS

The author would like to thank to Master student Andrei Preda for his assistance and support with the research facilities.

REFERENCES

[1]   Holmberg K., Andersson P., Erdemir A., Global energy consumption due to friction in passenger cars, Tribology International, Volume 47, 2012, p. 221–234.

[2]   **** ASTM D3233 – 93 (2009) , Standard test methods for measurement of extreme pressure properties of fluid lubricants (Falex Pin and Vee Block Methods).

[3]   **** ASTM D2670 – 95 (20109) , Standard test methods for measurement wear properties of fluid lubricants (Falex Pin and Vee Block Methods).

[4]   **** ASTM D5620 – 94 (2004) , Standard test methods for endurance (wear) life and load carrying capacity of solid film lubricants (Falex Pin and Vee Block Methods).

[5]   Gibson W., Jacobs L., The falling sphere viscosimeter, Journal of Chemical Society Transactions, Volume 117,1920, p.473-478.

[6]   Industrially Relevant characterization of Adhesion or Wear Resistance by Adapted ASTM methods, http://www.bluesphere.be/images/sites/114/editor/files/WhitePaper_2010-1_Pin_VeeBlockTestsCoatings.pdf

[7]   Brandao, J.A. et al., Comparative overview of five gear oils in mixed and boundary film lubrication, Tribology International, Vol.47, 2012, p. 50-61.

[8]   Maru, M. et al., Assessment of the lubricant behaviour of biodiesel fuels using Stribeck curves, Fuel Processing Technology, Vol.116, 2013, p. 130-134.

THE INFLUENCE OF PROCESS PARAMETERS ON MECHANICAL EXPRESSION OF SUNFLOWER OILSEEDS

A. O. Arişanu[1], Fl. Rus[1]

[1] Transilvania University of Brasov, Brasov, ROMANIA, e-mail: arisanu_ov@yahoo.com

Abstract: *Separation of oil from oilseeds is an important processing operation. The process employed has a direct effect on the quality and quantity of protein and oil obtained from the oilseeds. Basically, two methods are used for this purpose. One is the solvent extraction method in which a solvent, when brought in contact with the preconditioned oilseed, dissolves the oil present in the seed and the separated mixture is later heated to evaporate the solvent and obtain the oil. The other method used involves mechanical oil expression. In this process, the preconditioned oilseed is passed through a screw press where a combination of high temperature and shear is used to crush the oilseed to release the oil. The objective of this study was to investigate several process parameters (pressure, time, temperature and moisture content) and their influence on mechanical expression of sunflower oilseeds.*

Keywords: *sunflower oilseeds, mechanical expression, process parameters, oil yield, oil quality*

1. INTRODUCTION

Worldwide, especially in industrialized countries, there is increasing trend in human consumption of vegetable oils, which can be explained by the advantages they pose in terms of food, compared with animal fats:

- are more easily assimilated (predominantly unsaturated fatty acids as to the saturated ones);
- are nutritionally superior due to the presence of polyunsaturated fatty acids;
- contain less cholesterol for the human body;
- are more suitable for food products (mayonnaise, sauces, dressings etc.).

Obtaining quality vegetable oils greatly depends on the physical, chemical and biological characteristics of the raw materials, as well as on the extraction procedure. Although numerous experiments are being conducted with the aim of developing the vegetable oils extraction method based on the use of supercritical fluids, at industrial level, in Romania, the separation of vegetable oils from oleaginous raw materials is being achieved by two procedures: mechanical expression and solvent extraction, which can be applied independently or successively, depending on the oil content of the oleaginous raw material and the desired extraction degree [1].

The issue of vegetable oils extraction from oleaginous raw materials is quite topical nowadays, being a major concern both for specialists, who are permanently working on finding solutions intended to improve this process (increasing extraction efficiency, reducing energy consumption, introducing various environmentally friendly technologies), but in the present situation, when the humanity is faced with a serious economic crisis, potential consumers seem to grow more and more concerned about this issue [1].

2. MATERIALS AND METHODS

The oleaginous raw materials are numerous and most varied. Of more than 110 species of oleaginous plants, on the world market there are presently about 50, grouped in 15 important botanical families [3], [9], namely: compositae (sunflower), cruciferae (rape), leguminous plants (soya), malvaceae (cotton), papaveraceae (poppy), rozaceae (almond tree, hazel tree), peduliaceae (sesame), vitaceae (grape seed), jugladaceae (nut tree), palmae (oil palm, coconut palm, palm kernel), foleaceae (olive tree), linaceae (flax), cucurbitaceae (pumpkin seeds) leufobiaceae (castor oil plant) and solanaceae (tomato seeds, tobacco seeds) [9].

Due to their particular importance, oleaginous plants are being grown worldwide, the extent of each culture depending on the geographical area. Thus, if on a worldwide scale the palm tree holds the top position among the oleaginous raw materials, with 28.6% of the world vegetable oil production, in Europe, sunflower is ranked first, with 34.1% of the oil production, being closely followed by rape, with a share of 33.3%. In Romania, the mainly grown oleaginous plant is sunflower, with 76% of the domestic vegetable oil production.

As far as the areas under cultivation are concerned, in 2013 Romania was ranked first among the EU member states; however, the average yield per hectare remained by approximately 12% lower than the means on record in the other states of the European Union [1].

Considering the exceptional dietary qualities and the numerous industrial uses [3], [9], [12] of sunflower oil, and on top of that the prospect of using it on a large scale, together with the rapeseed oil, as a biofuel – biodiesel to be more precise [12], the materials which constitute the subject matter of this paper are the oleaginous sunflower seeds.

Sunflower seeds are achenes of variable dimensions (5...26 mm in length, 3...10 mm in width and 2...6 mm in thickness), generally in an elongated shape, pointed at the end attaching to the head. The chemical composition of the achenes is set forth in table 1, and they consist of a pericarp (hull), with a weight of 14...28% of the total weight of the seed (Table 2), of a ligneous consistency, ashy, white, black-coloured or striped, and the oleaginous kernel, which can store up to 60...65% oil content [2].

As may be seen from the data contained in table 1, besides the high oil content, oleaginous seeds also contain significant quantities of protein substances, which is why sunflower is classified as a high-protein oil content plant.

Table 1: Chemical composition of sunflower oleaginous seeds [2], [3], [5], [9], [12]

Designation	Value
Humidity, %	9...11
Raw oil, %	44...48
Gross proteins , %	18...20
Non-nitrogenous extractive substances, %	10...15
Cellulose, %	14...18
Ash, %	2...3

Table 2: Main characteristics of sunflower oleaginous seeds [2], [9]

Designation	Value
Hectolitre weight, kg/hl	38...42
Hull content, %	14...28
Equivalent diameter, mm	5.64
Density, kg/m^3	730
Bulk density, kg/m^3	438
Porosity index, %	40

3. RESULTS AND DISCUSSIONS

To separate the vegetable oil from the sunflower seeds, the combined process is almost always used in industry practice, namely the pressing of the oleaginous material, which enables an oil separation of up to 80...85%, followed by solvent extraction, whereby the remaining oil is being separated (99...99.5%) [3], [9].

The combined procedure (expression-solvent extraction) is applied in the processing of oleaginous raw materials with a minimum oil content of 30%. Raw materials with lower oil content are directly processed by solvent extraction, as the low yield of their expression does not justify the costs generated by this method [9].

Depending on the characteristics of oleaginous seeds, the degree of equipment of the processing facilities and the desired extraction degree, there are 5 main categories of oleaginous material which can be subjected to extraction by pressing: oleaginous seeds as such (having reached technological maturity, shelled and dried but unshelled, ground or hydrothermally processed), ground unshelled oleaginous seeds, ground unshelled but hydrothermally processed oleaginous seeds, ground shelled oleaginous seeds, ground shelled and hydrothermally processed oleaginous seeds [1].

Considering the above, we may note that the grinding of oleaginous seeds is an omnipresent operation in the vegetable oil processing technologies (except for the first situation, very rarely met in current industrial practice on account of the low quality of the obtained oil). As regarding the hulling of the oleaginous seeds and the hydrothermal processing of the ground seeds, although these perceived as optional operations, both the quality of oil and that of the resulting seed meal, as well as the enhancement of extraction yields, greatly depend on their application [4], [5], [12].

Due to their chemical composition characterized by a low botanical oil content (0.5...6%) and a high cellulose content (up to 60% in the case of sunflower seeds), the hulls of oleaginous seeds are an inert material in processing and unwanted in the composition of the seed meal. Therefore, as far as this is possible, hulls are partially eliminated by oleaginous seed shelling or hulling (a certain percentage of hulls – approximately 8...10% in the case of sunflower seeds – is not removed from the hulled material because it ensures optimal conditions for grinding and pressing). Considering that the hulling of oleaginous seeds involves two steps: cracking and separating the hull from the kernel, and separating the hulls from the resulting mixture, respectively, in order to obtain satisfactory results both from a technological and an economical point of view, only those oleaginous seeds with a high content of hull, which is loosely attached to the kernel (sunflower, soya, castor oil seeds, cotton seeds, etc.) shall be put through a hulling process.

In the case of sunflower seeds, the main advantages derived from the hulling thereof lie in the:

- improvement in the quality of seed meal, by reducing the cellulose content and increasing the protein content: the seed meal obtained from unhulled sunflower seeds contains approximately 25% protein substances and 25...28% cellulose, that obtained from partially hulled seeds (10...12% hull remaining in the hulled material) contains 35...37% protein substances and 18% cellulose, while the seed meal obtained from mostly hulled seeds (6...8.5% hull remaining in the hulled material) contains 40...42% protein substances and 12...14% cellulose [3];
- increase in the processing capacity of the grinding rolls, of the hydrothermal processing facilities, of the pressing equipment, as well as of the extractors;
- reduction of equipment wear, especially on the grinding rolls and presses, given that the sunflower seed hull contains silicon dioxide, which is an abrasive material;
- reduction of oil losses occurring during the expression and solvent extraction processes;
- reduction of the wax content in the raw press oil (waxes, due to their high melting point, give to the oil a specific cloudiness – white sediments on the bottom of containers – for which reason these are being removed during the refining process through winterization);
- salvage of hulls and their use as fuel, in the manufacture of various products (furfural) or as an ingredient in forage for ruminants (ground hulls easily absorb molasses) etc.

The hydrothermal conditioning of the ground oleaginous seeds is an integral part of the material preparation operations preceding oil extraction by pressing and sometimes even solvent oil extraction, and it involves the performance of the following 2 operations:

- moistening the ground seeds up to an optimal humidity, specific to each oleaginous raw material;
- heating and drying the ground seeds until obtaining a specific cellular structure, enabling an easier oil separation during the process of extraction by pressing.

During the first stage, simultaneously with the feeding of the moistening agent (water spraying or steam injection in the hydrothermal processing system) the ground oleaginous seeds are heated (in order to stop the enzymatic activity, favoured by the presence of water and who could cause and increase in oil acidity), operation which is continued in the second stage, during which the feeding of moistening agent is stopped.

Two phase result from the moistening of the ground oleaginous seeds: a solid phase (gel), consisting mostly of protein substances, with a marked absorbent character and a liquid phase, consisting of oil and water. Both components of the liquid phase are moistening the solid phase yet, given the fact that the superficial tension of water is higher than that of oil, it exerts a better moistening action, for which reason water intervenes between the solid phase and oil, thus reducing the forces retaining oil at the surface and in the open capillaries of the ground seed particles. At the same time, when soaked in water, the solid phase (proteins) increase their volume thereby causing a reduction in the diameters of capillaries and micro-capillaries of the ground seeds and thus forcing oil to surface. During the second stage, the heating of the ground seeds determines a drop in the superficial tension of oil but also a reduction in its viscosity, which makes it easier for the oil to be released from the closed capillaries of the ground seeds during the pressing process [12].

Although the hydrothermal processing of the ground oleaginous seeds is an extremely complex operation, which needs to be carried out under controlled conditions, through its application, due to the physical, chemical and structural transformations occurring in the oleaginous material, it becomes fit for extraction by pressing (forming an optimal cellular structure permitting to achieve maximum yields of oil on pressing).

After the grinding of the hulled oleaginous seeds and the hydrothermal processing of the resulting ground seeds, the technological process of vegetable oil extraction from sunflower seeds continues with the pressing of the hydrothermally processed oleaginous material. For that purpose, the oleaginous material heated at temperatures comprised between 60 and 95 Celsius degrees (in the case of sunflower seeds) can be transferred from the hydrothermal processing system directly inside the feeding vat of the pressing equipment.

The results of the experimental researches have pointed out that in this temperature range (60...95 Celsius degrees), a maximum extraction degree can be attained (Figure 1, 2, 3) and at the same time the obtained oil is of the highest quality, for which reason the ground oleaginous seeds are transferred from the hydrothermal processing system directly to the pressing equipment without any intermediate processing [5], 12]

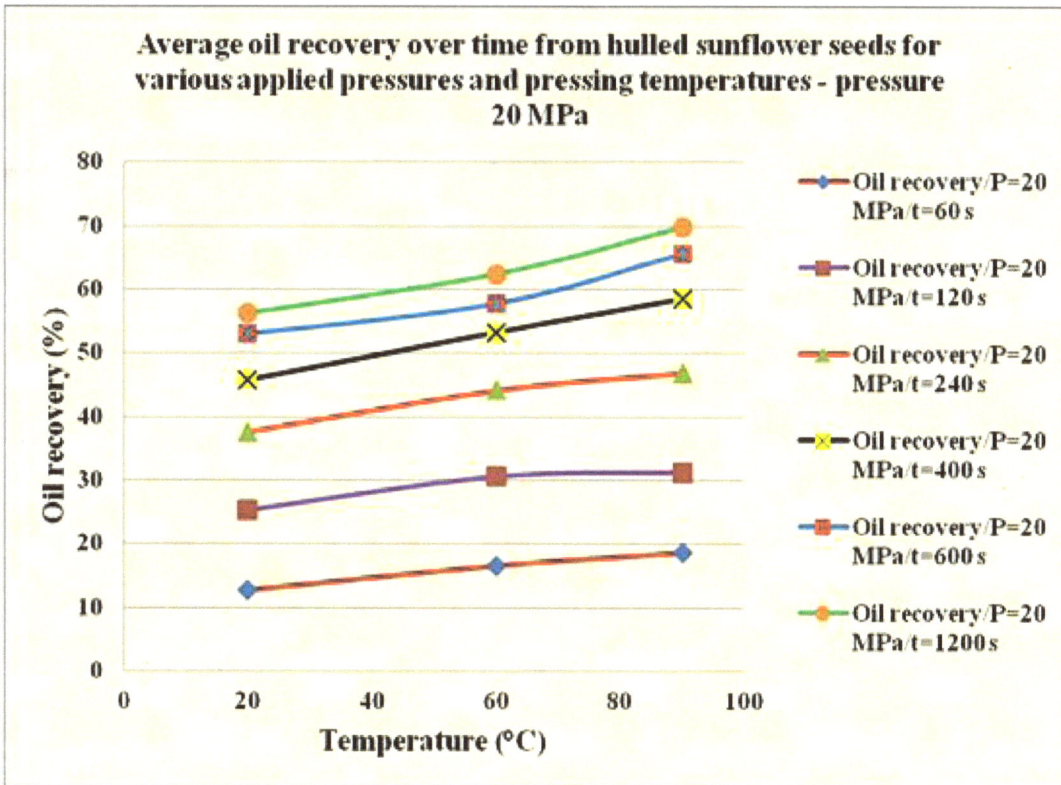

Figure 1: Average oil recovery over time from hulled sunflower seeds for various applied pressures and pressing temperatures – pressure: 20 MPa

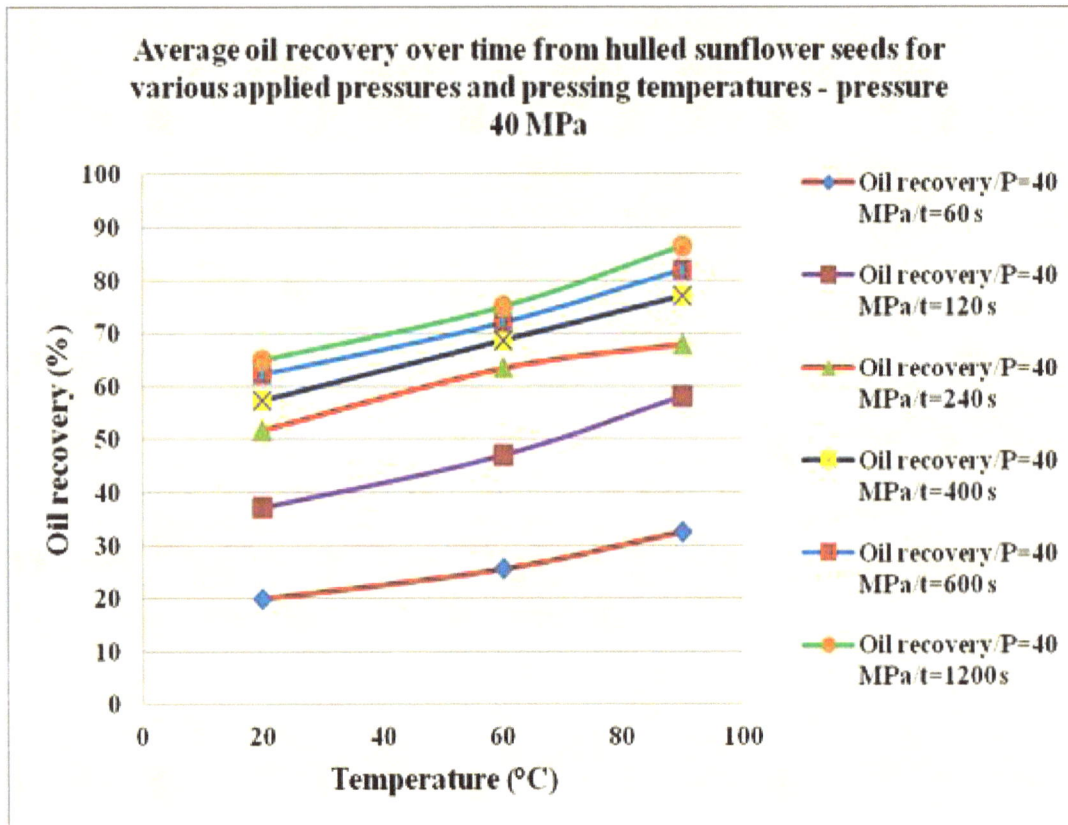

Figure 2: Average oil recovery over time from hulled sunflower seeds for various applied pressures and pressing temperatures – pressure: 40 MPa

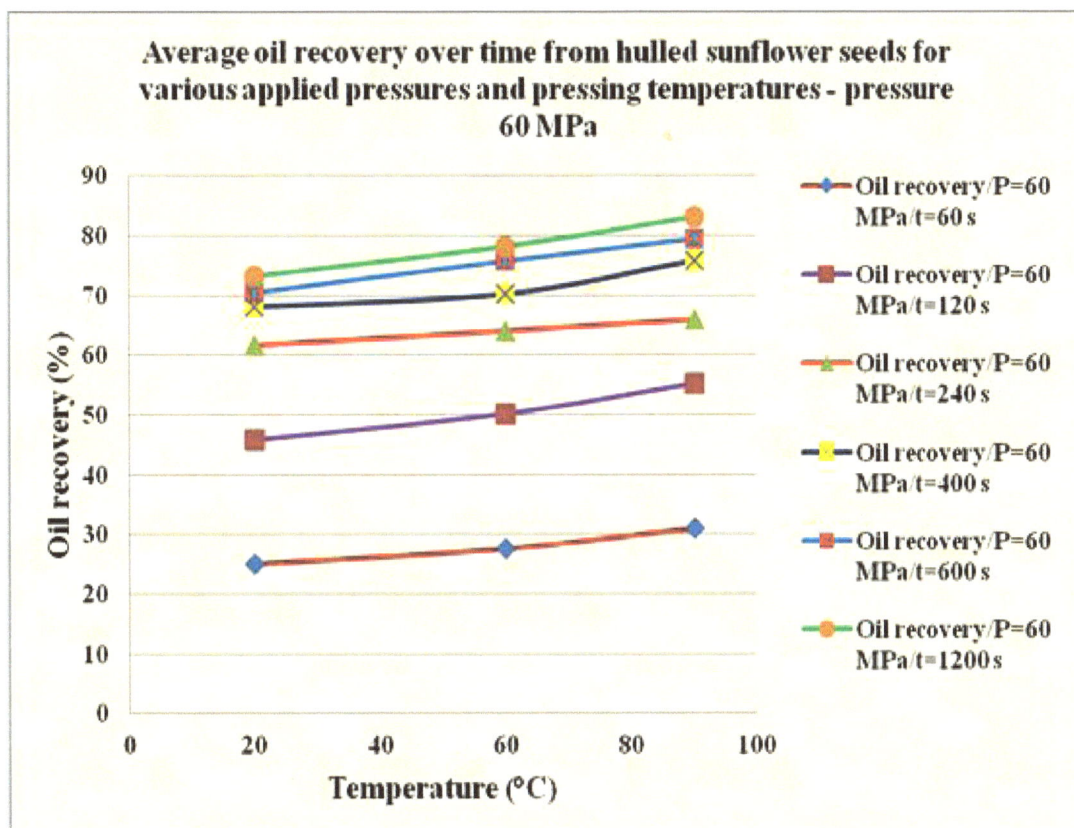

Figure 3: Average oil recovery over time from hulled sunflower seeds for various applied pressures and pressing temperatures – pressure: 60 MPa

Following the introduction of the oleaginous material into the pressing chamber (Figure 4), the first thing that occurs is the separation of oil (over a short period of time) without any exterior action, only through the effect of the gravitational field and of the pressure of material layers. This first phase unfolds as a sheer filtering process under the influence of a hydrostatic pressure. The actual pressing process is achieved through the action of an active organ (piston in our case), which initially achieves a compression of the oleaginous materials aimed at eliminating air pockets by evacuating the air existing between the particles of ground seeds. This is followed by the separation of the oil kept on the surface of the particles due to the surface forces of the molecular field, through the channels formed between the particles [4], [5], [8], [11].

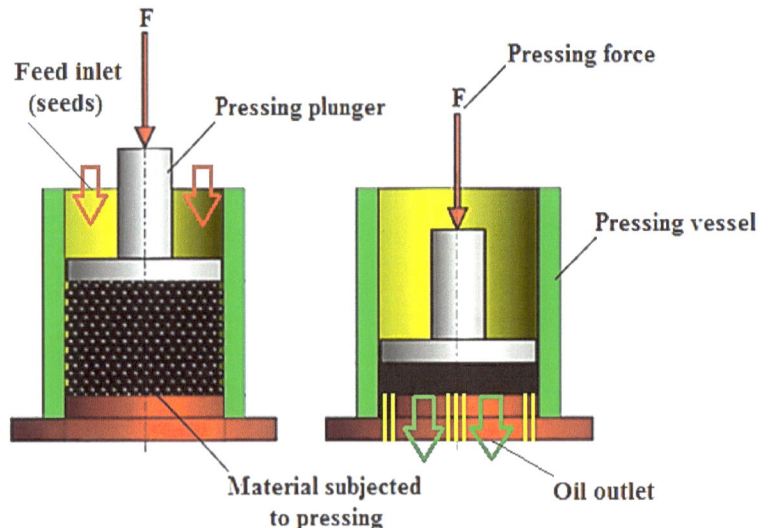

Figure 4: Schematic representation of the pressing process (hydraulic presses) [10]

The increase in the pressing forces engenders a decrease in particle volume, which causes the oil to be eliminated from the particle capillaries, at the same time as the separation of the oil existing on the surface thereof. The increase in the pressure exerted on the ground oleaginous seeds needs to be gradual, so that the finely ground particles do not obstruct the capillaries thereby blocking oil evacuation [11].

When the space between the surfaces of two particles becomes so narrow that the oil film is subjected to the retention forces exerted by both particle surfaces, the oil cannot be eliminated anymore, the film breaks in several places, the surfaces hit one against the other and the so-called oilseed cakes are formed.

4. CONCLUSIONS

Sunflower oil is an excellent edible oil which has come to be more and more appreciated in modern dietetics due to its high unsaturated fatty acids content (85...91%) mostly represented by the oleic and linoleic acid (up to 65%), one of the essential nutritive fatty acids. Unlike the other vegetable oils, sunflower oil ideally combines the high nutritive value with stability and a long shelf-life, owing to the absence of the linolenic acid. From this perspective, no other vegetable oil can stand comparison.
The results of the experimental researches have pointed out that in this temperature range (60...95 Celsius degrees), a maximum extraction degree can be attained and at the same time the obtained oil is of the highest quality.

Considering the structure of the ground and thermally processed oleaginous material and the manner in which oil extraction is achieved, the pressing may be defined as the physical process of partial separation, under the action exerted by outer forces, of the liquid phase (oil) from an heterogeneous solid-liquid mixture (ground oleaginous seeds). The essential requirement to be met by the materials to be put through the pressing process is that the skeleton of the solid substance of the phases system be compressible and that draining capillaries be formed, enabling the passing of the liquid phase.

ACKNOWLEDGEMENT

This paper is supported by the Sectorial Operational Programme Human Resources Development (SOP HRD), financed from the European Social Fund and by the Romanian Government under the contract number POSDRU/107/1.5/S/76945.

REFERENCES

[1] Arişanu A. O., Hodîrnău E., Particularities of oil extraction by pressing hulled and hydrothermally processed sunflower seeds, Journal of EcoAgriTourism, vol. 7, no. 2, p. 32-37, 2011.

[2] Banu C., Manualul inginerului de industrie alimentară, Vol. 1, Bucureşti, Editura Tehnică, 1998.

[3] Banu C., Manualul inginerului de industrie alimentară, Vol. 2, Bucureşti, Editura Tehnică, 2002.

[4] Bargale P. C., Mechanical oil expression from selected oilseeds under uniaxial compression, PhD Thesis, Saskatoon, University of Saskatchewan, Canada, 1997.

[5] Boeru G., Puzdrea D., Tehnologia uleiurilor vegetale, Bucureşti, Editura Tehnică, 1980.

[6] Brătfălean D., Cristea V. M., Agachi P. Ş., Irimie D. F., Improvement of sunflower oil extraction by modelling and simulation, Revue Roumaine de Chimie, vol. 53, no. 9, p. 881-888, 2008.

[7] Evon P., Vandenbossche V., Pontalier P. Y., Rigal L., Direct extraction of oil from sunflower seeds by twin-screw extruder according to an aqueous extraction process: Feasibility study and influence of operating conditions, Industrial Crops and Products, vol. 26, p. 351-359, 2007.

[8] Evon P., Vandenbossche V., Pontalier P. Y., Rigal L., Aqueous extraction of residual oil from sunflower press cake using a twin-screw extruder: Feasibility study, Industrial Crops and Products, vol. 29, p. 455-465, 2009.

[9] Ghimbăşan R., Tehnologii în industria alimentară, Vol. 1, Braşov, Editura Universităţii Transilvania din Braşov, 2000.

[10] Herak D., Kabutey A., Divisova M., Svatonova T., Comparison of the mechanical behaviour of selected oilseeds under compression loading, Not Bot Horti Agrobo, vol. 40, no. 2, p. 227-232, 2012.

[11] Kartika I. A., Pontalier P. Y., Rigal L., Extraction of sunflower oil by twin screw extruder: Screw configuration and operating condition effects, Bioresource Technology, vol. 97, p. 2302-2310.

[12] Rusnac L. M., Tehnologia uleiurilor vegetale şi volatile, Timişoara, Editura Universităţii Politehnica din Timişoara, 1995.

NOVEL IMPACT ATTENUATOR

Marian N. Velea[1], Simona Lache[2]

[1] Transilvania University of Braşov, Braşov, ROMANIA, marian.velea@unitbv.ro

[2] Transilvania University of Braşov, Braşov, ROMANIA, slache@unitbv.ro

Abstract: *The capacity of materials and structures to dissipate the impact energy generated by impulse forces (collisions, explosions, etc.) raises great interest in applications from vehicle industry – terrestrial, nautical or spatial – for increasing security of passengers and goods, as well as in applications from civil engineering – for protection of high security buildings.*

This paper presents a novel impact attenuator made of a multi-layered cellular structure - ExpaAsym. The analysis regards the main conditions an attenuator should fulfill in order to be efficient: the damage curve – which should be relatively constant during the deformation; the distance to compaction – which is expected to be higher, preferably more than 80% of the initial height; maximum initial load force – which should be as close as possible to the value of the average load force; the dissipated energy, respectively the area beyond the force-displacement curve – which is expected to be as large as possible. The results of the numerical analysis prove the feasibility of the novel proposed system.

Keywords: *cellular structure, energy absorption, numerical analysis*

1. INTRODUCTION

An efficient impact attenuator is the one that meets the following conditions, related to Figure 1: the crushing curve keeps itself constant during the whole process of deformation (> 80% from the initial height of the impact attenuator), the maximum initial peak load is as close as possible to the value of the average load, the quantity of the dissipated energy, respectively the area under the Force – Displacement curve, has to be as large as possible.

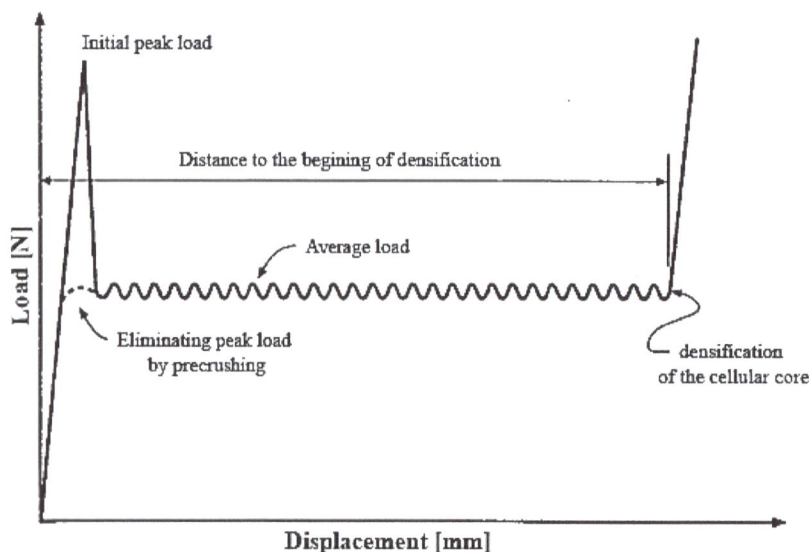

Figure 1: The behaviour description of the ideal impact attenuator [1]

The research in this field has demonstrated that the periodic cellular structures have a high potential to satisfy the above described conditions. Thus, several types of cellular structures used for attenuating the impact energy are known, such as: honeycomb structures, [2], [3], [4], single or multi-layered formed structures, [5], [6], or multi-layered corrugated structures, [7]. These cellular structures have several disadvantages: their construction implies high material consumption in order to dissipate a large quantity of energy; they have high relative density due to the material distribution in space; the plastic deformation takes place in shocks due to the buckling of the cells' walls; they possess a low number of variables that can be modified for the optimization of the absorption capacity; the manufacturing methods are relatively expensive, with a reduced degree of flexibility.

2. NOVEL IMPACT ATTENUATOR

According to the aspects mentioned in the previous section, it was considered the realisation of an impact attenuator by a simple process, having a low relative density, which should allow the dissipation of large quantity of energy as a result of an impact force. In order to uniform the reaction force, the plastic deformation of the constitutive elements should occur with low intensity shocks.

Due to the advantages that the expanded cellular structure ExpaAsym implies, [8], [9], its use for attenuating the impact energy may eliminate the disadvantages mentioned in paragraph 1 as follows: a multi-layered cellular structure 2 is constructed, Figure 2, consisting of two or more successive layers of ExpaAsym cellular structure rotated in plane from one layer to the other by 180°; this multi-layered structure is covered by two exterior face sheets 1 and 3.

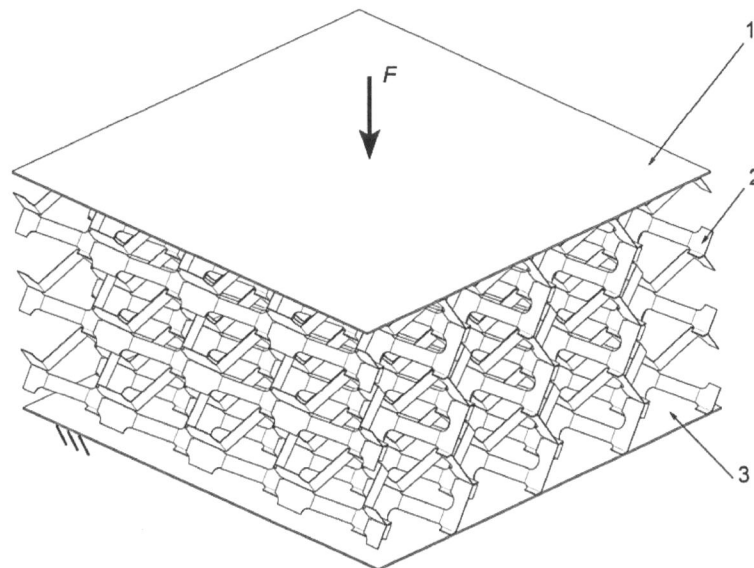

Figure 2: Impact attenuator containing ExpaAsym multi-layered cellular structure

3. NUMERICAL SIMULATIONS

Several quantitative criteria exist for the evaluation of the impact energy absorption capacity, [10], [11]. One of the most used refers to the determination of the volume specific impact energy absorption capacity (the absorbed energy at a unit volume). Thus, volume specific impact energy absorption capacity E_v

[J/m^3] of the impact attenuator obtained using the multi-layered cellular structure is calculated as the ratio between the absorbed energy E_{abs} and the initial volume V of the attenuator, Equation (1).

$$E_v = \frac{E_{abs}}{V} \qquad (1)$$

A preliminary numerical simulation was performed using Abaqus/Explicit in order to calculate the value of E_v, considering the geometric case when the internal angle has a value of 60°, the l/c ratio is 1.4 and the width b has a value of 10 mm, according to the notations made in.

The material properties introduced in the FE model correspond to a steel E = 210000 MPa, ρ=7870 Kg/m^3, σ_{yield} = 200 MPa for ε_p = 0% and σ_{yield}= 480 MPa for ε_p= 0.18%.

The impact attenuator consists of 6 layers of ExpaAsym cellular structure with the base material thickness of 0.5 mm, and one exterior face sheet with a thickness of 0.5 mm; this results in a total weight of 1,426 Kg (1,13 Kg - the weight of the cellular structure and 0.296 Kg - the weight of the exterior face sheet). The dimensions of the analysed attenuator are: width w = 317,3 mm, length t = 236,8 mm, height h = 203,9 mm.

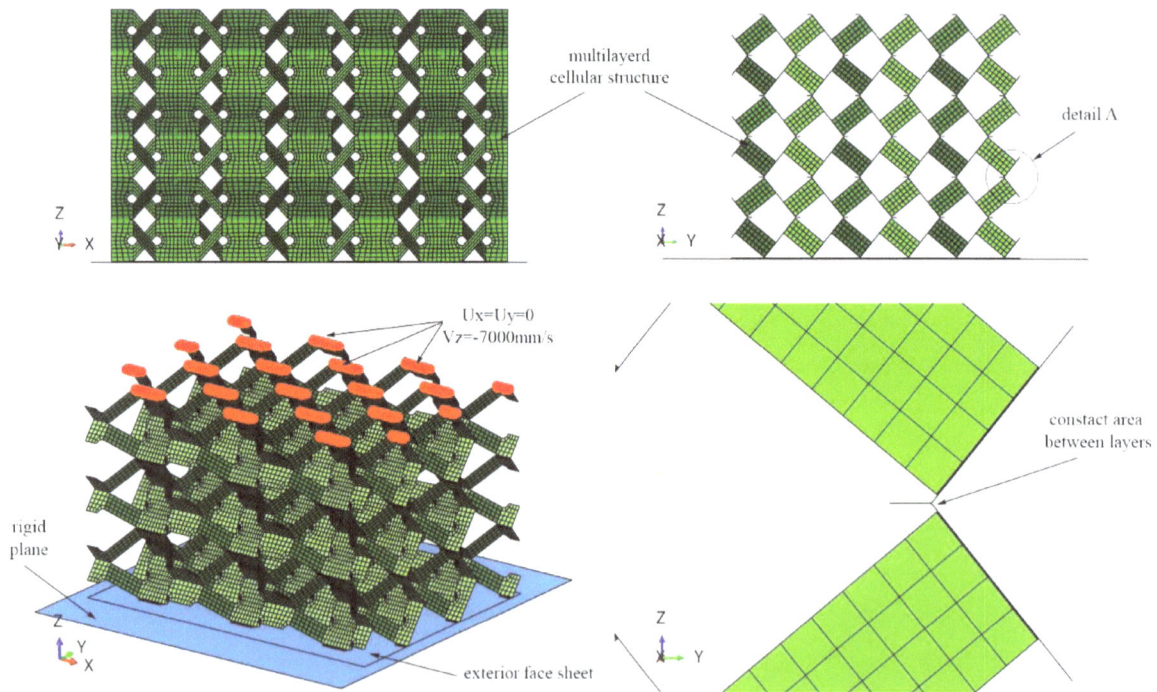

Figure 3: The FE model of the impact attenuator and the imposed boundary conditions

A concentrated mass of 300 Kg has been defined in a "master" node, positioned at 1000 mm along the Z direction, from the surface of the exterior face sheet that is in contact with the rigid wall. The superior nodes of the attenuator (marked with red in Figure 3), named "slave" nodes, are connected at the „master" node. An initial velocity of 7000 mm/s, Figure 3, was imposed to the master node which was implicitly transmitted to the "slave" nodes.

A great influence on the impact behaviour of cellular structures belongs to the deformation rate [10] $\dot{\varepsilon}$ which represents the rate of change of the specific deformation with respect to time. Its value can be calculated as the ratio between the impact speed v and cellular structure height h. According to the data presented in this analysis, the deformation rate has the value of 69.30/s. This value of the deformation rate, according to the classification of Ashby [10], is considered to be an *intermediate* value. As an order of size, the requirements of vehicles industry foresee the design of the elements required in impact considering deformation rates up to 40/s.

4. RESULTS AND CONCLUSIONS

The reaction force due to the impact is measured in the reference point that defines the position of the rigid plan, in terms of the distance of the nodes from the superior plane of the attenuator, and it is presented in Figure 4.

It can be observed that the value of the reaction force varies in lower limits around the value of 4000 N, Figure 4, allowing approximating that it is constant and thus fulfilling one of the conditions listed within section 1 for a high performance impact attenuator. Another requirement of an impact attenuator is also fulfilled, according to which the initial value of the impact force must be as close to the average load value. The energy dissipated at a specific strain of 0.5, respectively when the impact attenuator height value is halved (100 mm), Figure 5, has a value of 323453 mJ. Thus, considering Equation (1), for a specific strain of 0.5 and at a deformation rate $\dot{\varepsilon}$ of 69.30/s, the energy absorbed per unit volume E_v has a value of 4.31 mJ/mm^3, respectively 4310000 J/m^3.

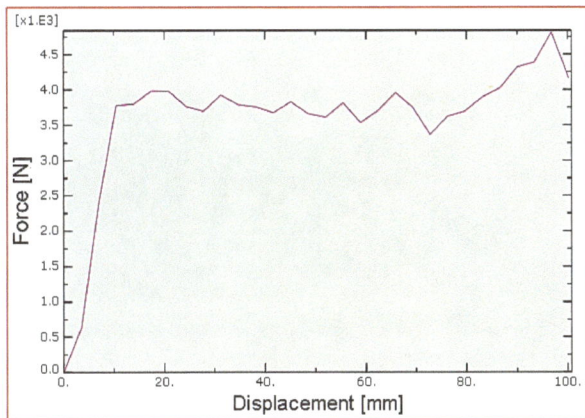

Figure 4: Reaction force registered at the deformation rate $\dot{\varepsilon}$ = 69.30/s

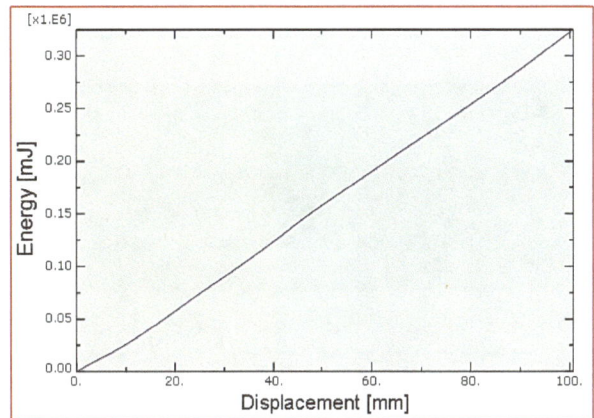

Figure 5: Dissipated energy by the plastic deformation of the cells' walls considering the deformation rate $\dot{\varepsilon}$ = 69.30/s.

Figure 6 illustrates the variation of deceleration up to a specific strain of 0.5, respectively when the height of the impact attenuator is halved (100 mm) being recorded a value of 15mm/s^2 (respectively 1.5g) at the beginning of the impact and then slightly increasing. This represents an important advantage in applications for passengers' transportation in order to increase their safety.

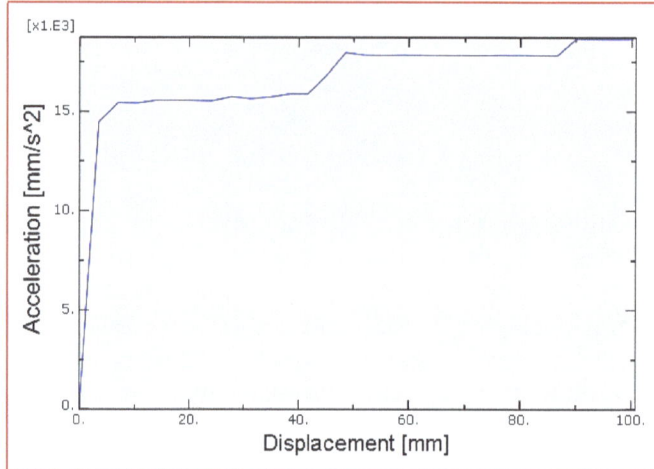

Figure 6: Variation of the deceleration within the impact scenario

Figure 7 shows the way in which the multi-layered cellular structure is deformed, and also the stress distribution in the material, expressed in MPa; it can be noticed the relatively constant distribution of stresses in the whole body of the cellular structure due to the impact loading.

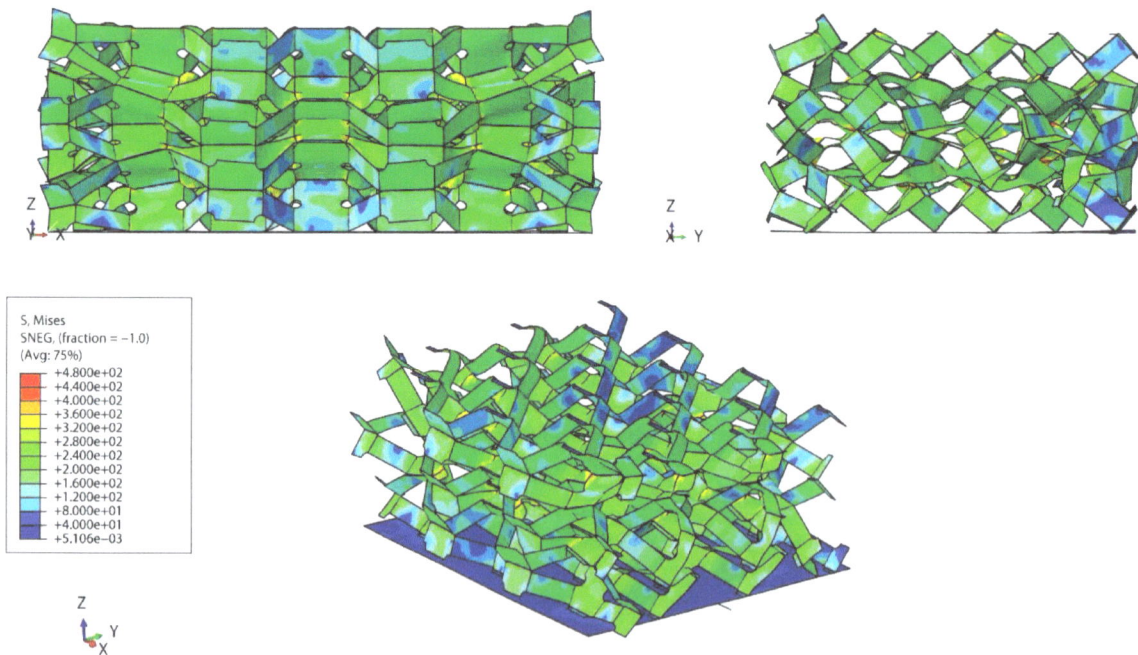

Figure 7: The deformations appeared in the structure at a strain of 0.5 and a deformation rate $\dot{\varepsilon}$ of 69.30/s

REFERENCES

[1]   Hexcell: HexWeb Honeycomb Attributes and Properties. 1999.
[2]   Arsentev A.S., Kosarev V.A., Fedotov J.V.: Impact Energy Absorber, Patent no. RU 2.246.646, 2005.
[3]   Jeyahsingh V.M.A.: Analytical Modeling of Metallic Honeycomb for Energy Absorption and Validation with FEA, Doctoral Thesis, Wichita State University, 2005.
[4]   Shepherd D.: Impact Energy absorber, Patent no. GB 2.323.146, 1998.
[5]   Schulz P., Gradinger R., Reiter J., Khalil Z., Birgmann A.: Deformable Body Having a Low Specific Gravity, International Patent no. WO 2007/009142. 2007.

[6]   Zupan M., Chen C., Fleck N.A.: The Plastic Collapse and Energy Absorption Capacity of Egg-Box Panels, International Journal of Mechanical Sciences, Vol. 45, No.45, 2003.

[7]   Bonnetain Y.: Shock Absorption Bumper for an Automotive Vehicle, Patent no. US 4.221.413, 1980.

[8]   Velea M.N., Lache S.: In-plane effective elastic properties of a novel cellular core for sandwich structures, Mechanics of Materials, Vol. 43, 2011.

[9]   Velea M.N., Wennhage P., Lache S.: Out-of-plane effective shear elastic properties of a novel cellular core for sandwich structures. Materials & Design, Vol. 36, 2012.

[10]   Gibson L.J., Ashby M.F.: Cellular Solids. Structure and Properties. 2 ed., Cambridge University Press, 1999.

[11]   Nagel G.: Impact and Energy Absorption of Straight and Tapered Rectangular Tubes [Ph.D. Thesis]: Queensland University of Technology; 2005.

RESEARCHES REGARDING THE CAUSES OF DEGRADATION OF ROOF SYSTEMS

E. Badiu[1], Gh. Brătucu[1]

[1] Transilvania University of Brasov, Brasov, ROMANIA, ed@cebb.net

Abstract: Roofing system durability is the result of many parameters, such as, but not limited to: local climate, local temperature which is correctly considered a result of changes in solar radiation, wind, rain, and other environmental influences. Other factors, such as the use or physical abuse to which the system will be exposed, are requirements mandated by the specific building in question.
Keywords: degradation, durability, roof system

1. INTRODUCTION

Contrary to popular opinion, the maintenance-free roof system is a misnomer. All types of roofs require a certain level of attention. In fact, from the moment of installation, the roofing system undergoes continuous deterioration. Extreme temperature fluctuations as well as snow, ice, hail and wind prevail upon the roofing surface. In short, the elements are the biggest deterrents to the roof system over its service life. Traffic on the roof and the installation of mechanical and other equipment can also cause physical damage that could lead to roofing failures.

There are a number of complex, systematic variables that cause distress in roofing systems. Rational planning for repair and replacement is necessary. It is important to note that distress of roof systems is not linear. The function of each system is defined in distinct terms, not on a continuum. If a roofing system's performance reaches a point of failure, its function and aging process change dramatically. From that point on, problems are not static, self-correlating, or reversible. A split in a roofing membrane amounts to an immediate failure to protect the building structure and its contents, as does a flashing failure, detachment due to wind forces, mechanical damage by man, or any change in the protective function of the roofing membrane [5].

2. STRUCTURAL PROBLEMS DUE TO THE DECK MOVEMENT

The type and stability of the deck are critical to the performance of the roof. Decks that are too light or that deflect contribute to excessive tensile forces upon the roof that can induce a form of cyclical fatigue stress that will eventually lead to the splitting of properly manufactured components of a roof system.

The stability of the deck can be affected by many factors, such as exposure to moisture, improper attachment of the deck to the structural supports, lack of proper joint spacing, inadequate structural support and the use of weaker or improper decks [5].

3. STRUCTURAL PROBLEMS DUE TO LACK OF VENTILATION

The most common problems caused by poor roof ventilation are due to two factors, heat and moisture [2].

The lack of proper ventilation can lead to excessive heat and moisture build-up which can adversely affect the short-term performance as well as the long term performance. Over time, excessive heat and moisture can lead to premature aging resulting in hardening of the coating asphalt, visible cracking, or fine, alligator-type cracking and the eventual splitting or cracking of the roof system components.

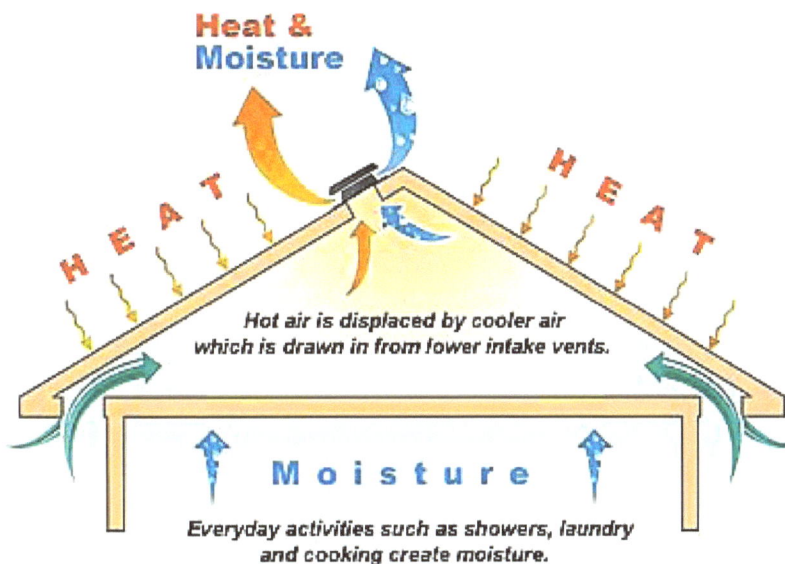

Figure 1: Problems caused by poor roof ventilation [2]

Trapped moisture and heat can lead to the following common problems:

- **Mold & Mildew:**
  A humid environment is the perfect place for mold or mildew to form. Mold can ruin stored items in the attic and cause health problems.

- **Rust:**
  Rust can begin to form on metal components like nails or other critical fasteners. Overtime it can rust the heads off of nails or cause plumbing or venting straps to fail.

- **Sagging or Spongy Decking:**
  When excessive moisture begins seep into the roof decking it can begin to dissolve the adhesives which hold them together and cause it to warp, sag between rafters or feel spongy when walked on. This can become a danger for anyone on the roof.

- **Roofing System Deterioration:**
  Not only can excessive heat and moisture ruin roof decking, it can also reduce the life of the underlayment and shingles themselves. Cracking shingles or premature loss of granules can be signs of improper roof ventilation.

- **Air Conditioner Replacement & Expenses:**
  As heat builds in the attic, air conditioners must work extra hard to keep the air inside the home cool. This undue stress on the unit can reduce its life and increase energy costs.

- **Frost:**
  Similar to how sitting in a cold car on a winter day will cause frost to form on the windows, the same can occur in a poorly vented attic. As the attic cools and warms with the day, frost formed inside the attic can melt and drip onto the ceiling.

- **Ice Dams:**
  Ice dams can form at the edge of a roof where trapped warm air can melt snow on the roof that then freezes as it cools. As the snow continues to freeze, melt and refreeze it creates a barrier, or dam preventing water from running off the roof. Once dammed, water and ice can creep back up under the shingles and underlayment resulting in leaks.
  Proper ventilation and the use of added insulation can help mitigate this melting and freezing process and eliminate ice dams [2].

4. PROBLEMS RELATED TO MANUFACTURING OF ROOFING MATERIALS

Materials used in roof construction vary in cost, design and longevity. The style of the residence or building, the desired color, and economic and ecological factors drives the selection of roofing materials. Understanding the variables of the numerous materials available for roof construction provides a clearer selection process. Whether choosing green (ecologically friendly) or choosing historically correct roofing, options for materials used in roof construction open new possibilities in the 21st Century.

- **Asphalt Shingles.**
  The variety of colors and the range in price from inexpensive 3-tab (three sections per unit) shingle to a more costly durable asphalt shingle remain popular. Easily repaired, petroleum-based asphalt shingles are environmentally unfriendly since they seldom get recycled and usually end up in landfills. Hot weather scars this roofing material while moss and mildew form during its short life span of 15 to 30 years [3].

- **Glass fiber-reinforced asphalt shingles**
  Glass fiber-reinforced asphalt shingles are the predominant roof material used to cover steep-slope roof systems in the United States and Canada. Many variations of this product are available, including three-tab and laminated shingles. Field performance of glass fiber-reinforced asphalt shingles ranges from outstanding to poor; deck conditions, wind, snow and ice, and extreme heat may shorten shingles' service life. Installation methods and fastening patterns may also affect field performance. A predominant field performance issue with glass fiber shingles is cracking. The cracking pattern on a three-tab shingle may be diagonal, vertical, horizontal or a meandering combination of the above. The mechanical and thermal loads a roof shingle experiences are many; extreme heat, large temperature swings, high winds, deck warpage and heavy rainfall are the primary agents loading a shingle. The sealant used to hold down the exposed lower portion of a shingle is vital to preventing wind uplift. If the sealant is too hard and not ductile, it prevents expansion of the shingle during extreme heat. Sealant location is also critical; applications of sealant that are close together prevent movement. Wider distribution allows for more gage length. Selfsealing turns multiple individual shingles into a unit. This will cause stress concentration to occur during temperature swings if nonuniform attachment is present either in the sealant or nailing. The performance of glass fiber-reinforced shingles has been studied and reviewed by many authors, including Cash, Ribble, et al., Noone and Blanchard, and Terrenzio, et al. Shingle cracking has been specifically addressed by Cash, Datta, et. al., Noone and Blanchard, Phillips, et.al., and Shiao. Although temperature extremes certainly occur on roofs, Rose and Cash have demonstrated that attic ventilation alone cannot control or significantly affect shingle temperatures. These authors separately concluded from field studies and mathematic models that attic ventilation is limited in controlling shingle temperature. Cash has shown that color has more effect on shingle temperature during solar load than attic ventilation. The measurement of a shingle's ability to resist cracking or splitting has been debated heavily [1].

- **Metal roofs**
  Metal roofs are great for any type of roof and are ideal in forested, moss prone, or heavy precipitation areas. Typically manufactured from steel, aluminum or copper, metal roofing offers homeowners the chance to choose from a multitude of colors and textures. Standing-seam steel roofing is the most popular residential metal roofing today. The term standing-seam describes the upturned edge of one metal panel that connects it to adjacent sections, creating distinctive vertical lines and a trendy historical look. But metal roofs can also be made to resemble wood shakes, clay tiles, shingles, and Victorian metal tiles. Aluminum or coated steel is formed into individual shingles or tiles, or into modular panels four feet long that mimic a row of shingles or tiles. Metal roofs are durable, fire retardant and almost maintenance-free. They are also energy efficient. Research by the Florida Solar Energy Center showed that metal absorbed 34% less heat than asphalt shingles, and homeowners switching to metal roofing reported saving up to 20% on their energy bills. Metal roofs typically have solar reflectance values between 0.50 and 0.70 but their overall efficiency are reduced by their low

emittance levels, which means they trap solar radiation and don't emit the heat. They perform better when combined with a polymeric coating that helps to offset the low emittance of the metal. These coatings, which are similar to paint, can be factory-applied. It can be manufactured in long panels, or in smaller pieces that more closely resemble tiles or shingles. The sound of rain on a metal roof, which some homeowners find unacceptable, can be reduced with the use of a foam underlayment. The cost of metal roofing is initially higher than that of composition shingles, but it has a longer life cycle and can significantly lower heating and air conditioning costs, making a metal roof a very good investment. Furthermore, metal roofs are made from recycled metals (60% or higher), so they provide an environmentally friendly option. The reduced weight is of particular importance in high seismic zones where roofs can experience severe vertical and horizontal forces during an earthquake. The lightweight metal roof significantly reduces the chances of catastrophic failure or collapse of the roof structure during a massive quake. It is also fire resistant, making it suitable for use in fire-prone areas, and can result in reductions in the cost of insurance coverage. Metal roofs are virtually maintenance-free. Periodic rinsing with a hose or pressure washer can help keep the surface clean and free of corrosive residue, such as bird droppings and acid rain. Although metal roofs can be walked on, care should be taken when walking on a roof with deep shake and tile profiles, to prevent damage to the contour of the ridges [4].

5. PROBLEMS RELATED TO APPLICATION/INSTALLATION

Roofing Contractors are a resource that always should to be valued and cultivated. Where defective design work contributes more than 50% of the roofing failures, defective workmanship accounts for about 30% of the roofing failures.

Most of these errors by contractors are due to ignorance about the consequences of their actions and absence or poor supervision. Skilled workers are usually supplemented with less experienced transient labor. Combining inexperienced workers, with inadequate supervision, with the emphasis on production rather than quality almost guarantees problems.

Among the most common problems are the following (which has often overlapped with matters of design):

- **Slope of the roof**:
  Low slope roofs is that category, which includes generally weatherproof membrane types of roof systems installed on slopes of 14 degrees or less. Such roofing systems have a weatherproof cover or a single membrane that prevents water from entering the host structure. Steep roof roofing is the categories that generally includes roofing discharging storm waters and are installed on slopes greater than 14 degrees. They are usually composed of individual parts or components installed shingles. These roof systems works with gravity to shed water from a place to another, ensuring drainage of roof surfaces.

- **Roof Drainage**:
  The next critical step, after the selection of the roofing system appropriate for the climate and the occupancy of the building, is to make sure that the roof drains promptly. Roof areas that promptly drain last at least twice as long as areas that don't drain promptly.

- **Flashing Details at eaves and valley construction**:
  Installation of a proper eave detail is always important, but in critical areas of high rainfall or where leaves, snow, ice, or where water dams are likely to accumulate on the roof. Installing a drip edge at the edge to protect the wood from possible damage is one of the most important conditions for a leak-free roof.

- **Flashing Details at roof penetrations and vertical walls**:
  Key areas of concern other than eaves and valleys are flashing details on roof penetrations and vertical walls, especially those associated with chimneys and skylights.

- **Proper Nailing to Structural Substrate**:

Whenever mechanical fasteners are in contact with the membrane, some problems can be expected. Proper nailing/fastening is a key to the roof's ability to stay in place. Nails/Fasteners must be installed per the manufacturer's installation.

There are specific repetitive elements; actions that might have been taken which might probably have prevented the failure from taking place, or at least minimized the event's impact [6].

6. CONCLUSION

- Roof systems are exposed to a variety of physical and chemical distress, which range from dramatic physical actions, such as wind forces, thermal loading, or mechanical damage, to slow, insidious chemical processes like photo-oxidation.
- The most common problems caused by poor roof ventilation are due to two factors, heat and moisture. The lack of proper ventilation can lead to excessive heat and moisture build-up which can adversely affect the short-term performance as well as the long term performance.
- The style of the residence or building, the desired color, and economic and ecological factors drive the selection of roofing materials. Understanding the variables of the numerous materials available for roof construction provides a clearer selection process.
- Roofing Contractors are a resource that always should to be valued and cultivated. Where defective design work contributes more than 50% of the roofing failures, defective workmanship accounts for about 30% of the roofing failures. Most of these errors by contractors are due to ignorance about the consequences of their actions and absence or poor supervision.

REFERENCES

[1] Dupuis R.M., The Effects of Moisture and Heat on the Tear Strength of Glass Fiber-Reinforced Asphalt Shingles, available at: http://docserver.nrca.net.
[2] http://www.trinityexteriorsinc.com.
[3] http://www.ehow.com.
[4] ***Green Affordable Housing Coalition Fact Sheet, No.18, p. 1-4, December 2005.
[5] www.buildings.com.
[6] www.roofingcontractor.com.

HEAT AND LIGHT REQUIREMENTS OF VEGETABLE PLANTS

C. Bodolan[1], Gh. Brătucu[1]
[1] Transilvania University of Brasov, Brasov, ROMANIA, bodolan.ciprian@gmail.com

Abstract: This paper addresses to the problem of knowledge requirements of each vegetable species to environmental factors (heat and light), which is of particular importance because the technology used for their control can intervene in closely aligned with requirements of different vegetable species. Although vegetable growing is one of the oldest human occupations, it has developed in modern times, because technological progress, thereby strengthening the independent science that deals with the peculiarities of agricultural plant culture.

This was the result of vegetable crops in protected areas having the climate controlled, which allows providing appropriate conditions for plant growth in out of vegetables season. The practice of modern technology, low pollution risk and low specific energy consumption, it tries increased productivity, which can benefit from economic point of view.

Keywords: climatic factors, gardening, heat, light greenhouses

1. INTRODUCTION

In vegetable production systems for a balanced supply of the market with fresh produce in season that they cannot be obtained under natural conditions in the field, are used greenhouses and solariums [1].

Table 1: The influence of environmental factors on plant growth
(after Laue, Forkel and Forberg 1968)

Environmental factors	Proportion %
Precipitations	5,74
Air relative humidity	2.03
The soil tillage	1.69
Soil micro-organisms	5.41
Nebulosity	4.28
The season of the year	4.28
Underground water	2.03
Irrigation and drainage	3.72
Soil	6.76
The fertilizers	3.72
The air pollution	7.66
Soil water available for plant	12.16
Soil nutrients available to plants	10.14
Physiologically effective electromagnetic radiation	12.84
	10.14
The content of CO2 in the air around the plant	7.40
Plant growth	

The proper knowledge of vegetation factors and appreciation that the proportion of their participation in plant growth and development processes allows the implementation of ecological systems models, methods of vegetable production models, especially in cultures where it is possible to automate the regulating of greenhouses environmental factors [2].

Knowing the plant relationships with the light is important in crop zoning, establishing optimal ages to set up cultures in protected areas to phased production. However, it is known that a decisive factor in seed germination, plant growth, chlorophyll formation, flowering, fructification and sweat is the heat [3]. This correlated with a strong intensity makes the process of photosynthesis to be high and during the night stops, while breathing is intensified. Knowing the vegetable demands has a practical importance, because it is the basis of establishment and conveyance culture technology [4].

Environmental factors such as light and heat are those together along with chlorophyll from plants directly influence the accumulation process in photosynthesis.

2. THE IMPORTANCE OF HEAT TO VEGETABLES PLANTS

Knowing the relationships between the heat as a growth factor and vegetable plants are important theoretical and especially practical. The heat is a factor that affects the whole range of plants vital processes [5].

Some data concerning the temperatures in the climate of our country shows that they records an diurnal, monthly and yearly variation, depending on the amount of solar radiation (Table 2).

Table 2: Annual temperature values in Romania, monthly and annual average, minimum and absolute maximum for a period of 59 years [°C]

Location / Month	București Filaret	Constanța	Craiova	Timișoara	Arad	Cluj-Napoca	Brașov	Iași
January	-2.9	-3.0	-2.5	-1.2	-1.1	-4.4	-3.9	-3.9
February	-0.8	-0.8	0.3	0.4	-0.3	-2.3	-1.8	-1.9
March	5.0	4.4	5.2	6.0	5.8	3.2	3.0	3.2
April	11.3	9.3	11.3	11.3	11.0	9.0	8.5	10.3
May	16.8	151	16.7	16.4	16.1	14.1	13.2	16.1
June	20.5	19.5	20.4	19.6	19.3	17.2	16.0	19.4
July	22.9	22.2	22.7	21.6	21.4	18.9	17.8	21.3
August	22.4	22.0	21.9	20.8	20.8	18.2	17.2	20.6
September	18.1	18.5	17.8	16.9	17.0	14.2	13.5	16.3
October	11.9	13.3	11.7	11.3	11.5	8.8	8.4	10.1
November	5.2	7.5	5.2	5.7	5.7	3.1	2.9	4.1
December	-0.1	2.6	0.1	1.4	1.4	-1.6	-1.6	-0.8
Annual average	10.9	11.2	10.8	10.9	10.8	8.2	7.8	9.6
Absolute minimum	-30.0	-25.0	-30.5	-29.2	-30.1	-32.5	-29.6	-30.0
Absolute maximum	41.1	38.5	41.5	40.0	40.0	36.8	37.1	40.0

Vegetable plants can be grouped according to heat requirements as follows:
- Exigent vegetable plants towards heat - include annual species, which consume fruits: tomatoes, peppers, eggplant, cucumbers, squash, melons, watermelons, beans, okra, etc. The minimum germination temperature is 10...14°C, the optimum germination is 20...25°C, it grows well at 25...32°C, supports a maximum of 35...40°C and a minimum of 10°C.
- Vegetable plants demanding average heat - annual and biennial species include: bulbs, roots, cabbage leaves, peas, beans, potatoes, etc. Minimum germination temperature is 2...5°C, the optimum germination and growth is 14...20°C and 22...25°C. Maximum growth. For a short-term support negative temperature between - 4°C and - 2°C.

- Vegetable plants resistant to cold - include perennial species as: asparagus, rhubarb, cardoon, artichoke, tarragon and sorrel. Generally requires a temperature similar to the previous group, or slightly lower, but over winter breeding organs that are found in soil supports a temperature down to - 20 ° C, and by protecting the right of - 27°C.

Seed germination, plant growth, flowering, and fructification, duration of the resting phase, assimilation, respiration, transpiration and other physiological processes take place in the presence of a certain temperature.

Regarding the temperature level of each biological species has three layers:
- a minimum when metabolic processes are slowed down and no accumulation occurs (F / R = 1);
- optimum when metabolic processes are intense and balanced, and perform the most intense growth rate and accumulation of reserve substances in consumer bodies (F / R> 1);
- a maximum when the intensity of metabolic processes is maximal, but F / R = 1, and by exceeding this threshold the plants are exhausted and dying (F / R <1);

Conducting mode of temperature for different vegetation stages vegetable plants can be seen in Table 3.

Table 3: The optimum temperature vegetable plantson growth period and phase [°C] by Stan, N. 1992

The species	The optimum temperature for vegetative growth	Periods						
		Seed	Vegetative growth		General growth			
		Phases						
		Germination	Appearance of the cotyledons	Seedling growth	Accumulation of reserve substances	Flowering	Fruiting and maturation of fruits
Cucumbers, melons, watermelons	25°C	32°C	18°C	25°C	25°C	23°C	32°C
Tomatoes, peppers, eggplant, beans and pumpkin	22°C	29°C	15°C	22°C	22°C	20°C	29°C
Beets, asparagus, onions, garlic, celery	19°C	26°C	12°C	19°C	19°C	17°C	26°C
Potatoes, salads, peas, carrots, parsley, parsnips, chicory, spinach, dill, sorrel	16°C	23°C	9°C	16°C	16°C	14°C	23°C
Cabbage, radish, horseradish	13°C	20°C	6°C	13°C	13°C	11°C	20°C

3. THE IMPORTANCE OF LIGHT TO VEGETABLES PLANTS

Light is the determining factor in the normal course of photosynthesis, acting through specific parameters such as intensity, duration, spectral composition, and determining the optimal value of other factors by species and stage of vegetation. Taking into account that the variations of these parameters cannot be directed only in small measure, the light is considered to be a limiting factor for crop production.

After Somos and collaborators, 1966 sowing in December when light conditions are poor, prolongs the period to harvest compared to sowing in February, with 19 to 48 days in different species (Table 4).

Table 4: The period until harvest, depending on the lighting conditions determined during the period of culture (Somos et al., 1966)

Culture	The period until harvest (days) on crops sowed on:		Diference (days)
	15 december	15 february	
Pepper	134	113	21
Tomatoes	168	134	34
Cauliflower	155	113	42
Early okra	133	85	48
Salad	100	71	29
Radishes	76	57	19

Depending on the light intensity vegetable plants are grouped in:
- Demanding plants (8000...12000 lx): solanaceous, cucurbits, beans, okra, are cultivated in the most favorable areas.
- Less demanding plants (4000...7000 lx): root, bulb, cabbage, leafy peas.
- Plants with small claims (1000...3000 lx) are grown for consumption extra-early and late (green onions, perennial onions, beet leaves).
- Plants that do not need light during growing of edible parts: cauliflower, chicory, asparagus, mushrooms.

Depending on Day length (photoperiod) vegetable species are classified as follows:
- long day plants - originating in northern areas requiring between 15...18 hours of daily illumination, but in lower intensity (3000...7000 lux).): Root, bulb, cabbage leaves, peas, green onions, perennial onions, beet leaves;
- short-day plants - originating in southern areas, requires between 10 to 14 hours of daily illumination, but with high intensity (8000...12000 lx.) Solanaceous, cucurbits, beans, okra,
- indifferent plants - adapted to different culture conditions (tomato, salad);
- in order to obtain large and quality productions, especially in plants which consume fruit and seed crops, it must be provided the lighting conditions required by plants.

Light intensity strongly influences the duration and progress rate of physiological processes and plant growth. As the light is more intense, speed up chemical reactions in cells and plant physiological processes take place faster.

Vegetable plants grow and develop themselves best at a light intensity of 20...30 thousand lux. In most species, as light intensity increases to the level of 50000 lux, the photosynthesis curve has an upward allure, and then remains constant up to 100000 lux and decreases sharply above this value called light saturation.

4. CONCLUSION

1. Knowing the vegetable plants requirements has a practical importance because it is the basis of establishment and conveyance of culture technology.

2. Environmental factors such as light and heat are those along with the chlorophyll from plants directly influence the accumulation process in photosynthesis.

3. The heat is a key factor that affects the whole range of plants vital processes.

4. Light is the determining factor in the normal course of the photosynthesis, acting through specific parameters such as intensity, duration, spectral composition.

5. The heat coupled with a strong intensity makes the process of photosynthesis to be high during the day and stops over the night, while breathing is intensified.

REFERENCES

[1]   Ceauşescu I., General and special horticulture, Didactic and Pedagogical Publishing, Bucureşti, 1984.
[2]   Ciofu R. et al., Treated of vegetable growing, Ceres Publishing, Bucureşti, 2004.
[3]   Goian M. et al., Horticulture, West Publishing, Timişoara, 2002.
[4]   Oancea I., Treated of agricultural technologies, Technical Publishing, Bucureşti, 1998.
[5]   Stan N. et al., Vegetable farming, Ion Ionescu de la Brad Publishing, Iaşi, 2003.

SOLAR ENERGY – AN ENERGETIC SOURCE FOR THE VEGETABLE AND FRUIT PRODUCTS DRYING IN BRASOV AREA

Gh. Brătucu[1], D. D. Păunescu[1]
[1] Transilvania University of Brasov, Brasov, ROMANIA, gh.bratucu@unitbv.ro

Abstract: *The dried vegetables and fruits can have multiple usages, if they keep much better their food and commercial features. For drying process an important amount of energy is consumed. In the paper is analyzed the possibility of the solar energy use for drying those products, with the condition that the process of water evacuation to be rigorous checked up, without damaging the nutritious elements which they contain. In Romania the method is recommended for the south and east zones where the solar radiation intensity is bigger, and the vegetables and fruits have an important weight in the agricultural production. The heat needfulness for drying it can be assumed with solar panels, and the energy for the fans with photovoltaic panels.*
Keywords: *drying conservation, fruits, solar energy, vegetables*

1. INTRODUCTION

The fruits and vegetables for fresh consumption or for different industrial food products satisfy the needs of alimentation by their qualitative value determinate by the taste, nutritious components, aroma etc.

The preservation of the vegetables and fruits through dehydration represent a possibility of supplementary capitalization of the production from this area, especially for their utilization in the recipes of different food products (instant soups), but also for direct consumption.

On this idea there have developed and utilized numerous constructive schemes of drying installations, as: discontinuous dryers Muger – of chamber type; convective dryers of tunnel type; convective dryers with overlap bands; fluidization dryers; column type dryers for cereals; conductive dryers through contact; contiguous dryers bellow pressure etc. [4].

The higher cost of the energy necessary for drying in these installations, as well as the requirements enforced to the products in the drying process made practically impossible their utilization by the small farmers. The use of solar energy for drying of agricultural products may represent an acceptable and accessible opportunity for all agricultural farmers [1].

For this purpose there have been studied and built many types of dryers, in which the solar energy solves completely the problems of drying, also at heat source and for the air motion inside the dryer. There are known the realizations of German firm Babcok AG, of the research workers from California Polytechnic University, of the research workers from the Hohenheim University, but also another old dryers or much more modern dryers like the cabinet dryer, the band dryer, the dryer with natural convection.

In all these cases, the food drying must carry out a series of requirements through which the chemical components, the vitamins, initial color, taste, smelt etc will be fully preserved. Obtaining these performances assumes conducting the drying process after rigorous criteria, so that the elimination of surplus water not to degrade the valuable parts of the vegetables and fruits [2].

2. MATERIAL AND METHODS

For the realization of a proper dryer project which uses the solar energy is needed to know the values of the climatic parameters, such as: multiyear monthly average of the air temperature, the parameters of the solar radiation, the speed of the wind and the average number of hours in which the sun glows in a month. For the computation of the solar global daily radiation (I_{za}) it is used the relation:

$$I_{za} = \frac{24}{\pi} \cdot C_s \cdot \left\{ \left[1 + 0.33 \cdot cos\left(\frac{360 \cdot n}{265}\right) \right] \cdot \left[cos\,\phi \cdot cos\,\delta \cdot sin\,t_r + \frac{2\pi \cdot t_r}{360} \cdot sin\,\phi \cdot sin\,\delta \right] \right\}, \quad (1)$$

were: I_{za} - the medium daily radiation out of the atmosphere, in W/m^2·day; C_s – the solar constant, $C_s = 1353$ W/m^2·day; n – the considered day of the year (the first day being 01 January); φ – the place latitude, in degrees; δ – the place the n day of the year declination, in degrees; t_r – the hour angle corresponding to the sunrise, in degrees.

For the drying installation design a series of dates about the vegetables and fruits which will be processed are needed, such as: the physical and chemical properties, the sorption isotherms, the water content, the heat needfulness etc.

The physical and chemical properties of the vegetables and fruits are adverted to their size, the specific heat, the firmness of the structure and texture, the color, the aroma, the taste, the water, vitamins and mineral substances content. Some properties from this category are specified afterwards [3].

The physical properties of a product are represented by weight, size and volume. The weight is expressed in grams and varies in large limits, depending on species, variety, climatic conditions, culture etc. (Table 1).

Table 1: The values of some physical properties of the main species of fruits and vegetables [5]

Species	Weight, g	Specific weight, kg/m^3	Volumetric weight, kg/m^3	Pieces/kg
Cherries	3...10	1.0060...1.0725	510...620	50...330
Apricots	15...60	1.0034...1.0547	490...560	17...66
Peaches	40...260	0.9312...1.0394	500...580	4...25
Plums	10...65	1.0016...1.0942	500...610	15...100
Apples	70...250	0.6572...0.9264	400...530	4...14
Pears	30...500	0.9843...1.0125	450...580	2...33
Potatoes	30...300	-	650...700	3...33
Onion	40...500	-	400...600	2...25
Carrots	25...200	-	500...650	5...40

3. RESULTS AND DISCUSSION

For the 45° north latitude which passes through the middle of the Romania, through the utilization of the relation (1) were obtained the values shown in Table 2.

Table 2: Medium daily solar radiation on the Romanian territory

Month	15.01	15.02	15.03	15.04	15.05	15.06
I_{za}, W/m^2·day	3278	4786	6823	9095	10789	11513
Month	15.07	15.08	15.09	15.10	15.11	15.12
I_{za}, W/m^2·day	11167	9799	7693	5443	3653	2883

For the dimensioning calculus is recommended to use the medium values of the daily global solar radiation measured on clear sky, on the horizontal plane (I_{gzs}) for the 15th day of every month, in W/m²·day (Table 3).

Table 3: Medium values of the solar radiation in Romania

Month	Measured values	%	Calculated values	%	Definitive values	%
January	2330	3.48	2240	3.60	2394	3.67
February	4080	6.10	3170	5.08	3459	5.41
March	6040	9.03	5145	8.27	4986	7.79
April	7235	10.81	6720	10.79	6711	10.49
May	8100	12.10	7715	12.38	8023	12.54
June	8480	12.69	8515	13.67	8595	13.44
July	8085	12.08	8040	12.90	8325	13.01
August	7230	10.80	6625	10.63	7254	11.34
September	6005	8.97	5400	8.66	5628	8.79
October	4220	6.31	3980	6.39	3947	6.17
November	3040	4.54	2765	4.44	2639	4.11
December	2070	3.09	1990	3.19	2061	3.22

From the analysis of the global daily solar radiation values measured on clear sky day on horizontal plan (I_{gzsm}), of the daily definitive solar radiation (I_{gzs}) and of the medium daily solar radiation from out of the atmosphere (I_{za}) is justified the adoption of the definitive values suggested in Table 2.

In Table 4 is presented the ratio between the global daily solar radiations measured on a clear sky day (I_{gzsm}), respectively the global daily definitive solar radiations (I_{gzs}) and the medium daily solar radiation from out of the atmosphere.

Table 4: Ratio I_{gzsm}/I_{za} and I_{gzs}/I_{za}

Day	15.01	15.02	15.03	15.04	15.05	15.06
I_{gzsm}, W/m²·day	2330	4080	6040	7235	8100	8490
I_{gzs}, W/m²·day	2394	3459	4986	6711	8023	8595
I_{za}, W/m²·day	3278	4786	6823	9095	10789	11513
I_{gzsm}/I_{za}	0.711	0.852	0.885	0.795	0.751	0.737
I_{gzs}/I_{za}	0.717	0.723	0.731	0.738	0.744	0.746
Day	15.07	15.08	15.09	15.10	15.11	15.12
I_{gzsm}, W/m²·day	8085	7230	6005	4220	3040	2070
I_{gzs}, W/m²·day	8325	7254	5628	3947	2629	2061
I_{za}, W/m²·day	11167	9799	7692	5443	3653	2883
I_{gzsm}/I_{za}	0.724	0.730	0.781	0.775	0.832	0.718
I_{gzs}/I_{za}	0.745	0.740	0.732	0.725	0.720	0.715

For the efficient utilization of solar energy is also needed to know the diffuse daily radiation of which values measured on the ground level around the parallel of 45°are shown in Table 5.

Table 5: Characteristics of diffuse daily solar radiation (I_{dzs}) and measured duffuse solar radiation (I_{dzsm})

Day	15.01	15.02	15.03	15.04	15.05	15.06
I_{dzs}, W/m^2·day	564	697	928	1186	1378	1452
I_{gzs}, W/m^2·day	2349	3459	4986	6711	8023	8595
I_{dzs}/I_{gzs}	0.240	0.201	0.185	0.177	0.171	0.169
I_{dzsm}, W/m^2·day	430	735	1015	1065	1160	1426
I_{gzsm}, W/m^2·day	2330	4080	6040	7235	8100	8490
Day	15.07	15.08	15.09	15.10	15.11	15.12
I_{dzs}, W/m^2·day	1410	1251	1035	787	583	515
I_{gzs}, W/m^2·day	8325	7254	5628	3947	2629	2061
I_{dzs}/I_{gzs}	0.167	0.172	0.184	0.199	0.222	0.250
I_{dzsm}, W/m^2·day	1255	1165	865	655	735	500
I_{gzsm}, W/m^2·day	8085	7230	6005	4220	3040	2070

Considering that the direct solar radiation is the difference between the global daily definitive solar radiation (I_{gzs}) and the diffuse solar radiation (I_{dzs}) are obtained the values included in Table 6.

Table 6: Values of the direct solar radiation on the Romanian territory

Daily sum, W/m2·day	Month											
	01	02	03	04	05	06	07	08	09	10	11	12
	1785	2762	4058	5525	6645	7143	6915	6003	4593	3160	2046	1546

For the conversion of the solar energy in Romanian conditions is recommended as the banking angle of the solar receptors against the horizontal line to be of 45°, which allows the appreciable decrease of the loads caused by the wind action and the decrease of the distances among the solar receptors. The average number of sunny days on the warm season is determined through multiplication of average hour duration of sun glowing with the number of days from the warm period of the year (01.04…01.11), respectively 156 days. In the warm season the average day used for dimensioning the solar installations is considered to have 9 hours and an intensity of solar radiation I = 580 W/m^2.

Knowing the features of the vegetables and fruits and tracing their sorption and desorption isotherms it can be drawn the energy balance through the utilization of solar panels as the source of energy for dryers with enforced production. On the strength of the sorption and desorption isotherms it can be conducted the drying process in such way that physical-chemical proprieties of the dried products to be much closer to the ones of fresh products.

4. CONCLUSIONS

- The preservation through drying of some vegetables and fruits present many economic interests, as much for the agricultural producers, and for the national economy. Difficulties in this process are generated by high energy costs required for drying.
- The utilization of solar energy for the drying of vegetables and fruits represent an acceptable alternative for the energy from unregenerate sources, on condition that the technical used-up equipment to be properly projected and exploited.
- An accordingly drying of every vegetables and fruits presupposes a better preservation of their physical-chemical features ensured as much through a properly sizing of the technical equipment, and through an appropriate management of the drying process.

- On the strength of the thermal balance of a dryer it can be sizing the heat sources, and the necessary (the flow) of the air which must be assured by the fans powered by photovoltaic panels.

REFERENCES

[1] Brătucu Gh., Agricultural Technology, Transilvania University of Brasov Publishing House, Brasov, 1999.
[2] Burtea O. et al., The Use of Solar Energy at the Fruits and Vegetables Drying, Ceres Publishing House, Bucharest, 1981.
[3] Epure G., Researches Regarding the Use of Tunnel Type Solar Equipment for the Vegetables Drying, PhD Thesis, USAMV Bucharest, 2004.
[4] Mitroi A. et al., Photovoltaic Operating for Agricultural Products Drying Solar Installations, in Mecanizarea Agricuturii Magazine, no. 8, p. 39-43, 1999.
[5] Udroiu A.N., Researches Regarding the Tunnel Type Solar Dryer Hohenheim Model for Agricultural Products, PhD Thesis, USAMV Bucharest, 2002.

SOIL PRESERVATION THROUGH THE PERFECTIONING OF ITS BASIC WORKS

Gh. Brătucu[1], I. Căpățînă[1], D. D. Păunescu[1]
[1] Transilvania University of Braşov, Braşov, ROMANIA, gh.bratucu@unitbv.ro

Abstract: *The classic technology of soil work for sowing presents many advantages, but also important drawbacks, regarding the destruction of the soil structure, the excessive subsidence, erosion etc., but also of big power consumption (the ploughing includes 30...35% from the total energetic consumptions in agriculture).*

The new technologies of sowing in un-plough lands or partial prepared, remove these disadvantages, but require the strict achievement of agro technical rules, for which the Romanian agriculture isn't prepared to accept them. Also, the optimization of the deep soil loosening work influence positive as much his features, and the economic results obtained to his cultivation. These aspects are analyzed in the paper on the strength of some authors' researches.

Keywords: *environmental protection, minimum tillage, no tillage, soil*

1. INTRODUCTION

Quality sowing has to ensure a good germination of the seeds, an explosive and uniform emergence, a normal growth and development of the plants, with an early covering of the land, offering to these ones the possibility to defeat in competition the weeds.

For thousand years, the sowing process had taken place in compliance with the same technology, that is the land was as a preliminary put in readiness by specific works of ploughing and stirring, subsequently the seed was incorporated at the necessary depth and covered, so that it should make contact with the soil in order to germinate and emerge under the form of plants as vigorous as possible [1].

As against this situation, deemed normal, during the last decades of the past century, in the context of the concept of durable agriculture, more and more criticisms rose as regards the classic sowing technology. The mechanical working of the soil by means of traditional methods is increasingly put under the mark of interrogation because of its great energy consumption and its continuous degradation by excessive compaction, erosion, destruction of the structure and of the texture, diminution of the humus etc; respectively the decrease of its fertility. In this framework there appeared and developed the conservative technologies for the preparation of the sowing, in which the plough is given up, the remains of the crop not being removed from the soil, forming a layer which protects and enriches it in humus and in other substances necessary for the development of the plants. The soil stirred without the turning of the furrow keeps its natural bedding; no less fertile soil is brought to the surface of the soil from the depths, with rocks or provided with salt. Moreover, the natural environment of the earthworms, whose action is beneficial for the soil as it contributes to the creation of the colloidal structure and of the soil particles undergoes almost no modification [2].

2. MATERIAL AND METHODS

The conservation technologies include no tillage, direct seed, minimum tillage, reduced tillage, and, as a general rule, the working systems which don't incorporate the vegetable remains form the previous crop, leaving them at the surface of the soil or the systems which work the soil upon the entire working width at a single pass.

These technologies are currently applied upon more than 50 million hectares, especially in the U.S.A., Brasilia, Argentina, Canada and Australia. Throughout Europe, the surface worked in a conservative system is estimated at about 1 million hectares, and in Romania this issue is timidly approached in the framework of certain scientific researches [3].

If the situation in this field in Romania could be explained by the destabilizations which have taken place in agriculture since 1990, it seems difficult to comprehend the European farmers' reticence towards the sowing technologies in non-ploughed land, farmers who cannot invoke the material hardships or the lack of information. This means that the application of these technologies is in connection with a series of other factors which, unless correctly solved, may annul the theoretical advantages of the system in question [4].

The basic factors which condition the application of the conservative technologies for the working of the soil and for the sowing may be:
- the technical factors;
- the agro-technical factors;
- the biological factors;
- the socio-economic factors etc.

The technical factors refer to the sowing equipment in non-ploughed land, which must wholly fulfill the requirements imposed by national norms, such as: the observance of the quantity of seed distributed upon the unit of area (maximum non-uniformity 2%); the degree of non-uniformity upon the working width of the sowing on a horizontal land (3%); the incorporation depth, which may vary with i 1 cm for the seeds buried at more than 4 cm and with ± 0.5 cm for the seeds buried at less than 4 cm; the distance between the beds must be rigorously equal, as the one of the seeds upon the bed. A modern broadcast seeder should be able to distribute quantities comprised between 1.5...400 kg/ha, according to the culture, to the biological and cultural value of the seeds, to the specific agro-technical requirements etc.

Moreover, the sowing outfit in non-ploughed land must make the soil loose upon a narrow strip (4...5 cm) at the necessary depth, the seeds must be incorporated, then covered and brought to safe contact with the soil. Often times, the land is covered with vegetable remains, which must be chopped finely so that they should not get in the way of the incorporation and covering of the seeds.

We have to notice that the technical aspect of the sowing in non-ploughed land is practically solved throughout the world, an important number of companies carrying out research and building competitive sowing outfits (NOKKA-TUME - Finland, HUARD, SULKY, KUHN - France, GASPARDO - Italy, JOHN DEERE - U.S.A., HOWARD, DUTZI and AMAZONE - Germany, BALDAN - Brasilia etc.). It has been ascertained that many companies producing such technical outfits are situated in Europe. In Romania the broadcast seeder SCN-17 for stalk cereals has been accomplished by the society SC MECANICA CEAHLĂU, S.A. - Piatra Neamţ, in collaboration with INMA - Bucharest.

3. RESULTS AND DISCUSSION

As a general rule, the broadcast seeders for non-ploughed land possess the same constructive structure as those for land worked by classic technologies (Figure 1), the main distinction referring to the attachment in front of the machines of some batteries with staggered disks (Figure 2) with a view to stirring the soil and to accomplishing some mechanisms for the incorporation of the seeds and for their covering, more solid than the ones for normal sowing.

The disk cutters may be flat, indented or corrugated (chamfered), the disks with intermittent bit penetrating more easily the soil, with a better cutting of the vegetable remains and with the elimination of the possibility of their pulling along in front of the disks. We also distinguish among outfits endowed with one, two or three disks for stirring the soil on the band which is to be sown.

Figure 1: Sowing outfit for non-ploughed land

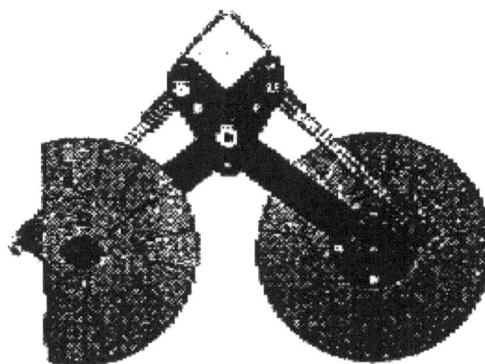

Figure 2: Battery of disks for stirring the soil

The outfit for directly sowing the soil with disk cutters stirs the soil in a small degree, and the bottom of the furrow is even slightly sunk. This aspect constitutes an advantage in the dry areas, due to the loss of a minimal quantity of water from the soil and to the driving of the water to the seed through the slightly sunk layer of soil from the bottom of the furrow.

The alternative to the system of stirring with disks are the chisel knives, which accomplish a stronger stirring and mixing of the soil, which leads to greater water losses. By means of a more intensive mixing, there also takes place a richer growth of the weeds, and in case of larger quantities of vegetable remains, the running of the equipment is no longer safe.

From the experimental research there has come out that the sliding-shares of the normal broadcast seeders don't possess a satisfying stability while they run, for which reason there is recommended their replacement with double disk shares, more solid and more reliable. For a safe covering of the seeds, very good results have been produced by the spur wheels fitted in the blades of the equipment for seed incorporation.

The agro-technical factors which may influence the sowing in non-ploughed land are the most numerous and the most important for the success of the activity and they take into consideration the compatibility of the soils with the respective cultures, the rotation of the cultures, the combating of the weeds and of the harmful insects etc. If we take into consideration the different types of soils (Figure 3) we may notice that in the soils which contain in the upper horizon rocks which overtop 30...50 mm in diameter, the activity of sowing becomes difficult, as the disks which make the soil loose are exposed to the deformations or to the breaks (a situation which is specific to the hill area, that is for 30...35% of the tillable surface of the country).

However, the most difficult problem is the scientific rotation of the cultures, so that, at the same time with the process of retaining the water and the nutritive substances within the soil, the weeds should be fought against as efficiently as possible. It has been acknowledged that one of the major advantages of the classic working of the soil consists precisely in the active combating of these weeds, which stands for an important disadvantage as regards the conservative works of the soil. In order to fight against the weeds in this latter situation, there is required an increased quantity of herbicides, a fact which contradicts the principles of the ecological agriculture (durable).

The integrated combating of the weeds stands for the optimal solution, but it implies the insertion in the system of rotation of the cultures of some vast and extremely vast surfaces so that the weeds eradicated during one year should not do over again in following year out of the seeds carried by the wind from the neighboring lands. However, it seems that until reaching this desideratum in Romania, and even in Europe, some time has still to pass. Moreover, the agriculture is obliged to produce what the market demands, to be efficient, an aspect which do not always harmonized with the scientific rotation of the cultures.

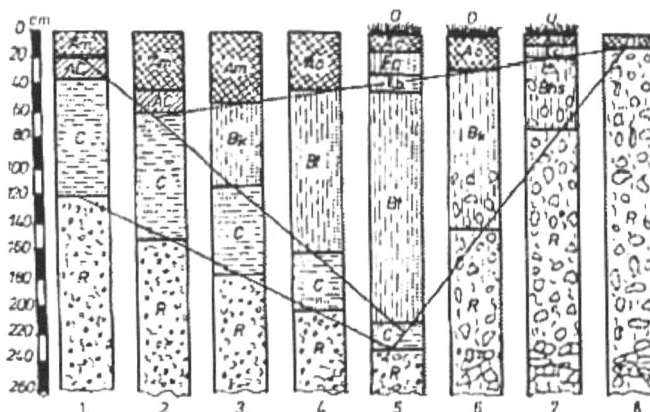

Figure 3: The section in the depth of the soil

Another aspect of an agro-technical nature belongs to the corresponding putting in readiness of the fields upon which the sowing on conservative principles is to be accomplished. This is about the insurance of the surface flatness, their settlement for irrigations and draining, so that the sowing outfit should be provided with the working conditions by whose means the basic requirements imposed by the norms should be fulfilled. Unlike in the framework of the well-developed European countries, in Romania this aspect practically fails to have been approached, and until it has been solved, a significant period of time has to elapse.

There is incumbent to the scientific research the role of determining the "suitability" of the soils with the sowing in a conservative system, and especially to convey the most efficient solutions for a rotation in compliance with the necessities of this system.

The biological factors, which have to be taken into consideration when adopting the conservative system belong to the possible different adaptability of the varieties of a certain species to the germination and afterwards the development within the soil which has been worked only upon narrow strips. It is also important that the vegetable remains from a certain culture should not hinder the process of working the respective strips by means of the normal equipment fitted in the broadcast seeders. For instance, a variety of the corn species, whose stem reaches in the lower part up to 35...50 mm in diameter will raise more difficult working problems than the variety of the corn species whose stem is of only 10...20 mm, sustaining at the same time an equal number of corn, cobs (and even greater) and being susceptible of being cultivated at a denser density of the plants upon the hectare. This means that the sowing in a conservative system needs varieties of a light germination, endowed with strong weeds and stems of reduced dimensions, characterized by an increased adaptability to the climate and soil of the respective area.

The socio-economic factors refer to support offered by the State to the promotion, at least in an experimental variant, of the soil working in a conservative system. Also, by the agricultural policies promoted by the State, the large-dimension farms have to be encouraged, as these ones are able to allow correct rotations of the agricultural cultures, so that the immediate economic advantage offered by the lack of tillage should be maintained by normal subsequent works, optimal plant densities and final competitive crops.

There has been specified that the system of conservative working of the soil is favored by the remaining on the soil of a part of the vegetation of the previous culture, which has nonetheless to be finely chopped and uniformly spread across the field. This means that the cropping machines have to be provided with chopping tools for the vegetable remains, an aspect which leads to the rise of their constructive complexity and to the cost of the work.

In conclusion, we may state that the issue of the soil working in a conservative system must be approached as a facet of the integrated management of the agro-systems, which deals simultaneously with distinct directions, correlated however among themselves, such as: biological management, ecological management etc., all of them related nonetheless to the classic management (of the production, financial, of the human resources, of the waste etc).

4. CONCLUSION

1. The traditional method of working the soil is increasingly criticized because of the high cost, of the excessive compaction of the fertile layer, of the predisposition to erosion, of the structure destruction, of the diminution of the humus content, of the useful micro- fauna destruction etc.
2. The method of working the soil in a conservative system partially eliminates these disadvantages, but its application is conditioned by a series of technical, agro-technical, biological, socio-economic factors etc.
3. At the present moment in Romania, only the technical aspect afferent to the sowing in non-ploughed land is almost solved, while some of the other factors have not been even approached so far.
4. The activity of working the soil in a conservative system has to be accomplished on the basis of the integrated management of the agro-systems, by the involvement of the scientific research of the upon the farmers, the producers of agricultural equipment and even upon the political factors.

REFERENCES

[1] Brătucu Gh., Agricultural Technology, Transilvania University of Brasov Publishing House, Brasov, 1999.
[2] Brătucu Gh., Drilling in Non Plough Land a Component for Sustainable Agriculture in Romania, in Bulletin of the 7th Conference BIOATAS, vol. 3, p. 869-872, Brasov, 2005.
[3] Brătucu Gh., Constructive Particularities of the Equipments for Drilling Cereals in Non Plough Land, in Bulletin of the 7th Conference BIOATAS, vol. 3, p. 873-878, Brasov, 2005.
[4] Gângu V. et al. Researches Regarding the Design and Working of the Drilling Machines for Cereals in Non Plough Land, in INMATEH 2004-r Magazine, p. 33-40, Bucharest, 2004.

STUDY ON RHEOLOGICAL BEHAVIOR OF BAKERY DOUGH

C. M. Canja, M. I. Lupu, V. Pădureanu
[1] *Transilvania* University of Brasov, Brasov, Romania, canja.c@unitbv.ro

Abstract: *Bread and bakery products represent basic food daily consumed of all the people. In this context, the quality characteristics of the bread have an important role regarding the consumer choice. So, in this study are specified the rheological properties of dough and its influence on bread.*
Keywords: *dough, rheological, bread, kneading*

1. INTRODUCTION

The dough is a complex colloidal medium that is formed during the mixing of the flour with the addition of water. The rheological characteristics of dough and the extensibility or elasticity are due to the most to the gluten, which is formed during the process of mixing from the protein of wheat flour gluten. The gliadins cause the extensibility of gluten and the bread volume, and the gliadins causes the extensibility of gluten and its tolerance to the mixing. By adding certain substances of oxidation to the dough, the protein network suffers some important changes, caused by the conversion of cysteine aminoacid in cystine, forming some disulfuric linear molecules in protein fiber [6].

The kneading is the technological operation after which is obtained a homogeneous mass of dough with a specific structure and rheological properties (strength, extensibility, viscosity, elasticity, plasticity), by mixing raw and auxiliary materials. The rheological characteristics of dough affect directly the quality of final product: the elasticity of the core and the peel, the volume and the form of the bread, as well as maintaining its freshness. When the dough has the elasticity and the extensibility sufficiently high, results loose bread, with developed volume and core which has pores with thin walls. If the dough is too tough, the bread that is obtained is undeveloped, with dense core and when the dough is too extensible, the bread is flatting and it has low volume and coarse porosity [6].

2. THE FORMATION OF THE DOUGH AND THE PROCESSES THAT OCCUR

During the kneading process there takes place a number of physical, chemical, and biochemical, colloidal and microbiological processes that cause significant changes of the substances in the dough mass.

The physical processes depend on the way of mixing the flour with water. Depending on the manner of conducting the operation, the forming of the dough is divided into two stages: mixing the raw and the auxiliary materials and proper mixing of the mixture thus obtained.

In the first moments of the mixing, the flour water absorption leads to the formation of small wet clumps separately. Following the contact with water, is developing the hydration heat, around 27 cal/g of flour.

Continuing the mixing, it is reached the development stage of the dough, when little wet clumps merge into an uniform mass, and the water from its surface disappears, becoming smooth and shiny and begins to manifest the elastic properties. The time for the optimum development of the dough is 2…25 minutes, depending on the quality of the flour, the water added and the type of mixer.

The next phase is the stability of the dough when it is subject to the distortion due to the velocity of the gradients that arise.

The last phase of mixing, which should be avoided is the softening of the dough, and is characterized by changing the rheological characteristics. The dough becomes soft, slightly elastic and highly extensible and finally loses cohesion, becoming sticky and just like a viscous liquid.

The phases of dough forming can be observed by tracing the farinograph curve.

During the kneading, due to the heat of hydration and the transformation of part from mechanical energy into heat kneading, the dough temperature increases. The increase of dough temperature accelerates its formation. Exceeding the optimum temperature of dough formation, 28°C, leads to the increase of enzyme activity, the dough viscosity decreases, which has negative influence on the rheological properties of dough and can occur even distortion of proteins.

The colloidal processes are represented by the hydration and swelling processes of the dough components.

Hydration of the flour is a complex meal. Flour components bound water in various ways. Although the protein and the starch bind the largest amount of water in the dough, an important role also has the pentozans.

The proteins from flour bind the water both by absorption and by osmosis. The osmosis leads to the swelling of gliadin and glutenin resulting the gluten. The water related through absorption forms around the proteins the film hydration [4].

At the formation of the gluten an important role is played by the amount of water used. Not enough water will not satisfy the necessary required by the gluten, its structure is not formed completely, and its quality will be poor [4].

Biochemical processes occuring under the action of enzymes and they tend to the degradation of macromolecular constituents of flour to form simpler compounds which modifies the rheological characteristics of dough [4].

As a result of amilolyses process, during the kneading the dextrins and the maltose increase in the dough. They, in particular β – dextrins contribute to the increase of the dough viscosity. Also, the dough begins to activate the lipoxygenase, which in the presence of oxygen oxidizes the polyunsaturated free fatty acids and their monoglycerides [1].

The nature of chemical groups on the protein structure leads to the formation of covalent bonds as: disulfide, as well as non – covalent bonds: hydrogen bonds, hydrophobic bonds, ionic bonds. The gluten is formed so as a result of the interaction between gluten protein. The main role in the formation of gluten plays the glutenin, which favors the interactions and the associations with other proteins and other constituents of flour. Due to its large molecule, the hydrated proteins can form films, and at the kneading, its ability to interact increases.

During the kneading in the dough is included some air. A part is dissolved in the aqueous phase, and the remaining air forms microbubbles. These bubbles contribute to the pore formation in the dough at the kneading, and the oxygen from the air takes part in the oxidation processes from the dough.

2.1. The factors which influence the dough formation

The formation of the dough and its rheological properties are affected by a number of factors, as they are represented in the following figure 1.

The kneading conditions are represented by the intensity of kneading, the amount of energy transmitted to the dough, the kneading duration. It decreases with the increase of the speed of the kneading arm. The kneading duration influence a lot the kneading properties, thus lead to an optimal development, or an incomplete development or to a development too high [1].

The flour quality: The dough obtained from flour of poor quality is different from the dough obtained from flour of good quality. The dough made from low flour the protein' films break easily, even before their uniform distribution in the dough. In the dough obtained from flour of high quality the hydrated proteins are elastic, and to a kneading too high, the protein films present just a few breaks. This stability to this kneading is one of the most important characteristics of flour required.

The amount of water. A higher or a smaller quantity of water different than is required to achieve the normal consistency extends the kneading duration. Dough is very sensitive to a kneading too high, contrary to the dough that has a sufficient tolerance.

The electrolytes, particulary the salt (NaCl). The addition of neutral salts modifies the nature and the intensity of the hydrophobic interactions between the gluten proteins. Increasing of the ionic strength in the dough following the introduction of salt reduces the water capacity of retention by proteins [1].

2.2. The constituent phases of dough

From the physical point of view, the dough is composed of three phases: solid, liquid and gas. The solid phase is composed of insoluble constituents and bound water: gluten proteins limited swollen, starch granules, bran particles and other solid ingredients.

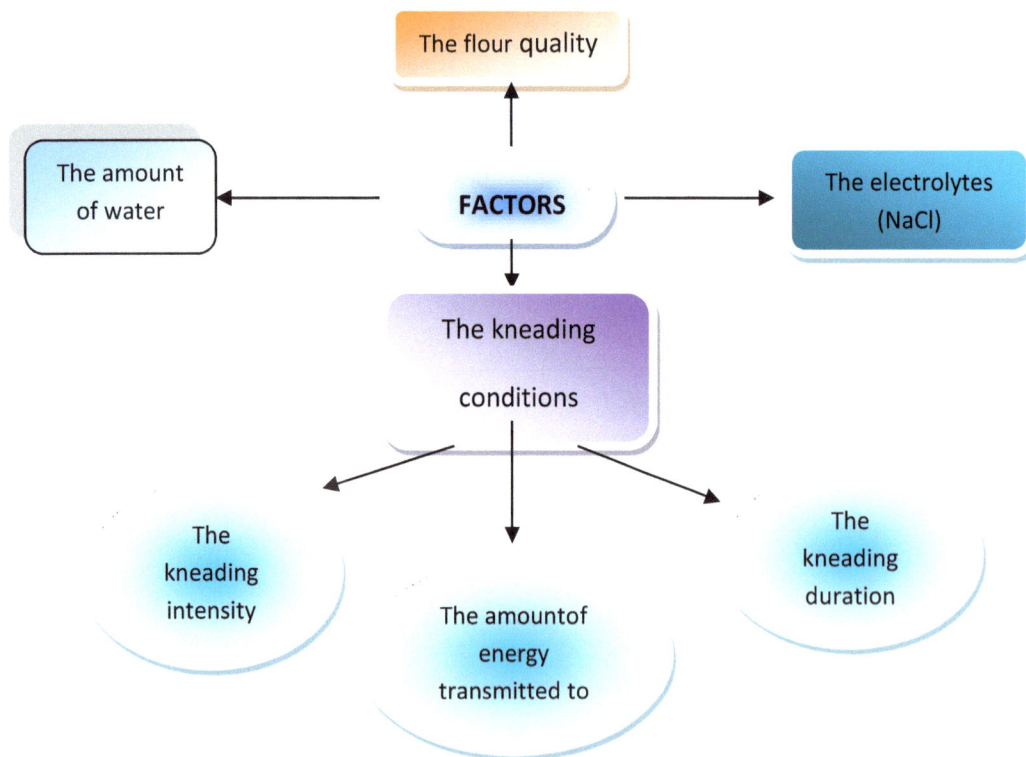

Figure 1: The factors which influence the dough formation

The liquid phase is formed in that part of water which is not bound by adsorption and there the soluble constituents of the dough which are dissolved: minerals, simple sugars, dextrins, water soluble proteins, polypeptides, amino acids. It is found partly in the form of thin films surroundings the elements of the solid phase and most of it is in the dispersed state, osmotic input by the gluten proteins in the swelling process. The liquid phase represents 8-37% by the weight of the dough. A big influence on the liquid phase has the flour quality and the kneading time. To a normal kneading it represents approx. 20%, and to a short kneading about 11% by the dough weight.

The gas phase is formed of air bubbles included in the dough while kneading. It is presented as an emulsion of gas in the liquid phase of the dough, and mostly in form of air bubbles included in the gluten protein which are swelling. To a normal kneading, the gas phase reaches 10% of the dough volume. To the kneading extension it can reach 20% [1].

3. THE DOUGH AND ITS RHEOLOGICAL PROPERTIES

As defined by The International Organization of Standardization (ISO Standard 5492.1.1977), the food rheology is the science that deals with the study of the deformation and the flow of raw materials, the intermediates and the finished products in the food industry.

The rheology accepts as old models the bodies with uniform properties, those whose behavior is described by linear law. The perfectly elastic solid (Hooke), perfectly plastic solid (St. Venant) and purely viscous fluid (Newton) are particular rheological bodies.

The rheological properties express the deformation of dough in time under the action of external forces exercised on it.

The dough prepared from wheat flour is a non-liner visco-elastic body. I t has properties which are characteristic to both solids and liquids, and therefore has an ideal behaviour intermediate between solids and the liquids: when it is stress, some of the energy is dissipated and the other part is stored [2].

3.1. The rheological properties of bread dough

The rheological properties of dough are: elasticity, viscosity, relaxation, creep. All these properties are largely due to the gluten which is formed at the mixing, but also how it interacts with the other components of the flour and the dough ingredients [3].

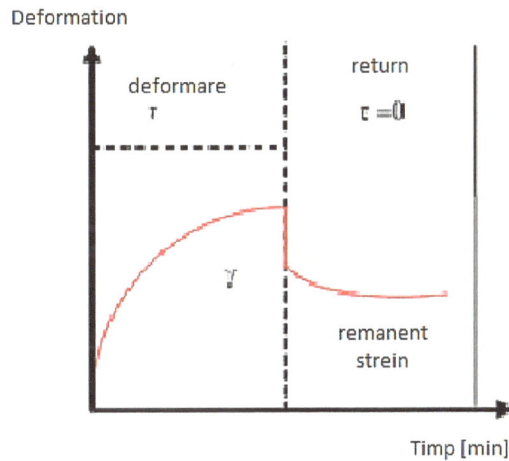

Figure 2: The deformation and its returning for a visco-elastic body
T- the applied voltage, γ-deformation

The elasticity is a property of a solid, deformable to reversibly store strain energy [5].

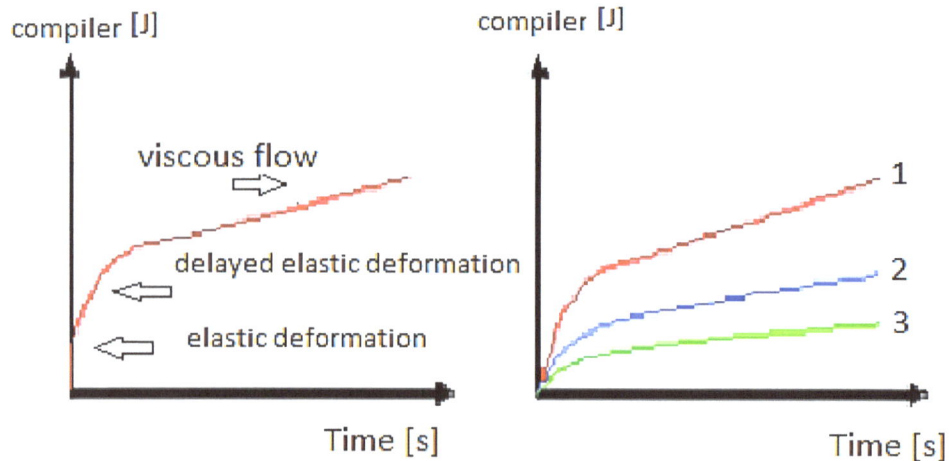

Figure 3: The typical curve of a viscoelastic material/ Deformation curves of dough from what flour (1-strong flour, 2-good flour, 3-weak flour)

The dough elasticity is provided by gluten, and particularly by glutenin, and that consists in that the dough deforms reversibly to a given applied force, then is it irreversibly deformed. The dough has an instantaneous elasticity that occurs upon the force application, and an elastic delay which occurs after the removal of the force.

The viscosity is the property of the bodies to resist to the deformation. The viscosity of the dough is an apparent viscosity, which, unlike the viscosity of the liquid, depends not only on the temperature and pressure, but also of a number of other factors such as the rate of shear that the dough has previously submissively.

The relaxation is the process of bone resorption, by decrease of the internal pressures of the dough, while maintaining the shape. The reabsorb of the pressures is made through the gradual elastic deformation in plastic deformation. The relaxation does not occur until the cancellation of internal tensions, but up to a limit determined, that is the limit of the elasticity, under the relaxation does not develop.

The relaxation time is the time when the tension from dough decreases of 2.7183 times, respectively with the base of natural logarithms e = 2.7183.

The creep is the property of a solid to flow slow and continuous under the action of a constant load [5].

The factors which influence the rheological properties of the dough

The rheological properties of dough play an important role in the production process, where the dough is subjected to the action of forces that realize the appearance of tensions and causes its deformation [3].

The quality of the flour, respective the protein content and the glutenin/gliadin report, has a great influence on the properties of dough. Thus, responsible of the dough viscosity are the gliadins which contribute to the dough extensibility, while the glutenins gives elasticity and resistance, increasing the resistance to breaking [2].

During the technological process, the dough is subjected to tensile and shear tensions.

The viscosity at the breaking by stretching and the breaking tension increases with the protein content of the flour, which explains the good performance of baking of the flour with high protein content [3].

The study of the dough behavior at shear with rotary viscometer showed that, at the request through shear, it increases its viscosity, proving its growth resistance. Increasing the viscosity of the dough to the severance occurs when it drops. The maximum viscosity to the shear decreases with the incresing of the protein content, but increases with the glutenin/gliadin ratio [1].

The amount of water; increasing the water content is accompanied by a reduction of the elastic properties of the dough and its viscosity.

The rheological properties of the dough, the elasticity and the viscosity, increase until certain values of water content, corresponding to the maximum swelling of the protein, after which the value decreases. The optimum consistency is achieved when the dough contains enough water for the flour swelling components. An optimal swell of the components influence favorably the shape stability of the dough and the bread quality [1].

The optimum temperature for dough is 28...32°C. During the mixing process, the dough temperature increases due to the heat released during the hydration of flour particles and to the pass of an energy quantity into thermal energy. Increasing the temperature above the optimum temperature leads to the elasticity worsening and consistency of the dough, due to the increase of the fermentative activity. Lowering the temperature under its optimal value shrinks the dough plasticity with negative consequences on products quality [1].

Due to the temperature influence on enzyme activity, on the microbiota activity and on the rheological properties of the dough, it is best to use a lower temperature to the process of the weak flours and to the strong flours a higher temperature.
The kneading time is influenced by: the quality of the flour, the water quantity and the speed of the kneading arm.

Depending on the flour quality used, the dough can be formed slower or faster. The dough prepared from flours with high extraction and big extraction is more sensitive to the kneading than those obtained from low extractions flours and high extraction.

The dough of low consistency is very sensitive to a high kneading, contrary to the consistent dough which has a big tolerance. The kneading time decreases with the kneading arm speed [2].

For the traditional kneading, the kneading duration is between 6 and 12 minutes, while in case of the indirect method, the yeast is kneading 6...10 minutes, and the dough 8...12 minutes [2].

The end of kneading; it is appreciated through sensory analysis. The dough well kneading should be smooth, tight, consistent, flexible and easy to detach by the arm mixer and by the box wall where has been kneaded. At the manually sample, stretched between thumb and forefinger, the dough should stretch into a thin strip, transparent and flexible without breaking. The dough insufficiently kneaded is homogeneous, but is sticky and viscous. The dough excessive kneaded is highly extensible, without tenacity and to the manual sample it breaks.

4. CONCLUSION

- The dough is a complex colloidal medium that is formed during the mixing of the flour with the addition of water.
- The rheological characteristics of dough and the extensibility or elasticity are due to the most to the gluten, which is formed during the process of mixing from the protein of wheat flour gluten.
- From the physical point of view, the dough is composed of three phases: solid, liquid and gas.
- Depending on the flour quality used, the dough can be formed slower or faster. The dough prepared from flours with high extraction and big extraction is more sensitive to the kneading than those obtained from low extractions flours and high extraction.

REFERENCES

[1] Bordei D., Tehnologia moderna a panificației, Editura A.G.I.R., București, 2004.
[2] Bordei D., Controlul calității în industria panificației. Metode de analiză, Editura Academica, Galați, 2007.
[3] Codină R., Proprietățile reologice ale aluatului din făina de grâu, Editura A.G.I.R., București, 2010.
[4] Ghimbasan R., Tehnologii în industria alimentară, Editura Universitatii Transilvania din Brașov, Brașov, 2000.

[5]   Rus Fl., Bazele operatiilor din industria alimentara, Editura Universitatii Transilvania din Braşov, Brasov, 2000.
[6]   Voicu Gh., Procese şi utilaje pentru panificaţie, Editura Bren, Bucureşti, 1999.

OPTIMIZING THE ENERGY CONSUMPTION TO MOULDS OF PALLETING MACHINES OF MIXED FODDERS

C. Csatlós[1]

[1] Transilvania University of Brasov, Brasov, ROMANIA, csk@unitbv.ro

Abstract: This paper aims to analyse the development of technological works of granules production from concentrate fodders and to insist upon the physical-mechanical principle of their achievement. The proposed mathematical model has few simplifying assumptions, which is why it can be used for the rational sizing of moulds of pelleting machines. The methodology of approach of this topic can be expanded with specific elements and also for granulation, and respectively for briquetting the fibrous fodders.

Keywords: moulds, granulation, mixed fodders

1. GENERAL CONSIDERATIONS

The granulation is the process of pressing of concentrated mixed fodders, of vitamins, of bioenergizers etc, in order to obtain some fodders with a high content of nutritive substances.

The granulation allows the increase of the mechanization and automation degree of the feeding process, with an important disposal of the dissipation.

The most important advantages of this process are:
- rational use of resources;
- the decreasing of storage by increasing its density;
- a high homogenization of the composition;
- a natural repose angle, with positive effects of the flow through the plant ducts of food distribution;
- also allows the air circulation through granules during storage;
- and do not make possible for animals to choose in a preferential way the components of the fodders.

The granulation process involves a series requirements, among the basic ones we mentioned:
- the ones which does not modified the organoleptic properties of fodders;
- those which realise granules with a high mechanical resistance and a low friability;
- the granules sizes should match with the zootechnical and biological needs of animals to which they are for.

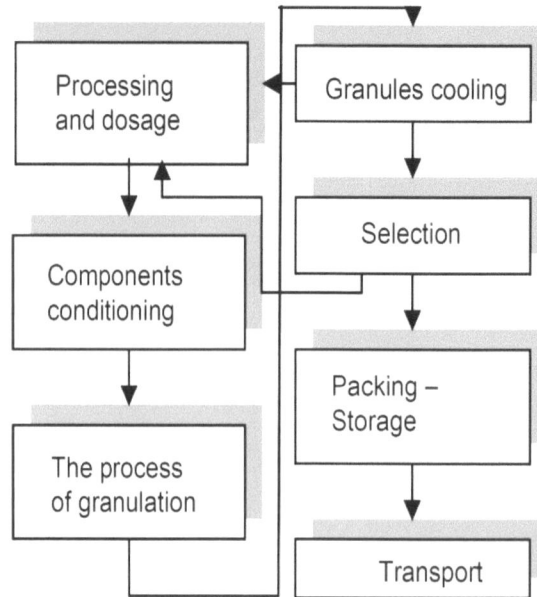

Figure 1: Block pattern of the technological process of fodders granulation

The conditioning of raw materials consists of shredding and of dosing the components in parallel with the adjustment of working temperature and it is carried out with the following purposes:
- particles agglomeration (with steam or sticker addition);
- the increase of mechanical resistance;
- an easy decomposition to digestion;
- the correlation of friction forces to pressing;
- the decrease of the sticker effects of the working to pressing.

The block pattern of the technological process is shown in Figure 1.

The marked returns indicates the bringing back of crushed grains or of those which do not have the requirements imposed in the processing sector, and they are once again placed on the stream.

We can se that the granulation still presents some technological difficulties, additional costs related to the investments in plants and high energy consumptions. However these difficulties are attenuated through the advantages presented at the beginning of the work.

2. THE MATHEMATICAL MODEL OF THE COMPRESSION PROCESS

The fodders pressing essentially represents their density increase, by raising up the pressure which acts upon them. The experimental researches show that the variation of pressure depending on density can be expressed as a relation under the form of [2]:

$$p = C \cdot \rho^{m} \tag{1}$$

where C and m represents the dependent coefficients of the material subjected to compression and ρ- the specific mass.

This variation is represented in Figure 2, and it can be touched by fodders humidity. It was observed that the pressing needed to granulation decrease at the same time with the increase of the humidity.

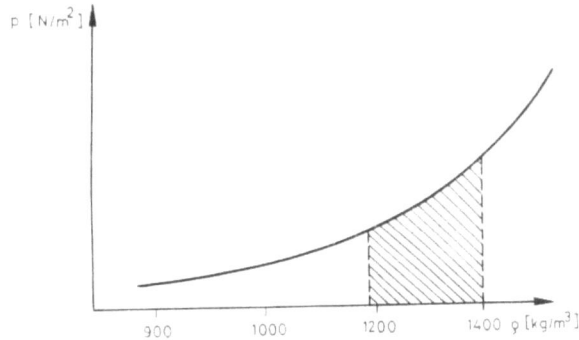

Figure 2: The change of granules density depending on the working pressing [2]

The physical principles of the granulation process with the symbolical representation of the active working organs are shown in Table 1.
Regardless the form of pressing chambers, during working, there are forming friction forces with sides. They

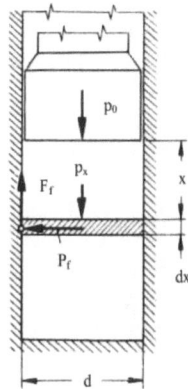

Figure 3: The physical model of pressing chamber

change the pressure in the direction of compression and the one perpendicular to it.

The mathematical model proposed is based upon the simplifying assumption of a cylindrical pressing chamber with inner diameter d, as shown in Figure 3.

It is considered that at the current distance x, the friction pressure p_f, perpendicular to the motion direction is constant. If you take an elementary volume of a cylindrical shape with the height dx and pressure given by the working organ p_0, we can write the equilibrium equation of forces after axial direction of motion:

$$\frac{\pi \cdot d^2}{4} \cdot dp_x + \mu \cdot \frac{v}{1-v} \pi \cdot p_x \cdot dx = 0 ; \tag{2}$$

where: p_x represents the the pressing on the elementary conssidered surface; μ - the friction coefficient with the inner walls and v - Poisson coeffient.

After simplification and separation of variables it results the following differential equation:

$$\frac{dp_x}{p_x} = -\mu \cdot \frac{v}{1-v} \cdot \frac{4}{d} d_x . \tag{3}$$

After the integration between the variables limits p_0-p_x, respectively $0 - x$ we obtained:

$$ln\frac{p_x}{p_0} = -\mu \cdot \frac{v}{1-v} \cdot \frac{4}{d} \cdot x , \tag{4}$$

which after the logarithmation is transformed into:

$$p_x = p_0 e^{-\mu \cdot \frac{v}{1-v} \cdot \frac{4}{d} \cdot x}. \tag{5}$$

This relation is valid in the case of compression of closed chambers.

Compression, in the case of granulator with moulds or spiral or with spur, takes place in pressure chambers realised under the shape of some open ducts.

In this case, the pressing forces are balanced on the friction ones between the material and the duct sides. Their size depends on the length of the duct.

If generally the pressure relation (5) is written down with x = l (duct length) and $p_x = 1$ (atmospherical pressure), results:

Table 1: Physical principles of the granulation process [2]

Granulation by aggregation	Granulation by drawing				Granulation by compression	
	With slug	With annular molds	With plat mold	With spurs	With sectional rollers	With cylindrica l chamber
p=0	p = 10...100MPa				p ≤ 1000 MPa	

$$1 = p_0 \cdot e^{-\mu \cdot \frac{v}{1-v} \cdot \frac{4}{d} \cdot l}. \tag{6}$$

After logarithmation we can extract the expression of the duct length for which is necessary a pressure p_0:

$$l = \frac{d \cdot (1-v) \ln p_0}{4 \mu \cdot v}. \tag{7}$$

This relation highlights the fact that the pressure of duct length is proportional with the natural logarithm of pressure.

The determination of power consumed during the pressing process can be realised with the following relation:

$$P_{granulation} = F_f \cdot v_m \cdot n_{cps}, \tag{8}$$

where: F_f represents the friction force of the material with the sides of the passing ducts; v_m – the average motion velocity of granules and n_{cps} – the number of ducts with simultaneous pressing of the blending.

The variation of friction force is calculated with the expression:

$$F_f = \mu k \cdot p_0^l p^l \tag{9}$$

where k is the coefficient of lateral displacement (k=0,4...0,45) and l_p – the perimeter of the pressing duct.

The average motion velocity of granules is:

$$v_m = \frac{l}{t_g}. \tag{10}$$

where t_g represents the bleeding time which is included between 16 and 18 s.

For the number of ducts pressed simultaneously we can write:

$$n_{cps} = n_c \cdot n_r \cdot \frac{\alpha}{2\pi}, \tag{11}$$

where n_c represents the total number of the mould ducts, n_r – the number of pressing rollers and α- the central angle of the pressing zone of the roller.

3. THE RUNNING OF THE MATHEMATICAL MODEL AND THE INTERPRETATION OF THE RESULTS

The mathematical model presented above has been rolled out in MATHCAD programme and there were used the following input data: the inner diameter range of the flute compression between 4 and 8 mm, length 25 ...50 mm, friction coefficient of 0.1 and Poisson's ratio of 0.85.

The graph obtained is shown in Figure 4 and represents the working pressure variation which depends on the length and of the diamater of compression duct.

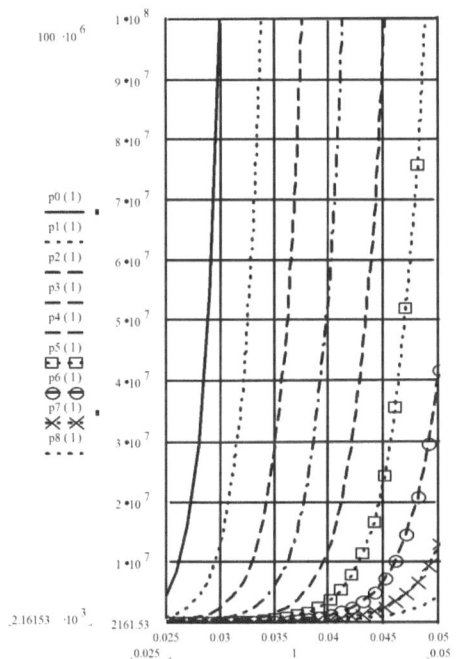

Figure 4: The dependence of the granular pressure depending of the length of the flute

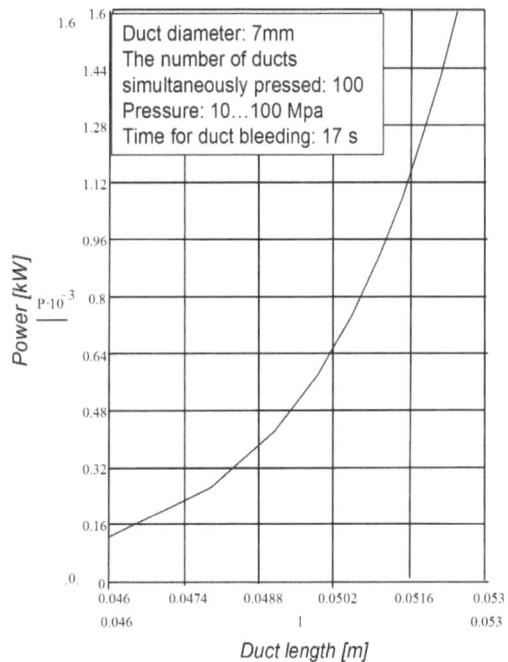

Figure 5: The dependence of the consumed power to granulation depending of the length of the flute

Knowing the variation of the length of the duct depending on pressure, it was studied the power consumption during the compression process, in the case of a hypothetical granulator with the hole diameter of 7 mm.

For this was required a variable pressure from granulation range (of 10 ...100 MPa), depending on which resulted a range of ducts lengths. These have allowed the study of friction varition, and in the end of the consumed power during the process of granulation. The results are graphically shown in Figure 5.

4. CONCLUSIONS

The analysis and modeling of the granular process presented in this paper allows the achievement of some forms and sizes for a wide range of moulds. The way how the process of pressing and making-up the granules makes possible the application of the method described and the sizing of the active elements of the pressing and briqueting equipments.

Thus we cand find an optimum variant between energy consumption, geometric dimensions of the end product and the biological requirements of the animals, the size of granules depending on species and age.

REFERENCES

[1]   Griba V. K., et al., Mehanizaţia jivotnovodstva, Minsk Uradjanîi, Minsk, 1987.
[2]   Mikecz I., Az álattenyésztés gépei, Mezőgazdasági Kiadó, Budapest, 1985.

RESEARCH ON IMPROVING COMFORTABLE CABLE CAR BY LATERAL DAMPING

M. Hodîrnău[1], E. Mihail[1], C. Csatlos[1]
[1] Transilvania University of Braşov, Braşov, ROMANIA, mariushod@unitbv.ro

Abstract: In the present article are presented the results of recordings made while traveling by cable car under disturbances caused by wind. This research trying to find a lateral oscillation damping solutions and their disposal gain due to lower pendulum arm consists of assembly cable, battery rollers, arm, and cabin. Below are the measurements obtained on a model 1:10 scale on which it was mounted an active suppression system for lateral play
Keywords: cableway, cable, cable transportation, dampening, oscillation

1. INTRODUCTION

Cable transportation is used for quick access to mountains equipped to ski slopes and connects the massive of mountains for tourist attraction. It also represents a means of travel and entertainment, having an important role in the development of mountain tourism (infrastructure tourist altitude resorts), but also for winter tourism.

Experiments remain present in the life of a cable transport installation and after its introduction into production, to check the stability of manufacturing technology, maintain quality and reliability, confirmed during certification (approval). Any changes to a product in mass production involve a review and approval based on appropriate tests.

Therefore, experimental research precede and accompany all stages of the existence of a facility, as an object of economic activity, giving original certificate of conformity with the requirements of the design project and later all certificates of maintaining technical and functional parameters to defaults, so its dynamics to maintain the desired set by the manufacturer and the customer.

2. INVESTIGATION OF VORTEX EXCITED CROSS-OSCILLATION OF BICABLE ROPEWAYS

Observations and reports of various aerial ropeways by operating personnel have time and again shown significant cross-oscillations of the cabins to occur and build up even during meteorologically calm periods in the absence of wind. This was described most recently in the case of bicable ropeways which were equipped with barrel-shaped cabins (circular cross-section). These oscillations always appeared only at reduced operating speeds. Similar behavior has been observed in the past for ropeways with cylindrical cabins; turbulence was clearly identified as the cause of these oscillations [3]. This led to the speculation that here, too, the forces which excite the oscillations are the result of periodic vortex forces acting on the cabin due to the relative wind. In order to confirm these suspicions, this was examined from a theoretical point of view, on the one hand, and by measurements taken of two different bicable ropeways while in operation, on the other [2]. As it is described later in greater detail, computer simulations have been carried out in the meantime regarding the behavior of the cabins when subjected to vortex effects [4].

3. ASPECTS FROM THE MEASUREMENTS ON THE CABLE CAR

As can it be seen in the article "Aspects of analog theoretical and experimental research on the dynamics of cable cars" [1] were measured accelerations which were then compared with a dynamic model with 5 degrees of freedom where the results were compared with data from reality.

4. ASPECTS FROM THE MEASUREMENTS ON THE CABLE CAR MODEL

To achieve scale model studied different patterns of similarity and concluded that it will take part in the study similarity as to obtain a perfect similarity is necessary to make a 1:1 scale model. Thus, the cable will be made of the geometrically in a scale of 1:10, in terms of weight will be 1000 times lower than in reality; used wire diameter will be 5 mm.

Winds will be a variable that clicks the wind distorted because the cable car to have the same effect (tilt angle) as in the real case. The calculations effectuated in Mathcad program resulted in a speed of 4.7 m/s, this speed during the tests had no effect on the cable car thus confirmed by experimental research wind speed of 4.7 m/s which reality scale model to obtain the same effect as the real model (angle of inclination of the cable car) at a wind speed of 15 m/s (Figure 5).

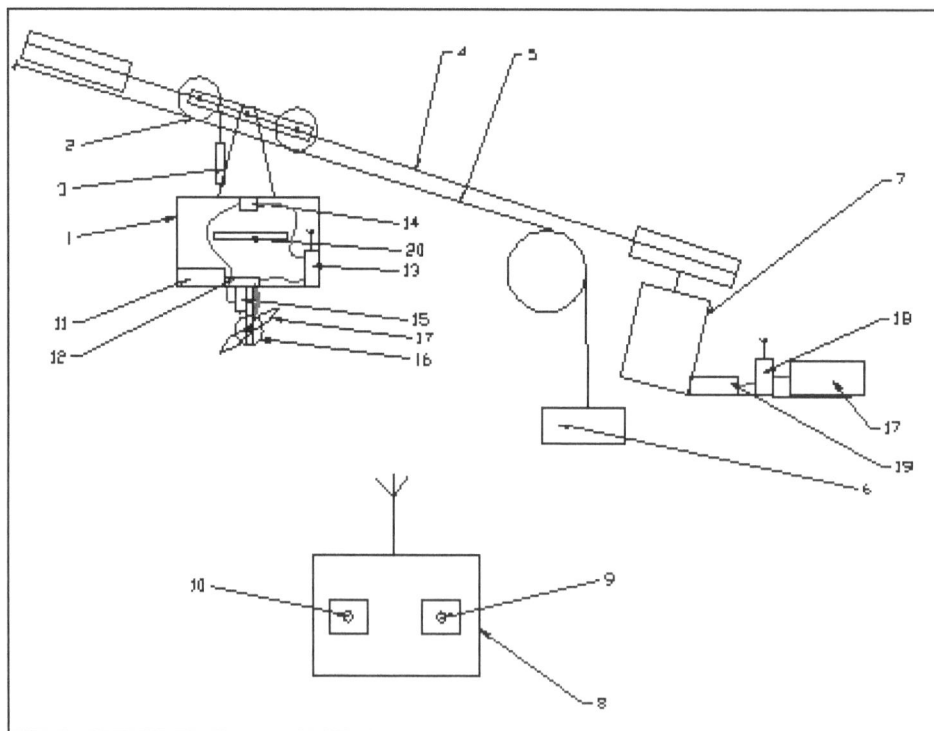

Figure 1: Functional diagram of experimental testing stand.

In Figure 1 we have the following components: 1 - cable; 2 - the battery roller; 3 - battery damper roller drawer; 4 - cable; 5 - cable carrier cable tension; 6 - weight carrier; 7 - cable drive motor carrier; 8 - control group transmitter; 9 - speed lever of the cable car drive motor; 10 - speed control lever of the motor 16; 11 - battery supply circuit in the cable car; 12 - motor speed controller of the motor 16; 13 - cable car radio control receiver; 14 - gyroscope control (rotation accelerometer); 15 - servo motor for variable pitch propeller; 17 - the propeller, 18 - radio receiver control group drive drawer cable; 19 - speed controller pulling group; 20 - data acquisition board.

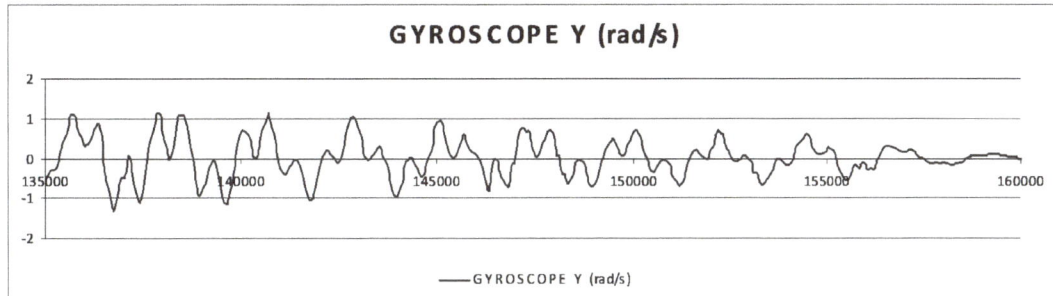

a.

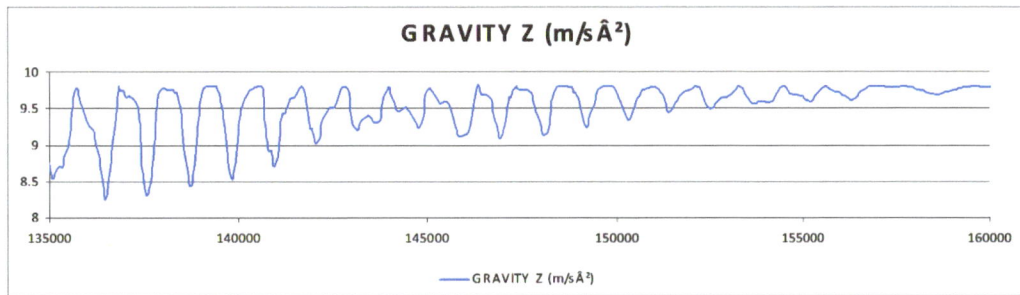

b.

Figure 2: Transverse oscillations without damping cable car in the conditions of a side gust of 5.2 m/s

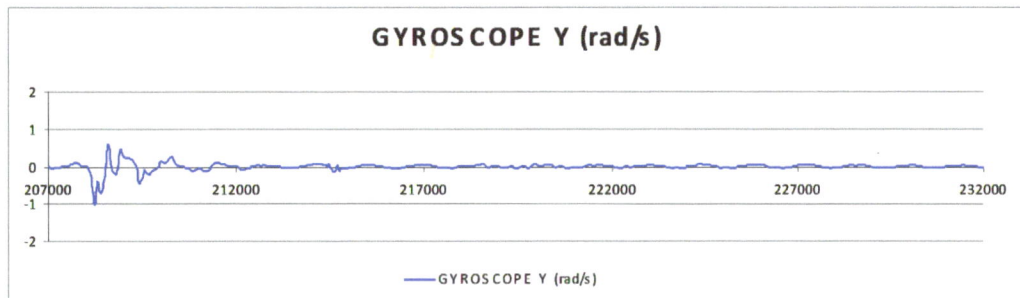

a.

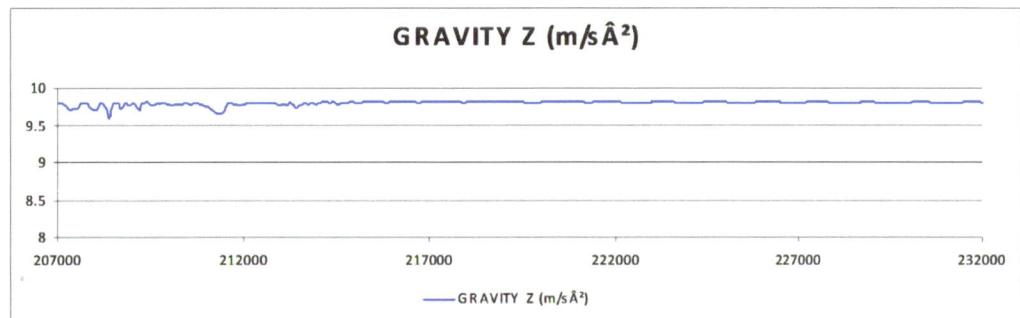

b.

Figure 3: Oscillations of dampened cable car in the transverse plane through controlled variable pitch rotor gyro sensor in terms of a wind of 4 m / s with gusts side of 5.2 m/s, the tension in the cable carrier 150N, 40N cable car mass, length of 20 m and a carrier difference at 5 m

In Figure 2 we can see that the decay time of oscillations cable car has just mounted a numbness between the roller and the arm cable car battery is 25 seconds, and if the cable car comfort but controlled variable pitch rotor gyro started, the timeout is 5 seconds (Figure 3) and discomfort to passengers is much lower. Figure 2, b can be seen as the gravitational acceleration varies passenger perception that it varies between 9.8 m/s^2 and 8.3 m/s^2.

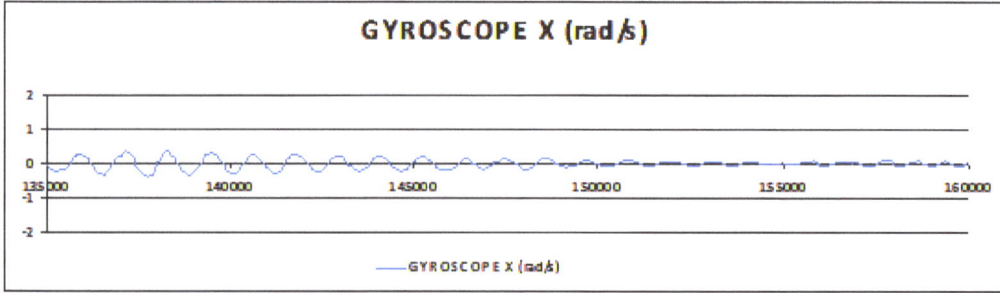

a.

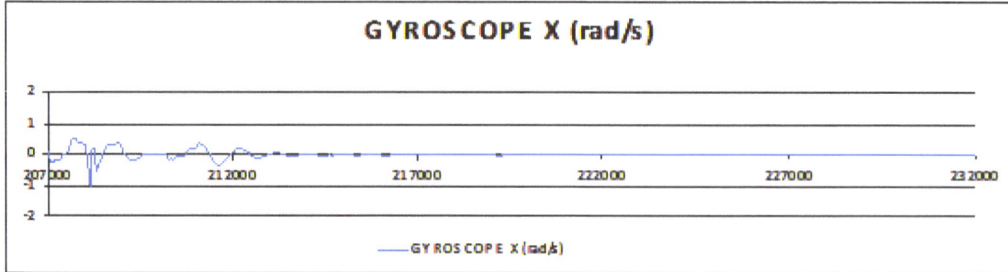

b.

Figure 4: Undamped oscillations in terms of lateral gusts of 5 m/s at the top of the figure and longitudinal oscillations damping cable car through the controlled variable pitch rotor gyro sensor on the same terms at the bottom

As can be seen in Figure 4 lateral damping plays an important role in the overall oscillation of the cable car because it helps to extinguish felt in the longitudinal oscillations as shown in Figure 2. Keep up the complete extinction of longitudinal balance occurs after 10 seconds and if no side damping Figure 2 below, the balance is off time of 20 seconds.

Figure 5: Aspects during testing with cable car crossing wind

5. CONCLUSION

Wide lateral oscillations of the cable car without stabilization device are switched off in 20 seconds and when it is on fire oscillations is done in 5 seconds.

Attempts were made to stabilize the engine and propeller step on the different fixed values to 0 or cabin but behaved as if a simple cable car, oscillation keeps extinguished in 20 seconds.

When crossing cable car equipped with damping device blast it If cable swing damping system balance without side pillars approach must slow down to each of them.

If cable car cable car system can prevent oscillation passes the pillars without slow down the speed of transport may increase, thus increasing the number of passengers that can be transported on an exchange.

ACKNOWLEDGEMENT:

This paper is supported by the Sectorial Operational Programme Human Resources Development (SOP HRD), financed from the European Social Fund and by the Romanian Government under the contract number POSDRU/107/1.5/S/76945.

REFERENCES

[1] Hodîrnău M., Mihail E, Csatlos C., Aspects of analog theoretical and experimental research on the dynamics of cable cars, in The 5th International Conference Computational Mechanics and Virtual Engineering COMEC 2013 24- 25 October 2013, Braşov, Romania.

[2] Hoffmann K. and Liehl R., Querschwingungen von ZUB-fahrzeugen, Internationale Seilbahnrundschau, No. 8, pp. 14-16, 2004.

[3] Oplatka G., Zum aerodynamischen verhalten von kabinen mit kreisförmigen grundriss, Internationale Seilbahnrundschau, No. 3, pp.4-5, 1998.

[4] Hoffmann K. and Petrova R.V., Simulation of vortex excited vibrations of a bicable ropeway, Engineering Review, Vol. 29, No. 1, pp. 11-23, 2009.

ENERGY OPTIMIZATION OF SMALL HOUSE PHOTOVOLTAIC PANEL BY INCREASING THE WORKING EFFICIENCY THROUGH ADDITIONAL MIRRORING OF SUN LIGHT ON THE PANEL AND RECOVERY OF THE HEATING ENERGY FROM THE PHOTOVOLTAIC CELLS

D. C. Ola[1], D. M. Danila[1], M. E. Manescu[1]

[1] University Transilvania from Brasov, Brasov, ROMANIA, danielola@unitbv.ro

Abstract: *Solar energy is considered a huge source of renewable energy and is regarded as one of the future solutions to the ever growing demand for energy. The industrial production of photovoltaic panels has recently increased hugely for households use. Solar cells efficiency vary under temperature conditions with direct impact on the power output. The paper presents a photovoltaic panel system designed to maintain a balanced working temperature of the solar cells and in the same time increase efficiency by over exposing the solar cells to mirror reflected solar radiation to increase the efficiency during cloudy solar diffuse light.*

Keywords : *photovoltaic panels, mirror, heat pump, microcontroller.*

1. INTRODUCTION

Since the price of the fossil fuel has escalated in the recent years so much, alternatives for independence towards the use of fossil fuels are being searched and investigated lately. The use of renewable energy sources of energy is the fore front in achieving this goal and solar energy is seen as the main energy source. The solar radiation can be converted through photovoltaic panel into power and this solution has been implemented in households that seek ways to reduce their energy costs.

A solar cell panel is basically a p-n semiconductor junction that when is exposed to light produces a dc current. Photovoltaic panels offer several advantages such as: high reliability, low maintenance cost, no environmental pollution, and absence of noise. The equivalent circuit of the Photovoltaic cell is shown in Figure 1.

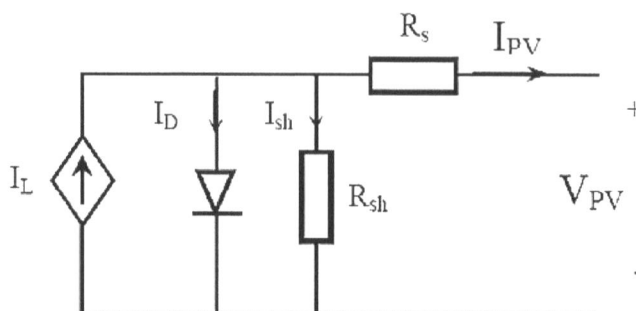

Figure 1: The schematic representation of an equivalent circuit for photovoltaic cell [2].

The mathematic model that describes the working functionality of the solar cells is presented in equation 1 and 2. These equations are used to plot the PV curves depending with the solar insolation and the temperature of the solar cell.

$$I_{PV} = I_l - I_0 \left(e^{\frac{q(V_{PV}+I_{PV}R_S)}{DKT}} - 1 \right) - \frac{V_{PV} + I_{PV}R_{SC}}{R_p} \qquad (1)$$

$$P_{PV} = V_{PV} \times I_{PV} \qquad (2)$$

where: I_{PV} is the photovoltaic current (A), I_l represents the current generated by the light (A), I_0 is the diode saturation current, q is the charge of electron (coulomb), D is the diode factor, K is the Boltzmann's constant, T is the temperature of the solar cell (K), R_{sc} is the solar cell series resistance (ohm), R_p is is the solar cell parallel resistance (ohm), V_{PV} is the module output voltage (V), and P_{PV} is the extracted photovoltaic power (W) [1].

The effect of temperature on solar cells is on of the major influences in the power output of the solar panels so that the voltage is inversely dependent on the temperature raise over the optimum level of temperature. The decrease in efficiency is higher with the increase of the temperature of the photovoltaic cells over the optimum temperature.

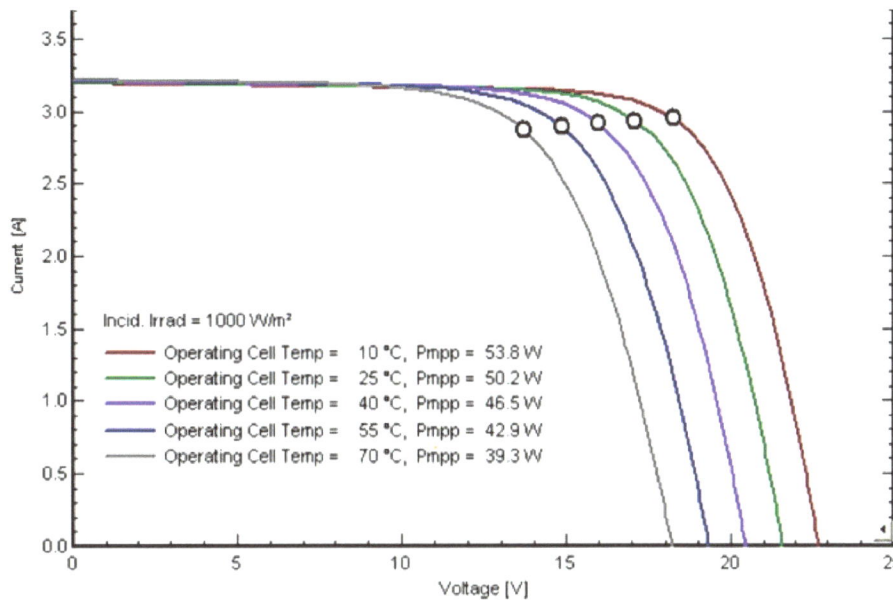

Figure 2: Output characteristics of a photovoltaic module for different temperatures [2].

In Figure 2 it is shown the effect of temperature on a photovoltaic panel I-V characteristic when exposed at constant radiation. It is well seen that for lower temperatures the voltage increases while at higher temperatures there is significant decrease [3].

The drop in energy production efficiency of photovoltaic solar panels is dependent on working temperatures of the panel, such that a drop of 1% of the peak output is considered for every increase in the temperature of the photovoltaic solar panel over the optimum 42°C. The experiments conducted by Rivers State University of Science and Technology at Port Harcourt, Nigeria found similar results for operating panel temperatures over 44°C [5].

2. THE EXPERIMENTAL STAND AND RESEARCH METHODS

By utilization of state of the art microcontroller technology for the automation [4] of the monitoring and regulation of the heat exchanger, it is possible to optimize the working temperature of the photovoltaic cells so that the electricity output of the panel will be optimized during the critical exploitation periods (ex.: high sun radiation or cold freezing winters) and in the same time gaining the excess heating energy to be used in other applications that exploit this energy source.

The experimental stand will use the mirror system to shine upon the photovoltaic solar panel to provide higher sun radiation intensity. The panel is also fitted with a heating exchange element that will maintain the photovoltaic cells at an optimum working temperature.

The working conditions will be those of lower sun light found in cloudy days as well as during the early hours of mornings and late hours of the sun set.

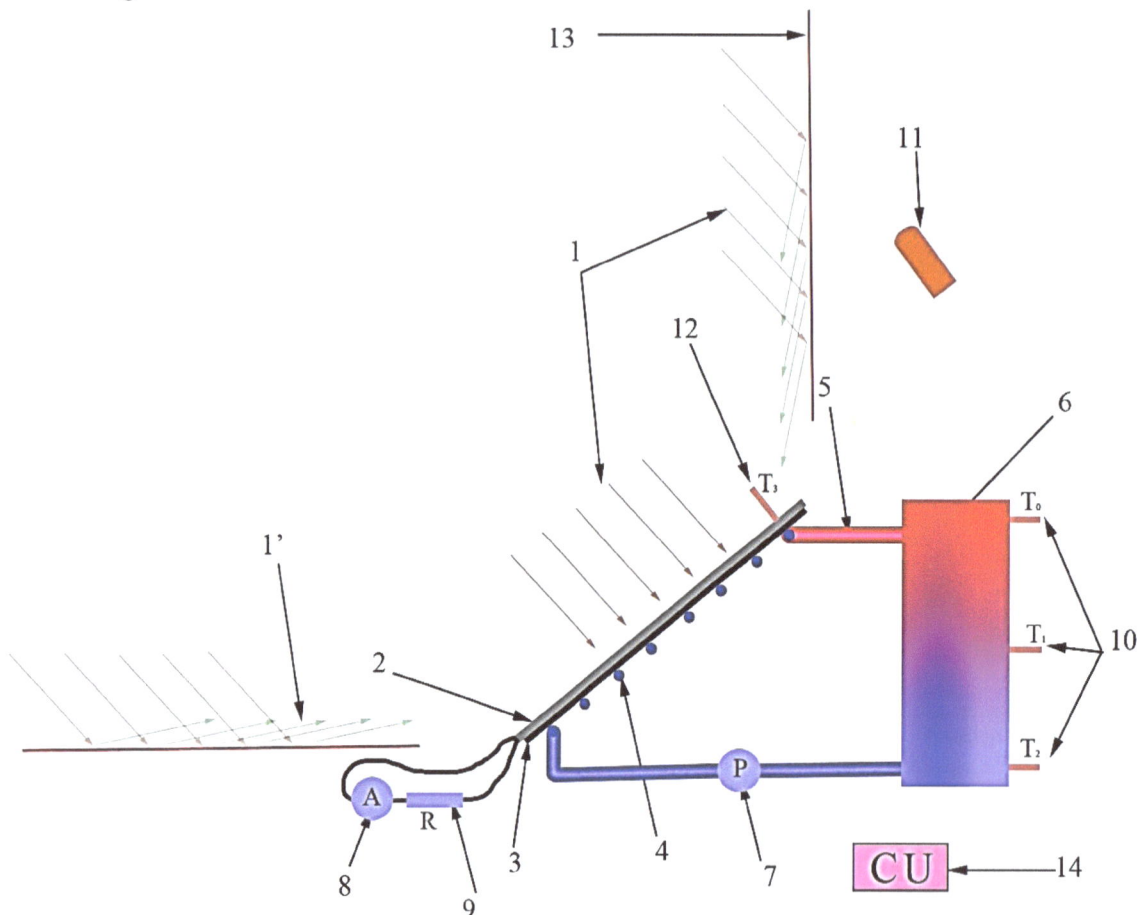

Figure 3: The schematics of the working principles.
1 – direct sun light; 1' – reflected sun light; 2 – photovoltaic cells; 3 – support plate for the photovoltaic cells; 4 - heat exchanger coil; 5 - return pipe hot water; 6 – water tank; 7 – coolant circulation pump; 8 – ampere meter; 9 – electric power resistive load; 10 – thermal sensors of the water tank; 11 – lux meter sensor measuring the light intensity; 12 – photovoltaic cells thermal probe; 13 – mirrors; 14 – control unit.

In Figure 3, the solar radiation (1) shines on the photovoltaic panel surface (2). In addition the mirrors (13) will reflect more light on the panel through the reflected radiation (1'). On the rear side of the photovoltaic cells (2) it is applied a heat sink sheet metal material (3) that is foreseen with proper insulation so that the functioning of the photovoltaic cells will not be hindered. On the other side of the element (3) it is welded a coil pipe (4) that will be used to extract the heat from the cells. The cooling coil (4) is connected to the circuit (5) made up from a storage tank (6) and a recirculating pump (7). The experimental stand will monitor the following three parameters: the electric current generated by the photovoltaic panel (2) by using a voltage sensor (9), the panel temperature T3 (12), the water temperature from the storage tank (10) on three locations (T0, T1, T2) and the intensity of the sun radiation by using the sensor (11).

The control unit (14) will run the automation program that will be responsible to the monitoring of the parameters and the positioning of the mirrors so that the panel will be exposed to more light and in the same time the cooling system will ensure an optimum working condition.

The experimental stand will be used in natural conditions so that the cooling system will maintain the photovoltaic panels at a temperature of 40°C during the over exposure to the sun radiation. The monitoring of the heating recovery will be also analyzed.

The results of the tests will provide a characteristic graph of the photovoltaic panel according with the efficiency of the output current when using the mirrors during diffuse light.

The experimental test will be done in natural conditions in order to cover the most representative conditions of diffuse light or lower sun radiation during the sun rise and sun set.

3. CONCLUSIONS AND EXPECTED RESULTS

Considering that the efficiency of the photovoltaic panels is directly dependent on the working temperature, the present paper aims to present an optimization to the photovoltaic panel by additional reflected sun light by a system of mirrors. The control of the operating temperature of the photovoltaic cells is made through a cooling system that will extract the heating energy and will make it available in household applications.

The advantage of adding more sun light radiation to the panel is beneficial especially on cloudy days and in the early morning and evening hours when the amount of sun radiation is lower. The photovoltaic panels are not foreseen with most of the time when high sun radiation is available during summer the output of the home solar panels will not provide the optimum output expected according with potential of the solar radiation.

The proposed system tries to exploit this disadvantage and harvest the potential heat energy through a heat exchanger in order to provide house hold running water (remote touristic resorts, poor access to other sources of energy etc.) or other sources of renewable energy production systems (steam micro-turbines, heat pumps etc.).

The experimental results will show the potential to optimize the energy efficiency of solar cells in periods of low output and in the same time to harness the heating energy that is directly linked with this panel.

The data collected will be used in designing a new type of house solar panel that could provide extra efficiency especially for remote locations were agro-tourism makes an impact upon the resources of the place.

REFERENCES

1. Azab M., Optimal power point tracking for stand-alone PV System using particle swarm optimization, IEEE Int Symposium on, in Industrial Electronics (ISIE), 2010, pp. 969-973.
2. Jafari Fesharaki V. et al., The Effect of Temperature on Photovoltaic Cell Efficiency, Proceedings of the 1st International Conference on Emerging Trends in Energy Conservation -ETEC,Tehran, Iran, 20-21 November 2011.
3. Patel H. et al., Maximum Power Point Tracking Scheme for PV Systems Operating Under Partially Shaded Conditions, IEEE Transactions on Industrial Electronics, Vol. 55, No. 4, pp. 1689-1698, 2008.
4. Thierheimer W., The theory of technical systems of agriculture and food industry, Transilvania University of Brasov Publishing House, ISBN 973-8124-76-X, 2001, pp. 30-33.
5. http://greenliving.nationalgeographic.com/effects-temperature-solar-panel-power-production-20500.html.

THEORETICAL RESEARCH TO IMPROVE TRACTION PERFORMANCE OF WHEELED TRACTORS BY USING A SUPLEMENTARY DRIVEN AXLE

V. Pădureanu[1], M. I. Lupu[1], C. M. Canja[1]
[1] Transilvania University, Braşov, Romania, padu@unitbv.ro

Abstract: Nowadays tractor manufacturers tend to reduce the specific weight of those. For that purpose, this work aims to improve traction performances of universal wheeled tractors by using a supplementary driven axle actuated by power timed plugs of tractor.
Keywords: driven axle, tractor, traction performances

1. INTRODUCTION

Lately, on a global level within the tractors industry it is observed a more and more accentuated tendency to reduce the constructive weight of tractors. This tendency has been generated on one side, by the necessity of cutting the tractor's price and on the other side, by the necessity of improving the traction and economic indexes of the tractor when it is exploited under different circumstances.

The traction and economic qualities of the tractor (tractive efficiency, productivity, fuel consumption etc.) depend on a large scale on the correlation of two parameters: tractor's weight at exploitation and the nominal power of the engine [1].

For the consumed power used for the tractor's motion to be minimal, the weight of tractor needs to be the smallest possible. On the other side, this weight needs to suffice in order to ensure a good adherence of the system of rolling up the soil. As tractors are exploited in a large range of velocities, they should have a variable exploitation weight: small for the transportation works with high velocities – when the drawbar load is reduced and great in the case of exploitation with low velocities – when the drawbar load is great. But most of the times it is difficult to modify the exploitation weight of the tractor, its traction qualities for the inferior velocity stages are not limited to the power of the engine. This way, for low working velocities the tractor's engine is not used rationally [2].

An effective solution for improving the utilization degree of the engine and, at the same time, for improving the traction and economic efficiency of the tractor regarding varied agricultural works, is using a supplementary driven axle.

In order to underline the efficiency of these systems, as follows, there is drawn up a comparative study of the traction qualities of the tractor 4x4, with and without supplementary driven axle.

The basic scheme of the system formed by the tractor combined with the supplementary driven axle mechanically activated by means of the tractor PTO (power take off) is presented in Figure 1.

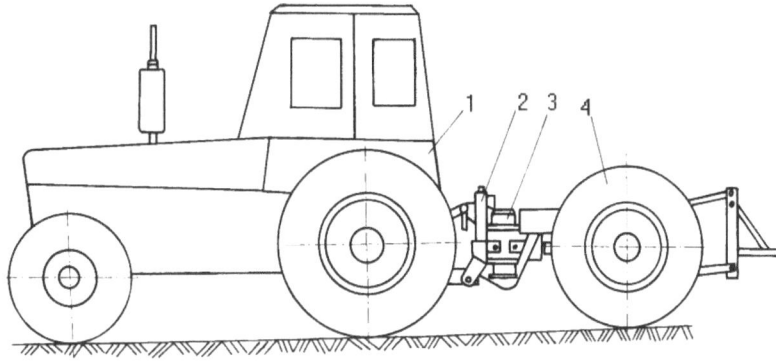

Figure 1: **The tractor with driven axle: 1 – tractor; 2 - support;**

2. TRACING THE TRACTION CHARACTERISTIC OF THE TRACTOR IN BOTH CASES

For tracing the theoretical traction characteristics only 5 stages from the gearbox - with values within 4.30 and 12.53 km/h – as this interval of velocities is the most used when effecting most of the agricultural works.

2.1. Mathematic model of the interaction between the roll up system and the soil

The **tractor slipping** δ is calculated as function of the specific driving force, which is defined by equation:

$$\phi_m = F_m / G \quad \text{for 4x4 tractors.}$$

$$\delta = \frac{A\phi_m - B\phi_m^2}{C - \phi_m}, \tag{1}$$

where: A, B, C are coefficients depending on the driving wheels tire dimensions, and especially on the road conditions.

The **efficiency of the tractor** can be estimated by means of the traction efficiency η_t defined by the equation:

$$\eta_t = \eta_{tr}\left(1 - \delta\right)\left(1 - \frac{R_r}{F_m}\right) = \eta_{tr}\left(1 - \delta\right)\left(1 - \frac{fG}{F_t + fG}\right), \tag{2}$$

where: η_{tr} is the total efficiency of tractor transmission; R_r - the rolling resistance force of the tractor, $R_r = fG$; f - the rolling resistance coefficient of the tractor on ground; G - the tractor weight in the center of gravity; F_m - the traction tangential force (driving force), developed by the engine; F_t - the drawbar load.

$$F_m = F_t + R_r. \tag{3}$$

By means of the relation we may trace the curve representing the traction efficiency depending on the traction force Ft representing, in fact, the potential traction characteristic of the tractor.

The **real moving speed** value of the tractor can be determined by the relation:

$$V = V_t\left(1 - \delta\right), \tag{4}$$

where V_t is the theoretical traveling speed of the tractor, in m/s;

The **driving force and a drawbar load.** In case the tractor rolls up on a horizontal area, in a stabilized system (V = const.), the traction balance of the tractor has the following form:

$$F_m = F_t + R_r = \frac{M_e i_{tr} \eta_{tr}}{r},$$

(5)

where i_{tr} is the transmission total reduction ratio;. R - the dynamic radius of the driving wheels, m; M_e - the effective torque of the engine, Nm.

The traction power P_t, in kW, is defined by the equation:

$$P_t = 10^{-3} F_t V,$$

(6)

2.2. Mathematical model of the engine's characteristic

In most of the papers the power curve is approximated to a third degree parabola:

$$P_e = P_n \left[\alpha_1 \frac{n}{n_n} + \alpha_2 \left(\frac{n}{n_n}\right)^2 + \alpha_3 \left(\frac{n}{n_n}\right)^3 \right],$$

(7)

and the curve of the torsion moment in the shaft is approximated, as a consequence, to a second degree parabola:

$$M_e = M_n \left[\alpha_1 + \alpha_2 \frac{n}{n_n} + \alpha_3 \left(\frac{n}{n_n}\right)^2 \right],$$

(8)

where α_1, α_2 and α_3 are determiners, for the above mentioned functions to approximate in the best possible way the external characteristic obtained experimentally. The values of these coefficients depend on the relations n_m / n_m and M_m / M_n

On the branch controlled by the regulator of the engine rotation characteristic, the dependency $M_e = f(n)$ is considered, generally, linear and, as a result:

$$M_e = M_n \frac{n_0 - n}{n_0 - n_n},$$

(9)

where: n_0 is the maximum no-load speed of the engine (the maximum speed of the crankshaft), $n_0 = (1,06...1,1)n_n$; n_n - the nominal rpm.

For diesel engines used for tractors, the specific fuel consumption curve is described more precisely by the function:

$$c = c_n \left[1,55 - 1,55 \frac{n}{n_n} + \left(\frac{n}{n_n}\right)^2 \right].$$

(10)

The **nominal specific fuel consumption** can be determine, approximately, by the relation $c_n = 1,05 \cdot c_{min}$, in g/(kW.h).

The hourly fuel consumption, in kg/h, can be calculated by the relation:

$$C = 10^{-3} c P_e.$$

On the linear branch of the engine characteristic, the hourly fuel consumption has the following equation:

$$C = C_g + \frac{C_n - C_g}{P_n} P_e,$$

(11)

where: C_g is the hourly fuel consumption at n_o; C_n - the hourly fuel consumption at n_n

In Figures 2...4 a comparative analysis of the traction and economic efficiency of tractor is presented, in the standard variant and in case a driven axle is used.

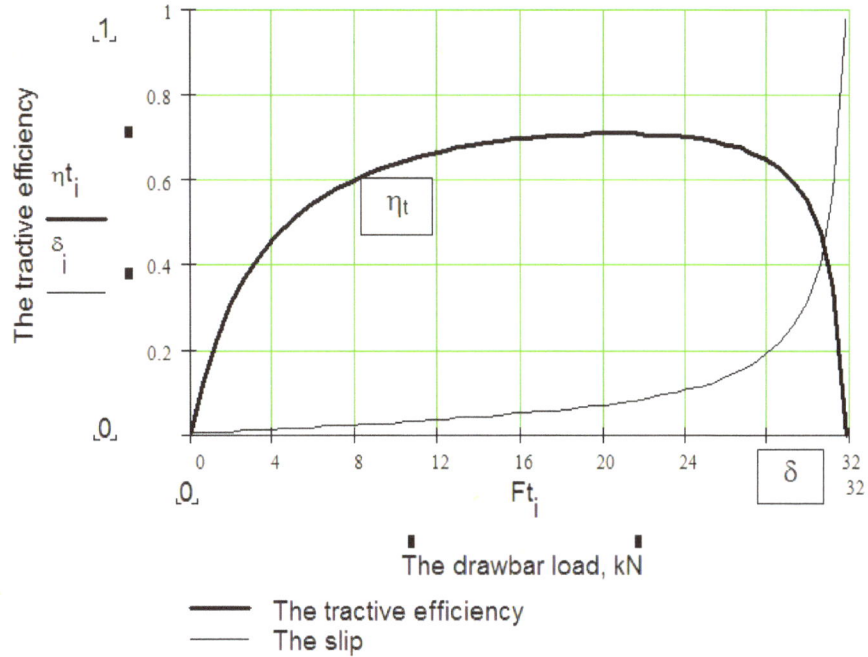

Figure 2: Variation of skidding and a traction efficiency with drawbar load:
a.   – standard tractor;

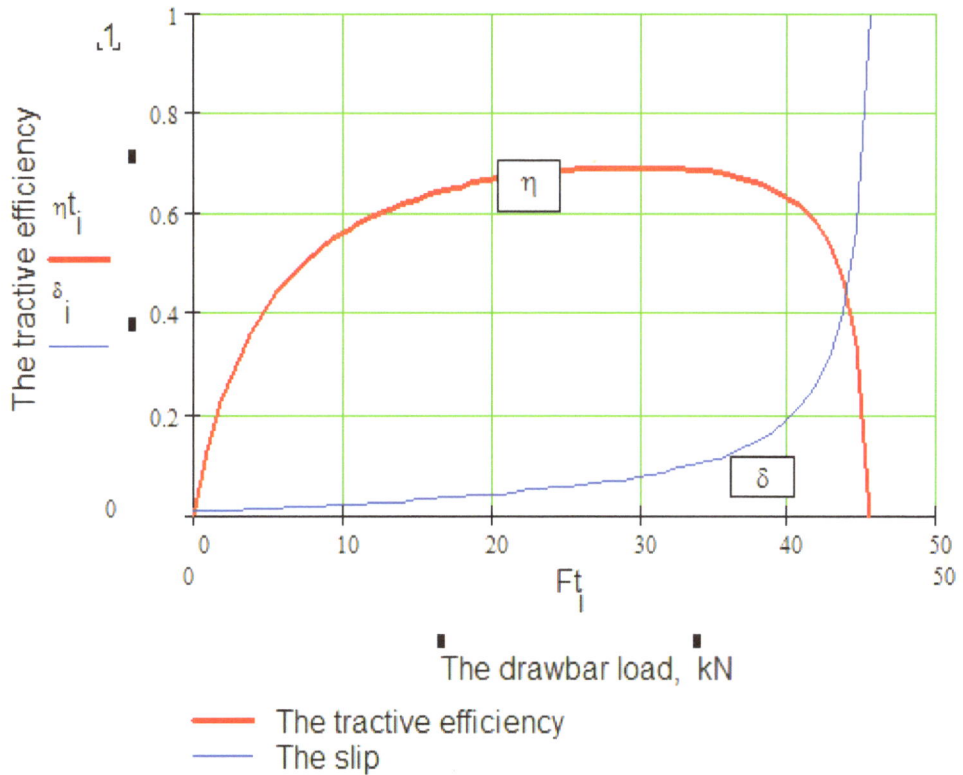

b. – tractor + driven axle.

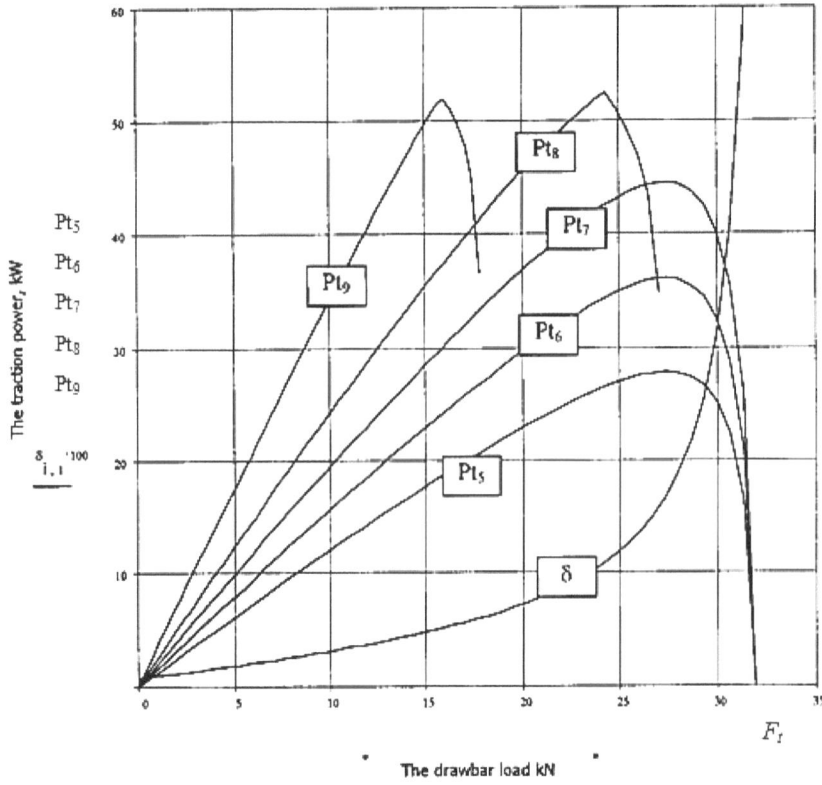

Figure 3: Variation of the traction power with drawbar load:
a. – standard tractor;

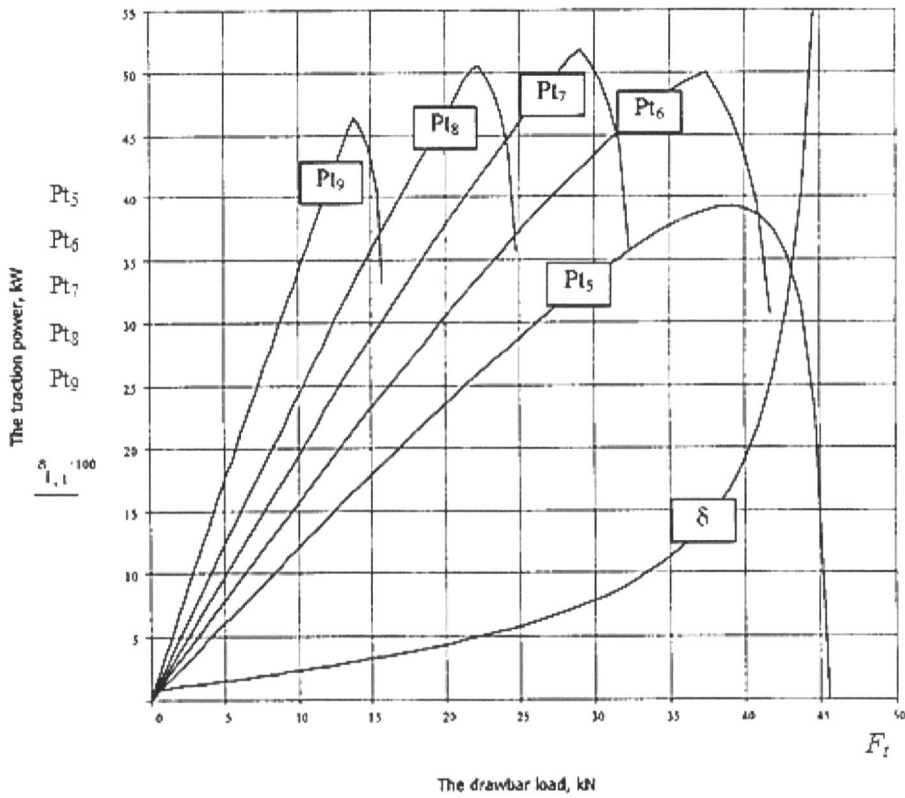

b. – tractor + driven axle.

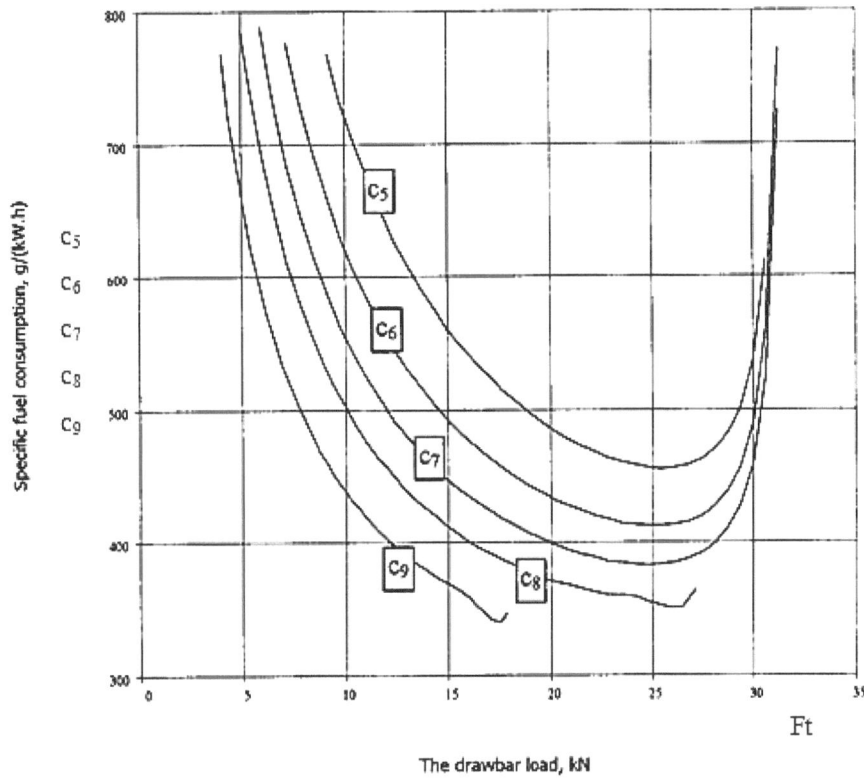

Figure 4: Variation of the specific fuel consumption with draw bar load:
a. – standard tractor;

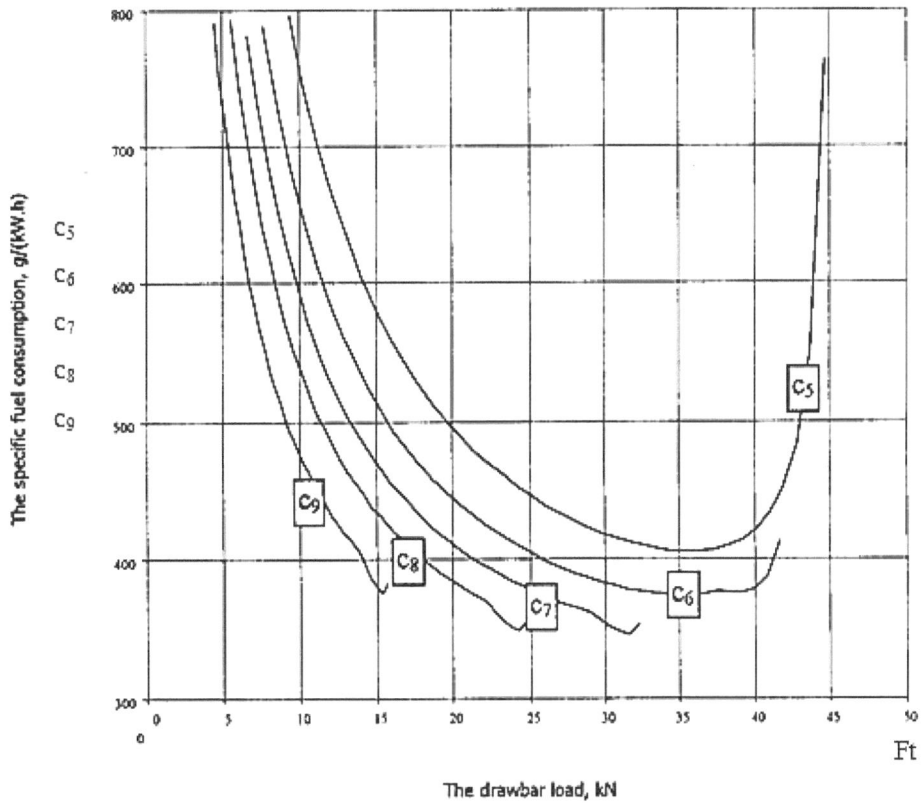

b. – tractor + driven axle.

3. CONCLUSIONS

- The limits of the traction forces in which the tractor has a high efficiency are increased. Thus, the traction efficiency of the standard tractor is above 0.6 in the interval 8kN....29.3kN (Figure 2, a), and the traction efficiency of the tractor with a supplementary driven axle is above 0.6 in the interval 12.1....41.9 kN (Figure 2, b).
- The traction force limited by the adherence increases 1.43 times more at the tractor equipped with supplementary driven axle, than at the standard tractor (Figure 2).
- Within the limits of the adopted velocities, at the tractor equipped with supplementary driven axle, the number of the stages of velocity to which the engine is used totally, increases from 2 to 4 (Figure 3).
- At the tractor equipped with supplementary driven axle the respective minimal fuel consumption is smaller and is moved to the area of the great traction forces (Figure 4).

REFERENCES

[1]   Nastasoiu M., Padureanu V., Tractoare. Transmisii ale tractoarelor, Editura Universităţii Transilvania, Braşov, 1999.
[2]   Nastasoiu M., Padureanu V., Tractoare, Editura Universităţii Transilvania, Braşov, 2012.

ENERGY EFFICIENCY ANALYSIS OF AGARICUS BISPORUS MUSHROOM PRODUCE IN FELDIOARA-BRASOV

H. Gh. Schiau[1], Fl. Rus[1]

[1] Transilvania University of Braşov, Braуov, ROMANIA, e-mail: horia.schiau@unitbv.ro

Abstract: *The purpose of this research was to assessment of the nature energy and specifically to measure and benchmark the effi- ciency for Agaricus Bisporus mushroom production in Feldioara, Brasov county. The data used in this study were collected poersonaly from October 2011 to November 2012 in the mushrooms farm KADNA BIONATURA one of the important mushroom producer in the area. In the investigated farm, the average and total energy were calculated, during both summer and winter season, considering the financial implications. Electricity and fossil fuel were found to be used in excess in mushroom production farm. The conclusion is that the total energy consumption can be reduced with 8% for mushroom production especialy by improving the thermal insulation of the building and use of heat recuperators.*
Keywords: *Agaricus Bisporus, Energy Equivalents Used, Energy Equivalent, Technical Efficiency*

1. INTRODUCTION

Currently, champignons are grown in 80 countries of the world. At present, of all the diversity of mushroom realm, more than 14 species are cultivated on a commercial scale. Production of mushroom has already crossed 7 million metric tons annually in the world and is expected to reach 10 million metric tons in the next 10 years [10]. The leading position in the global production belongs to white button mushroom (champignon); its annual volumes exceed 1.5 million tons. The champignon is followed by such wood-attacking fungi as Siitake (527,000 tons) and oyster mushroom (250,000 tons).

Besides the mentioned species, the following are also grown on a commercial scale: volvariella volvacea (200,000 tons), Jew's-ear fungus (120,000 tons), velvet foot (105,000 tons), nameko (30,000 tons), Coprinus comatus, honey fungus and stropharia rugoso-annulata (10,000 tons each). Torq, black truffle, agrocybe, and some other species are grown in insignificant volumes.

Presently, the scientific research work is underway to introduce new prospective species of edible fungi into cultivation. White button mushroom (Agaricus Bisporus) cultivation began in France two hundred years ago and has developed into a thriving industry not only in Europe, but world over [19].

Greenhouse production is one of the most intensive parts of the world agricultural production. It is intensive in the sense of yield and annual production, but also in sense of the energy consumption, investments and costs [22]. Efficient use of resources is one of the major assets of ecoefficient and sustainable production, in agriculture [4]. Efficient use of energy is one of the principal requirements of sustainable agriculture. The shares of greenhouse crops production were as follows: vegetables 59.3%, flowers 39.81%, fruits 0.54% and mushroom 0.35% [16].

The energy use in agriculture has been increasing in response to increasing population, limited supply of arable land, and a desire for higher standards of living.

Continuous demand in increasing food production resulted in intensive use of chemical fertilizers, pesticides, agricultural machinery and other natural resources. The development of energy efficient agricultural systems with a low input of energy compared to the output of products should therefore help to reduce the emissions of greenhouse gasses in agricultural production [3].

The energy use is one of the key indicators for developing more sustainable agricultural practices. Renewable energy sources coming from agricultural crops could play an important role to supply the energy requirement and in terms of environmental effects [19], [20].

The study was conducted from October 2011 to November 2012 in the mushrooms farm KADNA BIONATURA of Fledioara village, Brasov County. Farm has 4 spaces for mushrooms production, with area of 150 square meters, equipped with 4 rows of shelves with 4 levels. Developed culture surface is 256 sqm / room, respective 1024 sqm / farm.

The selection of button mushroom growing rooms was based on random sampling method.

2. AGARICUS BISPORUS (CHAMPIGNON) MUSHROOM PRODUCING

The mushroom culture is a cyclical process and involves several different operations, each of which must be carefully performed. The white button mushroom is more acceptable to the consumer and fetches higher prices. For its successful and profitable cultivation, careful attention must be paid [10].
Producing Champignon mushroom needs cool weather condition in temperatures ranged of 16.5-18.5°C or 18.5-21°C [11], [12], [24].

Most of farmers in Romania grown in winter but many important farmers work throughout the year, also in summer. These farmers have to create cool condition in their rooms in summer but would sell their product more expensive than winter growers. During the crop cycle, mushrooms are harvested in a series of breaks or flushes that occur at approximately 7 or 8 day intervals with hand carefully. After two flushes, mushroom production declines rapidly so that each successive flush produces fewer mushrooms. Usualy, romanian farmers harvest only 3 flushes to obtine an 35-36% productivity. In recent years large number of commercial units has been built to increase production, but a few effective unorganized were closed.

3. ENERGY EQUIVALENTS USED

The amounts of inputs (Champignon mushroom compost, electricity, human power, machinery, water and chemicals) used in the production of button mushroom were specified in order to calculate the energy equivalences in the study. The energy coefficients of inputs are the energy used from primary production to the end user. The energy equivalent of water input means indirect energy of irrigation consist of the energy consumed for manufacturing the materials for the dams, canals, pipes, pumps, and equipment as well as the energy for constructing the works and building the on-farm irrigation systems [19], [22]. The energy equivalent of human power is the muscle power used in growing room operations [22]. Transporting machines and other machines used by farmers is also calculate.

Chemicals energy equivalents include the energy consumption for producing, packing and distributing the materials and they are given on an active ingredient basis. Also in button mushroom production chemicals only contains insecticides (Dinilin, Mirage, Formalin), that used to control insects. The units in Table 1 were used to find the amount of inputs.

Table 1: Energy equivalents for different inputs in agricultural production

Inputs	Unit	Energy equivalent (MJ/Unit)	Reference
Human power	h	1.96	Mohammadi et al., 2008
Machinery	h	62.7	Mandal et al., 2002
Diesel fuel	l	47.8	Kitani, 1999
Electricity	kWh	11.93	Qasemi-Kordkheili et al., 2013
Chemicals	kg	101.2	Erdal et al., 2007
Water	m3	1.02	Khan et al., 2009

3.1. Button mushroom energy equivalent

Button mushroom contains 87.5-89.5% moisture and 12.5-10.5% dry matter [18]. Energy equivalent of 100 gr button mushroom dry matters is 1.55 MJ [2], so button mushroom energy equivalent is calculated as 1.6275 MJ/kg.

3.2. Button mushroom compost

Producing button mushroom compost is a separated procedure and has different formulas with different inputs in each country. In Romania there is no compost plant, it is imported from other European countries (Hungary, Nedreland, Italy).

From the specifications of suppliers, mushroom compost contains mainly wheat straw as dead organic matter and Farmyard manure about 310 kg and 320 kg per 1 ton respectively. Wheat seed, urea (46% N) and water are other inputs. Other input quantities for producing 1 ton button mushroom compost and their energy equivalents are given in Table 2.

Table 2: Energy equivalents and quantity of inputs for producing 1 ton button mushroom compost

Inputs (unit)	Energy equivalent (MJ Unit-1)	Quantity per 1 ton compost	Total energy equivalent (MJ)
Human power (h)	1.96	6.20	12.15
Diesel fuel (l)	47.80	14.20	678.76
Machinery (h)	62.70	6.00	376.20
Electricity (kWh)	11.93	209.50	2499.34
Wheat straw (kg)	12.50	360.00	4500.00
Farmyard manure (kg)	0.30	280.00	84.00
Urea (46% N) (kg)	78.10	10.00	781.00
Wheat seed (kg)	25.00	24.00	600.00
Water (m3)	1.02	8.00	8.16
Total			9539.61

Energy equivalent estimated for producing 1ton mushroom compost in Hungary is about 9540 (MJton-1). The amounts of inputs were calculated per hectare and then, these input data were multiplied with the coefficient of energy equivalent. The energy equivalences of unit inputs are given in megajoule (MJ) unit. The total input equivalent can be calculated by adding up the energy equivalences of all inputs in Mega Joule (MJ).

3.3. Data envelopment analysis

A non-parametric method of DEA was employed to evaluate the technical, pure technical and scale efficiencies of individual farmers. So, the energy consumed from different energy sources including: human power, machinery, chemicals, water, electricity and button mushroom compost, were defined as input variables; while, the button mushroom yield was the single output variable; also each farmer called a Decision Making Unit (DMU).

In DEA, an inefficient DMU can be made efficient either by reducing the input levels while holding the outputs constant (input oriented); or symmetrically, by increasing the output levels while holding the inputs constant (output oriented) [14]. The choice between input and output orientation depends on the unique characteristics of the set of DMUs under study. In this study the input oriented approach was deemed to be more appropriate because there is only one output while multiple inputs are used; also as a recommendation, input conservation for given outputs seems to be a more reasonable logic [6], [19], so the button mushroom production yield is hold fixed and the quantity of source wise energy inputs were reduced.

3.4. Technical efficiency

The technical efficiency (TE) can be expressed generally by the ratio of sum of the weighted outputs to sum of the weighted inputs. The value of technical efficiency varies between zero and one; where a value of one implies that the DMU is a best performer located on the production frontier and has no reduction potential. Any value of TE lower than one indicates that the DMU uses inputs inefficiently [14]. Using standard notations, the technical efficiency can be expressed mathematically as following relationship

$$TE_j = \frac{u_1 y_{1j} + u_2 y_{2j} + \cdots + u_n y_{nj}}{v_1 x_{1j} + v_2 x_{2j} + \cdots + v_m x_{mj}} = \frac{\sum_{r=1}^{n} u_r y_{rj}}{\sum_{s=1}^{m} u_s y_{sj}} \qquad (1)$$

where, u_r, is the weight (energy coefficient) given to output n; y_r, is the amount of output n; v_s, is the weight (energy coefficient) given to input n; x_s, is the amount of input n; r, is number of outputs ($r = 1, 2, ..., n$); s, is number of inputs ($s = 1, 2, ..., m$) and j, represents jth of DMUs ($j = 1, 2, ..., k$). To solve Equation (1), Linear Program (LP) was used, which developed by Charnes

$$Maximize\ \theta = \sum_{r=1}^{n} u_r y_r \qquad (2)$$
$$Subjected\ to\ \sum_{r=1}^{n} u_r y_r \qquad (3)$$
$$\sum_{s=1}^{m} v_s x_{sj} = 1 \qquad (4)$$
$$u_r \geq 0,\ \ v_s \geq 0,\ and\ (i\ and\ j = 1,2,3, \dots k) \qquad (5)$$

where, θ is the technical efficiency and i represent ith DMU, it will be fixed in Equations (2) and (4) while j increases in Equation (3). The above model is a linear programming model and is popularly known as the CRS DAE model which assumes that there is no significant relationship between the scale of operations and efficiency [19]. So, the large producers are just as efficient as small ones in converting inputs to output.

3.5. Pure technical efficiency

Pure technical efficiency is another model in DEA that introduced by Banker et al. in 1984. This model called VRS and calculates the technical efficiency of DMUs under variable return to scale conditions. Pure technical efficiency could separate both technical and scale efficiencies.

The main advantage of this model is that scale inefficient farms are only compared to efficient farms of a similar size [1]. It can be expressed by Dual Linear Program (DLP) as follows [14]:

$$Maximize\ z = uy_i - u_i \qquad (6)$$
$$Subjected\ to\ vx_i = 1 \qquad (7)$$
$$-vX + uY - u_0 e \leq 0 \qquad (8)$$
$$v \geq 0,\ \ u \geq 0,\ and\ u_0\ free\ in\ sing \qquad (9)$$

where, z and u_0 are scalar and free in sign. U and v are output and inputs weight matrixes, and Y and X are corresponding output and input matrixes, respectively. The letters x_i and y_i refer to the inputs and output of ith DMU.

3.6. Scale efficiency

Scale efficiency shows the effect of DMU size on efficiency of system. Simply, it indicates that some part of inefficiency refers to inappropriate size of DMU, and if DMU moved toward the best size the overall efficiency (technical) can be improved at the same level of technologies (inputs) [15] The relationship among the scale efficiency, technical efficiency and pure technical efficiency can be expressed as [19]:

$$Scale\ efficiency = \frac{Technical\ efficiency}{Pure\ technical\ efficiency} \tag{10}$$

In the analysis of efficient and inefficient DMUs the energy saving target ratio (ESTR) index can be used which represents the inefficiency level for each DMUs with respect to energy use. The formula is as follow [9]:

$$ESTR_j = \frac{(Energy\ Saving\ Target)_j}{(Actual\ Energy\ Input)_j} \tag{11}$$

where energy saving target is the total reducing amount of input that could be saved without decreasing output level and j represents jth DMU.

The minimal value of energy saving target is zero, so the value of ESTR will be between zero and unity. A zero ESTR value indicates the DMU on the frontier such as efficient ones; on the other hand for inefficient DMUs, the value of ESTR is larger than zero, means that energy could be saved. A higher ESTR value implies higher energy inefficiency and a higher energy saving amount [9]. In order to calculate the efficiencies of farmers and discriminate between efficient and inefficient ones, the Microsoft Excel spread sheet and Frontier Analyst software were used.

4. EFFICIENCY OF GROWING ROOMS

Results obtained by application of the inputorientated DEA are illustrated in Table 3. The mean radial technical efficiencies of the samples under CRS and VRS assumptions are 0.82and 0.97 respectively. This implies first, that on average, growing rooms could reduce their inputs by 18% (3%) and still maintains the same output level, and second, that there is considerable variation in the performance of growing rooms. Increasing the technical efficiency of a growing room actually means less input usage, lower production costs and, ultimately, higher profits, which is the driving force for producers motivation to adopt new techniques.

Table 3: Energy equivalents and quantity of inputs for producing 7.5 ton button mushroom

Equipments/items	Power Kw	Hours	People	Total KW	Tot hours	L diesel	Total energy equivalent (MJ)	Percentage (%)
Total	388.58	1311	60	2479	1484	46	34680	100 %
	Energy equivalent (MJ/Unit)			11.93	1.96	47.8	Total 20 t compost	Total/1 t mushrooms
	Total energy equivalent (MJ)			29573	2908	2198	34680.48	4624.06
	Unit price (Euro)			0.12 €	1.20€	1.27 €	Total 20 t compost	Total/1 t mushrooms
	Value (Euro)			293 €	1780€	58.55 €	2132 €	266.54 €

5. CONCLUSION

Due of the variable weather condition in Romania, especial in Braşov county, the energy consumption is determined especial by the temperature differences between winter and summer (-28 °C + 36 °C). Cold weather require high consumption for heating and in the warm season, major consumption is given by the plant cooling water.

The comparative analysis performed in the period 2010-2012, shows that energy consumption is the main HVAC (Heating system + Ciller + fans).

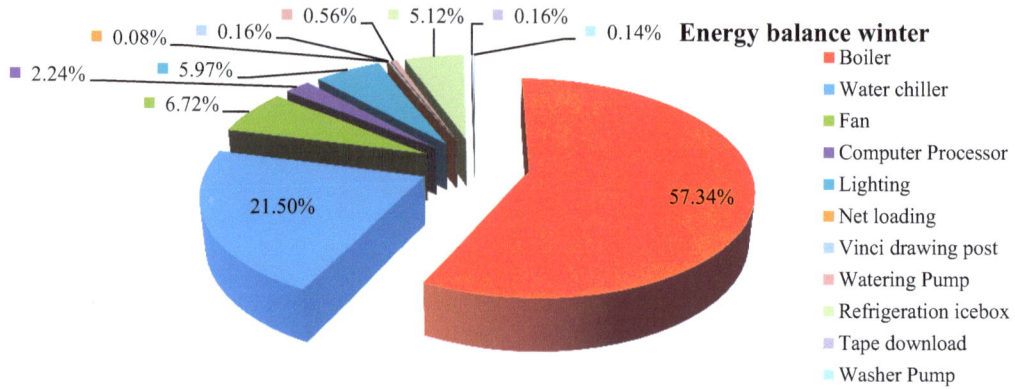

Figure 1: Energy balance for winter season

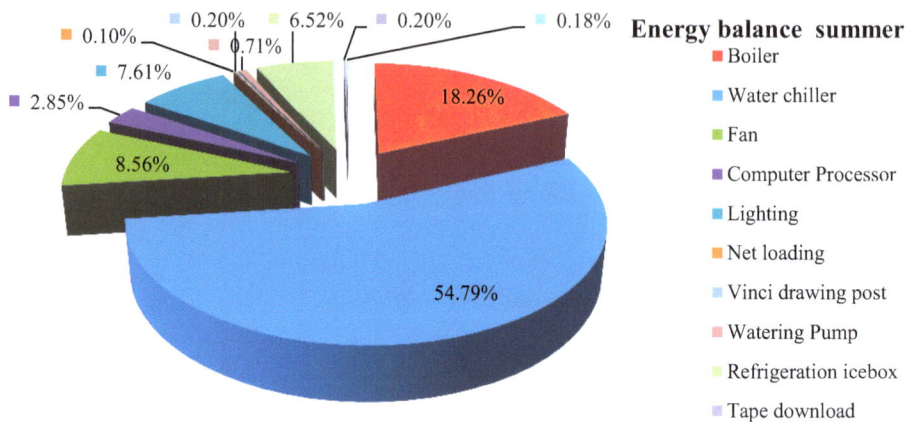

Figure 2: Energy balance for summer season

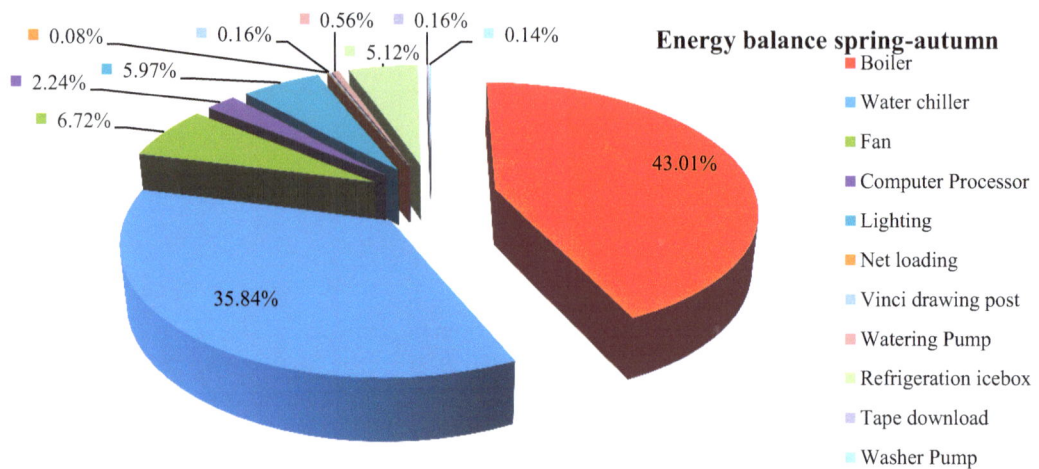

Figure 3: Energy balance for spring-autumn seasons

ACKNOWLEDGEMENT

This paper is supported by the Sectorial Operational Programme Human Resources Development (SOP HRD), financed from the European Social Fund and by the Romanian Government under the contract number POSDRU/107/1.5/S/76945.

REFERENCES

[1]   Bames A., Does multi-functionality affect technical efficiency? A non–parametric analysis of the Scottish dairy industry, Journal of Environ Manage, 80(4) :287–294, 2006.

[2]   Barros L., Cruz T., Baptista P., Estevinho L. M., Ferreira, L.,Wild and commecial mushrooms as source of nutrients and nutraceuticals, Food and Chemical Toxicology, 46: 2742-2747, 2008.

[3]   Dalgaard T., Halberg N., Porter J.R., A model for fossil energy use in Danish agriculture used to compare organic and conventional farming, Agric Ecosystem Environ, 87(1): 51–65, 2001.

[4]   De Jonge A.M., Eco-efficiency improvement of a crop protection product: the perspective of the crop protection industry, Crop Protect, 23(12): 1177–86, 2004.

[5]   Erdal G., Esengün K., Erdal H., Gündüz O., Energy use and economical analysis of sugar beet production in Tokatprovince of Turkey, *Energy,* 32(1): 35-41, 2007.

[6]   Galanopoulos K., Aggelopoulos S., Kamenidou I., Mattas K., Assessing the effects of managerial and production practices on the efficiency of commercial pig farming, Agric Sys, 88: 125-141, 2006.

[7]   Gibbons W.R., Button Mushroom Production in Synthetic Compost Derived from Agricultural Wastes, Bioresource Technology, 38: 65-77, 1991.

[8]   Houshyar E., Sheikh Davoodi M.J., Nassiri S.M., Energy efficiency for wheat production using data envelopment analysis (DEA) technique, Journal of Agricultural Technology, 6(4): 663-672, 2010.

[9]   Hu J.L., Kao C.H., Efficient energy-saving targets for APEC economies, Energy Policy, 35: 373–82, 2007.

[10]   Kumar A., Singh M., Singh G., Effect of different pretreatments on the quality of mushrooms during solar drying, J Food Sci Technol., DOI 10.1007/s13197-011-0320-5, 2011.

[11]   Lambert E.B., Ventilation requirements during cropping, In Proc. 1st Sci. Symp. on Cultivated Mushrooms and 6th Int. Congress on Mushroom Sci., Amsterdam, pp. 371-378, 1965.

[12]   Long P.E., Jacobs L., Some observations on CO 2 and sporophore initiation in the cultivated mushroom, In Proc. 2nd Sci. Symp. On Cultivated Mushrooms and 7th Int. Congress on Mushroom Sci., Hamburg, pp. 373-384, 1969.

[13]   Mohammadi A., Tabatabaeefar A., Shahin S., Rafiee S., Keyhani A., Energy use and economical analysis of potato production in Iran a case study: Ardabil province, Energy Conversion and Management, 49(12): 3566-3570, 2008.

[14]   Mousavi-Avval S.H., Rafiee S., Jafari A., Mohammadi A., Improving energy use efficiency of canola production using data envelopment analysis (DEA) approach, Energy, 36: 2765- 2772, 2010.

[15]   Nassiri S.M., Singh S., Study on energy use efficiency for paddy crop using data envelopment analysis (DEA) technique, Applied Energy, 86: 1320-1325, 2009.

[16]   Nassiri S.M., Singh S., A comparative study of parametric and non-parametric energy efficiency in paddy production, Journal of agricultural science and technology, 12: 379-389, 2010.

[17]   Omid M., Ghojabeige F., Delshad M., Ahmadi H., Energy use pattern and benchmarking of selected greenhouses in Iran using data envelopment analysis, Energy Conversion and Management, 52:153-162, 2011.

[18]   Pathak N.V., Nagendra Y., Maneesha G., Mushroom producing and processing technology, New Dehli: Agrobios (India), 179 page, 2003.

[19]   Peyman Q-K, Mohammad A.A., Morteza T., Mohammad SEA, Energy consumption pattern and optimization of energy inputs usage for button mushroom production, International Journal of Agriculture: Research and Review, Vol. 3(2), 361-373, 2013.

[20]   Piet W.J., Groot D.,Visser J., Leo J.L.D., Griensven V., Schaap P.J., Biochemical and molecular aspects of growth and fruiting of the edible mushroom Agaricusbisporus, Mycological Research, 102(11): 1297-1308, 1998.

[21]   Qasemi-Kordkheili P., Kazemi N., Hemmati A., Taki M., Energy consumption, input–output relationship and economic analysis for nectarine production in Sari region, Iran, International Journal of Agriculture and Crop Sciences, 5: 125-131, 2013.

[22]   Rafiee S., Mousavi-Avval S.H., Mohammadi A., Modeling and sensitivity analysis of energy inputs for apple production in Iran, Energy, 35(8): 3301-3306, 2010.

[23]   Schroeder M.E., Schisler L.C., Snetsinger R., Crowley V.E., Barr W.L., Automatic control of mushroomventilation after casing and through production by sampling carbon dioxide, Mush Sci, 9, 1974.

[24]   Singh H., Singh A.K., Kushwaha H.L., Energy consumption pattern of wheat production in India, Energy, 32: 1848-1854, 2007.

CONSTRUCTIVE SOLUTION, USING FINITE ELEMENT METHOD, FOR OPTIMIZATION STRUCTURE OF COMPOSITE MATERIALS

Anca Elena Stanciu[1]

[1] Transilvania University of Brasov, Braşov, Romania, anca.stanciu@unitbv.ro

Abstract: *The paper presents a theoretical approach regarding the mechanical behavior of glass fabric reinforced composite for liquid storage tank. The internal stress and deformation field is locally influenced by the relative difference between the constituents' properties, their size, shape and relative orientation as well as by the geometry of the repeating structures that form the composite material*
Keywords: *finite element method, layers, stress.*

1. INTRODUCTION

Finite element method (FEM) becomes more and more a general method used for solving different types of complex problems concerning both stationary and non-stationary phenomena from all engineering fields but also in other activity and research areas.

As far as the stress and deformation are concerned we may observe that the internal mechanical work is linked to three components of the stress in 2D coordinates, the normal plane component of the stress does not involve the canceling of other strains or stresses.

From mathematical point of view, the problem is very similar to that of plane stress and deformation analysis, this is why the situation may be regarded as two dimensional.

In order to control the complexity of the problem and "filter" the irrelevant aspects we need to accomplish a suitable mathematical model. This model should consider the fact that we are dealing with an anisotropic material, consisting of several layers and also that the loads and deformations along the contours are difficult to be obtained.

2. MATHEMATICAL MODELLING WITH FEM

The main part of the process is, as shown in the diagram, the mathematical model. This is mostly an ordinary equation or a differential one, developed in space and time. A discrete model with finite elements is generated by help of the variation form of the mathematical model. This stage is called meshing. The FEM equations are solved using an equation solver that will provide a discrete solution.

More relevant are the meshing errors, representing the extent to which the discrete solution does not check the mathematical model. The replacement in the ideal physical system might identify the modeling errors.

Then the model will be analyzed by help of MSC Nastran processor but before running the file we need to do some previous checking in order to validate the finite elements model, as follows:

- determination of the distance between two locations or nodes;
- determination of the angle between two directions determined by three point, one of them being considered as origin;
- identification of common points;
- identification of common lines;

- identification of common nodes and joining them;
- identification of nodes belonging to a selected plane, with the possibility of moving to this plane of the nodes from the adjacent area;
- identification of the common finite elements;
- determination of a finite element distortions;
- identification of the normal in a plane finite elements group and comparing them to a given direction;
- determination of mass properties for the finite elements;
- checking the geometric boundary conditions;
- determination of the loading forces sum in a node.

3. FEM ANALYSIS FOR LIQUID STORAGE TANK

The paper presents a theoretical approach regarding the mechanical behavior of glass fabric reinforced composite for liquid storage tank.

The model was achieved using MSC Patran preprocessor/postprocessor and MSC Nastran processor. In the preprocessing stage, the finite elements geometric modeling requires the finite element model, which will be finally solvable by help of the programs kit meant for this purpose.

A finite element modeling requires the material behavior modeling, selection and personalization of finite elements; finite elements structure generation, introduction of boundary conditions and loads.

The analysis and solution of the finite element model, elaborated during preprocessing requires the preliminary setting of the solving parameters and the execution of the specific program modules.

The geometric modeling previous the meshing requires the generation of closed contours consisting of lines for plane areas or surfaces. In figure 1 we presented the detailed model geometry.

Figure 11 Geometric model of the liquid storage tank

According to standards the water density is of 1000 kg/ m^3. Considering a safety factor of 1,25 the water density used in calculus is of 1250 kg/ m^3. The pressure was determined upon the 5 sectors obtained after model meshing, taking into account the length of each sector and the water volume.

Table 1: Succession of material layers

Layers direction	Layer(Ply)	Material type
Inferior side	1	MAT600
	2	MAT600
	3	RT800
	4	RT800
	5	RT800
	6	RT800
	7	MAT450
Superior side	8	MAT450

The structure is made of 5 different layers as shown in Table 2, the arrow representing the succession of the layers starting from the interior towards the exterior of the better structure.

Table 2: Succession of material layers

Layers direction	Layer(Ply)	Material type
Inferior side	1	MAT600
	2	RT800
	3	RT800
	4	RT800
Superior side	5	MAT450

In figure 7 and 8 is presented the maximum Von Misses stress distribution for all layers

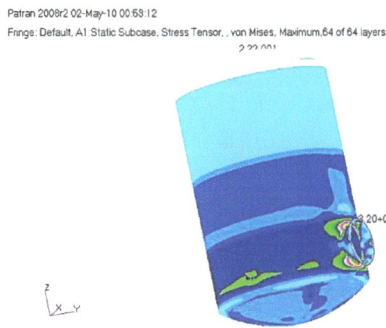

Figure 2: The maximum Von Misses stress distribution for 8 layers, MAT-Roving material

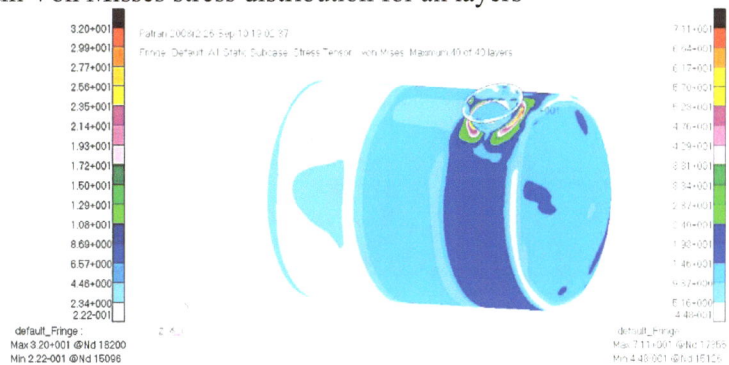

Figure 3: The maximum Von Misses stress distribution for 5 layers, MAT-Roving material

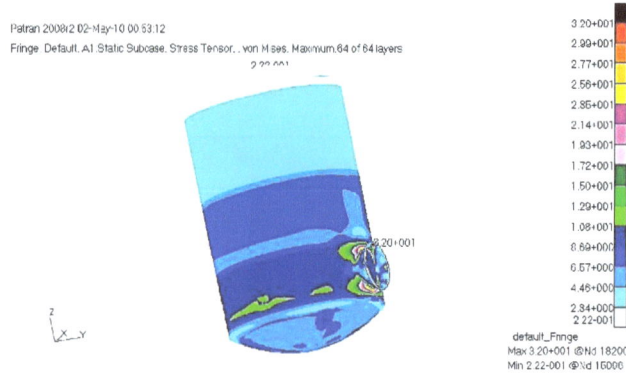

Figure 4: The maximum Von Misses stress distribution for 8 layers, Roving material

In figure 9 and 10 is presented the maximum stress distribution on all layers for X axis

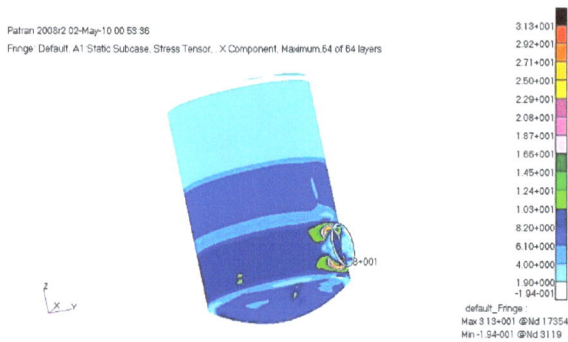

Figure 5: The maximum stress distribution on 8 layers for X axis, MAT-Roving material

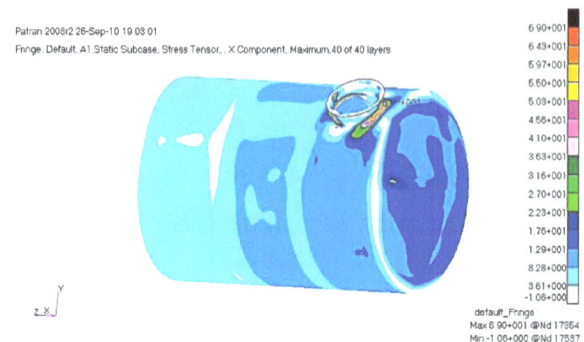

Figure 6: The maximum stress distribution on 5 layers for X axis, MAT-Roving material

In figure 11 and 12 is presented the maximum stress distribution on all layers for Y axis

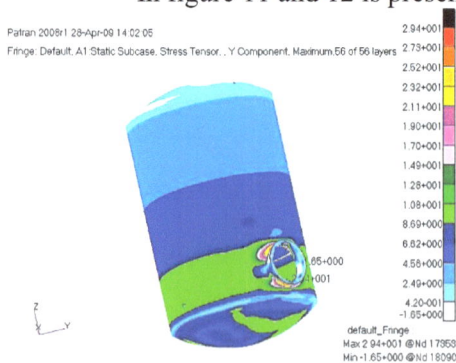

Figure 7: The maximum stress distribution on 8 layers for Y axis, MAT-Roving material

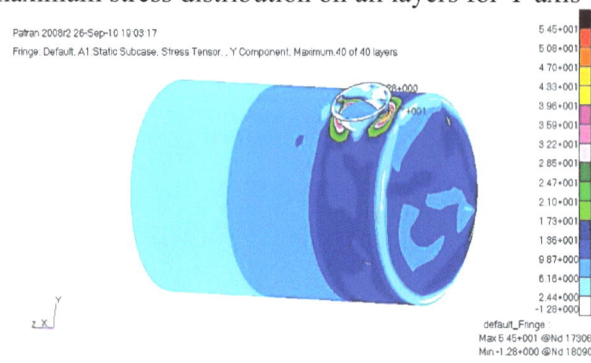

Figure 8: The maximum stress distribution on 5 layers for Y axis, MAT-Roving material

There was obtained a MAT-Roving optimized material with 5-layers, canceling from each type of material one layer from first model realized, where the maximum stress is 40% higher.

Table 3: The maximum stress for all composite materials used

Nr. Crt.	Composite materials	The maximum stress for X axis (σ_x), MPa	The maximum stress for Y axis (σ_x), MPa	The maximum stress von Mises ($\sigma_{ech\ v}$), MPa
1	MAT-Roving 8 layers	42.8	29.4	46.9
2	Roving 8 layers	31.3	21.1	32
3	MAT-Roving 5 layers	**69**	**54.5**	**71.1**

Following the experimental researches and also the studies based on FEM we conclude that the material thickness is oversized.

In this respect we decided that we may reduce the number of layers from 8 to 5, namely we may give up to one of each type of material.

4. CONCLUSION

MAT type material resists much better to the applied loads (considering all directions), the values are smaller in comparison to the efforts occurred in the roving.

This leads to costs diminishing, weight loss and not last to less exposure of the working personnel to toxic wastes.

REFERENCES

[1] Alămoreanu, E., Constantinescu, D.M., Design of laminated composite plates, Romanian Academy Publishers, Bucharest, 2005.

[2] Enescu,I., Stanciu, A., Finite elements, Transilvania University Publishers, 2007, ISBN 978-635-947-7.

[3] MSC/NASTRAN for WINDOWS, Version 2.0, Users manuals.

[4] Sims G. D., Broughton W., R., Glass Fiber Reinforced Plastics—Properties, Comprehensive Composite materials, vol. 2, ISBN 0-080437206, pp. 151-197.

[5] Stanciu, A. E., Bencze, A., Munteanu, M.V., Vlase, S., Purcărea, R., (2010), Constructive Solution, Using Finite Element Method For Optimization Of Tanks, 3nd Int. Conf. COMAT, 27-29 octombrie, ISSN 1844-9336, vol. II, pp. 273-278..

[6] Teodorescu, H., Basic concepts and mechanics of polymeric composite materials, Transilvania University Publishers, Brasov 2007.

THE COMPUTATIONAL MODELLING APPROACH A MULTI-SCALE OVERVIEW

C.I. Pruncu[1]

[1]Univ Lille Nord de France, F-59000 Lille, France / UVHC,
TEMPO EA 4542, F-59313 Valenciennes, France, CatalinIulian.Pruncu@univ-valenciennes.fr

Abstract: The revolutionary progress of Computational Methods (CM) involve efficient, user-friendly source to detect the evolution of material behaviour from a multidisciplinary point of view: physical, biological, chemical or mechanical system. Besides, this technique present high level of sustainability due to the cheap price of computational modelling within reasonable times acquired under modern powerful computer.

A significant advantage rise when this technique is employed at large scale in industrial platform. Thus, in the industrial field the CM transposes the experimental investigations on the numerical simulation, making the possibility to save as well main source of row material. The CM technique is applied in different sectors as: wind energy, aeronautical, naval, automotive, bio-medical, biomechanics industry, power plant assembly, building construction, electromagnetic, electronic field, weather forecasting and etc.

This paper present an summary of general benefit/challenging encountered in CM converted in principal improvement into industrial area, and concludes with an open problem that provides stimulation for further research.

Key words: Computational Methods (CM), multi-scale algorithm, contacts interaction;

1. INTRODUCTION

The scientific development of CM is recognized as a function of continuous growth of the number of papers published in Journal of specialities (i.e. Computational Mechanics, International Journal for Numerical Methods in Engineering, Theoretical and Computational Fluid Dynamics, Theoretical and Computational Fluid Dynamics, Computational Materials Science and etc.) connected to the industrial field.

The necessity to solve general problem encountered in nature and mostly applicable in industrial area entail the research organization (i.e. Universities and research centres) to imply more energy to obtain new knowledge for solving today's and tomorrow's problems raised from computational algorithm.

Departure algorithm for CM starts from knowledge of the reference coordinate. As location, the position of a particle relative to its coordinate system can be specified by a vector function of time – the position vector x(t). Then, it is normal to define an equation for x(t) to design the trajectory of this particle, and to find its trajectory during time [1]. This border can be considered in order to fix the coordinate of reference system for the stability of the local condition. Besides, if a system undergoing slow time variation in comparison to its time constants can usually be considered to be linear time invariant (LTI) and thus, slow time-variation is often ignored in dealing with systems in practice [2]. The issues arise when time change its sequence, case that is related to time-dependent problems, and planned to be solved with systems of partial differential equations in which the time derivatives are of first order. Moreover, the numerical methods can be used to solve partial differential equations containing higher-order time derivatives by defining new unknown functions equal to the lower-order time derivatives of the original unknown function and expressing the result as a system of partial differential equations in which all time derivatives are of order 1 [3]. The general form of second-order partial differential equation is present in equation 1.

$$\frac{\partial^2 \psi}{\partial t^2} + \psi \frac{\partial \psi}{\partial x} = 0$$

(1)

By setting these basic conditions in the CM, it is possible to go further to implement the algorithm of numerical analysis to detect the evolution of material behaviour from a multidisciplinary point of view: physical, biological, chemical or mechanical system.

In this paper, we present the general benefit/challenging encountered in modelling methods to further improve the efficiency and accuracy of CM within the applicability in industrial sector.

2. CONDITION AND SCENARIOS TO IMPLEMENT THE CM

In the CM, an important step is represented by the validation of the model that entails an accuracy level reliable with the requested application. This procedure can be acquired by a calibration method developed as well as computing code, which can involve an automate regression of the calibration procedure (optimization method) [4]. Figure 1 present the general model for a cycle of CM.

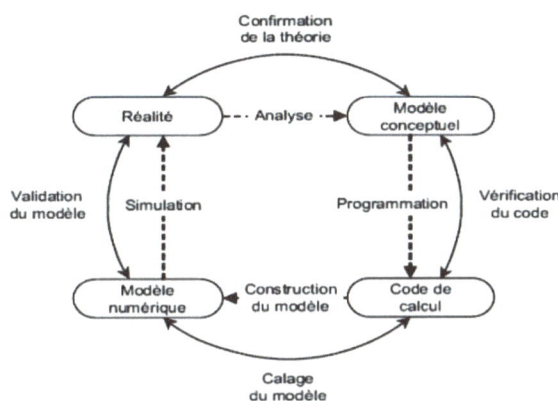

Figure 1. Components of a modeling terminology incorporating the stage of model calibration 4.

The calibration procedure can be very difficult task, for example at atomistic scale. The challenge can rise when the required input data represent properties of small-scale volumes, embedded and constrained by other phases, with complex geometry and some internal heterogeneity. Obviously, the properties cannot be directly measured or tested, or might be different from the properties of macro specimens (due to the scale and size effects, constraints by other components, effects of treatment regimes, etc.). The extraction of the data for the phases from the standard or modified tests of materials can become a nontrivial problem, which requires both rather complex experiments and an inverse analysis step [5]. A particular case, that confirms this situation, can by exemplified when is applied the generation of unit cells, as is present in Figure 2, where molecular distribution are very sensitive to CM procedure.

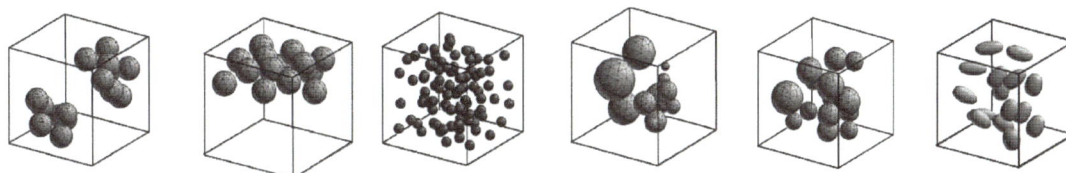

Figure 2. Examples of the unit cells generated with the use of the program Meso3d [5].

Furthermore, at the mesoscale levels are a wide range of configuration that encloses different heterogeneities. Such arrangement can involve a large number of small inclusions, so that a small parameter, the relative size of an inclusion, may compete with a large parameter, represented as an overall number of inclusions. To solve this kind of problem, the method of meso-scale asymptotic approximations is usually applied in order to obtain a uniform approximation to the solution of the Dirichlet problem in the multiply perforated domain [6].

An alternative approach could be considered homogenization techniques [7]. Through this method can analyze the local stress and strain in the constituent material. In this case the grain core (inclusion) and the grain boundaries (matrix) mechanism are engaged. The behavior of the grain core and of the grain boundaries are, respectively, elastic visco-plastic and elastic perfect–plastic.

Besides the heterogeneous nature of the compound, one must consider all kinds of interactions developing at even very different scales. This complex problem that requires careful depth treatment, with accurate and fast numerical methods (i.e. finite difference method, finite volume method, finite element method, mesh-less finite element methods, Discrete Elements Methods, Molecular Dynamics etc.), deal to the contact detection between these particles. In general, the configuration of contact particle is surrounded by two or more body (solid, liquid, gaseous) established in direct contact as a multi-layered structure comprised of different matters (metal-metal, metal-oxide, sol-gel film, metal-liquid, metal-polymer, polymer-concrete, polymer-textile, rubber/metal composites and polymer-human body) what involve different constraints. The atomistic and/or molecular scale the contact body is considered by the Van der Waals interactions and depends on the geometric configuration of the group of atoms encountered [8]. Figure 3 show the Van der Waals interactions, a schematic representation of the repulsive part of the interaction potential.

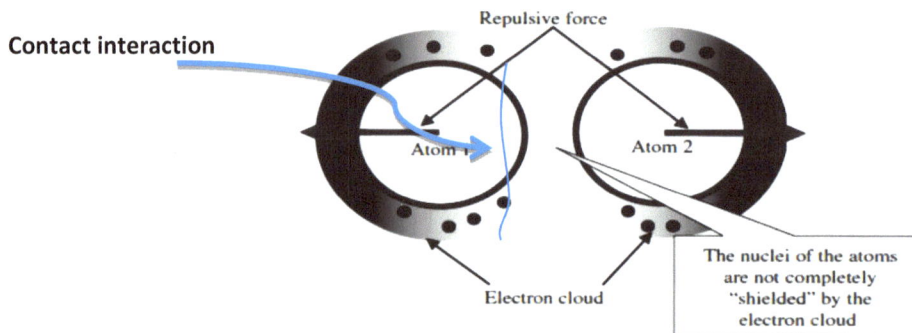

Figure 3. Van der Waals interactions-schematic of the repulsive part of the interaction potential [8].

At the macroscopic scale, the load transfer for interface field is analyzed using molecular structural mechanics method and the matrix deformation by the continuum finite element method. The van der Waals force between interacting atoms is written as:

$$F(r) = -\frac{dU(r)}{dr} = 24\frac{\varepsilon}{\sigma}\left[2\left(\frac{\sigma}{r}\right)^{13} - \left(\frac{\sigma}{r}\right)^{7}\right] \tag{1}$$

where r is the interatomic distance, □ and □ are the Lennard-Jones parameters. For carbon atoms the Lennard-Jones parameters are □= 0.0556 kcal/mole and r= 3.4 A˚.

Panin et al. [9], [10] applied the concept of structural scaling and structural-scaling transition to assess defect on subsystem for a deformed Nano solid. An important functional role in the response of the internal medium to external actions, in multilevel system, is played by the heterogeneous medium. The surface layer substrate interface may be described with fundamental effect of a chessboard-like stress strain distribution model. This effect implies the generation of strain induced defect on the surface of a loaded solid. Plastic flow propagation at meso- and macrobands is localized as well as, and show the behavior of solid under various external actions. The mechanical interface is generating by sinusoidal surface layer due to rotation of constraints successively in tensile and compression in the structure. Described by the following equation:

$$\sigma = A\sigma_y \sin\frac{x - l_x}{t\sqrt{2}} \tag{3}$$

where t, x, σ_y $\square\square$A, are the thickness of the coating, distance of crack propagation, stress coefficient and surface area covered by "paper", respectively.

$$l_x = \frac{1}{\sqrt{2}}\left(\frac{\pi}{2} + n\pi\right)t \qquad (4)$$

For macroscopic contact interaction, this complex problem can be discretizing as a function of contactor behavior (elastic, elasto-plactic, and perfect plastic mechanical behavior) and/or contactor shape that act to the specimen sample (i.e. from meso-micro-to macroscale). Generally, the shape of contactor is described by a curve or a segment. It is well know that the numerical approximation of contact interaction is allocated to node to segment surface. For example, the interaction between an segment surface and an disc curve present three situation as is exemplified in Figure 4.

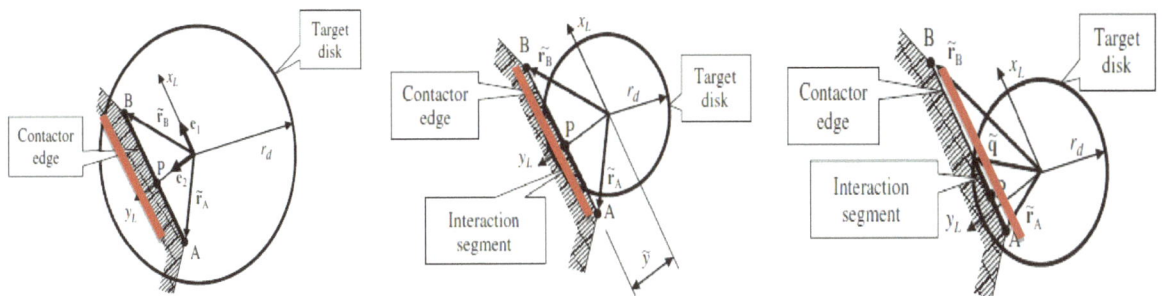

Figure 4. Schematic of interaction between an segment surface and an disc curve for different case a) both nodes of the contactor edge are inside the target disk, b) both nodes of the contactor edge are outside the target disk and c) Node A is inside the target disk while Node B is outside [8].

These types of contact problem showed above surround a large domain of application, from mechanical forming (cold, worm, hot) problem to biomechanics approach (interaction between different implants applied on human body).

Multiple-scale techniques, where the emphasis is made on the latest perspective approaches, such as the bridging scale method, multi-scale boundary conditions, and multi-scale fluidics. As a possible application of studies in bone failure may be detection of milder osteoporotic symptoms.

3. CHALLENGES

Considering the limit when a "numerical model" encloses structure from a unit scale at the quantum level to a model with multiscale processes (see Figure 5) at the Human scale, it is possible to be ideal calibrated to respect the powerful of length scale approach? Or this methodology applied under the philosophy of multiscale analysis has the effects on the upscale after modeling, simulation, and validation occurred at the length scale of interest?

The successfully integration of multiscale approach can lead to a comprehensive industrial solution.

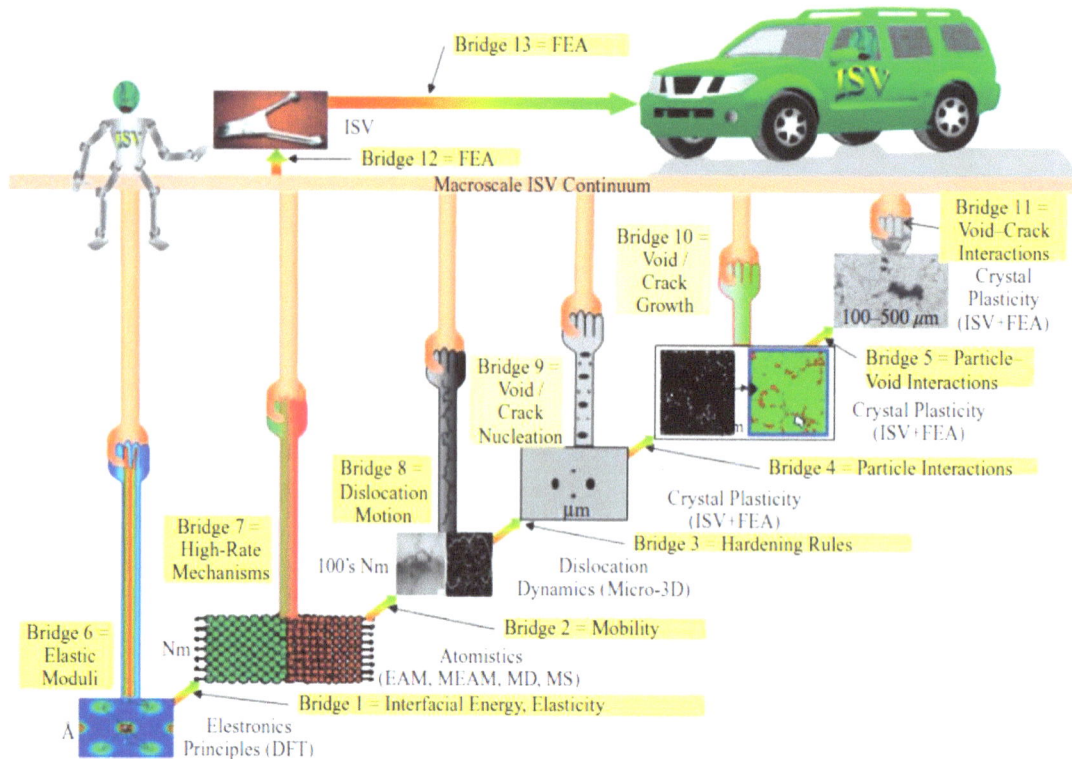

Figure 5. Multiscale approach applied in a metal alloy process for design an automotive component. The structured methodology present different length scale analyses used through various bridges within CM. ISV, internal state variable; FEA, finite element analysis; EAM, embedded atom method; MEAM, modified embedded atom method; MD, molecular dynamics; MS, molecular statics; DFT, density functional theory [11].

4. CONCLUSION

This paper cover the main issue generated when the CM are applied to predict the behaviour of material behaviour, and point out the boundary condition (knowledge the time and the reference coordinate) necessary to implement the CM processes. Besides, a multiscale approach is called for consideration as further application on the industrial management.

REFERENCES

[1] O.L. de Lange and J. Pierrus. *Solved Problems in Classical Mechanics.* ISBN 978–0–19–958252–5, Oxford University Press Inc., New York, 2010.

[2] DongBin Lee and C. Nataraj. Model-Based Adaptive Tracking Control of Linear Time-Varying System with Uncertainties. Mikhaylo Andriychuk. *Numerical Simulation – From Theory to Industry.* Janeza Trdine 9, 51000 Rijeka, Croatia, 2012.

[3] Dale R. Durran. *Numerical Methods for Fluid Dynamics.* ISSN 0939-2475, Springer Science+Business Media, LLC 1999, 2010.

[4] Denis Dartus. Feedback on the Notion of a Model and the Need for Calibration. [auteur du livre] Jean-Michel Tanguy. *Environmental Hydraulics Numerical Methods, Volume 3.* ISBN 978-1-84821-155-1 (v. 3), © ISTE Ltd, 2010.

[5] Leon L. Mishnaevsky. Computational experiments in the mechanics of materials: concepts and tools. *Computational Mesomechanics of Composites Numerical Analysis of the Effect of Microstructures of*

Composites of Strength and Damage Resistance. ISBN 978-0-470-02764-6, John Wiley & Sons Ltd, England, 2007.

[6] Vladimir Maz'ya, Alexander Movchan, Michael Nieves. *Green's Kernels and Meso-Scale Approximations in Perforated Domains.* ISBN 978-3-319-00356-6, © Springer International Publishing Switzerland, 2013.

[7] L. Capolungo, C. Jochum, M. Cherkaoui, J.Qu. Homogenization method for strength and inelastic behavior of nanocrystalline materials. International Journal of Plasticity 21 (2005) 67–82.

[8] Antonio A. Munjiza, Earl E. Knight, Esteban Rougier. *Computational Mechanics of Discontinua.* ISBN: 978-0-470-97080-5, Wiley Series in Computational Mechanics, November, 2011.

[9] V.E. Panin, A.V. Panin, V.P. Sergeev, A.R. Shugurov. Scaling effects in structural-phase self-organization at the "thin film - substrate" interface. Physical Mesomechanics, Volume 10, Issues 3–4, May–August 2007, Pages 117–128.

[10] V.E. Panin, V.A. Panin and D.D. Moissenko. Physical mesomechanics of a deformed solid as a multilevel system. II. Chessboard-like mesoeffect of the interface in heterogeneous media in external fields. Physical mesomechanics, 5-14 (20), 2007.

[11] Mark F. Horstemeyer. Integrated Computational Materials Engineering (ICME) for Metals Using Multiscale Modeling to Invigorate Engineering Design with Science. The Minerals, Metals & Materials Society, 2012.

DETERMINATION OF COEFFICIENT OF THERMAL CONDUCTIVITY ON GLASS FIBERS-REINFORCED POLYMER MATRIX COMPOSITES

Gheorghe Vasile[1], Bejan Costel[2], Sandu Veneţia[3], Lihteţchi Ioan[4] Eugenia Secară
[1] INAR S.A., Braşov, ROMANIA, ghesile@yahoo.com
[2] INAR S.A., Braşov, ROMANIA, cvbejan@yahoo.com
[3] Transilvania University, Braşov, ROMANIA, sandu@unitbv.ro
[4] Transilvania University, Braşov, ROMANIA, lihtetchi@unitbv.ro

Abstract: *This paper presents the determination of the coefficient of thermal conductivity for composite materials reinforced with fiberglass. The sheets tested on the stand were cut from a sheet made from 15 layers of fiberglass fabric.*
Keywords: *coefficient of thermal conductivity, composite materials, fiberglass fabric.*

1. INTRODUCTION

Thermal conductivity coefficient characterized the material properties to lead the heat flow. The coefficient of thermal conductivity is a physical constant that depends on the shape of the material, on the nature, temperature, aggregate state and it is determined experimentally.

The coefficient of thermal conductivity is numerically equal to the stationary conductive heat flow, passing through a unit area of a sheet of uniform thickness, when the difference of temperature between the outer surfaces is equal to unity. So:

$$\lambda = \frac{q \cdot \delta}{\Delta t} \quad \left[\frac{W}{mK} \right]$$

(1)

Experimental determination of these coefficients involves experimental measuring of q, δ and Δt.

2. TEST BENCH

The test bench for determination of the thermal conductivity coefficient of insulating materials with flat shape, homogeneous, microporous, with fibers or of particles is designed by Dr. Bock. The testing area is $\lambda = 0,029...1,977 \ W/(mK)$. The schematic diagram of this system is shown in Figure 1.

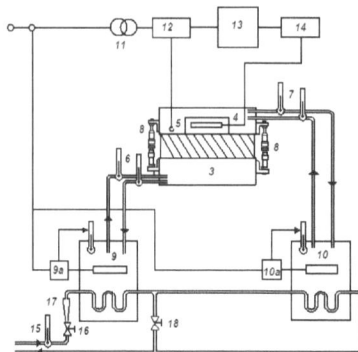

Figure 1: Schematic diagram of the stand for determination thermal conductivity coefficient
The determination of the thermal conductivity coefficient is based on the heating plate with one test body. The sample material (1) is placed between two flat metal plates, the upper (2) with a higher constant temperature, called heaters, equipped with an electrical resistance. The lower plate (3) with a lower constant temperature - cold called, that gives warmth. The hot plate is covered by another protective plate (4) to prevent loss of heat from the hot plate.

The protection plate temperature is kept constant through its connection to the heating circuit of the thermostat (10) with thermoregulator (10a). The constant temperature of the cold plate is achieved by the cooling circuit of the thermostat (9) with its thermoregulator (9a).

The cooling water circulating through the coil thermostats, in whose routes lie thermometer (15) and rotameters (17), reduce the thermal inertia of the water in the thermostatic hot plate and take the heat transferred to the cold plate and to the its thermostat.

With thermometers (7) determine the mean temperature of thermal agent in protection plate. The thermometers (6) determine the mean temperature of thermal agent in cold plate. With these thermometers can calculate the temperature drop of the sample.

In electrical resistance circuit is interposed a rheostat with twelve positions (14). The maintenance of a constant temperature over the entire surface of the top plate in contact with the sample of material is performed by the thermocouple (5) which is connected to the millivoltmeter (12).

The consumption of electrical energy give to the heating plate are recorded by an electric meter (13) located in the electrical resistance circuit. In the power supply circuit is located a variable transformer with seven positions (11).

During the measurement, the metal plates are surrounded by a protective box, which is designed to reduce heat loss to the outside.

With the help of four micrometers (8), fixed to the top plate, measure the thickness of the sample material.

3. PREPARATION OF MATERIAL SAMPLES

Samples of the materials are in the form of square or circular plate with plane and parallel surfaces, with an edge length or a diameter between 200 mm and 250 mm. The thicknesses of the samples are between 3 and 70 mm.

In this paper test sheets were cut from a sheet made from 15 layers of fiber-glass fabric (Figure 2). Plates size is 250 x 250 mm.

Figure 2: The plates used for determining the thermal conductivity coefficient

Figure 3: Milling fiberglass plates

Because the plate was free molded in an open mold, its thickness was not constant and one of its surfaces wasn't flat. For this reason the plates were made subject to a milling processing (Figure 3). By milling, obtain an uniform thickness of plates and both sides have become flat. In Figures 4 and 5 are shown the face and the edge of a plate which has been processed.

Figure 4: Milling face of fiberglass plate

Figure 5: Edge of fiberglass plate

4. MEASUREMENTS

After the system is entering in steady state operation, following determinations are made from 0.25 to 0.25 hours, which fall in the worksheet:

- hours and minutes indicated by the clock;
- indication of electricity meter;
- thermal agent temperature reading on thermometers (7) located at the entrance and to the exit of the upper plate (tci, tce) and to the thermometers (6) of the bottom plate (tri, tre);
- room temperature;
- cooling water temperature;
- thickness of the sample is measured at the end of the experiment with these four micrometers.

Data reading by measurement instruments and data calculated are recorded in a single worksheet. To each plate was carried out a worksheet. The worksheets for the two parts are attached down (Table 1 and Table 2).

Table 1. Thermal test result

Specimen no.	1	Matherial	Plastics				Date	
Read no..	1	2	3	4	5		6	...13
Time								
Room temperature TR [°]	20							
Meter reading E [kW]	556,7793	556,7919	556,8036	556,8142	556,8241	556,833		
Entering heating plate protective tri [°]	31,58	31,72	31,62	31,65	31,71	31,72		
Output heating plate protective tre [°]	31,55	31,58	31,53	31,52	31,6	31,6		
Entrance cooling plate trt [°]	22,24	22,65	23,05	23,4	23,62	23,78		
Output cooling plate tre [°]	22,05	22,42	22,84	23,19	23,35	23,55		
Power stage	9							
ΔE [kW]	0,0126	0,0117	0,0106	0,0099	0,0089	0		
$\Delta \zeta$ [h]	0,25	0,25	0,25	0,25	0,25			
$\Delta E/\Delta \zeta$	0,0504	0,0468	0,0424	0,0396	0,0356	0		

Micrometers		Measurement	Error	Corrected value
$\Delta 1$	[mm]	10,31	0,01	10,3
$\delta 2$	[mm]	10,55	0,01	10,54
$\delta 3$	[mm]	10,53	0,01	10,52
$\delta 4$	[mm]	10,48	0,01	10,47
$\Sigma \delta 1$-4	[mm]			41,83
$\Sigma \delta 1$-4/4	[mm]			10,4575
Λ	[kcal/(hmgrad)]			0,133835907834897

Table 2: Thermal test results

Specimen no.	2	Matherial		Plastics			Date							
Read no..	1	2	3	4	5	6	7	8	9	10	11	12	13
Time													
Room temperature TR [°]	20												
Meter reading E [kW]	556,9353	556,9464	556,9568	556,9656	556,9747	556,9833	556,9916	556,999	557,0055	557,0125	557,0181	557,0242	557,0294
Entering heating plate protective tci [°]	31,7	31,65	31,62	31,8	31,63	31,61	31,6	31,6	31,75	31,65	31,73	31,58	31,7
Output heating plate protective tce [°]	31,58	31,55	31,59	31,61	31,55	31,53	31,51	31,52	31,61	31,59	31,64	31,51	31,61
Entrance cooling plate tri [°]	22,2	22,48	22,7	22,9	23	23,2	23,4	23,57	23,74	23,88	24,04	24,22	24,2
Output cooling plate tre [°]	21,96	22,25	22,45	22,64	22,7	22,91	23,16	23,32	23,5	23,58	23,73	23,86	23,9
Power stage	9												
ΔE [kW]	0,0111	0,0104	0,0088	0,0091	0,0086	0,0083	0,0074	0,0065	0,007	0,0056	0,0061	0,0052	0
Δζ [h]	0,25	0,25	0,25	0,25	0,25	0,25	0,25	0,25	0,25	0,25	0,25	0,25	
ΔE/Δζ	0,0444	0,0416	0,0352	0,0364	0,0344	0,0332	0,0296	0,026	0,028	0,0224	0,0244	0,0208	0

Micrometers		Measurement	Error	Corrected value
Δ1	[mm]	9,52	0,01	9,51
δ2	[mm]	9,7	0,01	9,69
δ3	[mm]	9,53	0,01	9,52
δ4	[mm]	9,53	0,01	9,52
Σδ1-4	[mm]			38,24
Σδ1-4/4	[mm]			9,56
Λ	[kcal/(hmgrad)]			0,0911042736053292

5. ACKNOWLEDGEMENT

PROGRAMUL OPERAŢIONAL SECTORIAL
„CREŞTEREA COMPETITIVITĂŢII ECONOMICE"
„Investiţii pentru viitorul dumneavoastră"

Proiect: Dezvoltarea de componente din materiale compozite avansate cu aplicatii in industria auto civilă şi militară.

Proiect cofinanţat de Uniunea Europeană prin Fondul European de Dezvoltare Regională

REFERENCES

[1] Badea A., Necula H., Stan M., Ionescu L., Blaga P., Darie G., Echipamente şi instalaţii termice, Editura Tehnică, Bucureşti, 2003.
[2] Kreith F. Handbook of thermal engineering, CRC, Boca Raton, Florida, 2000.
[3] Pop M.G., Leca A., Prisecaru I., Neaga C., Zidaru G., Muşatescu V., Isbăşoiu E.C. Îndrumar. Tabele, nomograme şi formule termotehnice, Editura Tehnică, Bucureşti, 1987.

INFLUENCE OF TEMPERATURE ON MECHANICAL PROPERTIES OF POLYMER MATRIX COMPOSITES SUBJECTED TO BENDING

Gheorghe Vasile[1], Bejan Costel[2], Sîrbu Nicolae[3], Lihteţchi Ioan[4], Arina Modrea

[1] INAR S.A., Braşov, ROMANIA, ghesile@yahoo.com
[2] INAR S.A., Braşov, ROMANIA, cvbejan@yahoo.com
[3] INAR S.A., Braşov, ROMANIA, nica_sirbu@yahoo.com
[4] Transilvania University, Braşov, ROMANIA, lihtetchi@unitbv.ro

Abstract: *This paper presents the influence of temperature on the mechanical characteristics of a composite material. Results are presented in an attempt to rupture in bending of a lot of samples made from glass fiber fabric on one stand. Various attempts were made on samples heated at different temperatures.*

Keywords: *composite material, rupture in bending, glass fiber fabric, bending testing.*

1. INTRODUCTION

I   In this paper are presented the results obtained from rupture on test bending bench, applied of a lot of composite material samples.

2. THE METHOD USED FOR EXPERIMENTAL TESTS

The working principle is as follows:

The test piece is supported as a lever, between two suports and subjected to a bending constant speed rupture. During the test are measured the force applied to the specimen and its arrow deformation (displacement of a midpoint of to the test piese between the supports). These measurements are embodied in a force-arrow graphic.

The scheme of the three-point of bending of test specimen is shown in Figure 1.

As shown in the figure, the test piece is placed on the two cylinders at 80 mm rest against each other, and of the center of the upper is operating force F by the pusher. Under the action of F force by pressing, the test piese are deforms with *f* arrow.

The test piece is considered broken at the first fall of the force-arrow deformation graphics.

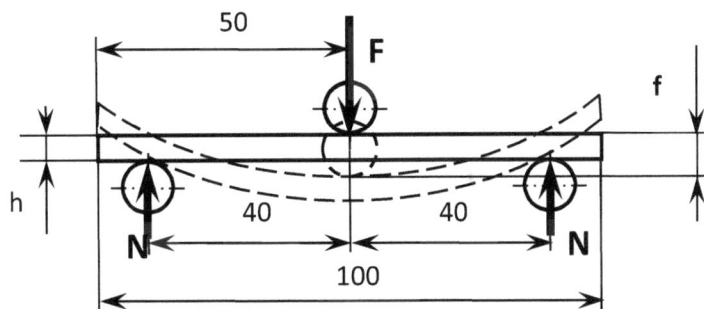

Figure 1: Application scheme in three-point bending of specimen

For the test sample of the specimen to rupture in bending was made a sheet of composite material with 15 layers of glass fiber fabric. This layer has a thickness of 0.8 mm and is shown in Figure 2.

The achieved plate thickness is 12 mm. From this plate were cut 20 samples of dimensions: (Figure 3.)
- Length 100 mm;
- Width 14,5 mm;
- Thickness 12 mm.

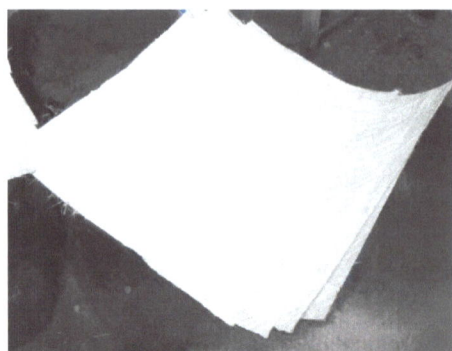

Figure 2: Glass fiber fabric

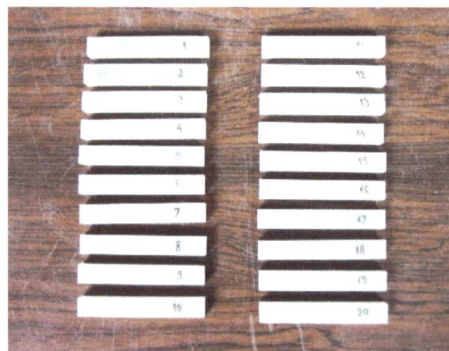

Figure 3: The specimens used

Then, were measured their gauge dimensions.

These test pieces (specimens) thus obtained were divided into four groups.

The first group contains the test pieces marked with numbers 1, 2, 3, 4, 5, and 6. These samples were tested to breaking in flexure at room temperature or 20 °C.

Test pieces of the second group, those marked with the numbers 7, 8 and 9, were heated to a temperature of 50 °C and tested in bending at this temperature.

The test specimens of numbers 10, 11 and 12, of the third group, were heated to a temperature of 65 ° C, after which they were tested.

The fourth group was made up of the test specimens with the numbers 13, 14 and 15, which have been heated to a temperature of 100 ° C and then was allowed to cool to ambient temperature (20 ° C), they were tested for breaking bending ..

The speed of pressure was 0.1 mm/s

In Figure 4, is shown the specimen on the stand during bending testing.

Figure 4: Testing glass fiber fabric on the stand

For each specimen tested was carried out a chart force-arrow. In figure 5 is shown chart of the registration bending test for No. 1 specimen and in Figure 6 is the force-arrow chart in aggregate for the bending test of the other six specimens.

Maximum values recorded during the application of bending test of the three samples tested, dimensions and values for the bending load [δ] are shown in Table 1.

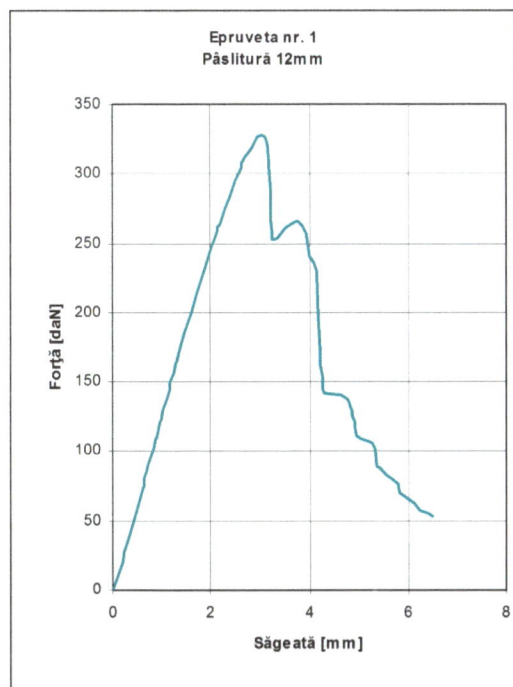

Figure 5: Arrow-force graph registered for specimen no. 1

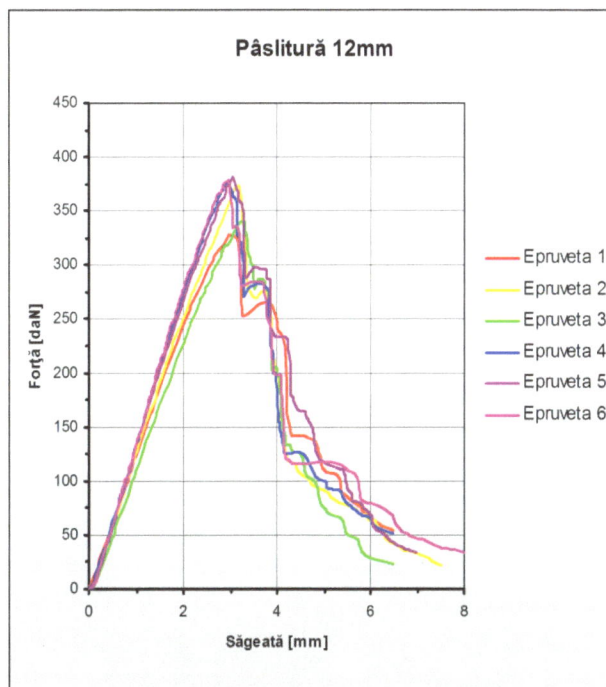

Figure 6: Arrow-force graph registered cumulated for bending test of all specimens

Table 1: Maximum values recorded during the application of bending test of the three samples tested

Glass fiber fabric	Downforce	Arrow	Specimen dimensions		Bending load
			Width	Thickness	δ
	[daN]	[mm]	[mm]	[mm]	[MPa]
Specimen no. 1	328,00	3,05	14,50	11,20	21,64
Specimen no. 2	372,00	3,19	14,50	11,40	23,69
Specimen no 3	340,00	3,31	14,50	11,10	22,84
Specimen no. 4	377,00	2,94	14,50	11,80	22,41
Specimen no. 5	381,00	3,05	14,50	11,90	22,27
Specimen no. 6	379,00	2,95	14,50	11,90	22,15

When the specimens in the second group reached 50 °C, were tested to the bending breaking. The temperature of the specimen was checked on the stand, with an electronic thermometer (Figure 7).

Figure 7: Checking on the stand, of the specimen temperature, using an electronic thermometer

Flexural breaking test at 50 ºC was done on the specimens with numbers 7, 8 and 9.
The sample was performed at the same speed push to the specimen at 0.1 mm/s.
In Figure 8. is shown the force-arrow chart, cumulated, for the bending test of the three samples.

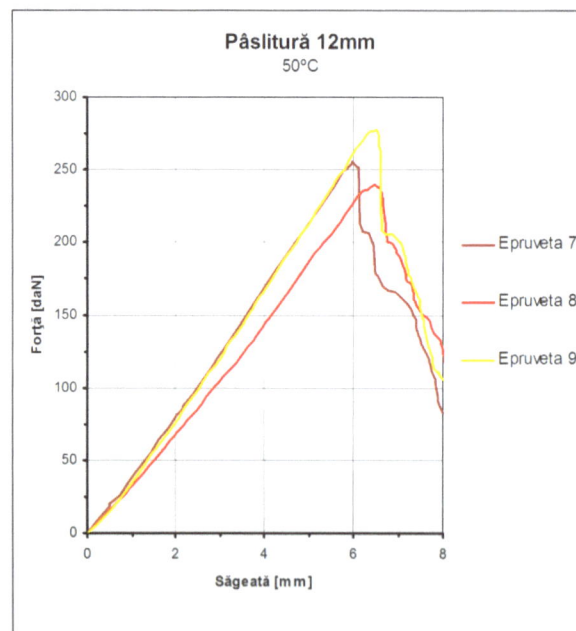

Figure 8: Arrow-force graph registered cumulated for bending test of all specimens

Table 2 presents the maximum values recorded during bending breaking test for the three specimens tested.

Table 2: Maximum values recorded during bending breaking test for the three specimens tested at 50ºC.

Glass fiber fabric 50ºC	Downforce	Arrow	Specimen dimensions		Bending load
			Width	Thickness	δ
	[daN]	[mm]	[mm]	[mm]	[MPa]
Specimen no. 7	255,34	5,96	14,50	11,10	21,64
Specimen no. 8	240,00	6,49	14,50	11,00	23,69
Specimen no. 9	278,00	6,52	14,50	11,60	22,15

Flexural breaking test at 65 °C was done on the specimens with numbers 10, 11 and 12 (third group). Specimen temperature was checked this time too, with an electronic thermometer (Figure 9).

Figure 9: Checking on the stand, of the specimen temperature, using an electronic thermometer

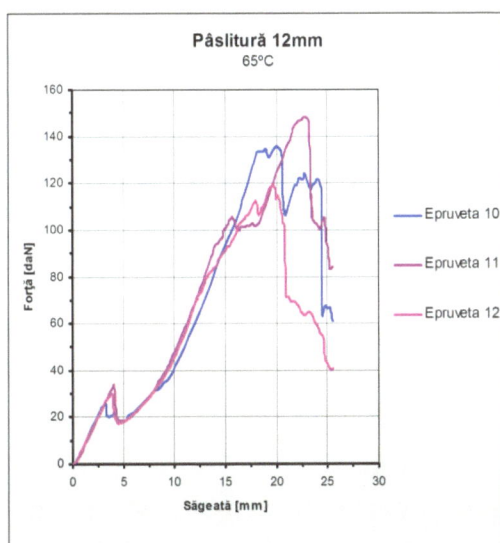

Figure 10: Arrow-force graph registered cumulated for bending test of all specimens

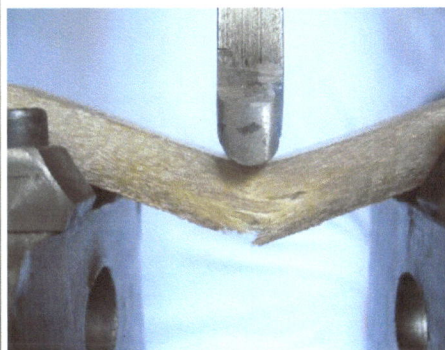

Figure 11: Test of glass fiber fabric specimen on stand

Specimen deformation rate was 0.1 mm/s.
In Figure 10 is shown the force-arrow chart, cumulated, for the bending test of the three samples tested at 65 ° C
Table 3 presents the maximum values recorded during bending breaking test for the three specimens tested.

Table 3: Values recorded during the application of bending test of the three samples tested at 65°C

Glass fiber fabric 65°C	Downforce	Arrow	Specimen dimensions		Bending load
			Width	Thickness	δ
	[daN]	[mm]	[mm]	[mm]	[MPa]
Specimen no. 10	25,84	3,17	14,50	12,00	1,49
Specimen no. 11	33,77	4,02	14,50	11,00	2,31
Specimen no. 12	30,13	3,84	14,50	10,80	2,14

The final three samples were heated to a temperature of 100 °C. They were allowed to cool to room temperature, after which the breaking bending tests were repeated. We wanted to see the influence of temperature on the behavior of this material.

Finaly, to the breaking bending test were take the pieces numbered 13, 14 and 15. The sample test was also performed at a push speed of 0.1 mm /s. In Figure 11, is shown the specimen on the stand during bending test.

In Figure 12 is shown the force-arrow chart, cumulated, for the bending test.

The values recorded during the application of bending test of the three samples tested, dimensions and values for the bending load [δ] are shown in Table 4.

In Figure 13 is shown the force-arrow chart, cumulated, for the bending test to one of four groups tested specimen.

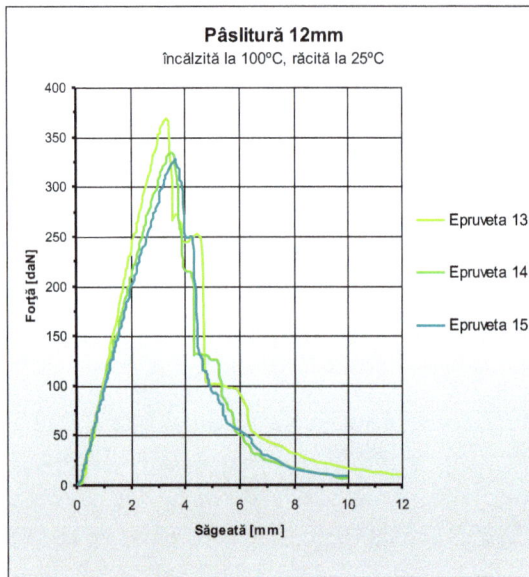

Figure 12: Arrow-force graph registered cumulated for bending test of all specimens

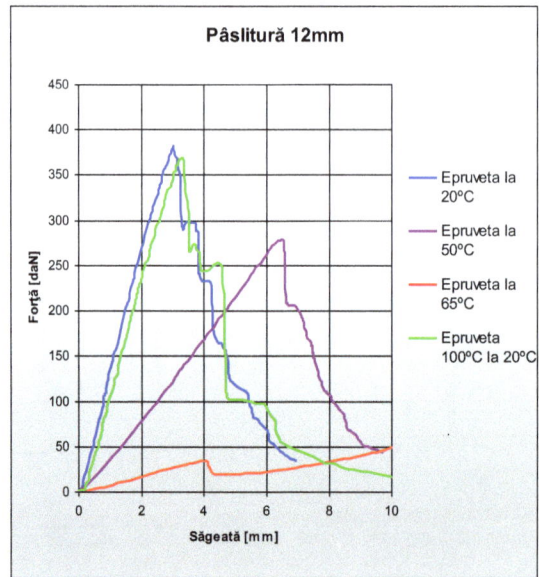

Figure 13: Arrow-force graph registered cumulated for the four groups

Table 4: Values recorded during the application of bending test of the three samples tested at 100°C

Glass fiber fabric 100°C	Downforce	Arrow	Specimen dimensions		Bending load
			Width	Thickness	δ
	[daN]	[mm]	[mm]	[mm]	[MPa]
Specimen no. 13	368,00	3,34	14,50	11,70	22,25
Specimen no. 14	333,92	3,53	14,50	11,90	19,51
Specimen no. 15	327,00	3,69	14,50	11,30	21,19

3. CONCLUSION

Temperature influences the characteristics of the composite material. As the temperature is higher, the performance of the material decreases. Note that even if heated to a high temperature, the composite material regained initial strength once the return to baseline temperature.

4. ACKNOWLEDGEMENT

Instrumente Structurale
2007 – 2013

UNIUNEA EUROPEANĂ

GUVERNUL ROMÂNIEI

PROGRAMUL OPERAŢIONAL SECTORIAL
„CREŞTEREA COMPETITIVITĂŢII ECONOMICE"
„Investiţii pentru viitorul dumneavoastră"

Proiect: Dezvoltarea de componente din materiale compozite avansate cu aplicatii in industria auto civilă si

REFERENCES

[1]   Huba, G., Iovu, H. - Materiale compozite - Editura Tehnica 1999.
[2]   Mihalcu, M. , Materiale plastice armate - Editura Tehnica Bucuresti ,1986.
[3]   R. Purcarea, V.Gheorghe, M.V. Munteanu Endurance tests on specimens from compozite materials sandwich type - The 4th International Conference - Advanced Composite Materials Engineering - COMAT 2012

THEORETICAL STUDIES AND EXPERIMENTAL RESEARCH FOR THE INCREASE OF THE WORK SAFETY AT GANTRY CRANES PART I – THEORETICAL STUDIES (ABAQUS). CONCLUSIONS

Gheorghe N. RADU[1], Ioana Sonia COMĂNESCU[2]

[1] Transilvania University of Brasov, Brasov, ROMANIA, e-mail: rngh@unitbv.ro
[2] Transilvania University of Brasov, Brasov, ROMANIA, e-mail: ioanacom@unitbv.ro

Abstract: *The need for cranes is extremely wide and various. The present research is focused on the gantry cranes, which are largely spreaded as destination and work conditions. The study of the above mentioned cranes is the subject of a classical analytical approach and also of an actual approach of high interest – the Finite Element Method (FEM) of the mechanical structures. An experimental approach which confirms the methods and the results obtained through other methods is also achieved.*

The analysis was performed on a particular structure – a 3D one consisting mainly on beams, loaded in static and dynamic regime. One emphasizes the fact that this above mentioned structure is a particular one, on which one tries to get some original contributions concerning the improving of the work safety.

The work safety of the mechanical structures subjected to the static and dynamic load's action represents a highly complex issue. The precise knowing and understanding through adequate techniques and technologies of the whole structure but of the components, too regarding the state of stress, stands at the base of the work safety. One established that the structure's answer analysis in static and dynamic regime, mainly expressed by the values of the displacements and stresses fields is essential for the work safety.

Keywords: *gantry crane, stress, deformation, main beam, windbracings*

1. PRESENT STUDIES

In a modern assumption, the cranes are complex hoist equipment, usually made of a metallic frame of variable shapes, designs and sizes; they are composed of a single or more mechanisms which are required to the uplift and to the loads displacement.

According to the action made or to the type of design the cranes may be classified by various criteria, as follows: fixed or movable on its own race, without own race, self – propelled or pulled.
The work area of the cranes mostly coincides with the activity area of the people.

Because of the particularities of the cranes work, these must satisfy many standards regarding the work security and safety. This is the reason why the present paper is focused on the improvement of the gantry cranes work safety.

The authors achieved an analysis, with help of the classical theory of strength of materials, of the main elements of the gantry crane (elements subjected to tension, compression, stability calculus – buckling, bending, the check of the trusses, the calculus of the main beam's strength – both fixed ends; one appeals at the classical calculus of the gantry crane, inclusively, which is recommended by the present standards. All the above mentioned calculations are made in order to achieve modern methods which should lead to the increase of the gantry cranes work safety.

2. THE APPLICATION OF THE FEM FOR THE DETERMINATION OF THE STATE OF STRESS AND STRAIN AT GANTRY CRANES

One starts from the assumption that the big dimensions structures may be studied on patterns (theoretical studies based on computational methods and experimentally for their validation, Figue 1. On the base of the above assumptions stands the similarity criterion. The concept of similarity is used meaning that the study methods on pattern (generally, laboratory research) can be extended at any other sizes of the gantry cranes.

Figure 1 Pattern of a gantry crane

One specifies further on some theoretical issue concerning the similarity criteria [6, 7].

1.   HOOK's similarity criterion

It goes from the general case where Hook's law states that in the case of elastic structures loads and displacements, stresses and strains, respectively, are directly proportional ($\Delta = \rho \cdot F$). Hook's similarity criterion is finally achieved, being written between the real body (subscript r) and its attached model (subscript m).

$$\varepsilon_r = \varepsilon_m; \quad \varepsilon = \frac{\sigma}{E}$$

or

$$\frac{F_r}{A_r \cdot E_r} = \frac{F_m}{A_m \cdot E_m}$$

where F is the internal axial traction force developed in the cross section A.

Taking into account the similarity criteria, one may write also the proportionality between the forces scale K_F and the length scaleλ, as follows:

$$K_F = \lambda^2 \cdot K_E.$$

Hook's similarity criterion given by the above equation, which states the static similarity, yields as:

$$\varepsilon = \frac{\Delta \ell_r}{\ell_r} = \frac{\Delta \ell_m}{\ell_m},$$

Which emphasizes that the deformations have the same scale as the lengths ($K_{\Delta \ell} = \lambda$).

From the above presented issues, two important conclusions are emphasized:
- the above number H has the same value for the real body and for the prototype, as well; practical, the two compared elastic bodies are likely in terms of external loads;
- Hook's similarity criterion (concerning the finite elastic deformations) is satisfied when the geometrical similarity between the model and the prototype is achieved.

2. CAUCHY's similarity criterion

Cauchy's criterion is a similarity criterion applied to the dynamic loads due to the vibrations of the elastic systems. This criterion yields from the "cross – vibrations equation" of the prismatic beams of changing cross section; the below form of the criterion yields:

$$E \frac{\partial^2}{\partial x^2}\left(I \frac{\partial^2 y}{\partial x^2}\right) + \rho \cdot A \cdot \frac{\partial^2 y}{\partial t^2} = 0,$$

where the meaning of the above quantities is:
- ρ - the material density;
- $\rho \cdot A$ – mass per unit length;
- $\rho \cdot A \cdot (\partial^2 y / \partial x^2)$ – force of inertia per unit length;
- I – axial moment of inertia of the cross section A;
- y – the deflection of the equilibrium position;
- x – current calculus distance (with respect to the origin);
- t – time.

As a result of some transformations (theoretical study), Cauchy's similarity criterion yields:

$$l_0 / t_0 \colon \sqrt{E_0 / \rho_0} = \frac{v}{c} = C;$$

where:
- v – represents the oscillatory motion velocity of the beam's particles;
- c – the propagation velocity of the (axial) longitudinal waves.

From dynamical point of view the phenomenon on the model and that on the real structure are likely only if Cauchy's criterion has the same value on the prototype (the real case) and on the model, as the above equations emphasize.

From the performed analysis yields that Cauchy's similarity criterion and the ratio of the inertia forces and the really applied external loads are proportional; with the increase of the studied structures stiffness, Cauchy's number decreases and the similarity leads to accurate results.

In addition, to the two above presented criterion there is also FREUDE's similarity criterion which is adequate for researches of the patterned elastic systems, where the gantry has a significant meaning.

The complex issues of the similarity criteria lead to the significant conclusions. One emphasizes that in the case of two materials of the same Poisson's ratio and the same characteristic curve (in dimensionless coordinates) the general stress – deformation behavior is the same. In case of the present research, the model and the structure are executed from the same material and identical cross sections of the beams. The above mentioned similarity criteria lead to the statement that the load - displacement and the stress – strain dependence, respectively, are the same for the real structure and for the model (both in elastic state of load and at the limit of the elastic – plastic state). As a result of the fore mentioned issues, a functional pattern of the gantry crane was achieved; at 1/10 scale (Fig. 1). One used advanced software – ABAQUS, with highly efficient and accurate possibilities of modeling. This one was achieved for the entire structure, but the main element on which the attention is focused is the main beam, profile I, with the dimensions 42 x 80 x 2660 mm; the main beam is reinforced by help of a truss. One used 2818 Beam 3D elements. Two constructive versions are studied: structure with and without windbracings. By graphic post processing, the Von Mises stresses fields and deformations are presented. The achieved graphs show the change of the deflection along the main beam. Some results are shown in Table 1 and Figure 2.

Table 1. State of stress and strain of the successively loaded crane in points P1, P3 and P4, 1444 N weight

Load Application points	Analysis points													
	P1		P2		P3		P4		P5		P6		P7	
	Si (MPa)	S (mm)	Si (MPa)	S (mm)	Si (MPa)	S (mm)	Si (MPa)	S (mm)	Si (MPa)	S (mm)	Si (MPa)	S (mm)	Si (MPa)	S (mm)
P1	0	-0.32	2.9	-0.14	1	-0.035	0.5	0.0097	0.25	0.031	0.1	0.042	0	0.045
P3	0	0.0095	0.65	-0.06	1.94	-0.11	0.6	-0.14	0.3	-0.105	0.05	-0.05	0	0.01
P4	0	0.019	0.175	-0.056	1.5	-0.137	4.5	-0.231	1.5	-0.137	0.175	-0.056	0	0.019

The load is applied in the points P1, P3 or P4 while the state of stress and strain of the main beam is achieved in the initial established points: P1, P2, …, P7. Theoretical analysis for the windbracings crane is performed.

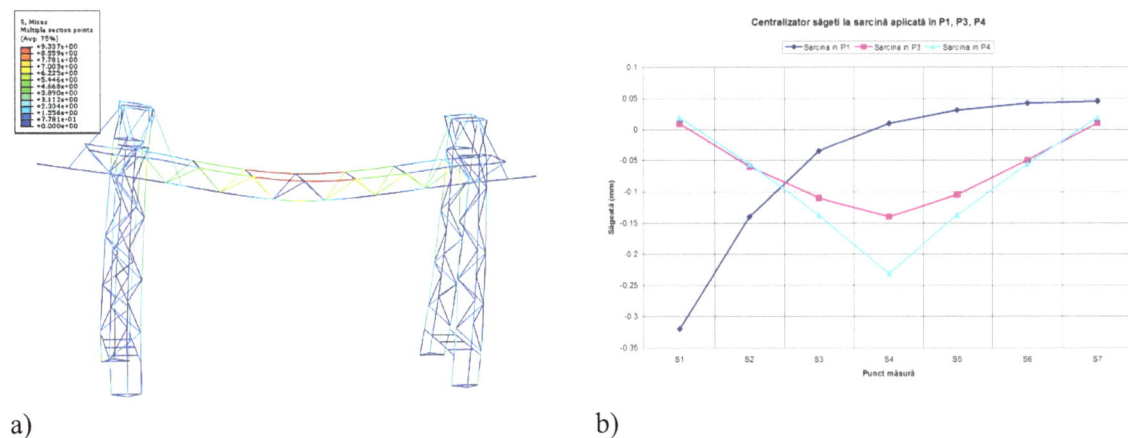

a)   b)

Figure 2 Stresses and strains - concentrated force in P1, P3 and P4

3. CONCLUSIONS (THEORETICAL STUDIES)

- In terms of the state of stress, analyzing the structures of the crane with windbracings and wingbracingless (stiffeningless), the two constructive solutions present no significant differences; the value of the stress in the middle of the main beam is higher for the crane with windbracings.
- In terms of deformations, analyzing the values of the deformations in the points P1 … P7, relative to the external load in P4, one concludes that the constructive solution with windbracings in more convenient. The windbracings ensures the stiffness of the structure and decrease the state of deformation of the main beam.
- The windbracing crane is a better constructive solution than the windbracingless solutions.
- If the elements used for windbracings are provided with pre – stressing systems, the behavior of the gantry crane is improved (in terms of deformations).
- The concepts and theory of similitude is used meaning that the study methods on pattern may be extended to any size of the gantry crane.

Figure 3 Measuring equipment for the highest loaded areas

From the performed theoretical studies the main failure areas of the crane are emphasized. These are:
- The main beam, profile I, in the grip area and in the area 1;
- The main beam, the straight longitudinal tube in the welding area of the support leg, Si2, Figure 3;
- The stand , in the welding area of the longitudinal straight tube – Si3, Figure 3;
- The main beam, the left longitudinal tube in the welding area of the stand Si4, Fig. 3;
- The stand, in the welding area to the left longitudinal tube, Si5, Figure 3.

4. REFERENCES

[1]   Manea, I., *Analiza modala experimentala*, Editura Universitaria Craiova, 2006.
[2]   Radu, N. Gh., Comanescu, I. S., Popescu, M., *The Study of the Behavior under static and dynamic Load of the Main Beam (profile "I") of a Portal Crane, Part I – The State of Deformation of the Main Beam under static Load, Part II – The state of deformation of the main beam in the "animation" working manner*, COMEC 2011, Brasov, 1991.
[3]   Radu, N. Gh., Comanescu, I. S., Popescu, M., *The Behavior under Load of the Portal Gantry Crane, referring to strength, stiffness and stability, Part I – Theoretical consideration, Part II – The Stresses and Deformations results by FEM analysis – ABAQUS, Part III – Experimental determinations of the field of stresses and strains*, The 9th International Conference OPROTEH 2011, ISBN 978 – 606 – 727 – 131 – 9, Bacau, Romania.
[4]   Radu, N. Gh., *Rezistenta materialelor si elemente de teoria elasticitatii,* vol. I (2002) si vol. II (2002), Ed. Universitatii Transilvania din Brasov, ISBN 073 – 9474 – 40 – 3.
[5]   Radu, N. Gh., *Capitole special de rezistenta materialelor*. Editura TEHNICA INFO. Chisinau., 2007, ISBN 978 – 9975 – 63 – 280 - 5.
[6]   Vasilescu, Al., Praisler, G., *Similitudinea sistemelor elastice*. Editura Academiei Romane, Bucuresti, 1974.
[7]   Vasilescu, Al., *Analiza dimensionala si teoria similitudinii*, Editura Academiei Romane, Bucuresti, 1974.
[8]   Nastasescu, V., Stefan, A., Lupuiu, C., *Analiza neliniara a structurilor mecanice prin metoda elementelor finite*. Editura Academiei Tehnice Militare, Bucuresti, 2002.
[9]   Gioncu, V., Ivan, M., *Bazele calculului structurilor la stabilitate*. Editura Facla, Timisoara, 1983.
[10]   Spinacovski, A. O., Rudenko, N. F., *Masini de ridicat si transportat*. Editura Tehnica, Bucuresti, 1973.

[11]   Vita, I., Sarbu, L., Nuteanu, T., Alexandru, C., *Masini de ridicat in constructii*, Editura Tehnica, Bucuresti, 1989.

[12]   ***SR ISO***Instalatii de ridicat. Organe de comanda. Reguli si metode de incercare. Cabine. Clasificare.

[13]   ***SR EN*** Securitatea ma;inilor. Principii ;i condi'ii tehncie. Principii generale de proiectare.

PLANNING THE TRAJECTORY OF THE SCORBOT-ER VII ROBOT

L. Predescu[1], M. Predescu[2]
[1] "Valahia" University, Târgovişte, ROMANIA, e-mail prelaur@yahoo.com
[2] "Valahia" University, Târgovişte, ROMANIA, e-mail zahariamirela@yahoo.com

Abstract: *The paper wishes to analyze the planning and the control of the trajectory of the movement of the end-effector of a rotating robot with 5 degrees of freedom (5R), imposing certain constraints in order to simplify the mathematical approach.*

By correctly modeling the direct and inverse kinematics, the positions of the end-effector can be determined in terms of time. Simultaneously, constraints can be imposed so that the robot should move on different trajectories within a previously established period of time and the final result would be the precise achievement of the given tasks.

The illustrated method allows the decoupled calculus of the positioning parameters of the end-effector from the orientation ones. This method simplifies and facilitates the analytical approach of the problem.

Keywords: *robot, kinematic model, trajectory planning*

1. INTRODUCTION

Robotics is usually defined as the science which studies the intelligent connection between perception and action. Referring to this definition, the robotic system is functionally complex, being made up of several subsystems.

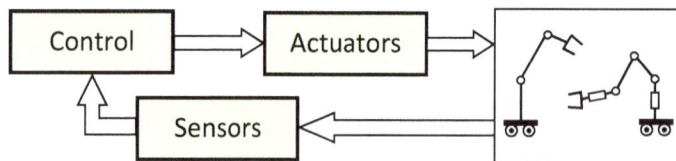

Figure 112 The robotic system components

The mechanical system together with the actuators, sensors as well as the control components make up the robotic system (fig.1). The mechanical system is commonly endowed with a motion apparatus (wheels, crawlers, mechanical legs) and a manipulation component (mechanical arms, end-effectors, artificial hands). In figure 1 the mechanical system is made up of two mechanical arms, either of them being set on a mobile vehicle thus covering as much of the work space as possible and generating its potentially limitless growth.
In terms of base mobility, robots can be:
- Fixed-base robots, also called manipulators;
- Moving-base robots, also called mobile robots.

The mechanical structure of a manipulator is made up of a chain of rigid elements (links), interconnected with different types of articulations (joints).

If this row is disposed in a serial manner to the base, it is called open kinematic chain manipulator, whereas if the links and joints support the task in parallel, it is called closed kinematic chain manipulator.

A manipulator's task is to follow a path in order to lead the end-effector to an established position.

In order to make a robot move along a trajectory between well determined target positions, the movement of different parts of the robot must be established. Consequently, constraints are imposed to reach these positions within a certain time limit, at a certain velocity, acceleration, etc. [1].

Its movement can be achieved through different configurations of its elements, and that is why the robot's movement within the work space must be studied [2], [3].

The kinematic analysis of the robots studies the motion of the composing mechanisms. Everything that has to do with the kinematic analysis, namely position, velocity and acceleration of all the elements will be computed in relation to a reference system which is considered fixed. As far as the kinematic analysis is concerned, the robot's actuating forces and torques are not considered.

In order to follow the functional trajectories, the kinematic model needs a program to generate these trajectories. The program can be either off-line or on-line.

Most of the manipulators are designed to achieve tasks in a 3D work space.
There are two different approaches related to the movement of the robot arm:

- specifying the end-effector's location in 3D coordinates;
- the individual movement of each articulation.

The manipulated part, tool or end-effector has to follow a planned trajectory.
The kinematic model supplies the relations between the position and the orientation of the end-effector and the space positions of the other elements, and articulations.

The kinematic modeling is divided into two problems: direct and inverse kinematics.
The task of direct kinematics is to determine the position and the orientation of the end-effector by giving values to the variables in the robot's joints.

Inverse kinematics focuses on determining the values for the variables in the joints which are necessary to move the robot's end-effector in a desired position and orientation.

2. THE INVERSE GEOMETRIC MODEL

In order to command the robot, the inverse geometric model is used. This model is based on determining generalized coordinates' vector (robot coordinates) $\bar{\Theta} = \bar{\Theta}(q_1, q_2, ..., q_k)$ in terms of the operational coordinates' vector $^0\bar{X} = ^0\bar{X}(p_x, p_y, p_z, \alpha, \beta, \gamma)$ (the coordinates of the characteristic point P and the angles necessary for the orientation of the end-effector in relation to the system{0}) [4], [5].

The common command algorhythms are made up of relations which express the engine movements in terms of the positioning parameters for the commanded body.

The geometric command modeling (the inverse geometric modeling) has the following vector expression:

$$\bar{\Theta} = f^{-1}\left(^0\bar{X}\right) \tag{1}$$

It is said that a robot can be solved for a range of tasks involving a $^0\bar{X}$ vector of the operational coordinates, if, by knowing the direct geometric model $^0\bar{X} = \bar{f}(\bar{\Theta})$, a unique mathematical solution can be obtained for the $\bar{\Theta} = \bar{f}^{-1}\left(^0\bar{X}\right)$ system.

The connection between the column vectors $^0\bar{X}$ and $\bar{\Theta}$ is thus achieved through the f operator:

$$^0\bar{X} = \left[^0X_j; j = 1 \to m\right]^T = \left[f_j(q_i; i = 1 \to n); j = 1 \to m; m \le n\right]^T \tag{2}$$

where m stands for the number of the kinematic parameters and n for the number of the degrees of freedom.

The equations in (2) represent a non linear transcending system of equations for which there is no calculating general algorhythm. Under particular conditions, connected to the relative position and orientation of the neighboring kinematic axes $\overline{k}_{i-1}, \overline{k}_i$, system (2) can be solved using either algebraic methods or methods belonging to the plane geometry. Unlike the geometric approach of the inverse geometric model which differs from one problem to another, the algebraic methods are based on reducing transcending equations to algebraic equation with a single unknown term and consequently they can be generalized. Equation (1) can be written as follows:

$$\left[q_i; i=1 \rightarrow n \right]^T = \left[f_i^{-1}\left({}^0X_j; j=1 \rightarrow m \right); i=1 \rightarrow n \right]^T \qquad (3)$$

Equations (3) express a certain configuration of the robot which satisfies the known position and orientation of the end-effector.

The great disadvantage of systems (2), (3), is that they are non linear. As we know, such systems can be solved through numerical methods which introduce unavoidable errors.

Any of the solving methods lead to multiple solutions for the generalizing coordinate q_i. Choosing the unique solution depends on the geometry of the mechanical structure of the robot and of its interaction with the environment.

3. THE KINEMATIC MODEL AND PLANNING THE TRAJECTORY OF THE SCORBOT-ER VII ROBOT

The robot used in this project is the **SCORBOT-ER VII** (figure 2). This robot, manufactured by the Israeli company Eshed Robotec Inc.[6], consists of a mechanical arm composed by 5(five) articulations (base, shoulder, elbow and wrist, this last one composed by 2 articulations) and 1 gripper (with 2 stages: open and closed), according to figure 2.

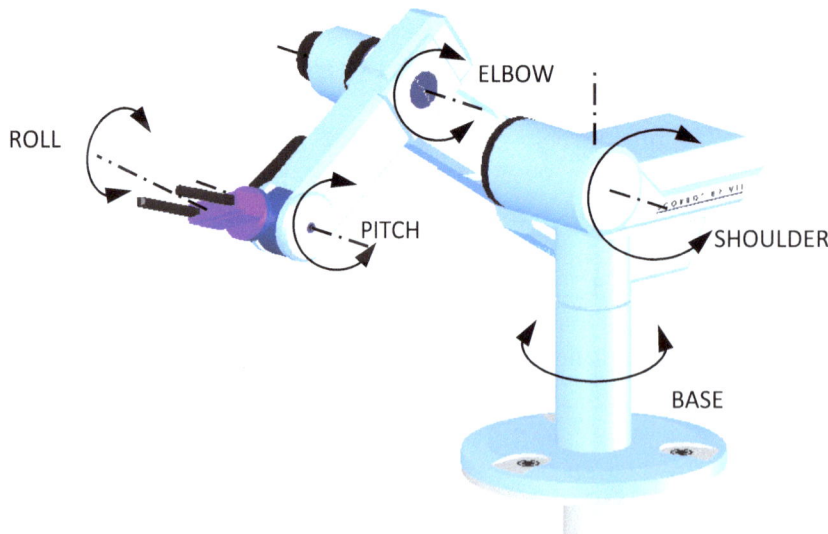

Figure 13 The elements and articulations of the SCORBOT ER VII robot

The kinematic chain in which the chosen reference systems are highlighted is presented in figure 3.

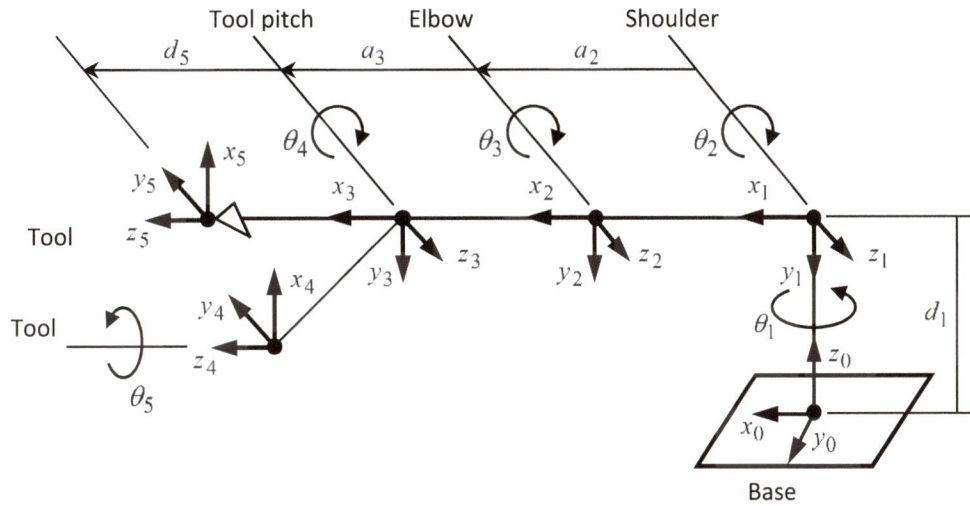

Figure 14 The kinematic chain of the SCORBOT ER VII robot

3.1. Simplified Manipulator Kinematics

Simplified manipulator kinematics has to do with imposing constraints to the kinematic relations. Such a constraint is that the end-effector should be parallel to the base, which results in a simplified model with a simpler solution.

Another simplifying method is the decoupled calculus of the geometric parameters of the robot, by separately analyzing the position equation from the orientation ones.

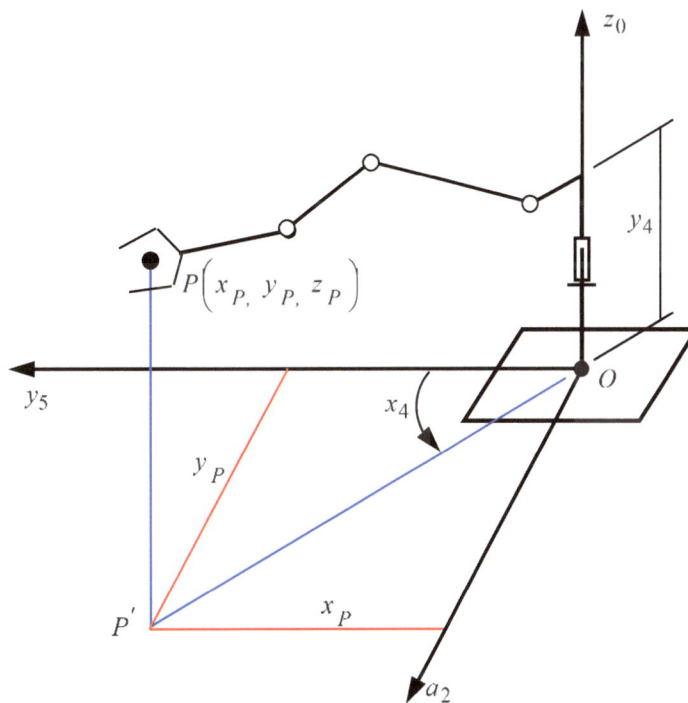

Figure 4 Determining variable θ_1

For a „Pick and Place" operation, the problem of determining a set of variables is considered, namely $\theta_1, \theta_2, \theta_3, \theta_4$, in order to satisfy the demand for position of the end-effector. The demand for orientation of the end-effector (namely θ_5) is neglected.

To calculate variable θ_1 in figure 4 the following formula is used:

$$\theta_1 = a\tan2\left(y_p, x_p\right) \tag{4}$$

In which the coordinates of point P represented by the values x_p, y_p, z_p as well as the lengths of the robot's elements (a_1, a_2, a_3, d_1 and d_5) are known.

Once θ_1 determined, the problem becomes a plane one, as shown in figure 5.

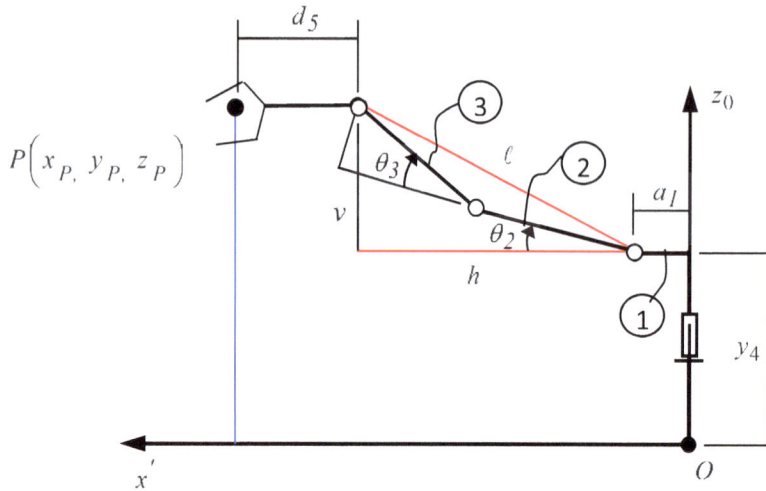

Figure 15 Graphic representation of the SCORBOT ER VII robot

Horizontal distance:

$$h = \sqrt{x_P^2 + y_P^2} - a_1 - d_5 \tag{5}$$

Vertical distance:

$$v = z_p - d_1 \tag{6}$$

$$\ell^2 = h^2 + v^2 \tag{7}$$

$$\ell^2 = a_2^2 + a_3^2 + 2 \cdot a_2 \cdot a_3 \cdot cos(\theta_3) \tag{8}$$

From (7) and (8) results:

$$\theta_3 = a\cos\left(\frac{h^2 + v^2 - a_2^2 - a_3^2}{2 \cdot a_2 \cdot a_3}\right) \tag{9}$$

If we annotate with α the angle between ℓ and element 2 (namely a_2) and with β the angle between ℓ and the horizontal h we obtain:

$$\alpha = a\tan2\left[a_3 \cdot sin(\theta_3), a_2 + a_3 \cdot cos(\theta_3)\right] \tag{10}$$

$$\beta = a\tan2(v, h) \tag{11}$$

$$\theta_2 = \beta - \alpha = a\tan2(v, h) - a\tan2\left[a_3 \cdot sin(\theta_3), a_2 + a_3 \cdot cos(\theta_3)\right] \tag{12}$$

The variable θ_4 was calculated keeping in mind the condition of parallelism between the end-effector and the base:

$$\theta_2 + \theta_3 + \theta_4 = 0 \qquad \text{The result is: } \theta_4 = -\theta_2 - \theta_3 \qquad (13)$$

3.2. Planning the Trajectory

The purpose of planning the trajectory is to generate input data in order to be able to man the movement control system so that the manipulator should execute a trajectory which is established within imposed constraints of velocity and acceleration [7].

A linear trajectory is imposed for the end-effector. It is defined by two points, M and N and the characteristic point P moves along it. The variation within a period of 120 seconds of the angle θ_2 ranging from 0° to 120° which means 0 and $\dfrac{2\pi}{3}$ rad, values which are situated within the functioning limits of the robot [6].

For a set of constant values $a_1 = 0.050\,m$, $a_2 = 0.300\,m$, $a_3 = 0.250\,m$, $d_1 = 0.385\,m$, $d_5 = 0.212\,m$, M(0.785, 0, 0.272), N(0.277, 0, 0.457) and for a sinusoidal variation of θ_2 variations of velocity and acceleration are obtained for point P, as shown in figures 7, and are expressed in $\left[mm/s\right]$ and $\left[mm/s^2\right]$. If we refer to a sinusoid variation of the angle which is similar to the one in [8], θ_2 have the structure (14) and the shape in figure 6.

$$\theta_2(t) = A_0 \cdot \frac{\omega_0}{2\pi}\left[t - \frac{1}{\omega_0}sin\left(\omega_0 \cdot t\right)\right] \qquad (14)$$

For this variation parameter $A_0 = 1\,rad$ and $\omega_0 = 1\,s^{-1}$ are considered.

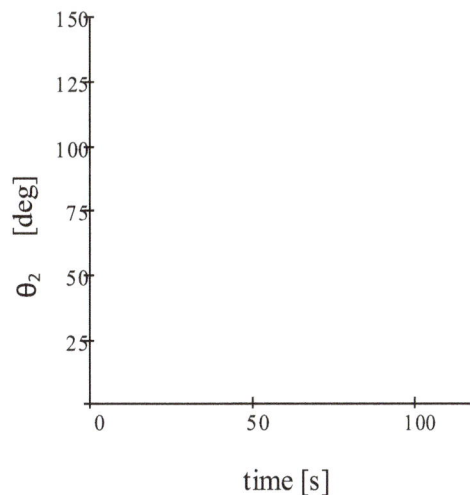

Figure 6 Sinusoid variation in terms of time of the θ_2 parameter.

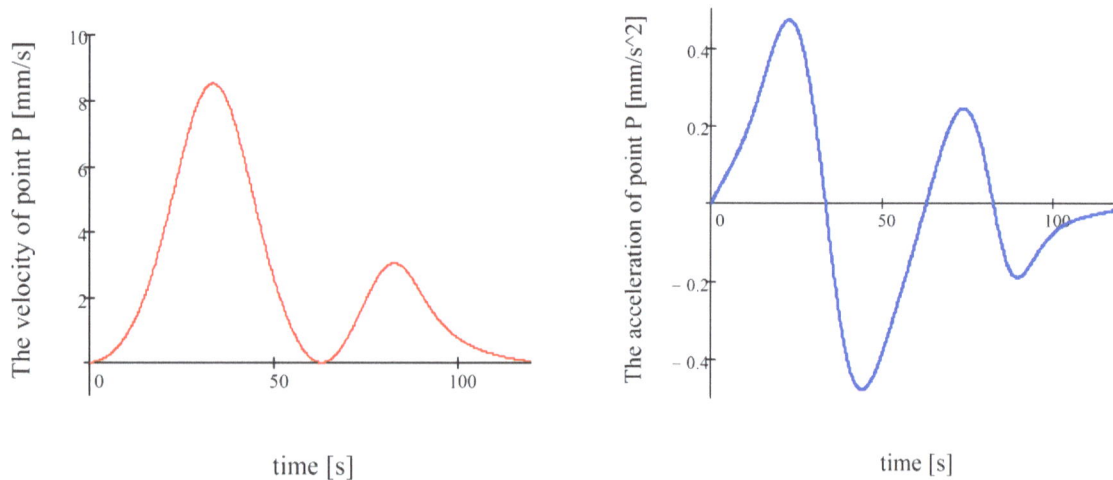

Figure 7 Velocity and acceleration history of point P for the sinusoid variation of the θ_2 angle.

5. CONCLUSIONS

This paper wishes to illustrate the stages of planning and the motion control of a robot with rotary couplers. In order to achieve complex trajectories, we can plan and control the movement, interpolating trajectories belonging to the straight line on a plane.

The method presented here allows the calculation of the geometrical parameters of the decoupled Scorbot-ERVII robot, analyzing separately the position equations and the orientation ones. Even if this procedure simplifies and facilitates the calculation effort, the analytical approach of the kinematic control solutions still remains a complex matter.

The base concept of the proposed approach is constituted by the fact that determining variables involves geometrical modeling of the robotic structure, which leads to multiple solutions, meaning that for a certain positioning, several configurations are obtained in which case it is necessary to intervene in the choice of the variable sets to generate the task.

The comparative study of the graphic representation generates the possibility of an optimal approach to the real work version in terms of the task imposed to the end-effector and its load.

REFERENCES

[1]   *B. Siciliano, L. Sciavicco, L.Villani, G. Oriolo,* Robotics Modelling, Planning and Control, Springer, 2009

[2]   *J.A. Snyman,* "On non-assembly in the optimal synthesis of serial manipulators performing prescribed tasks"J. Lenarcic and B. Roth (eds.), Advances in Robot Kinematics, Springer, 2006, pp. 349–356

[3]   R.N. Jazar, *Theory of Applied Robotics*, 2nd ed., © Springer Science+Business Media, 2010

[4]   *P.Popescu, I.Negrean, I.Vuşcan, N.Haiduc, R.Popescu*, Mecanica manipulatoarelor şi roboţilor, Vol.1,2,3 şi 4, Editura Didactică şi Pedagogică R.A., Bucureşti 1994

[5]   *V.Filip,* Modelarea manipulatoarelor robot. Calcul simbolic, 1. Geometria directă şi inversă. Cinematica directă, Editura PRINTECH, Bucureşti, 1999

[6]   *** User's Manual SCORBOT-ER VII, 2nd ed., Eshed Robotec, 1998

[7]   *S.R. Wang, Z.Z. Qiao and P.C. Tung,* "Application of the force control on the working path tracking" J ournal of Marine Science and Technology, vol. 10, no. 2, 2002, pp. 98-103

[8]   *I. Stroe, S. Staicu, A. Craifaleanu,* "Internal Forces calculus of Compass Robotic Arm Using Lagrange Equations" 11th Symposium on Advanced Space Tehnologies for Robotics and Automation, „ASTRA 2011", ESTEC, Noordwijk, The Nederlands, April 12-14, 2011.

TRANSILVANIA UNIVERSITY OF BRASOV
ROMANIAN ACADEMY OF TECHNICAL SCIENCES

THE 5TH INTERNATIONAL CONFERENCE PROGRAM

Computational Mechanics
and
Virtual Engineering

„COMEC 2013"

Thermique
Ecoulement
Mécanique
Matériaux
Mise en Forme
PrOduction
EMPo

24- 25 October 2013, Brasov, Romania

SCIENTIFIC COMMITTEE

CHAIRMAN
- **György SZEIDL, PhD, Prof. Habil. Eng, University of Miskolc, Hungary**

CO – CHAIRMAN
- **Petre P. TEODORESCU**, Prof. dr. hab.dr.h.c., Eng., member of the Romanian Technical Sciences Academy

MEMBERS
- **Vasile MARINA**, Prof.Dr.Hab, Moldavia Technical University, Moldavia;
- **Cornel STAN**, Prof. Dr.Eng. Habil. DHC, Zwickau, Germany;
- **Manfred BAU,** Dipl.-Ing.Dr.,Consaro Gmbh, Germany;
- **Damien SOULAT,** Prof., France;
- **Carmine PAPPALETTERE,** Prof.Dr. Eng., Politecnica di Bari, Italy;
- **Michael DEDIU**, Ph.D., DERC Inc., USA.
- **Constantin OPRAN,** Prof. Ph.D. MSc. Eng., Univ.Politehnica of Bucharest;
- **Mostafa KATOUZIAN,** Ph.D.Eng., Germany;
- **Károly JÁRMAI,** PhD, Prof. Habil. Eng., University of Miskolc,Hungary;
- **Mircea RADES,** Prof.Dr.Eng., Member of the Romanian Technical Sciences Academy;
- **Nicolae PANDREA**, Prof.Dr.Eng., University of Piteşti, Member of the Romanian Technical Sciences Academy;
- **Costică ATANASIU**, Prof.Dr.Eng. University Politehnica of Bucharest;
- **Tiberiu MĂNESCU**, Prof.Dr.Eng. University of Reşiţa, member of the RomanianTech.Sci. Academy;
- **Nicolae ENESCU,** Prof.Dr.Eng., Politehnica University of Bucharest;
- **Vasile NĂSTĂSESCU,** Prof.Dr.Eng., Technical Armee Academy, Bucharest;
- **Veturia CHIROIU,** Dr.Math., Institut of Solid Mechanics, Bucharest;
- **Iuliu NEGREAN,** Prof.Dr.Eng., Technical University of Cluj, corresp. member of the Romanian Technical Sciences Academy;
- **Ioan GOIA,** Prof.Dr.Eng.,University TRANSILVANIA, honorary member of the Romanian Tech.Sci. Academy;
- **Nicolae Victor ZAMFIR**, Dr.Phys., Member of the Romanian Acad., IFIN-HH;
- **Mircea MODIGA,** Prof.dr.Eng., University of Galaţi;
- **Zoltán MAJOR,** Univ. Prof. Dr. Mont. Johannes Kepler Universität, Linz, Österreich;
- **Tiberiu BABEU**, Prof.Dr.Eng., member of the Romanian Technical Sciences Academy;
- **Polidor BRATU,** Prof.Dr.Eng., ICECON Bucureşti;
- **Dinu TARAZA,** Prof.Dr. Eng., Waine State University, SUA;
- **Mircea LUPU,** Math.dr.ing., University TRANSILVANIA;
- **Lajos BORBÁS,** PhD., Prof., Univ. Budapest, Hungary;
- **Anton HADĂR,** Prof.Dr.Eng. Politehnica University of Bucharest;
- **Mihai HÂRDĂU,** Prof.Dr.Eng. University of Cluj-Napoca;
- **Stanislav HOLÝ,** PhD, Prof. Techn. Univ. In Prague, Czech Republik;
- **Milan Růžička,** PhD., Prof. Techn. Univ. In Prague, Czech Republik;
- **Eva Kormaniková,** PhD., Prof. Techn. Univ. Košice, Slovakia;
- **Luděk HYNČÍK,** Assoc. Prof. PhD., University of West Bohemia, Plzeň, Czech Republic

- **Nicolae ILIESCU,** Prof.Dr.Eng., Bucuresti;
- **Mircea MUNTEANU,** Prof.Dr.Eng., Udine, Italy;
- **Ion PIRNĂ,** Dr.Eng., Manager, INMA Bucharest;
- **Dan CONSTANTINESCU,** Prof.Dr.Eng. Politehnica University of Bucharest;
- **Nicolae HERISANU,** Prof.Dr.Eng. Politehnica University of Timişoara;
- **Marius CRACIUN,** Dr., University Ovidius of Constanţa;
- **Hans Karl LAERMANN,** Prof. Dr. Ing., M.d.h.c.Wuppertal, Germany;
- **Catalin PRUNCU,** Politecnica di Bari, Italy.

ORGANIZERS
- **Sorin VLASE,** Prof.Dr. Eng. Math. Head of Mechanical Engineering Departmentx, Transilvania University of Braşov, corresp. member of the Romanian Technical Sciences Academy;
- **Ioan Călin ROŞCA,** Prof.Dr. Eng., Dean of the Faculty of Mechanical Engineering, Transilvania University of Braşov.

ORGANIZING COMMITTEE
- **Ioan SZÁVA,** Prof.Dr.Eng.;
- **Gheorghe BRĂTUCU,** Prof.Dr.Eng.;
- **Ioan Călin ROŞCA,** Prof.Dr.Eng.;
- **Dumitru NICOARĂ,** Prof.Dr. Math.;
- **Radu Nicolae GHEORGHE,** Prof.Dr.Eng.;
- **Mihaela BARITZ,** Prof.Dr.Eng;
- **Vasile CIOFOAIA,** Prof.Dr.Eng.;
- **Camelia CERBU,** Assoc.Prof.Dr.Eng.;
- **Veneţia SANDU,** Assoc.Prof.Dr.Eng.;
- **Daniela ŞOVA,** Assoc.Prof.Dr.Eng.;
- **Mihai ULEA,** Assoc.Prof.Dr.Eng.;
- **Horaţiu TEODORESCU,**Assoc.Pr.Dr.Eng.;
- **Luminiţa SCUTARU,** Assoc.Prof.Dr.Eng;
- **Mihaela Violeta MUNTEANU,** Assist.Prof.Eng.;
- **Anca STANCIU,** Assoc.Prof.Dr.Eng;
- **Violeta GUIMAN,** Assist.Prof.Eng.;
- **Ramona PURCAREA,** Dr.Eng.

SECRETARY
Mircea MIHALCICA, Dr. Eng
- e-mail: **info-dimec@unitbv.ro**

Thursday, October 24th 2013
8^{00}-9^{00} Registration – *Transilvania* University, Conference Center
9^{00}-9^{30} Festive opening - *Transilvania* University, Conference Center
9^{30}-11^{00} Plenary Session – *Transilvania* University, Conference Center
11^{00}-11^{30} Coffee Break
11^{30}-13^{00} Technical Session - *Transilvania* University, Conference Center
13^{30}-15^{00} Lunch Break
15^{00}-15^{50} Plenary Session – *Transilvania* University, Conference Cente
15^{50}-16^{00} Coffee Break
16^{00}-18^{00} Technical Session – *Transilvania* University, Conference Center
19^{00}-22^{00} Gala Dinner

Friday, October 25th
10^{00}-17^{00} Social program

Thursday, October 24th
8.00 – 9.00
Participants welcoming and registration

9.00 – 9.30
Festive opening

9.30- 11.00
Plenary session UI3

CHAIRMAN: Prof.dr.eng G. Szeidl
1. Cătălin I. Pruncu,
   THE COMPUTATIONAL MODELLING APPROACH A MULTI-SCALE OVERVIEW
2. György Szeidl, László Kiss,
   VIBRATIONS OF HETEROGENEOUS CURVED BEAMS SUBJECTED TO A RADIAL FORCE AT HE CROWN POINT
3. József Farkas, Károly Jármai,
   MINIMUM COST DESIGN OF A RING-STIFFENED CYLINDRICAL SHELL LOADED BY EXTERNAL PRESSURE
4. Ligia Munteanu, Veturia Chiroiu, Ştefania Donescu, Ruxandra Ilie, Valerica Moşneguţu,
   ON THE EFFECTIVE MODULI OF SONIC COMPOSITES

11.00 – 11.30
Coffee break

11.30 – 13.00
Technical session

SECTION A
AULĂ UI2

CHAIRMAN: Prof.dr.eng. Ionel Staretu
Prof.dr.eng. Camelia Cerbu

1. **A. Coseru, J. Capelle, G. Pluvinage** - ON THE USE OF CHARPY TRANSITION TEMPERATURE AS REFERENCE TEMPERATURE FOR THE CHOICE OF A PIPE STEEL
2. **Aurora Potirniche** - COMPUTATIONAL DYNAMICS OF ELASTOMERIC-BASED ISOLATION SYSTEMS FOR RIGID STRUCTURES
3. **Camelia Cerbu** - MECHANICAL TESTING OF THE COMPOSITE MATERIALS BASED ON POLYPROPYLENE AND ITS APPLICATION IN AUTOMOTIVE PARTS
4. **Camelia Gheldiu, Mihaela Dumitrache** - THE DISCRETIZATION OF THE LIMIT OF A BOUNDARY VALUE DIFFUSION PROBLEM IN A PERFORATED DOMAIN
5. **Cristina Chilibaru–Opriţescu1, Amalia Ţîrdea, Corneliu Bob -** THE CHECKING OF THE SEMI-PRECAST R.C. FLOORS
6. **I. Milosan** - OPTIMIZATION OF SPECIFIC FACTORS TO PRODUCE SPECIAL ALLOYS.
7. **Ildiko Tulbure** - BIOFILTER MODELLING FOR ENVIRONMENTAL PROTECTION.
8. **Ionel Staretu** - STRUCTURAL SYNTHESIS FOR REDUNDANT INDUSTRIAL ROBOTS WITH MORE 6 AXES
9. **Iulian Girip, Rodica Ioan, Mihaela Alexandra Popescu, Ligia Munteanu, Veturia Chiroiu** - ON THE SONIC COMPOSITES WITH DEFECTS
10. **Mănescu Tiberiu Jr., Gillich Gilbert-Rainer, Mănescu Tiberiu Ştefan, Suciu Cornel** - EXPERIMENTAL INVESTIGATIONS UPON CONTACT BEHAVIOR OF BALL BEARING BALLS PRESSED AGAINST FLAT SURFACES

SECTION B
AULA UI3

CHAIRMAN: Prof.dr.eng. László Kalmár
Prof.dr.eng. H. Teodorescu-Draghicescu

1. **D. Stoica** - NONLINEAR DYNAMIC ANALYSIS FOR „OD" REPETITIVE BLOCK OF FLATS DESIGN PROJECTS.
2. **D. Stoica** - SEISMIC VULNERABILITY, RETROFITTING SOLUTIONS AND MONITORING FOR EXISTING BUILDINGS
3. **D.A. Micu, M.D. Iozsa, Gh. Frăţilă** - A ROLLOVER TEST OF BUS BODY SECTIONS USING ANSYS
4. **Daniel Condurache, Adrian Burlacu** - A DUAL VECTORS BASED FORMALISM FOR PARAMETRIZATION OF RIGID BODY DISPLACEMENT AND MOTION
5. **Daniela Şova, Bogdan Bedelean, Monica A. P. Purcaru** - EFFECTS OF SIMULATED WOOD DRYING SCHEDULES ON DRYING TIME AND ENERGY CONSUMPTION AT AN EXPERIMENTAL KILN
6. **Diana Cazangiu, Ileana Rosca, Yves Lemmens** - THE APPLYING OF AN AUTOMATIC

CONFIGURATION TOOL FOR THE INVESTIGATION OF THE UAV ELECTRICAL NETWORK

7. **Diana Cazangiu, Ileana Rosca**- THE CURRENT TRENDS IN STRUCTURAL HEALTH MONITORING IN AEROSPACE APPLICATIONS.

8. - **László Kalmár, Béla Fodor**, CFD Investigation of the Flows in One-Stage Blower Aggregate.

9. **Liviu Gaiţă, Manuella Militaru, Gabriela Popescu** - FRACTAL DIMENSION OF CHROMATIN REGIONS IN HISTOLOGICAL PICTURES REVEALS THE PRESENCE OF EPITHELIAL TUMOURS

10. **Liviu Gaiţă, Manuella Militaru** - OPTIMIZATION OF AN ARTIFICIAL NEURAL NETWORK USED FOR THE PROGNOSTIC OF CANCER PATIENTS

13.00 – 15.00
Lunch

15.00- 15.50
Plenary session UI3

CHAIRMAN: Prof.dr.eng **Luděk Hynčík**

1. **Ctirad Novotný, Karel Doubrava -** SOME PROBLEMS OF FEM MODELLING OF SANDWICH STRUCTURES
2. **Luděk Hynčík, Luděk Kovář -** TOWARDS PERSONALIZED VEHICLE SAFETY
3. **László Daróczy, Károly Jármai -** TOPOLOGY OPTIMIZATION BY A QUASI-STATIC FLUID-BASED EVOLUTIONARY METHOD

16.00 – 18.00
Papers presentation

SECTION A
AULĂ UI2
CHAIRMAN: Prof.dr.eng. S. Nastac
Prof.dr.eng. Ioan Curtu

1. **Marian N. Velea, Simona Lache** - NOVEL IMPACT ATTENUATOR
2. **Marian Truţă, Marin Marinescu, Valentin Vînturiş -** MULTI-SPECTRAL ANALYSIS OF THE SELF GENERATED TORQUE'S SIGNAL WITHIN A 4X4 AUTOMOTIVE DRIVELINE
3. **Mariana D. Stanciu, Dragos Apostol, Ioan Curtu** - EVALUATION OF STRESS AND STRAIN STATES BY FINITE ELEMENT METHOD OF PANORAMIC STRUCTURE MADE OF BARS AND PLATES
4. **Mariana D. Stanciu, Ioan Curtu, Dumitru Lica** - EVALUATING THE EFFICIENCY OF RECYCLING COMPOSITE PANELS MADE FROM ABS TO REDUCE THE TRAFFIC NOISE
5. **Marinică Stan, Petre Stan** - RANDOM OSCILLATIONS OF LIQUID IN THE U-SHAPED PIPE WITH PRONOUNCED RUGOSITY AND THE NON-LINEAR DAMPING FORCE

6. **Petre Stan, Marinică Stan** - RANDOM VIBRATION FOR THE DISK-SHAFT SYSTEMS WITH TWO DEGREES OF FREEDOM
7. **V. Roşca, C. Miriţoiu, I. Geonea, Alina Romanescu** - COMPARATIVE ANALYSIS OF THE CRACKING RATE FOR A STAINLESS STEEL LOADED AT 213K TEMPERATURE
8. **Sebastian M. Zaharia, Ionel Martinescu** - RELIABILITY AND ENVIRONMENTAL DEGRADATION OF COMPOSITE MATERIALS USING ACCELERATED METHODS
9. **Sebastian M. Zaharia, Ionel Martinescu** - RELIABILITY AND FATIGUE LIFE PREDICTION OF CYLINDRICAL ROLLER BEARINGS BASED ON FINITE ELEMENTS METHODS
10. **Silviu Nastac, Cristian Simionescu** - COMPUTATIONAL DYNAMICS OF HELICAL FLEXIBLE COUPLING WITH TRANSITORY CONTINUOUS REGIME
11. **Silviu Nastac** - ON NONLINEAR EFFECTS DUE TO THERMO-MECHANICAL BEHAVIOUR IN COMPUTATIONAL DYNAMICS OF ELASTOMERIC ISOLATORS

SECTION B
AULA UI3
CHAIRMAN: Prof.dr.eng. Venetia S. Sandu
Prof.dr.eng. H. Teodorescu-Draghicescu

1. **G. Dima, I. Balcu** - ACTUAL STATUS OF GUSSETED JOINTS OF AEROSPACE WELDED STRUCTURES
2. **G. Dima, I. Balcu** - NOTES ON EVOLUTION OF AIRCRAFT STRUCTURES LATTICED BEAM JOINTS
3. **Gabriel Popescu** - ON THE OPTIMIZATION OF THREE-DIMENSIONAL LINE CONTACTS INCLUDING SKEW AND MISALIGNMENT ANGLES FOR CAM-FOLLOWER TYPE CONTACTS
4. **Gigel Florin Capatana** - COMPLEX CONTINUOUS-LUMPED MODEL FOR SIMULATION OF VIBRATORY COMPACTION PROCESS
5. **Vasile Ciofoaia** - THE MECHANICAL RESPONSE OF TEXTILE COMPOSITE MATERIALS TO DYNAMIC IMPACT TESTS
6. **Vasile Ciofoaia** - THE STRENGTH OF SIMPLE BELT GUIDE UNDER TENSION
7. **Venetia S. Sandu** - MEASUREMENT OF COEFFICIENTS OF FRICTION OF AUTOMOTIVE LUBRICANTS IN PIN AND VEE BLOCK TEST MACHINE .
8. **L. Predescu, M. Predescu** - PLANNING THE TRAJECTORY OF THE SCORBOT-ER VII ROBOT

SECTION
Industrial Applications and Applications in Agricultural Machinery

CHAIRMAN: Prof.dr.eng. Gh. Brătucu
Prof.dr.eng. S. Popescu

1. **Filip Vladimir Edu, Carol Csatlos** - THE METHODS OF RESEARCH FOR MECHANICAL SORTING AND SIZING SYSTEMS
2. **A.O. Arişanu, Fl. Rus** - THE INFLUENCE OF PROCESS PARAMETERS ON MECHANICAL EXPRESSION OF SUNFLOWER OILSEEDS
3. **A.O. Arişanu** - MECHANICAL CONTINUOUS OIL EXPRESSION FROM OILSEEDS:

OIL YIELD AND PRESS CAPACITY

4. **E. Badiu, Gh. Brătucu**, RESEARCHES REGARDING THE CAUSES OF DEGRADATION OF ROOF SYSTEMS

5. **E. Badiu, Gh. Brătucu**, RESEARCH ON DEGRADATION BY CORROSION OF SOME COMPONENTS OF BUILDINGS ROOFS

6. **C. Bodolan, Gh. Brătucu**, HEAT AND LIGHT REQUIREMENTS OF VEGETABLE PLANTS

7. **C. Bodolan**, WATER, AIR AND SOIL REQUIREMENTS OF VEGETABLE PLANTS GROWN IN GREENHOUSES

8. **Gh. Brătucu, D.D. Păunescu**, SOLAR ENERGY – AN ENERGETIC SOURCE FOR THE VEGETABLE AND FRUIT PRODUCTS DRYING IN BRASOV AREA

9. **Gh. Brătucu, I. Căpăţînă, D.D. Păunescu**, SOIL PRESERVATION THROUGH THE PERFECTIONING OF ITS BASIC WORKS

10. **C.M. Canja, M.I. Lupu, D.W. Enache**, STUDY ON HYGIENE AND HYGIENE RULES IN BAKERY INDUSTRY

11. **C.M. Canja, M.I. Lupu, V. Pădureanu**, STUDY ON RHEOLOGICAL BEHAVIOR OF BAKERY DOUGH

12. **C. Csatlós**, OPTIMIZING THE ENERGY CONSUMPTION TO MOULDS OF PALLETING MACHINES OF MIXED FODDERS

13. **M. Hodîrnău, E. Mihail, C. Csatlos**, ASPECTS OF ANALOG THEORETICAL AND EXPERIMENTAL RESEARCH ON THE DYNAMICS OF CABLE CARS

14. **M. Hodîrnău, E. Mihail, C. Csatlos**, RESEARCH ON IMPROVING COMFORTABLE CABLE CAR BY LATERAL DAMPING

15. **D.C. Ola, D.M. Danila, M.E. Manescu**, ENERGY OPTIMIZATION OF SMALL HOUSE PHOTOVOLTAIC PANEL BY INCREASING THE WORKING EFFICIENCY THROUGH ADDITIONAL MIRRORING OF SUN LIGHT ON THE PANEL AND RECOVERY OF THE HEATING ENERGY FROM THE PHOTOVOLTAIC CELLS

16. **A.N. Ormenişan**, USING AUTOMATIC CONTROL SYSTEMS TO INCREASE DYNAMIC PERFORMANCE AND OPERATING ENERGY PLOUGHING AGGREGATES

17. **V. Pădureanu, M.I. Lupu**, C.M. Canja, THEORETICAL RESEARCH TO IMPROVE TRACTION PERFORMANCE OF WHEELED TRACTORS BY USING A SUPLEMENTARY DRIVEN AXLE

18. **H.Gh. Schiau, Fl. Rus**, ENERGY EFFICIENCY ANALYSIS OF AGARICUS BISPORUS MUSHROOM PRODUCE IN FELDIOARA-BRASOV

19. **H.Gh. Schiau**, AN INVESTIGATION OF THE AIRFLOW IN MUSHROOM GROWING STRUCTURES FOR MODELLING NEW STRUCTURES.

20. **Szilard Ilyes, Simion Popescu, Mihai Nedelcu**, COMPARATIVE ANALYSIS CONDUCTED ON A CONSTRUCTIVE-FUNCTIONAL BASIS ON TUBERS DISTRIBUTION SYSTEMS OF POTATO PLANTERS

21. **Szilard Ilyes, Simion Popescu, Mihai Nedelcu**, EXPERIMENTAL DEVICE FOR THE RESEARCHES ON THE PRECISION OF THE TUBERS PLANTING DISTANCE WITHIN THE ROW FOR DIFFERENTS POTATO PLANTERS

19.00 – Festive dinner

SECTION POSTER/
PAPERS IN CONFERENCE PROCEEDINGS

1. **C. Danasel**, MULTIOBJECTIVE AND MULTIDISCIPLINARY STRUCTURAL OPTIMIZATION OF A CONCEPT PART
2. **C. Drugă, M. Mihai**, EX-IN VITRO TESTING OF TOTAL KNEE REPLACEMENTS – FIRST PART
3. **C. Drugă**, EX-IN VITRO TESTING OF TOTAL KNEE REPLACEMENTS – SECOND PART
4. **Cornel Bit**, ON FATIGUE CRACKS MECHANICAL BEHAVIOUR
5. **Dumitru D. Nicoara**, OPTIMUM DESIGN OF SPINDLE-BEARING SYSTEMS.
6. **Enescu Ioan**, ROUGH SURFACE CONTACT – APPLICATION TO BEARINGS.
7. **Horaţiu Teodorescu-Draghicescu, Sorin Vlase**, STRESSES IN VARIOUS COMPOSITE LAMINATES FOR GENERAL SET OF APPLIED IN-PLANE LOADS
8. **Liviu Costiuc**, EXPERIMENTAL INVESTIGATION ON ENERGY DENSITY OF BIO-FUELS
9. **Călin Itu**, THEORETICAL ENGINE DESIGN SOLUTION TO MINIMIZE CONSUMPTION AND POLLUTION
10. **Maria Luminita Scutaru, Marius Baba**, MECHANICAL BEHAVIOR OF HEMP-BASED COMPOSITE SUBJECTED TO IMPACT TEST
11. **Maria Luminita Scutaru, Marius Baba, Janos Timar**, FLEXURAL RIGIDITY EVALUATION OF COMPOSITE SANDWICH PANEL OF CARBON-HEMP
12. **Sava Rodica, Lihteţchi Ioan,** AN EXPERIMENTAL DENSIFICATION METHOD BY COMPRESSING THIN VENEERS (0,3-1,2 MM)
13. **Traian Bolfa**, THE INFLUENCE OF PRINCIPAL FACTORS WITH WORKING CONDITIONS FOR HIGH SPEED BEARINGS
14. **Ungureanu I. Virgil-Barbu,** VARIABLE CONDUCTANCE HEAT PIPE MODEL FOR TEMPERATURE CONTROL OF PROCESSES
15. **Anca Elena Stanciu, Diana Cotoros, Ramona Purcarea, Mihaela Violeta Munte**anu, MICROSCOPIC ANALYSIS OF COMPOSITE MATERIALS TESTED
16. **Anca Elena Stanciu**, CONSTRUCTIVE SOLUTION, USING FINITE ELEMENT METHOD, FOR OPTIMIZATION STRUCTURE OF COMPOSITE MATERIALS
17. **Horaţiu Teodorescu-Draghicescu, Sorin Vlase**, PREDICTION OF ELASTIC PROPERTIES OF SOME SHEET MOLDING COMPOUNDS
18. **Vasile Gheorghe, Costel Bejan, Nicolae Sîrbu, Ioan Lihteţchi, Arina Modrea**, INFLUENCE OF TEMPERATURE ON MECHANICAL PROPERTIES OF POLYMER MATRIX COMPOSITES SUBJECTED TO BENDING
19. **Maria Violeta Guiman**, AN APPROACH TO MULTIBODY FORMULATIONS FOR BIOMECHANICAL MODELING
20. **Gheorghe N. Radu, Ioana Sonia Comănescu**, THEORETICAL STUDIES AND EXPERIMENTAL RESEARCH FOR THE INCREASE OF THE WORK SAFETY AT GANTRY CRANES
21. **Gheorghe N. Radu, Ioana Sonia Comănescu**, THEORETICAL STUDIES AND EXPERIMENTAL DETERMINATIONS FOCUSED ON THE INCREASE OF THE GANTRY CRANE'S WORK SAFETY, PART II – EXPERIMENTAL DETERMINATIONS, CONCLUSIONS AND RECOMMENDATIONS CONCERNING

THE INCREASE OF THE WORK SAFETY
22. **Vasile Gheorghe, Costel Bejan, Nicolae Sîrbu, Ioan Lihteţchi, Eugenia Secară**, Determination of coefficient of thermal conductivity on glass fibers-reinforced polymer matrix composites
23. **Mihai Ulea,** Consideration about Energy in Plastic Joints
24. **Mihai Ulea, Mihai Tofan**, Mathcad representation of Plastic Joints in Busbody Section
25. **V. Nastasescu, Gh. Ba**rsan, COMPARATIVE NUMERICAL ANALYSIS OF AN ARMOR PLATE UNDER EXPLOSION

Friday, October 25th

10.00 – 20.00
10.01 Social program

TABLE OF CONTENTS

www.ingramcontent.com/pod-product-compliance
Lightning Source LLC
Chambersburg PA
CBHW050759220326
41598CB00006B/66